1972

This book may be k

Foundations of Plant Geography

Foundations of Plant Geography

STANLEY A. CAIN

Professor of Botany
The University of Tennessee

(Facsimile of 1944 edition)

HAFNER PUBLISHING COMPANY
NEW YORK
1971

Published by
HAFNER PUBLISHING COMPANY, INC.
866 Third Avenue
New York, N.Y. 10022

Library of Congress Catalog Card Number: 72-152762

Printed in the United States of America

Contents

-»»><«<-

v

Contents

Part IV. EVOLUTION AND PLANT GEOGRAPHY

Part V. SIGNIFICANCE OF POLYPLOIDY IN PLANT GEOGRAPHY

Figures

Preface

---------------->>><<<---------------

To those who read these words before turning to the body of the book, I wish to say that I had no intention of writing a descriptive plant geography. This is rather an inquiry into the foundations of the science of plant geography. I have made an effort to survey the related fields of science for concepts and working methods which are useful in an interpretation of the phenomena of plant distribution. Many of these materials are from the fields of paleontology, taxonomy, evolution, genetics, and cytology. I have not deluded myself by an assumption that I am adequately conversant with the material and philosophy of these sciences. Neither do I accept a rather widely held educational fallacy that it is possible adequately to understand the generalizations of a field of learning without also being conversant with the facts on which such generalizations are founded. I have sought, rather, to cut the hedgerows between these fields of science, and to discover, for myself at least, some of the significance which one field has for another.

From the mixture and the synthesis there emerges, in part, a broader field of inquiry which forms the foundations of plant geography. The specialist may find his subject inadequately treated. If so, I plead space limitations. For errors, I admit ignorance and oversight, and suggest that this job needed doing, even imperfectly. The problem of selection of literature has been overwhelming. In many cases I have used those sources more familiar to me or more readily available—a complete search of the literature would have been unending. For the organization of the materials employed I ask understanding of the problem. Scarcely a subject is considered that does not anastomose with nearly every other subject, and one wishes for each topic a background formed by all other topics.

S. A. C.

Knoxville, Tennessee
February, 1944

Acknowledgments

———————————————— ·>>><<<· ————————————————

When I consider the pleasant obligation of expressing appreciation for assistance rendered in the preparation of this manuscript, one difficulty arises. In addition to those who have actively helped the progress of the work and who can be enumerated, there is a residue of numerous others who must remain nameless, for I can no longer tell from whom I got a certain idea, or even whether it is my own. To all these, sometimes for only a casual word, written or spoken, I express my appreciation.

There remains a considerable list of scientists to whom I know my obligation, for I have bearded them in their own dens and have later imposed on them from one to several chapters of manuscript for their critical consideration. Uniformly I have met with courtesy and a careful consideration of my problems, and of their busy days and deep experience they have given unreservedly. To list their names is inadequate recompense. If this book is worthy and proves to be of help in the field of geography, much of the credit is theirs. The errors, shortcomings, and peculiarities of emphasis of this book are in no way attributable to these friends. As a matter of fact, my mention that any one of them has read portions of this manuscript by no means infers that he is in total agreement with what is here written. Let them regret that I was not "converted," as perhaps, eventually, I shall so regret. I wish especially to thank Hugh M. Raup, Harvard University, for his many fine suggestions and enthusiastic help. To these other gentlemen, my thanks: W. C. Allee (University of Chicago), Edgar Anderson (Missouri Botanical Garden), W. H. Camp (New York Botanical Garden), R. W. Chaney (University of California), F. E. Clements (Carnegie Institution of Washington), T. Dobzhansky (Columbia University), Alfred Emerson (University of Chicago), H. A. Gleason (New York Botanical Garden), H. L. Mason (University of California), E. B. Matzke (Columbia University), Tom Park (University of Chicago), A. J. Sharp (University of Tennessee), A. B. Stout (New York Botanical Garden), and Frans Verdoorn (Chronica Botanica).

I am especially indebted to certain institutions and persons for necessary assistance. During the year from August, 1940, to August, 1941, I had the privilege of a fellowship with the John Simon Guggenheim Memorial Foundation. It was during this period that most of my material was assembled and the general form of this book developed. To Mr. Henry Allen Moe, Secretary General, and to the Committee of Selection, I express my deep appreciation. During this same interval I received a leave of absence from The University of Tennessee. For this consideration and many other practical aids I extend my thanks to Dean L. R. Hesler and the Administration. During the summer of 1939 I held a scholarship from the New York Botanical Garden, and for the first six months of 1941 was accorded staff privileges. For these favors I am deeply grateful. I wish also to extend especial appreciation to Mrs. Lazella Schwarten, Assistant Librarian, New York Botanical Garden, who has traced to its source every citation of literature available to her and has thus eliminated many errors from the bibliography. Mrs. Harvey Broome has typed the manuscript and been of assistance in ridding it of several errors, as well as having helped in the preparation of the index.

To F. E. Clements, I wish to express appreciation for his hospitality at the Alpine Laboratory, and for the many hours of his inspiring conversation. Anyone familiar with his Neo-Lamarckian concepts will realize the extent to which I disagree, yet I wish it known that I consider him a great ecologist of profound learning.

I am especially indebted to my wife, Louise Gilbert Cain, who has patiently worked with me in the preparation of the manuscript. Her meticulous care, good judgment, unfailing intuition, and humor have polished many a rough spot and contributed to the steady prosecution of the project. Her judgments have provided a litmus test for many statements and her acidulous reactions have often resulted in basic improvements. To my affection, I add appreciation.

PART ONE

Introduction

I.

Plant Geography
as a Borderline Science

Interpretive plant geography is a borderline science which depends for its materials and some of its concepts upon more specialized sciences, and it derives its distinction as a field of study from synthesis and integration. Interpretive plant geography is a second phase that follows naturally after descriptive plant geography.

Each field of biological science overlaps adjacent fields, contributing to and deriving from them. None is really independent except in its more descriptive phases. Modern genetics, for example, has turned to cytology on the one hand and to the phenomenon of distribution on the other, embroidering and elucidating its results with statistics. The new systematics is turning more and more to cytology and genetics for confirmation of classification and for criteria of relationship which the natural systems purport to show.

Static or descriptive plant geography, in its assembling of floristic and vegetational data, provides the materials and propounds the problems that interpretive plant geography seeks to understand. In so doing, plant geography differs from other more specialized sciences only in degree. When it leaves the static fields of description and searches for causes of distributional phenomena, both modern and historical, it finds its explanations in the material which is more particularly the province of special sciences. It is compounded of physiology and ecology, cytology and genetics, taxonomy and evolution, paleontology and geology, comparative morphology and floristics. It is the thesis on which this book is based that the principles of plant geography are derived from the findings of other more limited disciplines. Interpretive plant geography is, then, a synthesizing science.

3

Physiology, ecology, and genetics.—If we give to ecology a strict meaning, we can consider that it consists of an inquiry into the interrelations between the organism and its environment. The ontogeny of an organism, the very fact of living, is understood as consisting of readjustments between internal and external environments. From the zygote to reproduction, development consists of a series of time-coordinated structural changes which are the product of and the physical matrix for a series of physiological processes. Structure and function form an harmonious whole in a time sequence of a highly specific nature. The potentialities of an organism are firmly seated in its inherited genetic constitution. What an organism is and what it can become at any time in its development are strictly limited by the potentialities which it received from its parents or, in rare cases, which it develops by somatic mutation.

A species, in sexually reproducing populations, consists of a large number of biotypes differing slightly in their inheritance. It is well known that biotypes within a species differ not only morphologically but physiologically, with or without a strict correspondence between small structural and functional differences. Biotypes differ in respect to such physiological characteristics as winter hardiness, earliness, photoperiodicity, and disease resistance. Furthermore, gene mutations may cause physiological as well as morphological differences, changing an organism, for example, in respect to temperature optimum, rate of development, etc. The same holds true for larger mutations, such as interchanges, inversions, and chromosome doubling.

Any detailed consideration being omitted, it appears likely that genetic factors control the development of an individual through accurately balanced reactions in a normally definite time correlation under the influence of "formative substances" such as vitamins, hormones, enzymes. Experiments with *Hyoscyamus niger* (481),[1] for example, have shown this species to consist of annual and biennial races which differ by a single gene. If a branch from an annual plant of this species is grafted on a stock of a plant of the biennial race, the "natural" biennial comes into flower the first year. This change in development of the biennial appears to be due to a specific formative substance which is generated in the scion and diffuses into the stock.

The study of ecology has been spoken of as outdoor physiology, an expression that implies that intraspecific and interspecific ecological differences have an adaptational basis that resides ultimately more in function than in structure. There are, however, usually some morphological characteristics

[1] Numbers in parentheses refer to the "Literature Cited," at the end of the book.

that distinguish ecological races, whether or not they have a functional basis.

It is unfortunate that many physiologists and ecologists doing physiological work have apparently been so unaware of the diversity among the biotypes which compose a species that they almost never file vouchers of the plants they have studied. As a result, a considerable quantity of the data which have been obtained is completely useless because it is impossible to know which ecotype (or sometimes which species) has been studied. Although a species population contains genetically different plants, it is not without order. The experimental ecologist who combines the genetic and physiological points of view, the genecologist, has discovered that in nature a species population usually consists of local and regional groups of adapted biotypes which have resulted from natural selection. The ecotype, for example, is a group of closely similar biotypes which have been selected out of the species by a particular combination of environmental factors.

Genetics and taxonomy.—Taxonomy is based primarily upon the recognition of discontinuity. In the usual practical taxonomy an easily observable discontinuity of biotypes is demanded for the segregation of individuals into species groups. The taxonomist prefers one or a few completely discontinuous characters between closely related species, but frequently he has to content himself with a type of morphological discontinuity that results from a group of characters, no one of which is qualitatively different. The second qualification of biological classification, variation, is within the limits set by discontinuity. The great contribution of genetics to taxonomy lies in the description of types of biological variation and their causes and in the recognition of the several causes of discontinuity. When the taxonomist realizes that the barriers between species are of several types, it becomes quite clear that there are several categories of species in nature, as well as in systematics. Modern taxonomic treatments, where the genetic situation is known or reasonably inferred, are as much a product of genetics as of comparative morphology. The older evolutionary-geographical concept that the most closely related species of a genus occupy not the same region but contiguous ones separated by a barrier has been modified by the genetic discovery of intrinsic barriers to interbreeding. As long as two species are so closely related that there is no reproductive barrier between them, the maintenance of their distinctiveness depends upon their allopatric distribution. Related species can have sympatric distribution only when they are separated by intrinsic breeding barriers. The recognition of several isolating mechanisms in addition to geographical isolation has enabled taxonomists to clarify many problems of relationship and produce more natural treatments of related forms. The addition of cytogenetic methods to the older ones of comparative morphology and geography has greatly facilitated taxonomy and, since a sound taxonomic treatment is fundamental to a solution of geographical problems, the interrelations of the sciences are of mutual benefit.

Cytology and taxonomy.—The idea of the karyotype and genom analysis has contributed considerable information in support of taxonomic arrangements and revisions. The determination of chromosome numbers leads to the detection of multiple series of numbers and the conclusion that their origin was due to genom doubling. Whether the condition is auto- or allopolyploidy can sometimes be determined by a study of meiosis. If the genoms can be shown to be homologous, the condition is one of autopolyploidy and the duplication of genoms of one parent. If the genoms can be shown to be non-homologous, the polyploidy has had a hybrid origin; this phenomenon is of importance because it defines the relationships of the species involved. A genom analysis of hybrids, when several species have been crossed variously, even from different genera, can sometimes show the closeness of relationship of the species involved in the crosses. Cytology can go even further in a study of chromosome size, constructions, satellites, position of fiber attachments, inversions, segmental interchanges, etc., and sometimes permits an opinion concerning the constitution and relationship of different species. The ultimate of cytogenetic analysis has resulted in the prediction of the nature of the origin of certain natural species (and the subsequent reproduction of a natural species, as in the experimental work with *Galeopsis Tetrahit*), and in the prediction of a species not yet known. The only example of the latter, as far as I know, was the prediction of the existence in nature and the subsequent recognition of *Iris setosa* var. *interior* by Anderson (15). The contributions of cytology are to a natural taxonomy that reveals relationship through origin.

Phylogeny, taxonomy, and geography.—Taxonomy is the expression of relations between concrete or abstract items which are brought together by their similarities and separated by their dissimilarities. Taxonomic units are delimited by discontinuities which in practice frequently represent only a more or less sudden gradient of change in respect to the characters of the organisms classified. A classification is said to be more natural when the characters chosen for the basis of it are more important from the point of view of origin. Characters which a group of species have in common are considered to be natural because it is assumed that the species have evolved from a common stock and each received these unifying characters. This is not true, of course, where parallel evolution in different stocks has resulted in equivalent structures, and organisms are classified together because of these structures. Gregory (301) gives a good example of parallel evolution of fruit types in the Ranunculaceae.

Phylogeny introduces the factor of time and considers evolution from time level to time level. A classification at any one time level, as in paleontology or the arrangement of living forms, is taxonomy; phylogeny is an account of the changes with time. Taxonomic schemes are drawn at right angles, so to speak, to the flow of changes that constitutes phylogeny. Whereas taxonomy is based upon characters, phylogeny is based upon changes of characters, or evolution. Lam (407) has introduced the term genorheithrum for his concept of the stream of potentialities drifting in time. This concept is based upon the fact that characters are somehow materializations or functionalizations of genes or combinations of genes (and other mutations). The idea of a stream of potentialities is based upon the continuity of the ever-changing genoplasms. The more basic characters of a phylogenetic stock (the more persistent ones held in common by diverging forms) result, according to Lam, from the more solidly linked genes at the "center" of the genorheithrum, whereas the unlinked or loosely linked genes are at the "surface" of the stream.

Phylogeny can be investigated by cytogenetic methods, as we have seen, as well as by comparative morphology, geography, and paleontology; but there is frequently great difficulty in extrapolation from the small differences studied by cytogenetics to the larger differences, and to changes in the past (cf. 282). Purely geographical data are also largely understood through deductive processes because they are concerned with present distributions which have had a time sequence in origin. Paleontology, according to Lam, is the only science in which direct interpretations of data concerning direction and rate of evolution can be made (with a few exceptions).

In the two fields of biogeography, floristics is essentially static in its studies of species and areas, as is taxonomy. Historical biogeography, however, is dynamic, as is phylogeny, because it considers the migration of genorheithra and the dispersal of kinds.

Taxonomy attains a logical basis when the data of comparative morphology can be arranged in a geographical pattern that coincides with the probable phylogeny of the group and the history of the floras and climaxes in which it has been involved. The advantage, of course, is reciprocal for geography. The study by Gleason (275), for example, of the evolution and geographical distribution of *Vernonia* in North America leads to the assumption that migration and evolution have proceeded simultaneously. The interrelations of several sciences in taxonomy can be illustrated by Stockwell's study (614) of the Composite genus *Chaenactis,* where we find a combination of criteria employed in a taxonomic revision. The criteria used by Stockwell are quoted below. Those concerned primarily with comparative morphology and phylogeny are: (1) Within a family or genus, woody perennial species are more primitive than herbaceous annual species. (2) Large and indefinite numbers of parts in the floral whorls are indices of primitive nature. (3) The condition of asymmetry of the floral whorls indicates a more advanced state than symmetry. (4) In Compositae similarity of all flowers in the head is considered a more primitive arrangement than that indicated by the presence of distinctive marginal or ray flowers. The remaining criteria are based upon genetic and geographical conditions. (5) The greatest concentration of closely related species is near the point of origin, provided barriers of some kind are available for maintaining distinctions until genetic stabilization is attained. (6) Greater variability within local populations is to be expected in a region of speciation (active evolution) than elsewhere. (7) Old and static habitats of a genus or a species are marked by morphological stability of local populations and minor but well-marked differences between separated populations. (8) New and active habitats are marked by morphological variability of local populations and the recurrence of similar individuals at more or less widely separated points. (9) Points at which the ranges of two or more closely related species meet or overlap are often the centers of origin of great variation within these species. (10) At zones of contact between closely related species hybridization often occurs, resulting in new combinations that may perpetuate themselves. This is a convergent evolutionary trend likely to produce unstable and possibly aggressive new forms. (11) At the periphery of the range of a species conditions tend to vary most from those at the point of origin, and selective forces are most active in eliminating forms unsuited to the habitat. If conditions differ sufficiently, certain new types may receive nature's stamp of approval.

It was not claimed by Stockwell that these criteria necessarily apply to groups other than the one he revised. Also, I have no intention of attempting to evaluate these criteria, although some of them are discussed later; I have introduced them to illustrate how modern taxonomy is based on more than mere comparative morphology.

In conclusion, descriptive plant geography has built up an imposing body of data and has through the years been on a sound scientific basis because it has been consistently inductive in its method, advancing from observation to observation. Interpretive plant geography, especially physiological plant geography, is on a less secure footing because of the deductive nature of many of its fundamental concepts. It is the role of interpretive plant geography, however, to attempt to discover coincidental phenomena in the more or less basic fields of science which are parallel with the areal and historical facts of floristics and other types of descriptive investigation. Crossing the hedgerows, the geographer must pass from one scientific field to another in search of parallels and possible causes of geographic phenomena.

2.

Certain Previously Proposed Principles of Plant Geography

A perfect set of principles of plant geography has never been written and probably never will be, if for no other reason than the practical impossibility of defining the exact content of the science. No effort will be made to review here the history of attempts to enunciate a decalogue for plant geography, or to trace any particular geographical conclusion back to its original author, whether he be Darwin, Drude, Engler, Humboldt, Schimper, Wallace, Warming, or some other well-known geographer. This chapter will treat primarily two recent publications in the field. Good (287), a British plant geographer, published a brief list of principles, and Mason (454), of the University of California, wrote what is essentially a revision and amplification of Good's list. This chapter is a review of these papers. The remainder of the book consists, in part, of an amplification of these principles with illustrative data and discussion, and such additional materials as compose the framework of plant geography, as I see it, conceived as an explanatory science which attains its unity and justification by abstracting and synthesizing from the contributions of more specialized sciences.

The following outline is a reorganization of the principles of Good and Mason, together with a few additions:

    A. Principles concerning the environment:
        1. Climatic control is primary.
        2. Climate has varied in the past.
        3. The relations of land and sea have varied in the past.
        4. Edaphic control is secondary.
        5. Biotic factors are also of importance.
        6. The environment is holocoenotic.

B. Principles concerning plant responses:
   7. Ranges of plants are limited by tolerances.
   8. Tolerances have a genetic basis.
   9. Different ontogenetic phases have different tolerances.
C. Principles concerning the migration of floras and climaxes:
   10. Great migrations have taken place.
   11. Migrations result from transport and establishment.
D. Principles concerning the perpetuation and evolution of floras and climaxes:
   12. Perpetuation depends upon migration and evolution.
   13. Evolution of floras depends upon migration, evolution, and environmental selection.

1. **Climatic control is primary.**[1]—Good wrote, "Plant distribution is primarily controlled by the distribution of climatic conditions," and Mason added, "and in any given region the extremes of these factors may be more significant than the means."

Plant geographers have long recognized this fundamental principle in the broad parallel between temperature belts and vegetation, and in the subdivision of these belts on a moisture basis. Clements (150), in his exposition of the climax, formulated a thoroughgoing basis for the consideration of the vegetation of the major climatic regions. The climaxes are the major regional vegetational types that correspond to the principal climatic types. They are subdivided into associations, faciations, and lociations on a basis of climatic differences within the regions. The climax communities are considered to be the highest types of vegetation that can develop under the different aspects of climate, and are in dynamic equilibrium with the climate.

Mason's corollary, that the extremes of climatic factors are more important than the means in controlling vegetational patterns, will be discussed in more detail in later paragraphs. At this point it is sufficient to call attention to the fact that Clements' interpretation is diametrically opposed to Mason's. Clements has said,[2] "It is not the extremes of environmental factors which are of importance, but the means." I believe that these two statements can be understood if it is realized that the authors are looking at vegetation from different points of view. Clements refers to the fact that the great climaxes are the long-existing products of the great climates. Many, if not

---

[1] See Topics 6 and 7 for related discussions.
[2] In conversation at the Alpine Laboratory, Colorado, August, 1940.

all of them, have existed for millions of years. Annual weather conditions and short climatic cycles or trends produce changes within the climaxes, especially in ecotonal regions, but these changes are within the fabric of the climaxes and are not seriously inimical to them. Mason illustrates his point of view by reference to the disastrous effects of the cold winter of 1932 on the northern range of the coast redwood. It is true that recently, and probably at fairly frequent intervals in the past, the reproduction and survival or failure of individual redwoods at the periphery of range of the species have been good or poor according to the favorableness or unfavorableness of weather conditions. The resultant rather slight oscillations in frequency or occurrence of the redwood during the passing centuries have meant very little, for the redwood species and the redwood forest type have maintained their essential integrity during long periods of geological time. When consistent trends in weather have resulted after a long time in climatic change, the redwood forest has migrated in pace with the areal movement of the climatic type to which it is adapted. In the sense of Clements, then, it is not the extremes but the means or average conditions that determine the principal area of a species or of a vegetation type. The actual migration of a climax vegetation can take place only under the compulsion of a climatic change. The result is a cliseral movement. Only a few major cliseral movements have occurred in the history of modern vegetation, although there have been many lesser movements, largely of a fluctuating nature. Clements sees vegetation in the large, understands the unity of types and their long history, and does not deny the role of the individual organism or the influence of environmental conditions upon its welfare. Mason is looking at what happens to an individual during periods of unusual environmental stress, and he emphasizes that what happens to populations and to vegetation types is, in the last analysis, a question of what happens to individuals.

2. **Climate has varied in the past.**—Good said, "There has been great variation and oscillation in climate, especially at higher latitudes, during the geological history of the Angiosperms." Mason modified this statement merely by the substitution of "during the geological past" for "during the history of the Angiosperms."

The evidence for this principle is familiar to every student of earth history: it is apparent that the great climatic changes have wrought comparably great vegetational changes during the geological past, and animals, of course, have been associated with the changes. As Mason pointed out, "Climate seems to be the only significant variable that operates from one region to another in a manner which would stimulate the migration of floras." Since the causes of climatic change are usually not regional, and since one climatic change cannot occur without concomitant effects on various regions, the movements of vegetation have usually shown a regional parallelism and been cliseral. The great cliseres have migrated northward and southward and up and down in altitude, with temperature changes, and, in general, continentally or coastward with moisture changes. Local features

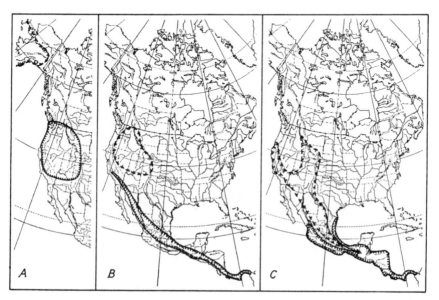

FIG. 1. Change in position of temperate (broken line) and subtropical (solid line) Cenozoic floras in western America, according to Chaney (130). *A*, Eocene; *B*, Miocene; *C*, Recent.

of topography and of relations between land and water masses, of course, have produced abundant deviations in the direction of movement. Certain climaxes are known to have moved a thousand or two miles, and prairie types in central North America have moved hundreds of miles eastward and westward. The biotic units of a clisere move together as they are simultaneously affected by changing climates, and the actual area of any one climax may also be compressed or expanded according to the conditions. During these great climatic changes, any one area has seen a constantly changing parade of vegetation back and forth if there have been great oscillations of climate, as during the Pleistocene.

3. **The relations of land and sea have varied in the past.**—Good says, "At least some, and probably considerable, variation has occurred in the relative

distribution and outline of land and sea during the history of the Angio-sperms." Mason contributed the same rewording as in the preceding prin-ciple.

Again, the evidence for this principle is completely familiar to geologists, geographers, and students of biotic history. The facts suggested by this

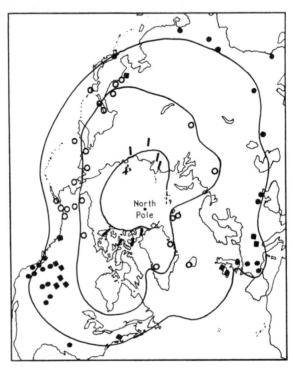

Fig. 2. Eocene isoflors. Black disks, subtropical floras; squares, intermediate floras; hollow circles, temperate floras; rods, cool-temperate floras. The Eocene isoflors (data after 135) in the northern hemisphere are assumed to indicate regions of essentially similar climate. Note, for example, that the subtropical isoflor swings northward along the west coasts of Europe and America and along the east coasts of Asia and America. The temperate isoflor behaves similarly, and swings even farther northward. The relations between floras seems fairly clear, but Chaney's drawing of isoflors appears to be rather free and not rigidly controlled by the data. In preparing this map, Chaney's data have been replotted from Mercator to Lambert's azimuthal equal-area projection; the present outline is based on Goode's copyrighted outline map No. 201PN, with permission of the University of Chicago.

principle have an importance for plant geography because of the connection between continental masses, oceanic basins, and climates. The disposition of land and sea is a primary cause of the climate of a region, and the climate is a primary cause of the vegetation. Heat, moisture, and wind conditions on land within the latitudinal belts result in a large way from the proximity to large bodies of water, relations to oceanic currents, the effects of mountain ranges, etc. The geological causes of such topographic changes are not of immediate importance to geography, but the fact that they have occurred and the nature of the changes they have induced are of importance.

Changes in the relative distribution of land and sea are also of importance to geography because of their relations to highways for and barriers to migration at various times in the past. It is of extreme importance to note that the establishment of the fact or probability of certain land connections in the past, whether by continental bridges or by continental displacement, is not sufficient for the student of biological history. Not only must a land connection have existed, but conditions suitable for the migration of particular types or organisms must also have been present. When two temperate floras show a strong genetic relationship and when they are today separated by a barrier, it is insufficient to show that the barrier did not always exist; it is necessary to show also that a temperate climate existed in the past.

4. **Edaphic control is secondary.**—Mason concurs with Good's statement, "Plant distribution is secondarily controlled by the distribution of edaphic factors."

The edaphic factors are those of the soil, in distinction to those of the atmosphere, which are climatic, but they are by no means independent sets of factors. Soil development is strongly under the influence of climate and vegetation in its broader aspects, and of topography in its smaller aspects. Accepting the concept of the climax and of the concomitant development of climax soils, we can expect only certain limited outcomes with the passing of time and in the absence of destructive changes. The ideal, hypothetical, essentially uniform integration of climate, vegetation, and soil, however, is by no means actual, for everywhere within a climatic region there are microclimates, and edaphic conditions vary almost from spot to spot. Out of this great heterogeneous mosaic of local conditions some order can be made, and it is apparent that climatic control is primary and that edaphic control in plant distribution is secondary. The thousands of kinds of organisms that are adapted in such a manner that they can tolerate the climatic conditions of a region cannot grow everywhere within that region because of the variable edaphic conditions which circumscribe their areas. On the other hand, rather similar edaphic situations exist under two or more climates, but similar organisms cannot occupy them because of the climatic limitations. For example, widely distributed weeds adapted to pioneer conditions in sandy soil of agricultural and other situations cannot grow successfully in all such situations because of climatic limitations. Species tend to be found within climatic areas everywhere that suitable soil conditions are found. This statement, of course, has to be qualified by the introduction of the element of time necessary for thorough dispersal, and by the element of chance which operates when time has been insufficient.

5. **Biotic factors are also of importance.**—In addition to climatic and edaphic factors, many ecologists and geographers employ other categories, including biotic, physiographic, catastrophic, and historical. Of these, the biotic factors compose a distinctive group that deserves separate considera- tion. Physiographic, catastrophic, and historical factors are largely if not completely climatic, edaphic, and biotic when analyzed, and approach to them from these points of view is largely a matter of emphasis.

All modern ecologists recognize the importance of biotic factors; Turrill (654), for example, has included a statement concerning them in his exposi- tion of the principles of plant geography. In Clementsian terminology, *action* consists of the effects of environmental factors on organisms; *reaction* con- sists of the effects of organisms on the habitat; and *coaction* is reserved for the interactions among organisms. According to the bioecologists (see 157), the role of animals is that of major and minor influents, and the primary characteristics of a biome (climax formation of plants and animals con- sidered together) are due to plants. Plant responses to climate appear usually to be more direct than in those of most animals, which are more closely related to the plant cover through their food and shelter requirements.

One of the best examples of the importance of biotic factors is the dis- climax—a long-enduring subclimax stage that is prevented from attaining cli- max condition because of animal disoperation. For example, the central North American short-grass plains were long considered to compose a climax as- sociation of the grassland formation in the same sense as the mixed-grass and tall-grass prairies farther east, but Clements and others have recently found that much of the plains are disclimax, held in the short-grass condi- tion by overgrazing. Through the action of biotic factors, which in the above example consist largely of changes due to overgrazing, the area of climax species and communities is diminished and certain successional forms are extended in area and density.

Other types of biotic factors include such relations as exist between hosts and parasites, obligatory food dependency, obligatory pollinating insects, etc. Among such phenomena are found clear examples of the action of biotic factors in the limitation of distributions within the complex of other limita- tions of a climatic and edaphic nature.

6. **The environment is holocoenotic.**—In a brief discussion of ecological

principles, Allee and Park (10) emphasize the fact that the factors of the environment act collectively and simultaneously, and that the action of any one factor is qualified by the other factors, i.e., that the environment is holocoenotic.

The concept of the action of a limiting factor is, of course, a familiar one to all ecologists doing any sort of physiological investigation, and I need only to refer to an example or two. The case of the redwood has already been mentioned—that although the mean temperature for 1932 was little different from the normal, the unusual freezing temperatures were evidently critical for the new redwood growth. Apparently such seasons occur sufficiently frequently to prevent the redwood from extending its area northward into what appear otherwise to be suitable territories. According to Mason, "This is clearly a case of the extreme being the controlling factor."

In another discussion of the biological results of the unusual year of 1932, Taylor (622) stated that the importance of such conditions and results requires a restatement of Liebig's Law of the Minimum (418) so as to bring it into line with the concept of critical time. Taylor's rewording of the law follows: "The growth and functioning of an organism is dependent on the amount of the essential environmental factor presented to it in minimal quantity during the most critical season of the year, or during the most critical year or years of a climatic cycle." Turrill (654) also discussed the application of Liebig's law to phytogeography. He stressed the importance of heat and moisture factors in relation to the "relic effects" of past climates and to soil development.

The point of view represented by the preceding authors requires a consideration of the actual situation by which ecologists and geographers are confronted, because physiological plant geography is rife with a priori explanations, deductive reasoning, and poorly founded conclusions as to causation.

Physiological processes are multi-conditioned, and an investigation of the effects of variation of a single factor, when all others are controlled, cannot be applied directly to an interpretation of the role of that factor in nature. It is impossible, then, to speak of a single condition of a factor as being the cause of an observed effect in an organism. Approximately the same result in plant development or distribution may be induced by combinations of different environmental factors or by different intensities of the same factors.

The principle of multiple causation was early recognized by Liebig, who supposed that when several factors affecting growth are present in different intensities, each less than the intensity most favorable to growth, only an increase in the factor present in the least amount would result in increased growth. Mitscherlich (468) believed that any deficient factor, when increased, will result in increased growth, and that the increase is greatest for a unit increase of the factor that is most deficient. That is, a unit increase of a factor will produce a greater effect when the factor is very deficient than when it is only moderately deficient, and a unit increase of each of several factors will have effects relative to the extent of their separate deficiencies. Furthermore, when a factor is increased, say by tenths of the amount necessary for maximum production, each succeeding increase results in approximately one-half the increase of production as does the preceding tenth. And when two factors are present in such an amount, for example, that each tends to result in one-half of the maximum production, the result is approximately their product, or one-fourth of the maximum production when both factors are present in sufficient amount. It is erroneous, then, to speak of a single condition of a single factor as being limiting; quite definitely, the environment *is* holocoenotic.

This situation is recognized by some physiologists but without sufficient emphasis, as I see its importance, and even less is it recognized by some ecologists and physiological geographers. Livingston and Shreve (421), however, went part way in the recognition of the interrelations of factors and the difficulty of integrating physiological studies of separate factors, but they were not willing to conclude that the ecological situation was so complicated that specific causation could not be analyzed. They say, "A large amount of laboratory experimentation of the most refined physical sort will be required before we shall ever approach an adequate knowledge of the influence of single conditions upon plants; the far more difficult study of the complex environmental systems of which single conditions are always component has already begun to attract attention." Klages (396), in his *Ecological crop geography,* has a reasonable approach to the problem. He says, "While classifications of habitat factors may not, and are not expected to, explain the very complex relationships of a plant during its various phases of development to its also changing environment as a growing season progresses, nevertheless, they may be of great help to the student in arriving at some conception regarding the processes involved."

Hartshorne (333) has made a searching inquiry into the historical development and modern concepts of geography that should be read by all investigators who, finding simple parallelism between certain environmental factors and certain chorological phenomena, assume that they have discovered the causes of the phenomena. As he points out, Humboldt, Ritter, and other early geographers were well aware that "in the great enchainment of causes and effects, no material and no activity may be studied in isolation." These great geographers tried to demonstrate their philosophical concepts not by deductive logic but by inductive reasoning based upon observations of nature. As Raup (519) has emphasized in his discussion of plant geography in which he takes up the thesis of Hartshorne, the complex interrelations among the environmental factors, and between them and the organism, with its complex physiological and morphological interrelations, are such as to defy solution in exact terms of causation. Ecological problems not only may be difficult of solution because of the interaction of factors and responses, but they may really be insoluble in a mathematical sense. I believe that it is worth saying again that the environment is holocoenotic and that organic responses are multi-conditioned. The geographer does well to adhere strictly to an inductive basis, hoping only to discover coincidences and parallelism between physiological reactions and ecological and geographical phenomena.

Two points in the preceding paragraphs need further emphasis. Taylor's emphasis on critical time is properly placed. As an organism goes through its ontogeny its physiological processes, its reactions, its requirements change. Environmental conditions that are critical at the time of germination or in the seedling condition may not be and likely are not critical later on. Conditions that are critical in flowering and pollination may not be critical during younger stages in the lives of plants. Ontogenetic conditions that are correlated with seasonal aspects may be adversely affected by unseasonal conditions at any time.

The second point is compensation. Although plants live under a basic climatic control and are, generally speaking, confined to the climatic regimen to which they are adapted, they may sometimes occur outside that climate because of compensation for the climatic difference through local edaphic, physiographic, or biotic conditions. Compensation must occur when, for example, plants of the tall-grass prairie live in the mixed-grass or short-grass associations to the west. In such cases it appears that the drier climate is compensated by a greater than usual availability of soil moisture, so that although the transpiration rate is higher, the supply of water is sufficient to produce a favorable water balance.

7. **Ranges are limited by tolerances.**—Mason stated this principle as follows: "The functions governing the existence and successful reproduction of plant species are limited by definite ranges of intensity of particular climatic, edaphic, and biotic factors. These ranges represent the tolerance

of the function for the particular factor." This statement is essentially a brief of a portion of Good's Theory of Tolerance (287). He said, "Each and every plant species is able to exist and reproduce successfully only within a definite range of climatic and edaphic conditions. . . . The tolerance of any species is a specific character subject to the laws and processes of organic evolution in the same way as its morphological characters, but the two are not necessarily linked. . . . Change in tolerance may or may not be accompanied by morphological change, and morphological change may or may not be accompanied by change of tolerance. . . . Morphologically similar species may show wide differences in tolerance and species with similar tolerances may show very little morphological similarity. . . . The relative distribution of species with similar ranges of tolerance is finally determined by the result of competition between them. . . . The tolerance of any larger taxonomic unit is the sum of the tolerances of its constituent species." These concepts will be scrutinized.

Originality is not claimed for these concepts by either Mason or Good, and the general idea involved has formed much of the warp and woof of geographical thought since the time of Darwin. They are related to the connotation of such terms as *Lebensfahigkeit* and vitality, and are comparable to Livingston and Shreve's Theory of Physiological Limits. The concept of limitation by tolerance is based on the following observations: that organisms do not live everywhere, but each lives only under certain more or less limited types of environments, and does better under conditions well within the extremes of environmental conditions that are tolerated; that in nature certain conditions of one or a group of factors appear to be limiting to the life or well-being of an organism; and that in the laboratory it is demonstrable that there are three cardinal points, the minimum, optimum, and maximum, with respect to the reaction of an organism to the variation of a factor. The operation of a condition of a factor in nature must be qualified by the multi-conditioned nature of responses, as pointed out in the preceding section; so it is misleading, if not erroneous, to state that "functions . . . are limited by definite ranges of intensity of particular . . . factors." That is scarcely demonstrable for individual factor conditions.

Good expresses the idea that the practical basis for the theory rests on the fact that organic evolution is slower than environmental change on the average, and hence migration occurs, and on the fact that changes in tolerance are specific characters conformable to and controlled by the laws of evolution. Good's statement is accepted by Wulff (718, 719), and he gives a series of instances of lack of harmony between some species and their present conditions. This lack of harmony shows itself in relic species, in lack of coordination between periodicity of growth and rhythms of climate, in incomplete life cycles, in the dying out of a relic species, etc. In many instances there may be a lack of harmony between communities and the general environmental conditions of an area. Such communities are relic as truly as species may be, and frequently consist of the preclimax and postclimax associations in the sense of Clements' system. If the plants of such communities can reproduce themselves, their lack of harmony is more apparent than real, for they must be living in regions of compensation. When further environmental changes make life increasingly difficult for relic plants and when there is no opportunity of further migration, they are doomed to extinction unless fortuitous evolution makes them better fitted for the new conditions.

There is, however, a different school of biologists who do not accept such specifically circumscribed hereditary limitations of ecological amplitude. These neo-Lamarckians believe that changing environments call forth favorable mutations, or even that adaptation and the character differences shown by related species of a phylad do not necessarily have a genetic basis. It seems to me that the burden of proof lies with those biologists who believe in adaptive morphological and physiological adjustments (other than the simple ecads) which do not depend upon genetic change. The opinions of Clements can serve to illustrate the neo-Lamarckian view. In referring to Good's opening statement, "Each and every plant species is able to exist and reproduce successfully only within a definite range of climatic and edaphic conditions," Clements[3] has said, "The word definite should be changed to read indefinite." This rather startling statement can be interpreted only in the light of his belief in the inheritance of acquired characters. Rather than believing with the geneticists that the characters of species originate with internal changes, Clements believes that, in some instances at least, the environment calls out the changes, causes them to occur. Clements has claimed[4] the reciprocal conversion of several pairs of Rocky Mountain species of plants by the simple experiment of reciprocal transplantation. When differences between two forms are purely modifications induced by the environment and when they are within the inherited amplitude of the type, i.e., when the forms are ecads, transplantation will overcome the differences. Clements, however, claims to have changed specific, generic, and even some

---

[3] In conversation at the Alpine Laboratory, Colorado, August, 1940.
[4] In several recent *Yearbooks* of the Carnegie Institution of Washington.

family characters through non-genetic manipulation. He admits that secondary species are easier to change than primary species, and that young species are easier to change than old ones; that is, the longer a "species" population has experienced a certain environment, the more difficult it is to change greatly. This is because the characters caused by the environment tend to become fixed. With such views, it is understandable that neo-Lamarckians would not believe that species have definite tolerances.[5]

One difficulty in applying to species the principle that ranges are limited by tolerances results from the fact that a species is biologically a very large population as a rule, and that it consists of thousands of biotypes arranged in a pattern of local populations, ecotypes, and geographical subspecies. Just as truly as a species consists of individuals which differ morphologically, so do they differ physiologically, in their tolerances. It is perfectly possible to speak of alpine species, salt-marsh species, etc., but within the alpine belt and within salt marshes there is a variety of conditions, and even such species will have their biotypes more or less sorted out on a combined habitat-tolerance basis. Other species are known to have maritime, montane, and even alpine ecotypes. With this phenomenon in mind, it seems to me that Good's statement that "the tolerance of any larger taxonomic unit is the sum of the tolerances of its constituent species" can have no real meaning. The tolerance of one organism cannot affect the tolerance of another except through inheritance. It would seem seldom if ever to serve any scientific purpose to add up the diverse tolerances, for example, of *Pinus* (with such species as *edulis, flexilis, banksiana, palustris,* etc.). There remain only such generalizations as that a certain genus is temperate or a certain family is tropical.

I wish to make one more point with respect to tolerances. Some species apparently have a wide adaptability, whereas others seem to be highly specialized. This is over and above the phenomenon of biotype richness or poverty. There are not very good cases to illustrate this point, but relics will serve my purpose. Some relics seem to be barely able to survive within their peculiar habitats, whereas others may be local but vigorous, as is the case with *Chamaecyparis Lawsoniana* in northwestern California, which is known as far back as the Oligocene.

---

[5] Although it is not within the scope of this discussion to develop the thesis for either side of the above argument, it should be stated that other work in experimental evolution fostered by the Carnegie Institution of Washington, and using transplantation as one of its major techniques, has recently resulted in publication (146). These investigators explain their results on a strictly genetic basis, recognizing that subspecies (ecotypes) and species (ecospecies) differ by inherited traits that cannot be overcome by transplantation.

8. **Tolerances have a genetic basis.**—This principle is inferred in the writings of Good and Mason, and the latter has decided, since his publication on geographical principles, to add a statement concerning the role of genetics. I quote from his letter to me written in February, 1941: "I am much interested in your conclusions as to the place of genetics in plant geography. Since you left Berkeley I have been doing a great deal of thinking along these lines and have formulated a statement that I have presented to my classes and added to my Principles of Plant Geography. I quote it to you. 'The capacity of the species to tolerate or to respond to its environment is governed by the laws of evolution and genetics, and the *range* of tolerance is the direct result of the genetic diversity of the species; that is, the mutation, combination and recombination of the genetic units within the species are responsible for variations in tolerance upon which the environment acts in the selection of its biota.' " There is no need to add any further comments at this point, for many of the subsequent chapters are devoted to the place of genetics in plant geography.

9. **Different ontogenetic phases have different tolerances.**—Mason states this principle as follows: "In the life history of the organism there are times when it is in some critical phase of its development which has a narrower tolerance range for a particular factor of the environment. The distribution of this intensity span of the factor limits the area in which the function can operate and, hence, governs the distribution of the species. The narrower the range of tolerance, the more critical the factor becomes."

In the development of the area of a species, migration is not accomplished by dispersal alone. Seeds must germinate and seedlings must become established. Establishment is usually the most critical phase of ordinary life because organisms usually have more narrow tolerances in juvenile than in adult stages. Mason says that the factors limiting establishment are among the most important of climatic barriers, and the same can be said for edaphic factors. Edaphic conditions, such as the water relations of *Taxodium distichum* in river swamps or of *Populus deltoides* on sand dunes, may control establishment, although offering no serious handicap to mature individuals. In fact, the relief from competition found in such situations may be a distinct advantage. In succession the role of ontogenetic differences in tolerance frequently appears. For example, the climax beech-maple forest (*Fagus grandifolia-Acer saccharum*) at Turkey Run, Indiana, contains several large, old, and vigorous white oak trees (*Quercus alba*), but the processes of succession have long since produced reactions that make the site unfavorable for white oak reproduction, and no young trees or seedlings are found in the forest.

Another type of illustration is found in the well-known fact that conditions unfavorable for pollination or for seed production may be excellent for the vegetative growth of certain plants.

10. **Great migrations have taken place.**—Mason concurs with Good's statement, "Great movements of floras have taken place in the past and are continuing to take place."

The paleontologists have provided abundant proof of this principle. It is especially certain when, as in the case of *Sequoia,* rocks of successive ages show the same fossils, but with a progressive shift of latitude through the Tertiary. Similar but more abundant evidence is being obtained from pollen analysis of Pleistocene and recent peat deposits. This whole subject is treated in detail in the succeeding chapters.

11. **Migration results from transport and establishment.**—Good said, "Species movement (plant migration) is brought about by the transport of individual plants during their motile dispersal phases." To this statement Mason added the all-important phrase, "and the subsequent establishment of these migrules."

With the exception of lower forms of aquatic life with actively motile stages, and limited amounts of spread due to growth and vegetative reproduction, plant parts play only a passive role in dissemination. The principal diaspores (migrules) are spores, seeds, and fruits, and the principal agents of dissemination are wind and water. There is only accidental cooperation between the two, although many plant structures provide facilitating devices; so it is purely fortuitous that a diaspore reaches a spot at a time when conditions are suitable for germination and establishment. Dissemination is but a necessary antecedent, and it is only upon establishment in a new locality that migration has occurred.

Mason has emphasized the fact that "the mechanism behind the establishment is the mechanism of stimulus and response wherein particular factors of the environment act upon particular functions of the plant." [6] This is no more true for juvenile stages than for later ones, but, as he points out, juvenile tolerances are usually more limited.

12. **Perpetuation depends upon migration and evolution.**—Mason says, "The perpetuation of vegetation is dependent first upon the ability of the species to migrate and secondly upon the ability of species to vary and to transmit the favorable variations to their offspring."

Migration, adjustment to new conditions, and extermination are the three processes which play a part in the development of vegetation under the

[6] See Principle 6.

duress of a changing environment. Except for the occupancy of bare areas, migration consists of emigration and immigration, which are simultaneous phases of a change caused by environmental change. Strictly speaking, adjustment to new conditions, or organic evolution, is not perpetuation. It is perpetuation only of a phylogenetic stock or of a community type, not of plants of a specific genetic constellation. Extinction and extermination, of course, result when neither of the first two processes is adequate to meet the changing conditions. The paleontological record contains abundant evidence of all three processes.

13. **Evolution of floras depends upon migration, evolution, and environmental selection.**—According to Mason, "The evolution of floras is dependent upon plant migration, the evolution of species, and the selective influences of climatic change acting upon the varying tolerances of the component species."

As one studies the fossil record of successive geological times, there appears a story of changing floras. A series from one locality represents what has lived in that area at different times. New plants have immigrated into the area and preexisting kinds have departed; other new plants have evolved in the area and some have become extinct. Immigration and evolution and emigration and extinction cannot be solved from a knowledge of a single area, but only when the larger geological story is known. Berry (62) is of the opinion that in any one area the geologically sudden appearance of a new flora is more a result of immigration than of evolution.

Some of the principles that have been briefly considered in this chapter will not receive more than incidental consideration in this book. This is specially true of those that belong more in the province of physiological plant geography. Others, those having primarily a chorological significance, will be dealt with in some detail. Some of the principles of Good and Mason are reworded and elaborated and additional ones are added as the subject is developed. The present chapter is considered primarily as an introduction and a means of obtaining a quick orientation in a field that seems naturally to tend to be diverse and ramifying.

PART TWO

Paleoecology

3.

Definition of Paleoecology and Limitations of the Method

The earth has been considered to consist of four spheres: the central lithosphere, a more or less superficial hydrosphere, an enveloping atmosphere, and a biosphere. Most life is restricted to a narrow horizon where the interactions of the other spheres are strongest, where it is subjected to the effects of a changing environment. Physical and chemical processes in the inorganic world continually and simultaneously bear upon organisms, adding their intricacy to the complicated dynamism of the forms which make up the biosphere. Life has been spoken of as a continual adjustment of organisms between their external and internal environments. Ceaseless change is an obvious phenomenon because of the reactions to the changing physical environment and the inherent specific ontogeny of the individual.

It is frequently not so apparent that plant and animal communities likewise undergo changes. This lack of obviousness is due to the fact that changes in the mantle of vegetation at any one place and in the composition and structure of communities, the development of vegetation, is on a different time scale from that of the individuals which compose the communities. The faster plant successions usually require decades and sometimes centuries for their completion. The great shifts of vegetational types are to be measured in hundreds or thousands of years at the least, and many involve larger units of geological time. The development of vegetation may await the slow processes of reaction and soil genesis. The shifting of climaxes awaits the slow deterioration or improvement of climates, which in turn may depend upon the slow geological processes of diastrophism or gradation.

Ecology, as a scientific discipline, comprehends the complex interrelations between organisms and habitats. It furnishes the points of departure which lead to past relations. Paleoecology, then, is the study of past biota on a basis of ecological concepts and methods insofar as they can be applied. No sharp line need be drawn between ecology and paleoecology. The principal difference is that the criteria of the latter are of necessity all inferential. The term, however, is useful because of the emphasis it gives and the restrictions it implies. Paleoecology bears the same relation to paleontology that ecology bears to biology. We may well accept the definition Clements (150) used to introduce the term: *The term paleoecology may be broadened to include the whole study of the interactions of geosphere, atmosphere, and biosphere in the past.*

Paleoecology differs from paleontology mainly in the breadth of subject matter and of evidence used in piecing together the history of life on the earth. In practice, the usual paleontologist has concerned himself mainly with the description, naming, distribution, and classification of fossils, together with the more immediately derivative subjects of stratigraphy, dating, and evolution. The paleoecologist must also be a biologist who is familiar with modern communities, climates, and their interrelations. He must be familiar with biological processes as exemplified in communities because any adequate understanding of history must be based upon processes now operating and conditions now existing. Starting where the paleontologist usually stops, the paleoecologist attempts to reconstruct an accurate picture of the life and the life conditions of the past by inferences derived through a study of modern biota that are comparable with fossil ones. He must combine the point of view, knowledge, and methods of the biologist and the geologist. With the statement that the science of paleontology is broad enough to include paleoecology there can be no quarrel, but the fact remains that in practice paleoecology has been the broader of the disciplines. The work of the paleoecologists has been largely concerned with botanical aspects. The reasons for that will become clear as this chapter develops.

At this point a further comparison may be useful. The relationship between paleoecology and paleontology is much like that between physiological plant geography and floristic plant geography (519). Floristic plant geography is largely an empirical and inductive science, progressing from observation to observation and finally, by weight of evidence, to conclusions. Frequently, however, it is satisfied by the discovery of coincidences and is less concerned with causes. Through a long history floristic geographers have found the environment to be too complicated in the interrelation of its factors to allow the statement of causal relations between factors and forms, floras, and areas. Physiological plant geography, on the other hand, commences with the assumption of adaptation and of direct relationships between causal factors and resultant biological phenomena. It arose with Darwinian natural selection and its inherent assumption of adaptation, and modern ecology is "heir apparent" to the philosophical method involved. Modern ecology has gone on to add its full quota of basic assumptions, many of them unproved or unprovable. Whatever scientific weaknesses ecology has, paleoecology also has.

Paleoecology has weaknesses, but let us remember that it attempts more than ordinary paleontology, and that it is making an effort to eliminate preconceptions from among its basic principles.

Basic assumptions of paleoecology.—According to Clements (150), the interpretation of past vegetations rests upon two basic assumptions from which naturally follows a third.

1. The operation of climatic and topographic forces in molding plant life has been essentially the same throughout the various geological periods.

2. The operation of succession as the developmental process in vegetation has been essentially uniform throughout the whole course of the geosphere.

3. The types of responses of animals to climate and to vegetation, both as individuals and in groups, have remained more or less similar throughout geological time.[1]

Without these basic assumptions the science of paleoecology could not exist. The universality of operation of physical and chemical laws and the continuity of cause and effect sequences cannot be doubted without discarding the whole structure of science. The basic assumptions of paleoecology would appear to be beyond question.

Basic paleoecological principles according to Clements.—The following outline of basic principles has resulted from an effort to extract, outline, and reorganize the principles expressed or implied in Clements' pioneer publication (150). In this skeletal form the ideas may not all be clear at first. They should be reread in the light of the discussion which follows, and with reference to the literature.

A. The present is the key to the past.

    1. Large topographic changes have always resulted in modification of climatic conditions, because of the effects of sea, land, and altitude on temperature, air movements, etc.

    2. Climatic forces have operated continuously on the development and structure of vegetation.

    3. The operation of the processes of succession was essentially the same

---

[1] This does not preclude the possibility that in the past there may have been temperate species in a genus which is now tropical, or vice versa.

during geological time as it is today, qualified as to details of expression by remoteness in time and evolution.

    4. The responses of animals to vegetation and to climate have remained of more or less similar order.

B. The vegetation is the key to the biome.

    1. The primary correlation is between the physical world and vegetation.

    2. A secondary but fundamental relationship exists between the animal and the plant worlds, with plants affecting animals more than the reverse.

    3. Vegetation must serve as the keystone because it is both an effect of topography and climate, and a primary cause in relation to the animal world.

C. The interrelations in the natural world are not strictly linear, but are usually mutual and complex.

    1. The fundamental sequence, however, is (a) topography; (b) habitat, in which the climatic and edaphic forces are direct factors; (c) vegetation; (d) animal communities, in which the main causal relations are essentially linear and progressive.[2]

    2. There are also complex cause and effect interrelations. Although vegetation is primarily the result of climate and soil, it is also affected by animals and man. At the same time that it exerts a causal effect upon these, it is also reacting in a critical manner upon the habitat itself. Likewise, while vegetation seems the decisive factor in the development and distribution of animal communities, physical factors operate directly and decisively upon them as well as indirectly through vegetation.

    3. Topographic change looks in but one direction, forward to its effects. Animal life, with certain exceptions, can only point backward to its causes because animals are never a principal control of large vegetational types or of large topographic changes. The evidences drawn from vegetation point in both directions, toward the originating physical causes and the resulting faunal effects.

D. The method of causal sequence places each fact in relation to its cause and effect in the primary sequence.

    1. Every fact to be used as evidence in paleoecology has its proper place in the primary sequence (habitat—vegetation—animals).

    2. Each bit of plant evidence can be read in two directions.

---

[2] Special cases in which this primary sequence is changed would include such as the following: earthworms—soil—vegetation, beavers—pond—vegetation, man—vegetation.

E. The method of succession is based on the phenomenon of development and stabilization.

    1. Climatic areas must have supported climax vegetation in geological times as they do now.

    2. Bare areas must have arisen within these climax areas and have been the focal points of seres.

    3. The progressive reactions of plants upon their environments must have resulted in shifting community structure, passing through initial, medial, and final associes to the relatively stable climax.

    4. The ecesic causes or processes, viz., aggregation, migration, ecesis, competition, invasion, and reaction, must have been universal, and their action essentially the same then as today.

F. Cliseral relations have always existed.

    1. Climatic zonation probably always existed, but was more or less pronounced at different times.

    2. The great climatic regions must have supported great climax formations (biomes, climaxes) which bore an orderly and inescapable relationship, spatially, to the climates.

    3. The movement of climatic-type areas, whatever the causes, must have caused corresponding shifts in the spatial positions of the climaxes.

        a. These great cliseral movements must have left behind, in places of suitable environmental compensation, relic colonies, even as they have in recent time.

        b. In regions of topographic diversity, and consequently of climatic diversity, there must have been alternations or interdigitations of climaxes.

        c. Two climatic climaxes cannot occupy the same territory simultaneously, but their elements can be more or less mixed in ecotonal or transitional regions.

G. The methods of causal sequence and succession have a concomitance.

    1. Every fact discovered, whether climatic, topographic, geological, botanical, or zoological, has its place in the causal sequence as well as in the successional development.

    2. The most critical part of an analysis consists in coordinating and harmonizing the evidence thus obtained.

        a. Evidences are considered especially trustworthy if those of the paleontological record agree with the facts as to climate and topography as revealed by the geological record.

        b. The converse must likewise be true; geological indications of a dry climate, for example, are more reliable when substantiated by biotic evidence.

As mentioned in the paragraph closing the discussion of the scope and characteristics of paleoecology, ecological thought has built an elaborate structure upon certain basic assumptions that may not be wholly true. The fact that the science is predicated upon these assumptions requires of an ecologist that he be wary of accepting them as universal phenomena and of allowing them to color his conceptions of observational facts. I refer to the assumptions of adaptation, of the organismal nature of vegetation, and of the climax.

Although the fundamental fact of adaptation—that organisms can live only under certain conditions—can readily be accepted, the difficulties of understanding adaptation are numerous, complex, and almost if not entirely insurmountable. Studies of the effects of single factors are almost useless, except as coincidences may be discovered, because organisms have relations to the totality of simultaneously operative factors. The conditions of maximum, optimum, and minimum for an organism with respect to intensity of a factor may change as any other factor of the total environment changes. In the last analysis, the environment is holocoenotic. Furthermore, the problem is complicated by the individuality of organisms and the fact that many if not most species are constituted of genetically (and physiologically) different populations.

The concept of vegetation as being universally developmental has led both to the holistic concept of the formation as being like an organism (or a quasi-organism or superorganism), and to the monoclimax hypothesis. Although the phenomenon of vegetational change is recognized as being universal, it is not proved to everyone's satisfaction that succession is always progressive, or that it always results in a single climax within a climatic region. The analogy with an organism and an insistence upon the monoclimax hypothesis are both assumptions that can and do lead to conclusions beyond the evidence of observation and thus lose their usefulness. The climatic climax in its largest expression, the formation or biome, such as the eastern American deciduous forest, is a recognizable and definable unit. On the other hand, the concrete community or association-individual is a reality and can be delimited in time and space, but the conceptions of abstract associations or community types may be only convenient classificatory units in a world of merging environments and communities.

These remarks are intended not to prejudge the science of ecology as practiced by the majority of American students, or its extension into paleontology, but merely to recognize the hypothetical basis of much of ecology and to provide a caution with respect to some of the ensuing discussion.

In the following pages it will be seen how the above ideas form the warp and woof of paleoecological thought. For the sake of ease in presenting the subject, the discussion is organized under a series of topics. Each topic begins with a few sentences which are considered to be concise statements of working principles, and is followed by a discussion designed to illustrate the concepts and methods. In this way it is hoped that the science of paleoecology will emerge clearly with respect to its bases, methods, and accomplishments, and that the importance of the field for plant geography will be stressed. Naturally, the topics are not independent; the principles and materials used for one topic apply there especially, but not necessarily exclusively.

Limitations and sources of error in the paleoecological method.—If paleontology consisted merely of the description and naming of fossils for purposes of cataloguing and stratigraphic correlation—and much of it has been little more than that—it would be a static although useful science. An interest in evolution has given paleontological taxonomy a meaning and a dynamism, as it has the taxonomy of modern organisms. The introduction of the ecological point of view and method, however, has added further interest and possibilities to the attempt to interpret the story of the past. Topographic, climatic, vegetational, and animal history can better be reconstructed by the addition of the paleoecological method than from a wholly geological approach. Much of the method hinges on the very important ability to recognize in modern floras the species that are equivalent to or most nearly like those represented by the fossils. A failure in this basic reference produces no simple mistake, but one that is far-reaching because the significance of the plant fossil is interpreted from its central position in the primary sequence of interrelations: habitat-plant-animal.[3] The paleoecologist reads from the fossil backward in the causal sequence suggestions as to the community which it represented, the climate which controlled the community, and the topography which controlled the climate. He reads forward from the plant fossil to its probable animal associates in the biome, because of their close food and shelter relations.

A false reference to subtropical species in a fossil flora, which is otherwise temperate, presents problems which really do not exist and results in conclusions which are entirely false. The false assignment of certain fossils to subtropical genera in far-northern floras (Greenland, Spitzbergen) has caused the supposition that a subtropical climate existed at very high latitudes. This in turn has played a role in theories concerning the migration of the north pole, in theories of continental drift, of oceanic continuities, and of land

[3] Secondary sequences have to be expected.

bridges. The paleoecological method requires not only a wide familiarity with fossils and geology, but also a wide familiarity with modern plants, especially with modern vegetation types. Sometimes ecologically diverse plants have leaves which resemble one another. The paleoecologist must search for those plants which not only resemble the fossil morphologically but match it in the minute details of shape, texture, venation, etc., and which, above all, present no great ecological disharmony with the major composition of the

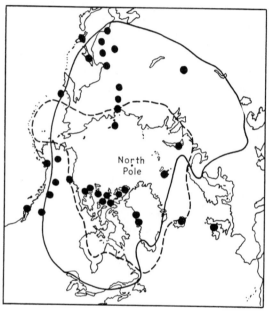

FIG. 3. The location of northern Tertiary floras of cool temperate nature. Contrary to widespread opinion, there were no warm temperate or subtropical floras in the Arctic region during the Tertiary. The +10° midsummer isotherm (broken line) and the −10° midwinter isotherm (solid line) of the present time are added for comparison. Data from (63, Figs. 4–6). Base map after Goode's copyrighted map No. 201PN, with permission of the University of Chicago.

fossil flora. The paleoecological method, it must be emphasized, opens up tremendous possibilities for the correct interpretation of history, but it also requires extreme care in the identification of fossils and their reference to modern types. This is the greatest single source of error and limitation of the method.

Although biological processes and ecological relationships are uniform (in the sense that natural laws know no limits of time or space), the paleoecological method is increasingly limited in application as one goes back further into geological history. The method is essentially inapplicable beyond the

Tertiary (or the Upper Cretaceous, at the most) because modern vegetations are so different taxonomically and in community structure. Down through the Cenozoic the method becomes increasingly accurate until in the Pliocene and Pleistocene its conclusions may be drawn both finely and with great accuracy. The limitations of the paleoecological method must be kept firmly in mind in studies of the early Cenozoic in order that gross errors of interpretation do not enter the story.

The possibility of a changed physiology (ecology) without a detectable morphological change should be kept in mind when conclusions are being made with respect to the significance of an element in a fossil flora. One or a few species of a distinct element should always be viewed with suspicion. When, however, several species compose an element their significance may be relied upon, for it is scarcely within the bounds of probability that several species would have undergone physiological evolution of the same type without some revealing morphological changes.

The application of the paleoecological method is slow, arduous, and expensive. As great a quantity of fossils as are available should be examined. This may at times mean reference to collections that are housed in widely separated institutions. An extensive and widely representative herbarium of modern plants should be available for reference in the identification of equivalent species. Furthermore, a familiarity with modern vegetation in the field is an absolute necessity, and this may involve travel and intensive studies. The paleoecologist should be not only a geologist and paleontologist but also a biologist (botanist and zoologist), a geographer, an ecologist, and a climatologist. Paleoecology is no field for the dilettante, but its compensation in results is wholly consistent with the time, work, and knowledge that are required. What has been said of paleoecology is true of plant geography as a whole. In fact, it is partly true of any science; but the borderline sciences that depend for their results upon synthesis from and coordination of numerous more specific sciences are especially difficult and hazardous.

4.

Identification of Fossils

1. *The identification of fossil organisms can best be made by comparison with living forms, searching among the members of natural ecological associations for equivalent species if the fossils are not too ancient, or for clues to their generic identity.*

2. *The ecological method helps to prevent the false inclusion, in lists of fossil biota, of taxonomic forms far removed climatically and distributionally from the nature of the biota as a whole; but final decision as to identity must depend upon morphological considerations.*

3. *The taxonomic considerations of a fossil flora are inseparable from the ecological considerations.*

4. *An examination of living equivalent species provides diagnostic characteristics, frequently with an emphasis on vegetative structures, and some measure of the limits of variation within a species, thus enabling a sound disposal of fossil materials.*

5. *The generic possibilities for a Tertiary flora were wider than for modern associations. Early associations apparently consisted of genera that no longer grow together, and consequently their counterparts cannot always be found in any one place today. This must be borne in mind when searching for modern representatives in the process of botanical identification.*

The following pages provide a discussion of the above principles together with selected illustrative materials. The whole method of paleoecology depends primarily upon the correct identification of the fossils which constitute an ancient biota. Without assignment of the fossils to correct genera the paleoecological method fails. Fortunately, the genera of Late Cretaceous and Tertiary plants appear frequently to be represented in modern floras, and the further one's studies progress up through the Cenozoic the closer the fossil species resemble modern species. The correspondence between fossil and modern species depends not only upon the age of the fossil flora but also upon the intensity of the study in some cases of both the modern and fossil forms.

Reid (522) found that several species of the western European Pliocene seed floras are apparently extinct. The extinct species are largely members of the element which is now referred to as Chinese-North American, absent from Europe, and their numbers decrease up through the Pliocene. Reid and Chandler (523), in a monograph on the London Clay flora of Eocene age, found that they could refer nearly all families to living ones, but they called no genus by the name of a genus of living plants. Their defense of this procedure rested on the fact that many genera have been proved to be extinct (in small families where all the living representatives are well known) and on the inadequate knowledge of seed characters of living or fossil material, or both. Nevertheless, close living relatives can usually be pointed out, and the use of distinctive generic names sometimes obscures the closeness of relationship. Seed floras are difficult to work with, however, and Berry (57), in a study of Upper Cretaceous floras composed mostly of leaf fossils, found that apparently most generic types were well differentiated from the beginning of this record on down. Many of the Cretaceous genera extended into the Eocene, but about 50 per cent are not known at all in modern floras; all species were extinct by the Eocene.

Although the species of a genus show considerable variation with respect to habitat requirements, a genus, at least among higher plants, can usually be referred to as temperate, tropical, etc. We know certain modern genera in which all species but one or a few are tropical or temperate; such anomalies probably existed among the species of fossil genera or the genera of families. In this connection, however, life form is important. A truly temperate member of a tropical family is likely to be characterized by an appropriate life form, and as we know a priori that its physiology has to be "temperate" no matter where its taxonomic relatives live, the problem of identification is facilitated.

Returning again to the Eocene London Clay Flora (523), we find in Table 1 an interesting comparison of living tropical families and those of the fossil "rain forest" flora. This close parallel gives confidence both in the climatic indications of the fossil flora and in the accuracy of the references of fossil to modern forms. Another aspect of the question can be approached through a consideration of the ranges of modern families. Reid and Chandler have determined that about 20 per cent of the families are either tropical or temperate and have been unable to adapt themselves to any except special

TABLE I.—A comparison of living tropical families and the families of the Eocene London Clay Flora (Data from 523)

	All Living Tropical Families (*per cent*)	London Clay Families (*per cent*)
Exclusively tropical	15.0	11
Mainly tropical	32.5	32
Equally tropical and extra-tropical	32.5	46
Mainly extra-tropical	20.0	11

conditions. About 50 per cent of the families are tropical or extra-tropical, with only a few species or genera out of the main area of the type. Only about 30 per cent of the families appear to be equally divided between the tropics and extra-tropical regions. The weight of evidence, consequently, does not allow for any general adaptability for families, and they think it would be rash to attribute to past life a greater adaptability than exists today. Such relations as exist for families must be even more pronounced for genera, and species are still more circumscribed in their ecological amplitude. The value of such a treatment of family characteristics depends to a considerable extent upon the delimitation of families taxonomically. With a narrower conception of families or of genera, they are found to be more completely restricted to one zone or climatic type than when they are broadly conceived.

When fossil species can be matched with equivalent species from modern floras, the method of paleoecology is on a firm basis. The likelihood of equivalent species having the same or closely similar requirements as the fossil species is considerable.

Among the specimens which compose the materials of a fossil flora there are usually a number of widely known and easily recognized species. Determinations of these should be checked first of all, irrespective of the formation, age, or geographical locations which they represent. An additional number of the fossils can likely be matched with named species from other studies and paleontological publications. The paleoecological method, however, provides a means of increasing confidence in the determinations thus obtained and of suggesting the relations of the remainder of unknown specimens.

Chaney (126) suggests two procedures. First, the modern species of the vicinity can be examined for suggestions relative to the fossils. Then the search can be extended to regions farther and farther away. If the climate of a region has changed much since the time of the sedimentation containing the fossils—and this is frequently the case—the local flora will probably not contain many of the same species and genera as the fossil flora. If, for example, the known species suggest a western American temperate flora, it is logical to look for matching species among the western temperate plants of today. Still lacking matched species, the search can be extended to eastern American temperate floras, to comparable Eurasian vegetation, and even to other climatic zones.

The second method is more valuable. When some of the dominant genera of the fossil flora have become known, a modern flora should be sought which contains as many as possible of those genera. If a truly representative modern vegetational association can be found, it will serve two very valuable purposes: (1) genera likely to match with the fossil specimens will be suggested and equivalent species may be discovered; and (2) tentative determinations, which now appear unlikely, can be discarded in favor of genera and species which are found to match better and were more likely to have been associated ecologically with the named kinds.

Chaney (132), in a review of some of the methods of paleoecology, especially as applied to Cenozoic plants in western North America, says, "The first adequate consideration of the systematic aspects of an American Tertiary flora was made by Berry as recently as 1916 in his monograph on the Wilcox flora." The Wilcox flora is composed of many genera now characteristic of forests of low latitudes, especially the Caribbean region, and their proper allocation would scarcely have been possible had not Berry (56, 62) made extensive use of large herbarium collections and also had some firsthand knowledge of the Caribbean. The handling by Chaney and Sanborn (140) of the Goshen flora, which is related to the Wilcox flora, develops the method of studying modern forests that contain equivalent species in order to ascertain the probable nature of the fossil vegetation.

At the beginning it should be realized, however, that sometimes floristic assemblages of the Tertiary cannot be exactly duplicated by modern vegetation. According to Chaney (132): "The generic possibilities for a Tertiary flora must be considered to be wider than for any one of its modern equivalents; such a consideration must be borne in mind when the ecological method is employed as a means of determining the likely alternative in a doubtful case." For example, several common genera of broad-leafed deciduous trees of eastern United States and Asia, now absent from western America, also occurred there during the Tertiary. Clearly, the paleoecologist must search the temperate floras of all continents for possible matching species. For a large number of genera, the history has been one of loss of territory during the later Tertiary and the Pleistocene. Berry (55) pointed out that in the Late Mesozoic and Early Cretaceous there was frequently an intermingling of forms now segregated. For example, willows and walnuts grew with figs, eucalyptus, laurels, and araucarias. Today such associations, according to Berry, are most likely found in that rather loosely characterized formation, the temperate rain forest of Schimper, in such places as Chile, Japan, Australia, and New Zealand. In New Zealand, for example, occur such genera as *Aralia, Laurus, Cinnamomum, Magnolia, Sterculia, Quercus, Fagus, Dryopteris,* and *Dicksonia,* together with both narrow- and broad-leafed conifers, coriaceous-leafed dicotyledons, and tree ferns.

In some cases the reduction of the generic richness of a flora was due only to geographical restriction of areas, but in other cases there appears to have been an actual climatic segregation of elements. The latter is by no means proved, however, and there are good theoretical arguments against the possibility which will be presented shortly.

The segregation of floras is shown in Berry's study of the Latah (61), which contained numerous members of the holarctic Miocene flora. The redistribution of generic types, once associated, resulted from climatic changes and migrations due mainly to the uplift of the Sierra Nevada, the Cascades, and the coast ranges. Among the Latah genera, the following survive in western America only in the valleys and humid western slopes of the mountains: *Tumion, Libocedrus, Platanus, Cercis, Sequoia,* and *Umbellularia.* Many genera are now absent from western America, of which *Ginkgo, Glyptostrobus, Paliurus,* and *Porana* survive in eastern Asia but nowhere in North America, and the following still exist in eastern North America: *Arisaema, Hicoria, Castanea, Fagus, Ulmus, Carpinus, Liriodendron, Magnolia, Liquidambar, Hydrangea, Celastrus, Aesculus, Tilia, Sassafras,* and *Nyssa.*

The value of the paleoecological method in the identification of species is well illustrated by the work of Mason (453) on the Pleistocene flora of the Tomales region. The disappearance of the soft parts of fruits frequently renders the fossil material difficult of determination because it looks so different from familiar fresh material. Compression, too, adds difficulties. In these connections Mason says:

Fortunately a method of identification has been devised that limits the field of search considerably, so that modern representatives of the fossil seeds can easily be found. The material which is most easily recognized is sorted and a preliminary notion of the fossil plant association is gained. A modern equivalent of the association is then sought and is studied in detail in the field. Collections are made of all parts of the plants and these are treated in such a manner as to remove the soft parts in order that they will simulate the conditions of fossil specimens. These are then used for comparative purposes. In general, it will be found that a great majority of the material can be positively identified in this way. Very frequently many species of a genus are so similar in their hard parts that it is impossible to differentiate between them. According to the ecological interpretation, the species selected is the one most in accord with the habitat of the fossil flora. . . . A strict adherence to this procedure may still involve some errors in identification, but on the whole it may be said that it is infinitely sounder in principle than the "hit or miss" comparison of fossil materials to a random collection of members of the plant kingdom from the four corners of the earth, or even from so small an area as that of the average state.

In paleobotanical work not employing the ecological method one sometimes finds reported mixtures of temperate and tropical species. Especially at higher latitudes, these are strange bedfellows. In his study of the redwood flora of Bridge Creek, Chaney (126) gives an example of the ecological method. *Cinnamomum bendirei* had been described from the Bridge Creek flora. Chaney pointed out that the genera associated with the redwood flora are not tropical, as is *Cinnamomum*, but consist of such temperate genera as *Acer, Corylus, Pseudotsuga,* and *Cornus.* He says, "It seems reasonable to suppose that the genera associated with the Tertiary equivalent of this forest should have also been temperate. . . . The use of this ecological method, involving as it does the climatic and distributional requirements of the modern relatives of fossil plants, may then serve to eliminate the unlikely genera. . . . The genus *Cinnamomum* is confined to eastern Asia and the East Indies. Consequently, the recording of a species of *Cinnamomum* in an assemblage of temperate species in the Tertiary of central Oregon is at once open to suspicion."

Following the method suggested above, it was discovered that the leaves of the supposed *Cinnamomum* match very well with *Philadelphus,* a genus which is common in many parts of the West and which is also associated with the modern redwood forest. In fact, it was found that both the large and small leaves are almost identical with leaves of the sterile and fertile shoots of the modern *Philadelphus lewisii.* The resemblance of the fossil leaves to *Cinnamomum* was principally in nervation. With *Philadelphus lewisii* the leaves match in shape and outline as well. Chaney adds, "It should be pointed out that while the ecological probabilities have in this case given the initial suggestion of referring these fossil leaves to a temperate rather than to a tropical genus, the fundamental basis for doing so is their morphological resemblance to the leaves of *Philadelphus lewisii.*"

A similar case is that of *Grewia* (once thought to be one of the most widely ranging forms of the Tertiary), which is now confined to Asia and there mostly to low latitudes. In 1927 Chaney suggested that the supposed Bridge Creek *Grewia* might be referable to the temperate *Cercidiphyllum* because the leaves are essentially identical to the living katsura, *C. japonicum*. Final proof came when Brown (96) reported katsura (*Cercidiphyllum*) seeds, previously unknown, from the Bridge Creek shales.

Many similar illustrations from modern paleobotany could be presented. The point which needs emphasis, however, is that the taxonomic considerations of a fossil flora are inseparable from the ecological considerations. Known facts concerning the nature of modern associations, their floristic composition, habitat requirements, and geographical arrangement serve as one of the most important bases for the identification of fossil plants.

As paleontology has progressed it has become clear that many of the studies which reported mixtures of temperate and tropical species at high latitudes, such as Heer's studies of the Arctic fossil floras (337), were faulty in allocating certain of the fossils to definitely tropical genera (63). In a general way, at least, the principle holds that a mixture of species of diverse floral and climatic regions is usually highly improbable.

A quarter of a century ago Clements (150) wrote, "It seems impossible that genera which we now know as boreal, temperate, and tropical should have existed in the most complex and uniform mixture through a vast region characterized by a warm climate, and then have completely differentiated by later climatic changes . . . into three great forest climaxes." There may be, however, situations in which less extreme mixtures do occur. For example, under the compulsion of a climatic shift, two or a series of contiguous plant formations tend to move together in a common direction. Such climaxes have more or less broad ecotones in which the characters of two climates are not distinct and in which some species of the two climaxes are inter-mingled. The cliseral movement may result in the extension or compression of the area occupied by any or all of the climaxes and their ecotones. Such movements also usually result in the production of a mosaic of communities representing the two climaxes. In a southward movement, for example, the more southern of two climaxes will leave behind relic stands in its former territory. Such relics become surrounded by the invading vegetation. They are not in harmony with the new prevailing climatic conditions. It is obvious that they must survive in regions of compensation where the local conditions of microclimate or soil allow them to resist, for a time at least, the climatic pressure and the competition from the invading vegetation. Such southern relics are likely to occupy the most favorable sites in a region, at least with respect to temperature and moisture conditions.

Should there be a climatic reversal and a northward migration, the processes are the same and the causes and effects are similar, but the direction is reversed. Under these conditions northern relics will be left behind in a region occupied now by a prevailingly southern type of vegetation.

Should a fossil flora enter the sedimentary record in an ecotonal region or in a region where relic colonies exist, it is obvious that the species may represent two climatic zones. The extent of the climatic differences which may thus come to be represented by the plants in a fossil flora will depend upon the vegetational history, the recentness of change, and the local diversity in sites which would allow a juxtaposition of diverse floral elements.

Another situation which would permit the entrance of plants from two climatic types of vegetation into the record of a fossil flora is that presented by the development of altitudinal belts with corresponding climates and climaxes. In the Goshen flora, Chaney and Sanborn (140) conclude from a wide variety of evidence that the flora consists of a mixture of tropical rain-forest and temperate rain-forest species which grew in a coastal situation. A rather obvious conclusion would be that the two elements, tropical and temperate rain-forest species, occupied a zonal arrangement, with the temperate element forming a forest belt at higher altitudes than the more coastal tropical forest. Leaves from the upper forest would then be carried down into the sedimentary basins, causing the apparent mixed-forest conditions. Such an altitudinal disposition of rain-forest types is the characteristic pattern today in some Central American regions. If the transport of the temperate elements seems improbable because of their abundance and condition of preservation, the sedimentary basin may have been in a transitional region between the two forest types.

Chaney and Sanborn, however, tentatively suggest another conclusion. They say, "The relationship of the Goshen flora both to the tropical rain-forest of Panama and to the temperate rain-forests of Costa Rica and Venezuela is consistent with its composition, including as it does an almost equal number of species whose modern equivalents occupy the lowland and middle slope forests. It may be concluded that the Goshen flora represents a forest of a type intermediate between the modern tropical rain-forest and the temperate rain-forest. It is possible that there is no such forest now in existence, because of climatic changes since the early Tertiary which have restricted the more temperate genera to the upper elevations, leaving only the tropical types in the original habitat near sea-level. Or it may be that a forest of the Goshen type will be found at some point in Central or South America not yet visited by the authors; there is some reason to believe that the temperate rain-forest of Chiriqui, the highest mountain in Panama, may show a closer similarity to the Goshen flora than any of the modern forests we have studied."

This situation leads to a possibility which presents the greatest obstacle to the application of the ecological method to paleobotany. Briefly, the situation and the problem are these. The history of vegetation, as shown by current interpretations of fossils, appears to be one of change from a relatively small number of types (climaxes) throughout most of the Tertiary[1] to the development of an increasingly larger number of vegetational types in the Late Tertiary and the Pleistocene. From a few widespread and relatively homogeneous types of vegetation have been derived a much larger number of more restricted types. The broad pattern is now one of heterogeneity on continents and between continents, with homogeneity only within relatively local vegetational types. According to the suggestion of Chaney and Sanborn quoted above, the Goshen flora represents an intermediate condition. From it has been derived, through segregation of elements, the Central American tropical rain-forest and the temperate rain-forest.

For this to be true, certain assumptions must be made. First, the environment of the Goshen flora must have been moderate, relative to the requirements of the now differing elements which compose the equivalent species —so moderate, in fact, that the limits of tolerance of no species of the mixture were surpassed. If that is true, it should be possible to find some situation today with an equivalent climate in which such an admixture of species exists. Such a situation should exist, at least on the ecotonal region between the lowland tropical rain forest and the temperate rain forest of middle elevations. The second assumption would be that the tolerances of the species of the flora have diverged through evolution in a limited number of directions.[2] If that is true, they can no longer live together as a single group because the conditions which are best suited for a portion of the flora are unsuited for the remainder. Segregation on a climatic basis is the only possibility for the survival of both elements. It follows, then, that the modern environmental conditions of either one or both of the elements are not equivalent to the environmental conditions of the Goshen flora.

With respect to the characteristics of the Goshen flora just discussed, a communication from Clements[3] expresses his opinion: "The conclusion as to

---

[1] This does not assume a lack of vegetational zonation, as supposed by some paleontologists, for there is good evidence of latitudinal belts of vegetation.

[2] This evolutionary improbability is discussed elsewhere.

[3] By letter, December 14, 1940.

the dual nature of the Goshen forest transgresses my understanding of natural (ecological) law and is equally contrary to the concepts of climax and of origin by adaptation. I cannot entertain it until a climax of this character is found."

Since there seems to be considerable evidence that the history of vegetation has been the breakup of wider, more homogeneous climaxes into narrower climaxes of more limited tolerances, it is necessary to make one or two conclusions concerning the nature of the species. If early species were characteristically of wide tolerances, and consequently of wide distribution, the modern descendants are of a relatively more limited occurrence because of (1) habitat limitations or (2) an evolution which has reduced their ecological amplitude and adaptation.

In the majority of Tertiary floras the most abundant fossil material, and sometimes the only type, is composed of leaf fragments. In a discussion of Heer's studies of Arctic Tertiary floras, Gregory [4] said that "most of Heer's determinations were based upon leaves, which give no data for generic identification." Of this opinion Berry (63) says, "Such a statement of genera such as *Liquidambar, Betula, Corylus, Ulmus, Platanus, Sassafras, Liriodendron, Acer, Potamogeton, Cornus,* and *Nymphaea,* to mention but a few of those recorded from the Arctic Tertiary, is the height of misunderstanding." The identification of leaves is frequently difficult; but faced with this situation, the paleobotanists have sought and found diagnostic characteristics among the vegetative features of which modern taxonomists of living plants may not be aware because they have not been driven by necessity to study the finer details of nervation, margins, and texture.

A most critical consideration in connection with identification is the matter of variation. Paleobotany, like all branches of taxonomy, has known two extremes of treatment of variable materials. Many of the fossils of the Bridge Creek redwood flora were early examined and named by Lesquereux (412). Many similar fossils were described by him under several different generic names: *Quercus, Alnus, Betula, Carpinus.* Newberry (486) lumped them all under a single polymorphic species, *Populus polymorpha.* Knowlton (398), reviewing the situation, took an intermediate position and placed certain of the forms under *Betula heteromorpha.* He recognized that a contrast of individual specimens which are extreme in form would warrant the establishment of separate taxonomic entities, but that in many cases a complete series of intergrades can be arranged and that it is impossible to draw a separating line. This would appear to be a most confusing situation and at

[4] J. W. Gregory, *Congrès Géol. Intern. Compte rendu* Xème *Session,* Mexico, 1907. Quoted from (63).

first it would seem impossible to resolve it. What confidence can an ordinary botanist have in the paleobotanical literature when the same fossils are referred variously to *Quercus, Alnus, Betula, Carpinus,* and *Populus?* There is, in fact, no way around such problems without reference to modern species and especially to the breadth of variation shown by leaves and other structures of modern representative species. It must always be remembered, however, that extinct species are to be expected and that it would be an equally bad mistake to lump diverse fossils with living representative forms under the same specific epithet.

In his comparison of the Bridge Creek redwood flora with the modern redwood forest, Chaney (126) says that Lesquereux's method is impossible, for "with the wealth of material available, there would be scores of such species or subspecies. If the second alternative is sound, the correctness of assembling all of these types under one species can be tested by a study of the variation in a single living species of a likely genus." In the case referred to above, Chaney found *Alnus* to be the genus, and he described the materials under the name of *Alnus carpinoides.* This eliminated from the Bridge Creek flora the following species names: *Betula bendirei, B. heterodonta, B. heteromorpha, B. angustifolia, Alnus macrodonta, Populus polymorpha, Quercus pseudo-alnus,* and probably *Quercus oregoniana.* Among modern species of *Alnus, A. rubra* is a member of the redwood flora today, and it is a very close relative of *A. tenuifolia.* It is to the American forms of *A. alnobetula* (*A. sinuata* or *A. sitchensis*) that the fossil material bears closest resemblance. By cleverly matching leaves from *A. alnobetula* and *A. tenuifolia,* Chaney was able to duplicate by fossil material (*Alnus carpinoides*) the forms which, because of leaf size or shape, apex or margin characters, had been assigned to diverse genera.

Further illustrations of the value of a study of variation of leaves in a modern species in the identification of fossil specimens are found in the paper by Chaney (127) on the Mascall flora. Fossil *Quercus pseudo-lyrata,* he says, appears to be closely related to the living California black oak, *Q. kelloggii,* as well as to several species of the living eastern oaks. With the variation in the leaves of these species in mind, it seems proper to consider *Q. merriami* and *Q. ursina* as variants of *Q. pseudo-lyrata.* In the Crooked River flora (128, Plates 17 and 18) several leaf variations of maple are all referred to *Acer osmonti* because each variant can be matched by variations in leaves of its modern representative, *A. glabrum.*

Although leaves are usually the most abundant fossil material of a flora, no possible aid should be overlooked in the identification of constituent members. Buds, twigs and wood, when present, should be used to substantiate or supplement the evidence for the presence of certain genera in a fossil flora. A single twig with the characteristic clustered terminal buds is sufficient to suggest whether certain leaves belong to *Quercus* or not. In some cases entire floras are based on seeds and fruits (522, 523) or on wood (679). Wood characters, together with the ecological characteristics of representative species, are used to advantage by Webber in the study of a small Pliocene flora called the Last Chance Gulch flora. The pine wood was found to possess thin, smooth-walled ray tracheids and thick-walled ray parenchyma cells with small half-bordered pits on their lateral walls. Three species living today possess such structure: *Pinus aristata, P. balfouriana,* and *P. cembroides* (with its varieties *monophylla, edulis,* and *parryana*). It is practically impossible to tell these forms apart on the basis of wood structure alone; but, according to Webber, environmental conditions are favorable for considering that the fossils are *P. cembroides* or, at least, that this species is the modern representative.

Another very important source of determinations and suggestions is fossil pollen. Cain (107) has published a general review of pollen analysis as a paleoecological research method. The special technique required in the study of fossil pollens has not been widely employed in the United States except for postglacial peat deposits in the glaciated region. Some progress is being made with intra-Pleistocene deposits, but little has been done with Tertiary materials. The sole study of Tertiary pollens in this country seems to be that by Wodehouse (709, 710) on the Eocene Green River flora. Brown (94), using data from Wodehouse, was able to demonstrate a definite temperate, "Miocene" aspect of the upland Eocene vegetation about the Green River basin. The work of Wodehouse is of such interest and promise for paleoecology that a brief description of it follows. In his first paper he prepared a key and descriptions of pollens of the modern representative species (or species of representative genera) of the Green River fossil flora. The purpose of this study was to facilitate identification of pollens of species thought to have been present because of macrofossils. His second paper describes and names a considerable number of pollens from these shales, many of which represent genera previously unknown from the deposits. All species, based on pollen, are described as new, although several are similar to living species or to ones described previously from Europe. The main reason for doing this is that the pollens are not found in association with other structures.

The names which are based on pollen alone are easily distinguished from names of other fossil species or of modern species by Wodehouse's practice of using the suffix *-pites,* which is a contraction of the form genus *Pollenites.* When the genus is certain and the reference is to a modern species the suffix is added to the specific name, as in *Pinus strobipites;* when the genus is uncertain and a reference is still desired, the suffix is added to the generic name, as in *Smilacipites molloides.*

There seems to be no doubt that a wider search for and study of fossil pollens among Tertiary deposits would add valuable information concerning the vegetation of these horizons. Because of their apparent ease of fossilization, their durability, and their wide dissemination, at least in comparison with macrofossils, they enrich a fossil flora greatly. Also, pollens seem likely to bring into a deposit a representation of forms farther from the sedimentary basin and more upland to it which otherwise would be unknown. Recent work with Carboniferous spores has revealed dozens of species and the possibility of their usefulness in stratigraphy, in both European and American coals (53).

Fossil materials from Pleistocene and recent deposits are frequently sufficiently complete, sometimes with leaves and wood and even flowers and fruits in association, that their identity with modern species is certain. The Willow Creek flora of Santa Cruz Island is made up of 9 species which are all still living in California (138). The 18 species of the Pleistocene flora from San Bruno, San Mateo County, California, are all living species, according to Potbury (508). The Carpinteria flora (139) consists of 25 species, of which only *Pyrus hoffmanni* is not known among living plants. The relatively rich Tomales flora of 53 species is closely related to modern closed-cone pine forests of the central mountains of California. The attitude of Mason (453) on nomenclature of these forms is expressed as follows: "A conservative attitude has been followed toward the erection of new fossil species and varieties. Absence of clear evidence of differentiation from the modern species is considered sufficient cause for retaining the name of the modern plant. Some of the species show characteristics which are represented in the modern plants only by extremes of variation. These are a record of evolutionary change, but not of adequate extent to justify establishing new species."

Apparently Pleistocene plants have passed into the modern floras with relatively few evolutionary changes of a morphological nature. The passage of time is suggested more by floral segregations, the development of new floristic assemblages and new areas, than by the development of new species. The practice of applying modern species names to Pleistocene fossils, except when the differences are obvious, appears to me to be a sound one.

According to Chaney (131), most students of the Tertiary assign fossil conifers and angiosperms to modern genera when they can, but give them specific names which are different from those of modern equivalent species. For example, the Tertiary redwood, on the basis of leaves, twigs, wood, and cones, appears to be identical with the living coastal redwood, *Sequoia sempervirens,* but the name *Sequoia langsdorfii* is maintained because of the implied antiquity.

In many cases proof of the identity of older fossil plants with modern species is lacking or impossible. Here the best procedure may be the application of a new specific name. The relationship of the fossil species to modern ones is then emphasized by an arrangement, parallel with the floristic list from a Tertiary flora, of modern representative or equivalent species. Whether the paired species are identical does not appear to be the most important consideration. They are frequently the two most closely related species known from a phylogenetic stock, sometimes apparently more closely related than any two living species of the genus. Especially in the early Tertiary, fossils are frequently encountered which cannot conservatively be given generic status among modern generic concepts. In such instances they are given a completely distinctive name or are assigned to *form genera* which suggest a probable relationship, as in the case of *Cinnamomites* and *Cinnamomum, Dalbergiites* and *Dalbergia, Alangiophyllum* and the paleotropical genus *Alangium.* Sometimes, in especially difficult cases, only a family relationship can be suggested by the use of such form family names as *Leguminosites* or *Laurophyllum.*

Somewhat along a different line, but still related to matters of identification and nomenclature, is the treatment of two or more fossil types of a single genus. If, for example, leaves and fruits of *Acer* are found in the same flora and the variations among the leaves and among the fruits do not seem to surpass the limits of a single species, they could be referred to a single species or to two. Which disposition is correct, when the materials are not discovered attached to the same twig, can be decided only by comparison with modern species. If the fruits are of a type normal to maples with leaves of the type, the logical procedure is to refer both fossils to a single species. If the leaves are of a red maple type and the keys of a sugar maple type, then obviously two specific designations must be made. Chaney (132) also points out that it is misleading to list two fruit-type species and two leaf-type species when there are two related living species with which these four remains may be paired.

In his study of the large Miocene Latah flora Berry (61) was aware of the probability that the species list contained several duplications that resulted from giving separate names to leaves and fruits in such genera as *Pinus, Alnus, Acer, Ulmus Liquidambar,* and *Quercus.* He recognizes that *Pinus latahensis,* based upon leaves, and *P. monticolensis,* based upon seeds, match well with the existing western white pine, *P. monticola,* and probably represent a single botanical species; but he considers it conservative to maintain their separate status in the absence of proof to the contrary. In the oaks, however, he reduces *Quercus elongata, Q. Chaneyi,* and *Q. praenigra* to synonymy under *Q. simulata.* Of this case he says, "There are a large number of specimens of a simple, prevailingly entire oak in the recent collections made around Spokane. . . . The similarity in texture and venation has convinced me that the citations in the above synonymy represent nothing more than the variants of a single Miocene species, comparable with and showing less variation than such recent species as *Quercus chrysolepis* Libermann. . . . I am sure that anyone who has handled a series of specimens of the latter will agree that these Latah leaves represent a single botanic species."

In concluding this subject, it can be reiterated that much of the usefulness of paleoecology depends upon correct reference of fossils to existing species. Ecology proceeds most soundly on a floristic basis. This does not deny that growth form has significance; it has been said that an ecologist could tell pretty much the nature of the environment of a place (assuming that he does not know his location) without knowing the name of a single species. He would do so, presumably, by examining the growth forms of the surrounding plants and drawing on his knowledge of the relations between environment and growth form. In the same way but to a lesser extent, the paleobotanist can judge the environment in which a fossil flora lived by deduction from the characteristic growth forms of the fossil plants. However, it is well known that many plants overcome apparent morphological handicaps and live in environments where their life form is exceptional. It follows, therefore, that the taxonomic reference is more certain than the morphological one, for it is unreliable to reason from structure to function.

5.

Determination of Dominance in a Fossil Flora[1]

1. *When a quantity of fossil material is available for study the significance of a list of species can be greatly increased by some estimation of the dominance of the different kinds of plants in the flora.*

2. *In a general way, the frequence of fossils of each species is useful, but the numerous variables that influence the quantity of leaves and other materials that get preserved can be resolved only by recourse to the paleoecological method of comparison with conditions in modern representative vegetation.*

3. *When specimens are available from several plant-bearing beds of the same age in the same general region, a high percentage of presence in the different localities is a better indication of the characteristic nature of the fossil for the flora and its probable dominance than its chance high abundance at one or a few localities.*

In the earlier days of ecology, quantitative descriptions of modern communities were seldom found in the literature. Frequently one read no more than a list of species from a community "arranged in order of their importance." In other instances, the quantitative relations of the various species of a community were designated by the use of such terms as "abundant," "occasional," "rare." Paleontology has had a similar history and one seldom finds a scientific attempt at quantitative description. The difficulties to be surmounted in a quantitative study of a fossil flora are tremendously greater than in a corresponding modern flora. In the first place, no quantity of material is available in many floras. Nevertheless, it is of importance to make such studies whenever possible.

[1] In paleontology dominance is of necessity largely determined on the basis of abundance, although it is admittedly a more complicated question than simple numbers, since it involves life form, coaction, etc.

Statistics on the percentage composition of the species represented in the whole collection of fossils are of some significance when the collections are representative of the fossil bed, but they are not necessarily representative of the quantitative relations of the plants that composed the community contributing materials to the sedimentary basin. That is another matter. In a modern community, the most significant factor is not the number of leaves, for example, produced by the plants of each species, but the dominant control over the community that is exerted by certain species in each of the life-form groups (layers or synusiae) which compose the community.

The paleoecologist must attempt to find the relation between the number of fossils of each species and the extent to which the species entered into the composition of the ancient community. Since a fossil community is represented only by those materials which happen to get preserved, it is of paramount importance to consider the factors which determine the abundance of fossils of a species in the deposit. Chaney (125, 126) and Chaney and Sanborn (140) listed several factors that influence the abundance of leaves in sedimentary deposits.

1. The distance of the plant from the basin of deposition where the leaf lodges and enters the sedimentary record will determine the abundance of leaves of that kind in the record.

2. The thickness and durability of the leaf are of importance because of their relation to its capacity for transportation without destruction.

3. The size and shape of the leaf are of importance in connection with its transport by air or water.

4. The habit of the plant in shedding its leaves is important. Leaves which are abruptly deciduous have a good chance of entering sedimentary basins, but those which are persistently withering, as in most herbs, have only the slightest chance of entering a fossil record except when such plants actually grow in or at the edge of the sedimentary basin.

5. The length of the erect plant stem, involving its habit as an herb, a shrub, a small or a tall tree, determines its exposure to agents of dissemination and consequently the likelihood of its leaves reaching a sedimentary basin by being transported a relatively greater distance if released from a greater height.

6. Plants of different life form, such as trees contrasted with shrubs, produce different quantities of leaves per plant and consequently have different chances of being represented in the fossil record.

7. Related to this is the fact that compound leaves are frequently represented by separated leaflets.

When it is realized that any or all of the above factors can operate to produce a discrepancy between the number of fossil specimens recovered and the relative number of plants from which they were derived, the task of producing a quantitative description of a community of geological age may well seem insurmountable. That some progress has been made in this direction is largely to the credit of the methods of paleoecology which depend heavily upon comparisons with modern representative communities.

An outstanding paper on quantitative relations of a fossil flora is Chaney's study of the Bridge Creek (125). He says, "In the study of the plant life of the past, exact data concerning the relative abundance of the species making up a flora are as important to a paleo-botanist as are similar data to a student of modern plants. Both the botanical and geological aspects of the study make it essential to determine which are the dominants and which are the accidental species in a fossil flora. To the geologist, the dominant species are of the greater value in correlating fossil-bearing deposits of two regions. . . . The dominant species give to the botanist a clue as to the general character of the vegetation which enables him to reconstruct it in terms of modern plant associations with similar dominants."

TABLE 2

Bridge Creek Fossil Flora	Redwood Forest Equivalent Flora
Pinus knowltoni Chan.	Pinus ponderosa Laws.
Tsuga sp.	Tsuga heterophylla Sarg.
Sequoia langsdorfii Heer.	Sequoia sempervirens Endl.
Torreya sp.	Torreya californica Torr.
Myrica sp.	Myrica californica Cham.
Corylus macquarrii Heer.	Corylus rostrata Ait. var. californica A.DC.
Alnus carpinoides Lesq.	Alnus rubra Bong.
Quercus consimilis Newb.	Quercus densiflora Hook. & Arn.
Berberis simplex Newb.	Berberis nervosa Pursh
Umbellularia sp.	Umbellularia californica Nutt.
Philadelphus sp.	Philadelphus lewisii Pursh
Crataegus newberrii Cock.	Crataegus rivularis Nutt.
Rosa hilliae Lesq.	Rosa nutkana Presl.
Fraxinus sp.	Fraxinus oregona Nutt.
Acer osmonti Kn.	Acer macrophyllum Pursh
Rhamnus sp.	Rhamnus purshiana C.
Cornus sp.	Cornus nutallii Aud.

The methods employed by Chaney are of such importance that they will be given detailed consideration. It had elsewhere been recognized that the Bridge Creek flora suggests a relationship to the modern redwood forest (126). The extent of the floristic similarity between these fossil and modern floras can be seen from Table 2.

The relationship between these two floras is striking, but the absence of quantitative data leaves much to be desired. To supply this deficiency, Chaney and an assistant counted 20,611 fossil specimens which they split from 98 cubic feet of fossiliferous shale from the type locality for the Bridge Creek flora. For the percentage composition data thus obtained they sought comparable information concerning the living redwood forest. Turning to the forest of Muir Woods National Monument, Marin County, California, they identified and counted the leaves found on square-foot sample plots laid out in 42 basins of accumulation along a stream. The numbers of leaves and fruits thus obtained was 8422. In Table 3 enough of Chaney's data are

TABLE 3

Bridge Creek Flora: *Percentage of Total* *Fossils Recovered*			*Muir Woods Flora:* *Percentage of Total* *Materials in Basins*
Alnus carpinoides	53.59	27.36	Alnus rubra
Sequoia langsdorfii	15.07	39.35	Sequoia sempervirens
Quercus consimilis	8.96	5.46	Quercus densiflora
Umbellularia sp.	8.82	13.27	Umbellularia californica
Acer seed (A. osmonti?)	0.68	2.86	Acer macrophyllum
Quercus fruit	0.34	0.45	Quercus fruit
Alnus, female ament	0.13	1.47	Alnus rubra, female ament
Rosa hilliae	0.05	0.01	Rosa nutkana
Cornus sp.	0.05	0.03	Cornus nuttallii
Alnus, male ament	0.03	0.01	Alnus rubra, male ament
Corylus macquarryi	0.01	1.38	Corylus rostrata
Equisetum sp.	...	...	Equisetum telmateia
Myrica sp.	...	0.02	Myrica californica

included to show strong similarity between the composition of the ancient and the modern forest.

The similarity of the two floras is emphasized when only the apparent codominants are considered. *Alnus carpinoides, Sequoia langsdorfii, Quercus consimilis,* and *Umbellularia sp.* constitute 86.44 per cent of the Bridge Creek fossils counted. In the representative modern forest of Muir Woods, *Alnus rubra, Sequoia sempervirens, Quercus densiflora,* and *Umbellularia californica* constitute 85.44 per cent of the fossilizable units sampled in the basins.

Correlations were made between leaf and tree numbers in the Muir Woods in order to develop a basis for estimating the probable number of trees of each important species in the Bridge Creek forest. These correlations were worked out between the leaves of each species found at the 42 stations in Muir Woods and the number of individual plants of these species within a 50-foot radius of each station. For several of the species the following correlation values were obtained:

Umbellularia californica	.649
Quercus densiflora	.569
Sequoia sempervirens	.493
Alnus rubra	.486
Cornus pubescens	.477
etc.	

Chaney says, "The correlation values for the leaves . . . are sufficiently high to indicate relationship between the numbers of leaves in the stream deposits and the numbers of adjacent trees. These correlations make possible a series of predictions as to the numbers of trees of the various species which were present in the Bridge Creek forest."

Without going into the statistical procedure, which can be sought in Chaney's publication, it is interesting to note the results with respect to the four species that compose 86.44 per cent of the Bridge Creek fossils. The prediction of the number of plants of different species within a radius of 50 feet of the average site of deposition is as follows:

Alnus carpinoides	9.4 plants
Quercus consimilis	7.3 plants
Sequoia langsdorfii	5.4 plants
Umbellularia sp.	4.8 plants

Even should these numbers be greatly different from the actual abundance of the plants of the species, the relative number of each species is still likely to be correct.

Qualitative comparisons between the redwood forest of Bridge Creek time and the modern equivalent of Muir Woods indicate that the former was more diversified, but the quantitative studies show that the dominants were closely similar. A botanist walking through the Muir Woods today might easily imagine himself at home in the Oligocene and Miocene forests of Oregon.

Chaney and Sanborn (140), in their consideration of the quantitative relations of the Goshen flora, made counts of the fossils of each species (based upon 1000 specimens) and determined the percentage composition. They carried the estimates further, however, by applying correction factors which allowed the indication of the probable number of plants. They say, "In showing the estimated numbers of individuals of each species, the following corrections have been applied to the figures based on leaf representation only: (1) The number of leaves on a tree has been assumed to be twice that on a shrub or vine, and the figure indicating the number of leaves of a tree species has therefore been halved to determine the number of individuals; (2) a thin leaf [membranaceous] has been estimated to have only half as good a chance of entering the fossil record as a thick leaf [coriaceous], and the figure indicating the leaf number of thin-leafed species has therefore been doubled to determine the number of individuals; (3) for species with compound leaves, the total number of leaflets is divided by the number of leaflets per leaf, and the quotient is then further changed as may be required for the other corrections."[2] The results of their two tables are combined in Table 4, which includes only the more important species and some of those that have changed percentages as a result of the correction factors.

On the basis of the recalculations with factors for life form, several changes are to be noted. For example, *Allophylus wilsoni* falls from second to sixth place, and the combined percentage of the three leading species is raised from about 30 to 40.

As already indicated, questions of overrepresentation and underrepresentation of fossil leaves in proportion to the numbers of plants they came from are difficult to resolve. Chaney and Sanborn (140), for example, have arbitrarily disposed of species with compound leaves by dividing the number of fossil leaflets recovered by the number of leaflets composing a leaf in order to make their representation proportionate to species with simple

---

[2] The assumptions of Chaney and Sanborn are admittedly arbitrary. It can be suggested that their effort to introduce quantitative correction factors is futile because of the unknown variables. With respect to (1) it would seem that most trees have on the average many more than twice the number of leaves of most vines. And the third correction, concerning compound leaves, seems especially subject to criticism. It would appear that the *leaflets* of one species are to be compared directly, without a factor of correction, to the *leaves* of a simple-leafed species. That is to say, it does not seem likely that on the average the plants with trifoliate leaves will have three times as many leaflets as a similar-sized plant with simple leaves would have leaves. My criticism, however, is also vulnerable. We do not have data on these points, and it depends, anyway, upon what plants are being compared.

TABLE 4

Species in the Goshen Flora	Per Cent of Total Based on Specimens	Corrected Results	
		Number of Individual Plants	Per Cent
Meliosma goshenensis	13.3	133	17.8
Allophylus wilsoni	9.1	31	4.1
Nectandra presanguinea	8.8	88	11.8
Ficus quisumbingi	8.3	83	11.1
Tetracera oregona	5.8	58	7.8
Anona prereticulata	4.0	20	2.7
Magnolia reticulata	4.0	20	2.7
Quercus howei	4.0	20	2.7
Lucuma standleyi	2.6	18	2.4
Aristolochia mexiana	3.5	35	4.7
Ilex oregana, etc.	2.2	22	2.9

leaves. In the Goshen flora *Allophylus wilsoni* was represented by 91 fossil leaflets. Its nearest living relative, *Allophylus punctata,* has three leaflets to a leaf, so they conclude that "the Goshen material may actually represent only 31 leaves." This is a minimum figure. It is highly probable that more than 31 leaves are represented because it is inconceivable that all leaflets of a leaf would get preserved, especially as they are usually found separated from the rachis.

Of the six most abundant species in the Goshen flora, only *Anona pre-reticulata* has a living equivalent which is a large tree. "If the assumption is made," they say, "that a Goshen individual of this species had twice as many leaves as the small tree or shrub of *Meliosma goshenensis* . . . it is clear that the numerical ratio of individual plants of the two species in the Goshen forest was not 133 to 40 in favor of *Meliosma goshenensis,* but 133 to 20. Such ratios are, of course, no more than approximations, but the general conclusion may be reached that the Goshen species whose modern equivalents are large trees were relatively less abundant than their leaf representation would indicate." Not only are larger plants more productive of leaves, but they also are more subject to the action of the wind as a dispersal agent. Tree leaves may be expected to carry farther and consequently enter into the record not only in greater abundance but from a greater distance than leaves of smaller plants.

The habit of the plant in shedding its leaves is of importance in any consideration of a species' representation in a fossil flora. In all of their work (Chaney, *et al.*), *Sequoia* leaves are not considered units, but the characteristic branchlets with their numerous leaves are. Considering the size of these dominants, it is likely that *Sequoia* is usually underrepresented. The individual small leaves are not suited for any considerable wind dissemination. The branchlets, on the other hand, are too heavy for any appreciable transport except by water. Herbaceous plants are usually absent from Tertiary floras or represented only by an occasional fossil. This is likely due not to their absence from Tertiary forests but to their characteristics. In the first place, their leaves are usually thin and delicate and consequently more easily destroyed before they have a chance to enter the sedimentary record. Probably a more important factor is their withering persistence. The leaves are not shed, but dry up or are killed by frost while still attached to the plant (126).

When it can be shown by comparison with modern representative vegetation that certain species probably grew in close proximity to the basins of sedimentation, it is likely that such species are overrepresented in the fossil record. And the converse is also true.

One of the most extensive statistical studies of fossil plants, in connection with the determination of numerical dominance, is that of Davies (178) on 29 horizons of the Westphalian, Staffordian, and Radstockian series of the Coal Measures. A total of 389,983 plant records covering 323 species were made. From these data Davies computed percentage dominance by classes (Equisetales, Lycopodiales, Filicales-Pteridosperms, etc.) and by genera, and was able to draw conclusions concerning the ecological conditions (dry upland, moist lowland, and swamp types) and land surface elevation and subsidence.

Berry (62) does not have much confidence in the attempts to determine dominance by counting the fossils of each kind because differences in the extent of study of different outcrops, differences in the preservation of various species from the different outcrops, and the differential effect of maceration and trituration preclude the keeping of any reliable statistics of the individual abundance of species at the different localities. He goes on to say that any method based upon the relative abundance of species at a single outcrop or a few outcrops is likely to give unreliable results regarding the dominance of certain species in a flora. For example, *Oreopanax oxfordensis* is known in the Wilcox from two out of 132 localities. These localities are about 300 miles apart, and the species is very abundant at one station.

Berry believes a more reliable method than the intensive study at one or a few stations is to consider those species commonest in a fossil flora which occur at the largest number of localities, particularly if these localities are widely scattered geographically. He says that the weight to be attached to the factors affecting the chances of preservation at any one locality becomes negligible when a particular species is found from several localities. For the Wilcox flora, Berry presents a list of the 46 species (out of 543 species) that are represented at 10 or more localities. The following eight species had the highest percentage of presence:

Sabalites grayanus	from 40	localities
Nectandra pseudocoriacea	" 40	"
Dryophyllum tennesseensis	" 39	"
Sapindus linearifolius	" 32	"
Evonymus splendens	" 31	"
Apocynophyllum sapindifolium	" 28	"
Sophora wilcoxiana	" 27	"
Ficus mississippiensis	" 24	"

Studies of this type also permit the recognition of facies in dominance. For example, *Ficus myrtifolia* has been found at 15 localities and *Capparis eocenica* at 18 localities on the eastern shore of the Mississippi Embayment. Neither of these species, however, has been found at any of the 38 localities of Wilcox age known from the western shore of the Embayment. They must have been absent or very rare on the western shore and widely distributed on the eastern shore.

The differences between the methods of Chaney and of Berry, and the results they obtain, are not just two points of view concerning the same phenomenon of vegetation structure, as Berry supposes. Actual dominance of a species is frequently a local phenomenon and a characteristic of stands, or association individuals. Such species may occur as codominants or minor associates in stands over a much wider area than those in which they are dominant. On the other hand, a vegetational type, such as a climax, usually has several species of wide occurrence that are nowhere of very great abundance or only rarely, under peculiar conditions, dominant. Such species may be very important to recognize as "binding species." Berry's Wilcox species of a high degree of presence at different localities may be such. To know whether they are either local or wide dominants, one would have to employ Chaney's methods.

6.

Determination of the Living Conditions of a Fossil Flora

1. *The environment under which a fossil flora lived can be inferred from the environment under which an equivalent flora lives today. That is, the characteristics of living equivalent species can be used for inferences concerning the fossil species with respect to the mutual aspects: the tolerances of the species and the nature of the physical environment.*

2. *The indications of environmental characteristics which a single species offers are greatly increased when the fossil flora includes a group of ecologically similar species which today form a natural ecological association and require approximately the same environmental conditions; it is the bulk of the flora that is most significant.*

3. *The presence of two or more elements in a fossil flora (groups of species belonging to different climaxes) indicates that one or more of several possible conditions prevailed. It is first necessary to rule out probabilities of redeposition and mixing of fossils, and of long-distance transportation of materials to the sedimentary basin. Such possibilities eliminated, the fossilized materials may have come from an ecotonal region, from a region containing relic colonies, from a region of diversified microhabitats and interdigitating communities, or from an association later segregated into two or more associations.*

4. *Even without botanical identification, the morphology[1] of fossilized plants is useful in indication of environmental conditions.*

5. *Evidence concerning the topographic situation in a region of deposition*

---

[1] I would include here not only life form, such as the tree habit, grass form, broad sclerophyll leaves, etc., but also anatomical features such as growth rings in wood, cutinization, stomatal characteristics, and attenuated leaf tips. It is necessary to caution, however, that there is no inviolate relationship between morphological characteristics and physiological adaptations.

can be used to suggest the environmental conditions of a fossil flora. For example, a mountain chain would be expected to have produced altitudinal belts and, possibly, a rain shadow on its lee side; a maritime site would be likely to have had an oceanic climate.

6. *The nature of inorganic sediments may offer evidence concerning the environment under which the sediments were deposited. Such indications of climatic conditions are secondary to fossil evidence; and in case of contradiction, the biological evidence concerning climate takes precedence over the inorganic evidence.*

Paleoecology, like modern ecology, must focus its attention on the habitat, the community, and the developmental processes that go on in communities. While the habitat comes first in the causal sequence, the interpretation of past habitats must wait on a knowledge of the biological aspects, because the biological effects are about all that are left. In this connection Clements (151) says, "It is the biome, or mass of plants and animals of a particular area or habitat, on which attention must first be fixed. The direct outcome of this is to reveal the successional movement, and on this as well as on the adaptive features of species and genera must be based our assumptions as to geological climates and soils." The occurrence of *Stipa* in the Miocene of Florissant, according to Clements, using the method of causal sequence, indicates not merely the existence of prairie but also, of course, a grassland climate and a grazing population; on the other hand, fossils of grazing animals even in the absence of botanical evidence suggest the presence in the region of both a grassland vegetation and a grassland climate.

Paleontology, in the absence of the ecological method, has been able to make only general inferences concerning the environment of a fossil biota. A modern representative ecological association having been established, it becomes possible to infer a great deal concerning the probable environment of the vegetation that contributed to the fossil flora.

If one is satisfied that a certain modern association is truly representative of a prevailing vegetation type of the past time, the conclusion that the ancient biota lived under approximately similar conditions to those of today is almost inescapable. The only objection that can be raised to such an assumption arises from the possibility that the species of the association have changed their tolerances and requirements and are living today under conditions dissimilar to those of the past. With respect to this point, it is known from modern investigations that related species, subspecies, and even strains usually have physiological differences; but in flowering plants, at least, these differences are frequently correlated with structural differences. It must be admitted, however, that this is a biological field which has received only scant attention. It is also a well-known fact that organisms have certain but often wide tolerances in the sense that there is a minimum and a maximum of any factor beyond which it is impossible for the organism to live. The actual environmental limits within which an organism can live depend upon the interaction of many factors so that the importance of a single factor toward limitation may vary both quantitatively and qualitatively. The importance of a particular environmental condition will also vary with the ontogenetic stage of the organism.

An additional aspect of the question of tolerance (the word tolerance, although somewhat vague and inconclusive, is convenient as a general term) is competition. The actual limitation of a species population seldom results from the action of climatic and edaphic factors alone; competition is also involved. That is to say, the potential physical boundaries of a species are seldom reached because of the limitations imposed by competitive species which are nearer their optimum living conditions. The importance of competition is abundantly shown by examples of cultivated plants which do well far beyond their native milieu provided only that they are relieved of competition.

Furthermore, it is known that there can be evolution which changes tolerances without changing the morphological characteristics of a species. Any one of the species of a fossil flora represented today by lineal descendants of the same species population (as in the western American redwood forests or the California coastal closed-pine forests) could have changed its tolerance without undergoing any detectable morphological change. That an entire association would do so, however, i.e., that the dominants and many of the associated species would evolve together and in the same direction so as to retain their integrity as an association and be living today under a different environment, is scarcely conceivable. In this connection Mason (453) says, "It may be true that in the course of time and through their evolutionary development, species may make adaptive changes to varying climatic conditions; but in view of the fact that each species reacts differently to the same stimulus and that the various species of an association are being affected at different points in their physiological limits for the particular stimulus, it is highly improbable that an association, as such, could continue through time intact."

Mason's conclusion can scarcely be disagreed with if he means that an association could hardly stay in the same place and remain intact under changing conditions. Some associations have retained their essential integrity through a considerable portion of geological time,[2] but this appears to have resulted from the migrations of their members, keeping pace with the shifting position of the climatic regimen to which they belong. That an association will more likely migrate than stay in one place and evolve—so far as the majority of its species, and especially its dominants, are concerned—seems probable from the fact that the dominants (for each synusia of the association) are physiologically more alike than different. It is conceivable that a number of different plants may together be isolated from their congeners through climatic or other changes, and that their isolation may result in similar evolution. It would seem, however, that this could be only because their previous selection and partial isolation—as in a position peripheral to an association—had already given them similar adaptations. In other words, it is my conviction that the preponderance of evidence supports a belief that mutational changes are generally at random. It is thus inconceivable that such diverse organisms as comprise an association could all, by pure chance, produce the appropriate adaptive mutations at the time they were needed because of the deterioration of the living conditions to which they were earlier adapted. Associations persist because conditions do not change to any considerable extent or because the associations migrate when conditions change. It would be an entirely gratuitous assumption to suppose that an association, at any other time or in any other place, lived under radically different conditions from those characteristic of it today.

It is with considerable confidence, then, that one can investigate the environmental conditions under which a modern representative association lives and therefrom reconstruct the probable environmental conditions of the fossil flora. In his study of the closed-cone pine forest of the Tomales flora, Mason (453) says, "Considered in terms of moisture and temperature, as well as the occurrence of fog, the climatic conditions represented by the fossil flora differ only slightly from those of the same region today. If anything, they suggest slightly more northerly conditions." He charted the precipitation tolerances of the modern species and found the total range to be from a minimum of about 10 inches to a maximum of about 100 inches. These are not the significant limits for the association. The significant limits are set by the species of the association having the highest minimum (about 23 inches) and the species having the lowest maximum (about 35 inches). Mason makes the tentative conclusion, therefore, that the rainfall during Tomales time in the closed-cone pine forest ranged between 23 and 35 inches

---

[2] All associations, if for no other reason than evolution of their constituent species, must have had a long development and only gradually attained what is called a characteristic composition and structure.

for the area contributing to the flora represented. If this seems like a wide range, it must be noticed that today the amount of precipitation varies even more within very close localities in the coastal region.

Mason found also that the mean temperatures varied between 45° and 57° for the component species, with a mean of about 51° F. for the association. He further emphasized that summer (dry season) fogs appear to be critical for many of the species and that in the modern range of *Pinus radiata* at least 35 per cent of the days are foggy, with most of them occurring in the dry season.

Chaney and Mason (139) studied the closed-cone pine forest flora from the asphaltic sands of the Carpinteria deposits and concluded that they lived under conditions very similar to those prevailing today on the Monterey Peninsula, 200 miles to the north. At Monterey the precipitation varies from about 19 inches at sea level to over 25 inches on the top of Huckleberry Hill. This precipitation is supplemented by an unmeasured amount of moisture condensed during the frequent fogs, which, they say, may amount to several inches annually. The mean annual temperature at Monterey is 56.6° F. In the Carpinteria region today the precipitation at sea level is about the same, but fogs are rare. Also, the mean annual temperature is higher (about 60° F. at Santa Barbara). The climate during Carpinteria time was probably similar to the modern climate at Monterey rather than to the modern climate in the region where the deposit exists.

The importance of vegetation in the primary causal sequence has been pointed out in the earlier section on basic paleoecological principles. Vegetation is both an effect of climate and topography and a cause in relation to the animal world. The nature of the vegetation would seem to be sufficient evidence of the nature of the environment. However, all sources of evidence concerning past conditions should provide facts which fit together into an harmonious pattern. The value of corroborative evidence is always to be recognized. Just as in the identification of fossil plants, determinations based upon leaves become more certain when they can be supplemented by wood, flower, or fruiting structures, so, in the determination of past environments, supplementary lines of evidence are of value. The members of the Carpinteria biota are an excellent illustration of this point because there is the rather unusual situation of an extensive fauna and flora being preserved together.

Chaney and Mason (139) conclude that the Carpinteria plants represent a somewhat open woodland or savanna in a coastal hill region dominated by *Pinus radiata, P. muricata,* and *P. remorata.* In the closely related Upper Pleistocene deposits at Rancho La Brea and McKittrick, the mammalian and avian fauna point to a condition of non-forest vegetation, probably grass-land. But at Carpinteria the presence of certain animals and the absence of others indicate an open forest or savanna environment (the groups studied include rodents, mammals, birds, insects). There were oaks, pines, junipers, etc., suggesting a condition of open woodland and grass comparable with that of the intermontane valleys in the Tchachapi ranges.

An excellent illustration of the details of information concerning the probable conditions of a fossil vegetation which can be obtained by studying modern representative forests is found in the study by Chaney and Mason (138) of the Willow Creek, Santa Cruz Island, flora. Although this small flora consists of only nine species, it contains three conifers of the modern closed-cone pine forests of the coastal regions and coastal islands of California: *Pseudotsuga taxifolia*, *Pinus remorata*, and *Cupressus Goveniana*. Several species of the closed-cone pine forest today grow together only in one locality near Fort Bragg, which is 440 miles north-northwest of Santa Cruz Island. The fact that this juxtaposition of the species is now so circumscribed enables a very close delimitation of the probable conditions under which the Willow Creek flora lived. With an average rainfall of about 39 inches that falls chiefly from October to May, a mean temperature of 52° F. with a low daily and annual range (the latter being only about 10° F.), and subject to almost daily fogs, particularly through the dry summer, this climate is in strong contrast with that of Santa Cruz today. The rainfall is considerably higher and the temperature is considerably lower. The conclusion is reached that the Willow Creek flora represents a southward extension of the northern forest that is probably correlated with one of the glacial epochs of the Pleistocene.

When several floras are placed in relation to one another, their climatic suggestions may reveal a broad relationship of considerable significance. Such an instance is illustrated by Chaney (127) in his study of the Mascall flora. In the following list the climatic suggestions from the upper Oligocene to the present are summarized:

	Rainfall Indicated
Present flora of the John Day Basin	15 inches
Alturas flora, Upper Pliocene	15 inches or more
Santa Clara flora, Upper Pliocene	20 inches or less
Sonoma tuff flora, Lower or Middle Pliocene	20 inches or more
Mascall flora of the John Day Basin, Miocene	30 inches
Bridge Creek flora of the John Day Basin, Upper Oligocene	40 inches

Chaney decided that "if the conclusions regarding the rainfall of these fossil floras, as based on the moisture requirements of their living equivalents, are correct, and if the floras are at all typical for the Tertiary of the western United States . . . there was a progressive trend toward aridity in the Great Basin and perhaps to some extent in west-central California during the Tertiary."

With respect to these claims, Berry (61) suggests that it is highly hazardous to conclude that the Mascall flora lived under an environment closely comparable to that of the modern oak-madrone forest when species of the Mascall no longer represented in the Pacific American floras are excluded from consideration. He says, "Their absence in the recent flora is the most conclusive evidence that the Miocene physical conditions differed from those of the present in that region. Otherwise there is no reason to account for their lack of survival." It is difficult to resolve these two opinions. In the first place, data are not provided concerning the conditions under which the segregated members of the flora now live in Asia and eastern America, but it would seem likely that they are of the same general order, so neither Chaney nor Berry can assert more than opinions. In the second place, it would appear certain that whatever the Mascall environment, it did not exceed the limits of tolerance (present limits) of the oak-madrone species. Berry was also uncertain that the progressive trend toward aridity characterized more than the Pliocene because the upper Miocene Latah flora was definitely mesophytic.

The closeness with which topographic conditions can be reconstructed with some assurance from even a small fossil flora is illustrated by Webber (679), who worked on the woods of the Ricardo assemblage. She says, "The present habitats of related trees and the growth rings in the fossil specimens suggest that the Pliocene assemblage grew in a steep-sided valley in a region characterized by low rainfall and temperature approximating that which distinguishes the Upper from the Lower Sonoran Zone. At the present time the closest approach to these environmental conditions is found in the San Jacinto Range between San Jacinto Peak and Santa Rosa Mountain, where palms are found living in the lower stretches, with pines and oaks occupying the slopes of the upper portions of the valleys. The suggested Pliocene climate differs from the present climate of the Mohave Desert in that the former is characterized by less extreme aridity."

Realizing full well the reciprocal relations between topography (and resultant climates) and vegetation, Chaney and Hu (137) have given extensive consideration to all aspects of the problem in their study of the Shanwang flora. Their concluding statement is worth quoting:

Summarizing the topographic evidence based upon inorganic data, the Shanwang formation is believed to have accumulated in upland lakes and valleys. A range of mountains to the southwest would have provided the coarse sediments which make up the yellow sands, and doubtless included the volcanoes from which boulders of basalt and showers of ash were carried to the Shanwang basin. No traces of the volcanoes have remained, but the mountain mass to the west now rises 1000 meters or more above Shanwang, and doubtless there was still greater difference in elevation during Miocene time.

The topographic indications of the Shanwang flora are completely in accord with conclusions based upon inorganic evidence. Modern forests with essentially the same composition have been studied in upland situations in China, Korea and Japan; at no place in Asia has a similar forest been found in a region of low relief or low altitude. The rather unusual mixture of climatic types which characterizes the Shanwang flora is considered to reflect topographic diversity, and to represent a forest of a sort found today only at middle altitudes in regions where the climate at sea level is sub-tropical. Mention of climatic relations at this point is an illustration of the close interaction of climatic and topographic factors in controlling the development of vegetation.

In the Mount Eden beds near Beaumont, California, Axelrod (31) found fossils from several vegetation types. He says, "From a distributional standpoint, the modern equivalents of the Mount Eden flora may be considered to consist of several units: (a) an assemblage on the north slope of Sierra Peak, 35 miles west of Beaumont, constituting 63 per cent of the flora; (b) a desert-border element on the edge of the Colorado Desert, some 15 miles east of the fossil area; (c) a digger pine forest 100 miles to the north, and (d) a Sonoran element now existing in the Southwest and northern Mexico." An analysis of the modern equivalents of the Mount Eden flora indicates that the flora is distinctly interior and, in the main, xeric. Rainfall in the region increased with elevation from about 10 inches (for the desert-border plants) to probably 25 or 30 inches (as suggested by *Pseudotsuga* and *Pinus Coulteri*). It is quite likely that summer precipitation was inappreciable. Axelrod concludes, "The habitat conditions suggested by the modern equivalents of the flora, and as further substantiated by both geologic and vertebrate evidence, are indicative of a low-lying basin occupied by shallow lakes and marshes, with adjacent highlands."

Berry (55) concluded that it was really remarkable to what an extent the Claiborne flora of the Grovetown estuary of Georgia agreed in indicating the character of the habitats. The flora was principally a tidal *Rhizophora* swamp, with sand-beach and rain-forest elements. The present winter isotherm at Grovetown is 48° F., but none of the representative species today flourishes north of the 52° F. winter isotherm, and most do not occur north of 60° F. The rainfall was probably 60 to 80 inches and it is unlikely that any of the species could have withstood freezing. The modern representative flora is so typical in area—as characterized by *Rhizophora*—and the conditions of its growth so uniform that there is little uncertainty in visualizing the coastal conditions of Claiborne time.

Plant geographers have long recognized general correlations between the physiognomy of vegetation and the characteristics of climate. In fact, it does not require a geographer to note the characteristics of rain-forest vegetation and to reach a conclusion as to the precipitation and temperature charac-

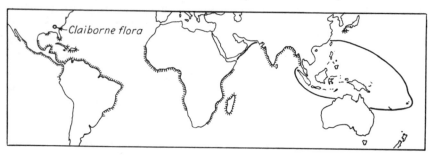

Fig. 4. The distribution of the existing species of *Rhizophora* within the limits of 32° N. lat. and 30° S. lat.; after Berry (55, Fig. 11). The Claiborne flora (Eocene) consisted of several species whose modern allies are all rather uniform in their distribution; the distribution of *Rhizophora* will serve for all the genera as an indication of the climatic conditions that probably prevailed at Grovetown, Georgia (shown by circle in southeastern United States), at the time the Claiborne flora was fossilized. Base map after Goode's copyrighted Map No. 101, with permission of the University of Chicago.

teristics. The same is likewise true for other principal world vegetation types such as desert scrub, winter-green forests, summer-green forests, temperate grasslands, taiga, and tundra. There have been scientific efforts to classify vegetation and climates both separately and simultaneously on the basis of the physiognomy of vegetation. Among recent well-known efforts along these lines is the work of Raunkiaer (516) on life-form classes and leaf-size classes as indicators of environmental conditions, especially climate.

Berry (62)[3] pointed out, "Plant fossils have this merit aside from any question of botanical identification, and this feature seems to have been lost sight of by numerous critics of paleobotanical practise: that the size and form of leaves, their texture, the arrangement and character of stomata, and the seasonal changes in wood, afford criteria that are quite as valuable climatically even though the species or genus to which they belong remains undetermined." In a discussion of Cretaceous floras, Berry (55) employs

[3] Also earlier in *Proc. Amer. Phil. Soc.*, 1922, 61:8–9.

interesting morphological observations. Noting the coriaceous-leafed forms, he concludes that they are due not to aridity but to insolation and "pseudo-xerophytism" of the coastal swamp habitats. The plentiful nature of the rainfall is assumed from the development of "dripping points" by leaves in several genera. This feature, supposedly characteristic today of regions of high precipitation, was especially emphasized in the Tuscaloosa flora of Alabama and in the South Carolina representatives of *Ficus crassipes*. The Cretaceous flora was a unit and only little changed from Alabama and New Jersey to the west coast of Greenland, but the frequency and size of dripping points were greater southward within the same species.

In their work on the early Tertiary Goshen flora, Chaney and Sanborn (140) have employed several approaches to the problem of indicating the low-latitude nature of the flora. Not only have they made comparisons with the species of Panama and other neotropical and paleotropical regions, but they have employed morphological life-form and leaf-character comparisons.

In the Goshen flora they found, by comparison with the life forms of the living equivalents, 18 species of trees, 23 species of shrubs or small trees or both, and 4 species of vines. The percentage ratio of tree to shrub species is 40 to 60, which is rather high compared to most related modern forests. Recalculating data of Sinnott and Bailey (580) to include only trees and shrubs, they provide data for the comparisons shown in Table 5.

TABLE 5.—A comparison of tree-shrub ratios of the Goshen flora with certain modern ones

Forest Flora	Percentage Ratios	
	Trees	Shrubs
Low-latitude floras		
Manila	44	56
Goshen fossil flora	40	60
Nicaragua	30	70
West Indies	27	73
Hong Kong	27	73
North temperate districts		
Average of eleven districts	18	82

Chaney and Sanborn conclude that the high tree-shrub ratio in the Goshen flora relates it to forests of low latitudes. They believe also that the presence of four species of lianes, and possibly a fifth, is wholly consistent with the other indications of a subtropical habitat for this Eocene Oregon flora.

They also analyzed the Goshen flora on the basis of the following leaf characters: length (whether under or over 10 cm.), organization (whether simple or compound), nervation (whether pinnate or palmate), margin (whether entire or non-entire), dripping point (whether present or absent), and texture (whether thick or thin). They then made a statistical comparison between the Goshen flora and a similar analysis of a portion of the Panama forest, and between the Goshen flora and the Bridge Creek flora, which is the temperate redwood forest of the same general region in Miocene time. As a result of these studies, Chaney and Sanborn come to the conclusion that "the possibility is indicated that with adequate data on fossil and modern leaf characters, the climatic conditions of a region or of a geologic period may be determined on the basis of the leaves of fossil or of living plant assemblages."

On the basis of their studies of the Goshen flora, Chaney and Sanborn are in full accord with the following conclusions of Sinnott and Bailey:

1. There is a very clearly marked correlation between leaf margin and environment in the distribution of dicotyledons in the various regions of the earth.

2. Leaves and leaflets with entire margins are overwhelmingly predominant in lowland-tropical regions; those with non-entire margins in mesophytic cold-temperate areas.

3. In the tropical zones, non-entire margins are favored by moist uplands, equable environments, and protected, comparatively cool habitats; in the cold-temperate zones, entire margins are favored by arid environments and other physiologically dry habitats.

4. In moist tropical regions the leaves are of comparatively large size, and these leaves are commonly semi-xerophilous (xeromorphic) in structure.

The significance of the ratio between entire and non-entire leaves is further investigated in the light of the third conclusion of Sinnott and Bailey above. Chaney and Sanborn compare the Eocene Wilcox flora of the southeastern United States and the Goshen flora of Oregon with certain modern floras. They conclude that it is clear that the Goshen flora has a representation of entire-leafed species which indicates its relationships. It corresponds closely with the upland forests of Simla and Hawaii. The larger number of non-entire species, as compared with the lowland forests of Panama, the Gangetic Plain, and Hawaii, may be interpreted as indicating a relatively moist habitat of a somewhat cooler type than occupied by the modern lowland forests of the tropics, according to Chaney. The suggestion of this type of habitat for the Goshen flora is in line with other types of evidence. The

TABLE 6.—A comparison of species with entire and non-entire leaves in certain fossil and modern floras

| Floras | Percentage of Dicotyledons with Leaves | |
	Entire	Non-Entire
Fossil floras		
Wilcox flora	83	17
Goshen flora	61	39
Modern floras		
Panama	88	12
Simla	58	42
Upper Gangetic Plain	71	29
Hawaiian upland	56	44
Hawaiian lowland	76	24

flora occupied a habitat in which the temperature was lower at the same elevation or the altitude higher than that of the related Wilcox flora of the southeastern United States.

Both the ecological relations of the equivalent vegetation of the Goshen flora and the general morphology of the equivalent species and the forests to which they belong indicate the lowland tropical and temperate rain-forest character of the Goshen flora. It is of special interest that the lithological and stratigraphic relations of the shales containing the Goshen flora also indicate that they were laid down near sea level. According to Chaney and Sanborn, "From the standpoint of the lithology of the Goshen sediments there is no basis for postulating their deposition in or adjacent to a region of high relief, since they are all relatively fine textured and show no evidence of having been laid down by streams of high gradient. The occurrence of marine invertebrates in the volcanic sediments at no great stratigraphic or geographic distance from the Goshen locality adds weight to the assumption that the plant-bearing sediments were laid down near sea-level. It seems possible to conclude, on the basis of floral composition, the lithology and the association of marine invertebrates in nearby deposits, that the Goshen flora occupied a region close to sea-level."

Having dwelt at length on the Goshen in order to emphasize the relations between working methods and paleoecological conclusions, let us turn to an entirely different region and a flora of much greater age. Davies (178) has made an extensive statistical study of the plant records of the Coal Measures of East Glamorganshire, with the result that he could determine in a striking way the relative elevation of the land about the sedimentary basins and the ecological condition of the flora represented in the different horizons. His data are based on nearly 390,000 plant records of 323 species and 29 horizons. The lower Westphalian series comprises 18 horizons, the Staffordian series 8, and the upper Radstockian series 3 horizons. The Lycopodiales (with 95 species) are accepted as indicating a wet type flora, the Filicales and Pteridosperms (with 141 species) a dry type of flora, and the Cordaitales and Equisetales (with 79 species) an intermediate or moist flora. There was a consistent recurrence of these classes—with the dominance of one over the other, or the reaction of wet and dry floras—which showed that the physical conditions of elevated and depressed land surface were the direct cause of the vegetational differences recorded in the shales. Not only was there a tendency at different times for a certain type of flora to dominate, but a certain genus tended to dominate in each individual type. Also, the significance of the change from one horizon to another was increased by the determination, in one case at least, that the characteristics of a horizon are not a local phenomenon but are preserved at different stations for a horizon. Furthermore, the fact that the changes are not erratic but are generally according to trends lends credence to the statistics.

A few more examples concerned with the climatic indications of certain fossils or fossil types will serve to show that terrestrial vegetation is the best guide to geological climates, and that caution is necessary and dogmatism unwarranted. It has been pretty widely believed, on the basis of certain fossils found at high latitudes and identified with groups now of the equatorial zone, that tropical and subtropical climates have extended to very high latitudes in the past. Berry (63) has reviewed the facts, and the following brief epitome of northern floras is taken from his work. During the Devonian most of the types extended northward from 45° or 50° lat. to as high as 75°, and the chief controlling factor was moisture rather than temperature. Lower Carboniferous floras extended to lat. 81° N., and were composed of palustrine types (as in the Devonian) providing no conclusive evidence as to climate. More than half the known fossils are Lepidophytes which seem to have shown little response to temperature. Triassic floras are known to about 70° N. lat. and are mostly cosmopolitan Rhaetic types of cycads and ferns that indicate a temperate climate and plentiful moisture. During the Jurassic there was a predominance of conifers with pronounced seasonal growth rings, suggesting a cool temperate climate reaching to 80° N. lat. Lower Cretaceous floras extended to lat. 70° N., with few differences from those of lower latitudes. The climate was humid, but temperatures were not high. Upper Cretaceous floras also extended at least to lat. 70° N. and appear to have had a definite temperate stamp. During the Tertiary, from Upper Eocene to Oligocene, temperate floras extended to 81.5° N. lat., being of cool temperate aspect northward.

Such interpretations are conservative and more in line with the facts than are the claims of tropical floras at high latitudes. It is true that at the latitude of southern England the London Clay flora of Eocene time was of a tropical rain-forest type, but nothing of the sort is known at higher latitudes. Reid and Chandler (523) found that the London Clay flora is now best represented in the East Asian lowland tropics, particularly of the Malay Islands where 73 per cent of the representative genera occur.

The authentic identification of *Artocarpus,* now a genus largely of Indo-China, from the Upper Cretaceous Atane bed of West Greenland raises an important question because the flora as a whole is preponderantly temperate. It is reasonable to assume that the temperate majority are less likely to have altered their requirements than that *Artocarpus* was always a completely tropical genus, but, as Berry says, "this falls short of actual proof." There is no reason why *Artocarpus* could not always have been primarily a tropical genus and also have included in the past one or more temperate species, just as certain genera or families do today. For example, *Diospyros* extends about 12° farther north than the bulk of the Ebenaceae and, in the Lauraceae, *Sassafras* and *Benzoin* are temperate members of a largely tropical family. Another possibility occurs to me. It is entirely possible that a strictly tropical genus insofar as we now know may have been larger and more variable and have lost temperate species it once contained.

One important cause of misinterpretations is the failure to distinguish between the equatorial belt and the truly tropical (hot, humid) regions within it. The modern representative of some fossil species may live in the tropical belt, but inland from the coast or upon a plateau or mountain where the climate is truly temperate. It is necessary for the paleoecologist to distinguish between geographical location and the actual environment where representative forms live. For example, the davalloid ferns of the Frontier formation (60), or tree ferns in general, are usually considered to indicate tropical conditions, but actually they find their optimum conditions in temperate rain forests, not in tropical lowlands. Berry (61) notes that the Miocene Latah flora contains an undoubted member of the fig genus (*Ficus*) and, although it is today confined to warm-temperate and tropical regions, many species of the large genus are not out of place in a humid but strictly temperate environment or such as Washington state in Latah time.

A quotation from Berry (61) will serve to sum up the problem. He says:

I am now convinced that most paleobotanists, nearly all of whom have been dwellers of the Temperate Zone, have been similarly misled.[4] Since 1917 I have had an opportunity to see the living flora in the Antilles, in Central America, and particularly in tropical Bolivia and Peru. My study of the upland floras of these parts of Bolivia and Peru that border the basin of the Amazon disclosed some remarkable facts. Plants that are elsewhere coastal tropical types, such as *Dodonaea, Swietenia,* and *Sapindus saponaria,* grow on the lower slopes of the eastern Andes, as if they had been stranded there by the withdrawal of a Pliocene sea. I noted also—and this fact is more significant to the student of paleo-ecology—that many of the genera and even species of the lowland tropics extend to elevations that carry them in effect well out of the Tropical Zone, and I came back from that region with the conclusion that none of the fossil floras of the Temperate Zone that paleobotanists have termed tropical, are in the strict sense of the word "tropical." . . . These facts apply with particular force to the attempted interpretations of the environment of the Claiborne, Jackson, and Vicksburg floras, and although that environment may have been almost tropical, I would modify the published statements concerning it.

[4] Berry refers to his earlier conclusions concerning the nature of the Wilcox flora. Cf. (56).

7.

Determination of the Structure of the Vegetation of a Fossil Flora

---

1. Vegetation of the past must have shown an organization similar to that which equivalent vegetation shows today. This organization is shown in the internal structure of communities and in the spatial arrangement of different communities, both successional and climax, on a combined topographic-climatic-edaphic basis.

2. Because the ecological adaptations of some species are strongly limited, they become good indicator species. It is possible to suggest many details concerning the major outlines of the community structure of a region from a relatively fragmentary fossil flora if it contains good indicator species.

There is every reason to believe that the communities of times past showed structural patterns similar to those known to exist today. Since the organization of the vegetation of a region depends upon the essential adaptations of the available species and the nature of the available habitats, and since these relations are inescapable, it is frequently possible to reconstruct the general pattern of vegetation when only a fraction of the total species is known.

For the sake of illustration, suppose one were handed specimens of the following modern plants (leaves, etc., analogous to fossils) and were asked the general nature of the vegetation and the situation where they were obtained: *Quercus macrocarpa, Salix nigra, Juglans nigra, Corylus americana, Andropogon scoparius, A. furcatus,* and *Calamagrostis canadensis.* Specimens of seven living species are not very many to perform the above task, but one familiar with the vegetation of the United States can make a pretty good guess. The oak suggests the western limits of the eastern deciduous forest where the burr oak (*Quercus macrocarpa*) is most frequent and, together with the tall prairie grasses (*Andropogon* spp.), suggests ecotone between woodland and grassland, possibly savanna. The hazel (*Corylus*) is not strange to the picture so far conceived since it occurs as an undershrub in open woods or at the forest edge. The willow (*Salix*) suggests either fluvial or marsh conditions and, together with the marsh grass (*Calamagrostis*) and the walnut (*Juglans*), would make one incline toward the assumption that the collection of plants came from a stream-side situation with *Calamagrostis* along marshy borders, *Salix* at the water's edge, and *Juglans* on the floodplain. All together, the species suggest a stream in the eastern American prairie (of Illinois, perhaps) with a low upland grove of oak near by.

To play this game again, let us suppose that one were asked the same question concerning the following plants: *Fagus grandifolia, Acer saccharum, Viburnum acerifolium, Carpinus caroliniana, Ostrya virginiana, Hepatica americana, Hystrix patula,* and *Thuidium delicatulum.* These species consist of two dominant climax trees (*Fagus* and *Acer*) of the northeastern deciduous forest and a group of other species of various life form which are characteristic under the cover of forest in mesophytic situations. The conclusion is rather safely arrived at that the plants represent a single forest type and that the associated species suggest layers of subordinate vegetation: *Ostrya* and *Carpinus* a small-tree layer, *Viburnum* a shrub layer, *Hepatica* and the grass *Hystrix* the herbaceous element, while the moss (*Thuidium*) grew on the soil or even on logs and tree butts. This is again a very small number of species, but it enables the partial reconstruction of the synusial complexity of a forest of a definite type.

The above illustrations are offered as a suggestion of the problem that confronts a paleoecologist in his efforts to reconstruct the vegetational pattern of a region when his fossil materials represent only a few of the many species which undoubtedly were present. It is obvious that the first necessity is that the person confronted with such a problem know the ecological characteristics of the species concerned, or of the modern equivalent species in the case of a study of fossil plants. Because the flora of a locality is not a heterogeneous mixture of hundreds of species but a highly organized phytocoenosis on a life-form—tolerance—habitat basis, usually with limited dominants, it is possible to reconstruct the essential outlines of the vegetational pattern.

The complexity of vegetation derives from the fact that plants of several different life forms live together in a single association (composing arborescent, frutescent, herbaceous, and cryptogamic synusiae), and from the fact that several different communities may occupy separate terrains and ecological niches within a rather circumscribed area. These different communities, representing different ecological situations, may be associes in one or more successional series, or they may be a mosaic of climax types, the prevailing climax with preclimax and postclimax relics and lociations. Nevertheless, many of the elements of a region can be detected and their proper relationships be approximately deduced (climatically, edaphically, topographically, and developmentally) on the basis of referring the fossil materials to modern equivalent materials. Furthermore, the modern equivalent species need not exist together today in the region where the fossil flora exists, or in any single region elsewhere, although the majority are likely to be associated somewhere.

The working of the paleoecological method in the problems of community structure of a fossil flora will be illustrated by a few selected examples.

The completeness with which a fossil flora can sometimes be matched with modern vegetation is illustrated by Chaney (127) in his study of the Mascall flora. The close relation between the Miocene Mascall forest and the California oak-madrone forest is shown by a comparison of the dominants. Cooper (161) lists the dominants of the oak-madrone forest as shown in Table 7. The equivalent Mascall species are found in parallel arrangement.

TABLE 7.—A comparison of equivalent dominants of the Mascall and the modern oak-madrone forest (Data from 161 and 127)

Dominants of the Modern Oak-Madrone Association	Equivalent Species of the Fossil Miocene Mascall Flora
Myrica californica	Absent
Castanopsis chrysophylla	Castanopsis chrysophylloides
Quercus agrifolia	
Quercus chrysolepsis	
Quercus Engelmanni	Quercus convexa
Quercus Wislizeni	
Quercus Kelloggii	Quercus pseudo-lyrata
Quercus lobata	Quercus duriuscula
Pasania (Quercus) densiflora	Quercus consimilis
Umbellularia californica	Umbellularia sp.
Arbutus Menziesii	Arbutus sp.
Acer macrophyllum	Acer merriami

Only *Myrica,* of the dominants listed by Cooper, is absent from the known Mascall flora. *Quercus convexa* of the fossil flora probably represents more than one species, possibly all four of the modern live-oak species grouped together in the table.

The dominant species of the oak-madrone forest listed by Cooper represent the regional association, not a local expression of it. It is of particular interest to see how well the Mascall flora matches a particular of the oak-madrone forest type. Chaney makes this comparison, using as a basis the common trees of the forest found at Jasper Ridge near Palo Alto, California. Here there is an admixture of redwood, which also occurred in the Mascall. By reference to Table 8 it is seen that every common tree of

TABLE 8.—A comparison of the common trees of the modern forest at Jasper Ridge, Palo Alto, California, and the matching species of the Miocene Mascall flora

Jasper Ridge Flora	Mascall Flora
Sequoia sempervirens	Sequoia langsdorfii
Salix laevigata	Salix varians
Salix lasiolepis	Salix varians
Populus trichocarpa	Populus lindgreni
Alnus rhombifolia	Alnus sp.
Quercus Kelloggii	Quercus pseudo-lyrata
Quercus agrifolia	Quercus convexa
Quercus lobata	Quercus duriuscula
Umbellularia californica	Umbellularia sp.
Arbutus Menziesii	Arbutus sp.
Acer macrophyllum	Acer bolanderi
Aesculus californica	Absent

the Jasper Ridge forest has an equivalent species in the Mascall flora. It is worthy of note, also, that the above list of Mascall species includes all but one of the wide-ranging species of that time.

With such a surprisingly close parallel between a modern forest and a fossil flora, it would seem entirely safe to make conclusions from the characteristics of the former concerning the latter. Not only is it likely that the climate that the Mascall flora enjoyed was similar to the modern climate of the oak-madrone forest (about 30 inches annual rainfall, 32 inches at Jasper Ridge), but also one would strongly suspect that the associated plants (especially flowering herbs, grasses, ferns, mosses) which are absent from the fossil record were similar both taxonomically and ecologically insofar as they existed at that time, and that the organization of the various elements in the community was also similar.

Studies by Mason (453) of the Pleistocene flora of the Tomales Bay area illustrate well the operation of the ecological approach to the reconstruction of community structure and the local community types of an area. Using the basic assumptions of paleoecology, Mason reconstructed the probable nature of the plant communities that contributed materials to the fossil flora. Comparing the species of the total flora with their modern equivalents, he decided that the prevailing major vegetation of Tomales time was a closed-cone pine forest with three life-form layers (a dominant tree layer, a shrub layer, an herbaceous layer) much like the modern Bishop pine (*Pinus muricata*) forest on Inverness Ridge along the adjacent coastal range. The closed-cone pine forest is probably to be considered subclimax, in the sense of Clements, to the montane climax dominated by *Pseudotsuga,* despite the long duration of the pine type. Other ecologists, adhering less strictly to the monoclimax concept, refer to the coastal closed-cone pine forests as climax.

The following enumeration suggests the probable structure of the forest.

The major forest cover: the closed-cone pine forest.

Trees: *Pinus muricata, Pinus radiata, Pseudotsuga taxifolia, Quercus agrifolia, Arbutus Menziesii, Cupressus Goveniana, Myrica californica.*

Shrubs: *Adenostoma fasciculatum, Arctostaphylos columbiana* (?), *Arctostaphylos uva-ursi, Baccharis pilularis, Ceanothus rigidus, Ceanothus thyrsiflorus, Corylus rostrata* var. *californica, Garrya elliptica, Prunus emarginata, Rhus diversiloba, Rubus vitifolius, Sambucus glauca, Vaccinium ovatum.*

Herbs: *Daucus pusillus, Eriophyllum artemisiaefolium, Fragaria californica, Galium, californicum, Pteris aquilina.*

In addition to the pine forest of the uplands and prevailing in the region in general, Mason recognized elements of two forests of moist to wet soil. On a successional basis, these forests would be considered associes, probably subseral to the closed-cone pine forest.

Forest of moist swales: swamp forest type.

Trees: *Myrica californica, Quercus agrifolia, Umbellularia californica.*

Shrubs: *Amelanchier alnifolia, Baccharis pilularis, Cornus californica, Prunus emarginata, Rhus diversiloba, Rubus parviflorus, Sambucus glauca, Symphoricarpos albus.*

Herbs: *Calandrinia caulescens, Fragaria californica, Galium californicum, Pteris aquilina.*

Riparian forest: stream-side forest type.

Trees: *Acer macrophyllum, Alnus rubra, Myrica californica, Picea sitchensis, Torreya californica, Umbellularia californica.*

Shrubs: *Amelanchier alnifolia, Cornus californica, Corylus rostrata* var.

californica, Prunus emarginata, Rhus diversiloba, Rubus parvi-florus, Rubus spectabilis, Symphoricarpos albus.

Herbs: *Calandrinia caulescens, Carex* spp. *Datisca glomerata, Oenanthe sarmentosa, Rumex salicifolius.*

Two additional communities not dominated by trees were recognized. One type consisted of the fresh-water marshes, with *Camassia Leichtlinii, Carex* spp., *Montia fontana, Montia Howellii, Montia siberica, Oenanthe sarmentosa,* and *Rumex occidentalis.* These marshes were probably bordered by the shrubs mentioned above in the forests of the moist soil type. The remaining community was the salt-water or subsaline marshes which contained *Atriplex hastata, Carex obnupta, Rumex occidentalis, Rubus salici-folius, Ruppia maritima,* and *Scirpus* spp.

From the nature of these five communities (the closed-cone pine forest, the swamp forest, the riparian forest, the fresh-water marsh, and the salt-water marsh), it is apparent that all could have existed within the compass of a few acres and definitely under a single climatic type. They represent the sort of vegetational mosaic that is everywhere encountered today in coastal regions with stream valleys and hills where the relief need not be over a few hundred feet at the most.

Sometimes a fossil flora contains much more diverse elements that are suggestive of rather strongly contrasted communities. Such a flora is found in the Pliocene Mount Eden beds reported on by Axelrod (31) from southern California. The Mount Eden flora is listed in Table 9 by associations with the modern representative species in the right-hand column. The desert-border element, the digger pine forest, the chaparral, and the *Pinus-Pseudotsuga* element, all probably represent climax vegetational types. The other two groups of species indicate communities of a successional nature and rather definite edaphic situations.

Axelrod sums up the probable situation with respect to the physical conditions and the disposition of the communities about the regions of sedimentation as follows:

The habitat conditions suggested by the modern equivalents of the flora, and as further substantiated by both geologic and vertebrate evidence, are indicative of a low-lying basin occupied by shallow lakes and marshes, with adjacent highlands. Around the lake borders and in the marshy areas was a typical hydric element of *Typha* and *Equisetum.* Riparian species of *Platanus, Salix,* and *Juglans* were also present and extended well up into the adjacent hills along stream courses; *Populus* was probably confined to cooler canyons. On the lower portions of alluvial fans, in dried water-courses, and on lower exposed slopes was a desert-border element of *Ephedra, Prosopis, Prunus* spp., and *Sapindus.* A

TABLE 9.—The Mount Eden Pliocene species and their modern representatives classified according to associations and associes (data from 31)

Mount Eden Species	Modern Representative Species
Desert-border element:	
Ephedra sp.	Ephedra spp.
Lepidospartum sp.	Lepidospartum squamatum
Prosopis pliocenica	Prosopis juliflora glandulosa
Prunus preandersonii	Prunus Andersonii
Prunus prefremontii	Prunus Fremontii
Quercus pliopalmeri	Quercus Palmeri
Sapindus Lamottei	Sapindus Drummondii
Savanna-woodland association: Digger pine forest:	
Arbutus sp.	Arbutus xalapensis
Juglans Beaumontii	Juglans californica
Pinus Piperi	Pinus sabiniana (Digger pine)
Quercus Hannibali	Quercus chrysolepis
Quercus lakevillensis	Quercus agrifolia
Quercus orindensis	Quercus Douglasii
Chaparral element:	
Arctostaphylos preglauca	Arctostaphylos glauca
Arctostaphylos prepungens	Arctostaphylos pungens
Ceanothus sp.	Ceanothus cuneatus
Ceanothus edensis	Ceanothus divaricatus
Cercocarpus cuneatus	Cerocarpus betuloides
Fraxinus edensis	Fraxinus dipetala
Quercus pliopalmeri	Quercus Palmeri
Rhus prelaurina	Rhus laurina
Coniferous element:	
Cupressus preforbesii	Cupressus Forbesii
Pinus Hazeni	Pinus Coulteri
Pinus pretuberculata	Pinus tuberculata
Pseudotsuga premacrocarpa	Pseudotsuga macrocarpa
Riparian element:	
Fraxinus edensis	Fraxinus dipetala
Juglans Beaumontii	Juglans californica
Lepidospartum sp.	Lepidospartum squamatum
Platanus paucidentata	Platanus racemosa
Populus pliotremuloides	Populus tremuloides
Salix coalingensis	Salix lasiolepis
Salix sp.	Salix exigua
Sapindus Lamottei	Sapindus Drummondii
Lake-border or marsh element:	
Typha Lesquereuxi	Typha latifolia
Equisetum sp.	Equisetum sp.

savanna of *Quercus* spp., *Pinus Piperi,* and to a lesser extent *Juglans,* existed over the rolling hills and plains, and on more favorable north slopes these same species formed woodland associations. The savanna-woodland was locally interrupted by chaparral species of *Arctostaphylos, Ceanothus, Cercocarpus, Rhus,* and *Fraxinus.* A coniferous element existed above the savanna-woodland, and was at lower elevations so that cones could be readily transported to sites of deposition. *Pseudotsuga* reached optimum development on cooler north slopes, while the other species were characteristic of drier exposures.

Questions concerning the altitudinal relations of the species of a fossil flora are sometimes difficult to resolve, but the use of the ecological method is illuminating. For example, Dorf (194) found that the Weiser flora of Idaho contained species that apparently represented several life zones. By studying the life-zone relationships of living equivalent species, he was able to conclude that the flora was definitely of the Transitional Zone, including both arid and humid elements. Only two out of 24 species whose equivalents live in the West today are absent from the Transitional Zone, although many of the species range downward into the Sonoran or upward into the Canadian Zone. Furthermore, the five species which are not represented in western America today are represented in eastern America by species that occur in the Transitional (Carolinian and Alleghenian) Zone.[1]

In his study of the Tehachapi flora of southern California, Axelrod (32) found that several vegetational types were represented and that they were probably distributed on an altitudinal basis. The first point of his summary bears directly on the relationship. "In all the areas where modern vegetation shows a relationship to the Tehachapi flora, the vegetation is distributed altitudinally. The different plant formations—the arid subtropical scrub, desert scrub, chaparral, and woodland—normally succeed one another in the order listed in regions of high relief. The large representation of the different elements in the flora suggests not so much transportation of leaves and other structures from upper to lower regions as an overlapping of formations like that found today in areas of high relief." In regions of relief, climates change rapidly with increasing altitude because of such factors as increased precipitation and reduced temperature. Furthermore, microclimatic conditions related to topography tend to permit an interdigitation of vegetational types because of the disturbance of altitudinal zonal arrangements due to differences in slope and exposure, in cold-air drainage, and the like. Strongly contrasting types of vegetation approach one another on al-

[1] Dorf was probably unfortunate in turning from climaxes to Merriam's life-zone concept, for life zones are frequently much more difficult to interpret than climaxes.

ternative north and south slopes and in intermediate valley situations. In such localities, also, there commonly exist relic colonies of various climaxes not now prevalent in the region, surviving in regions of compensation under a general climate that is "foreign" to them.

In conclusion, it appears to be possible to reconstruct with considerable accuracy several aspects of the vegetational pattern of a region represented now only by fossils of a small portion of the whole flora. This is possible from a knowledge of modern vegetational structure and the assumption that diverse life-form, climatic, and ecological elements must have held similar relations to each other in the past as they do now. Together with questions of dominance considered earlier in this discussion, it is possible to gain some idea of the structure of vegetation with respect to (1) the prevailing type of community in a region; (2) the organization of a community into societies, or life-form layers; (3) the presence of successional communities, associes; (4) the arrangement in altitudinal belts in a region of high relief; and (5) the intermingling of diverse climaxes on a basis of topographic-microclimatic contrasts.

8.

Migration and
Evolution of Vegetation

1. *With the passing of time, new floras and communities arise in a given area from preexisting ones through the loss of species by emigration and extinction, and through the acquisition of species by immigration and evolution.*

2. *All vegetation types, being products of environment, are subject to evolution, migration, or extinction under the compulsion of environmental change.*

3. *Migration appears mainly to be a mass phenomenon, with the climaxes of a clisere tending to move together in a direction and to a distance determined by the nature and extent of the climatic change. However, on analysis, migration is seen to be a matter of the dissemination and establishment of individual propagules; the mass aspect results from the concert of similar responses to identical causes.*

Modern vegetation has had a relatively ancient history. The apparent great outburst of angiospermous types in the Cretaceous, including not only the simplest but many of the more advanced families of flowering plants, simultaneously in many parts of the world, indicates that they were developing long before this closing period of the Mesozoic. It was in the Cenozoic, however, that the resemblance to living plants became increasingly close.

In speaking of the Wilcox flora, Berry (62) said that it was clear in the Mississippi Embayment region that during the close of the Cretaceous there was a certain amount of evolution of new forms, a considerable dying out of the older Mesozoic types, and an immigration of warmer temperate types from the south. He does not believe that the Upper Cretaceous should be characterized as essentially Cenozoic, because the great modernization of floras came in the Eocene. For example, 83 genera of the Eocene Wilcox flora are not known anywhere from the Upper Cretaceous. Berry's comment concerning the source of the new elements is interesting: "The evidence does not permit a conclusion as to whether the major factor in this modernization was evolutionary processes or whether it is only apparent, resting simply upon invasions into known areas from unknown areas, such, for example, as Asia on the one hand or tropical America on the other. Both of these factors were doubtless operative, but I am inclined to think that immigration was and is always the more important in enabling us to recognize seeming high and low points in the continuously unfolding drama of life."

Early Cenozoic plants consisted, in the main, of genera still extant. The older Cenozoic in western America seems to have been characterized by a mixture of temperate and subtropical types. In the Middle Cenozoic (extending through the Pliocene) there developed an increasing similarity between fossil and modern species, with a predominance of temperate types. For the Later Pliocene, most species appear to be similar to modern species. Starting with the Pleistocene, the Late Cenozoic plants are in the main definitely referable to existing species. Throughout the study of the Cenozoic it is possible to utilize the paleoecological method with increasing accuracy and significance for progressively younger periods.

In a broad way, the history of vegetation has had a dual character: (1) the expansion of phylogenetic stocks, the evolution of new species, and (2) the evolution and migration of vegetation types. Both of these process complexes have always borne an intimate relation to climate, which, in turn, has largely been a product of geological conditions and processes. A critical modifying factor in the evolution and migration of species and communities is isolation. The completeness of isolation, and especially the length of time isolation has been operative, have an important effect upon the course of the paleoecological history of two regions.

Aside from the problems of speciation, the evolution of floras and communities depends primarily upon the cause and mechanism of migration, modified by competition and isolation. It is pertinent first to consider certain aspects of migration, confining ourselves to the problem with respect to plants.

It can be assumed that throughout the Cenozoic the surface of the land available for occupancy by plants has been fully stocked by plants.[1] It can

[1] This is aside from relatively minor bare areas which everywhere are produced by disturbances of one sort or another, and which are rapidly colonized by plants and become the foci for successional development.

also be assumed that all kinds of plants in the past, as at present, have had reproductive capacities far in excess of the spatial possibility for germination, establishment, and maturation. Furthermore, all plants have some capacity for the dissemination of their propagules (spores, seeds, and vegetative productive structures). This capacity for dissemination differs greatly from species to species, but time is long, relative to the life span of an individual, and they all (except the most recent species, or those whose areas have most recently been disturbed) have become dispersed within limits that are set not by their capacity for dispersal but by other considerations, internal and external. This being the case throughout recent botanical history, it would seem apparent that a considerable biological pressure has existed in every area and been applied to every community. Migration of an individual plant—there is no other kind of migration—must ordinarily consist of dispersal followed by establishment in a "closed" community. It would appear, then, that there could have been little migration going on at any time in the past, except under conditions that would increase the competition pressure on one kind of plant and relieve the pressure on another kind. It would appear also that migration would consequently tend to be a mass phenomenon, similar organisms tending gradually to replace another group of organisms with different requirements.

Such mass migrations would appear to be possible only under the compulsion of a changing environment as first developed by Forbes' Theory of Climatic Migrations (253, 254). Changing environments are known for the past and involve such processes, for example, as the gradual desiccation of a climate, a shift from evenly distributed precipitation to a concentration of winter rainfall, a reduction of temperature maxima and minima. We need not enter into the causes of climatic change, but must examine briefly the way in which a climatic change causes vegetational migration.

The surface of the earth is characterized by different types of climates in different regions. Climatic characteristics change with latitude, with altitude, with distance from the oceans, etc.[2] The great climatic regions of the earth are characterized by great vegetational types, the climax formations (climaxes, biomes). It is obvious that there is an intimate relation between the great climatic types and the great climaxes. The great climaxes are evident from their physiognomy. Some are dominated by trees, some by shrubs, and others by the grass form. The trees may be deciduous or evergreen, broad-leafed or narrow-leafed. Each climax shows a certain amount of life-form

[2] Thornthwaite (625) has recently published an article on climatic types, presenting a new classification.

homogeneity, and one can conclude that there is some kind of causal relationship between environment and vegetation, however difficult it might be to analyze (202, 516, 678).

A more detailed consideration of the problem reveals that the available species in a region are not all members of a particular vegetation type. Only those kinds constitute any one community which are sufficiently well adapted for life under the conditions controlling the community to compete for space within it with other similarly adapted kinds of organisms. On the basis of information largely provided by physiological studies (coupled with genetic considerations), it appears that an organism can live and develop only within certain circumscribed limits (287, 454). Although these limits vary with the ontogeny of the individual, they appear to be real. The capacity of an individual for normal functioning, or even for survival, within certain environmental limits can be referred to as the tolerance of the organism, its ecological amplitude, or its adaptation. The area which a species population can occupy (assuming its availability) is seen, then, to be governed by three interacting sets of factors: (1) the physiology of the individuals that make up the species population, as interpreted in terms of tolerance; (2) the limitations of environment for the particular kinds of organisms; and (3) the success of the organisms in competition with other similar species. A possible fourth factor, the historical, is in reality mainly an important portion of the second, as geological and climatic history determine present environments and areas.

A type of vegetation, a community occupying a certain area, consists of a mixture of species populations that are associated because of the conjunction of certain environmental conditions and of kinds of organisms with certain tolerances. Because of the dynamics of vegetation, every region of the earth tends to become occupied by characteristic vegetational types that are in more or less stability with the prevailing conditions. The factors most likely to change such a dynamic equilibrium are the environmental ones, and inevitably a change in climate must mean a change in the nature of the life that occupies the climatic region.

Climatic change produces a pressure upon vegetation from which there is no escape except through migration or evolution. Failing in evolution, in finding a suitable migratory route, or in having an adequate migratory capacity, an individual, a species population, or a vegetation type is faced with extinction.

At this point it can be indicated that a migratory route is in its essence a matter of the continuity of suitable environmental conditions. Migratory highways are usually relatively narrow topographic features such as a river system for certain riparian species, a coastal strip of beach, dunes, and marshes for certain littoral species, a mountain chain for certain alpine and subalpine types. The continuity of such a highway consists in the regular recurrence of special habitats with no intervals greater than the capacity of the species concerned for at least occasional dispersal across the gaps. Continuity, then, is relative to the species concerned. Under the compulsion of a climatic shift, however, the species of a climatic type migrate coincidentally with the movement of the climate. This may mean the change of area of a climax expressed simultaneously through a southward movement along a coast, a lowering of altitudinal limits, a southward extension in a mountain system, and a gain of territory on continental interior plateaus. Under the climatic oscillations connected with the Pleistocene, climaxes moved southward and then northward along hundreds of consecutive miles of front, all of which constituted a highway for migration because of the breadth of territory affected by the climatic changes, and downward and then upward in altitude in mountainous regions.

From these statements it follows that a barrier to migration consists of an interruption of the continuity of suitable environments. Barriers are usually thought of as mountain chains, large bodies of water, deserts, and the like; but the existence of a barrier to the migration of a species population consists of *any* factor that becomes limiting to the establishment of new individuals of that species. The barriers mentioned are obvious and dramatic ones, but many others exist: length of day, length of growing season, frequency of certain high or low (this is relative) temperatures, low winter temperatures, frequency of fogs, etc.

The nature of the vegetation of a region has changed historically (and no vegetation type has remained unchanged through the Cenozoic) because of four happenings: (1) the evolution of new species, or subspecific groups; (2) migration, the immigration of new species and the emigration of pre-existing species; (3) extinction; and (4) segregation (through migration, of course) of what was one vegetational type into two or more vegetational types.

The unity of a climax is derived from (1) the essential life-form unity of its dominants; (2) a certain taxonomic unity, expressed in "binding" genera which are characteristic throughout the climax, and sometimes in dominant species which are widely distributed through the climax; and (3) the common historical origin of the vegetation type. Climaxes are large, wide ranging, and floristically heterogeneous. These characteristics frequently are not realized by critics of the climax conception. It is not required of a climax that it be floristically uniform throughout its extent. On the contrary, the climax shows a considerable complexity which is a direct reflection of climatic subtypes that exist from one limit of the climax to the other. The major subdivisions of a climax are the associations. The associations, in turn, geographically show faciations coincident with the lesser climatic differences, and lociations on a microclimatic basis (108, 154, 504, 678).

Keeping in mind the broad nature of the climax, we can say that the paleontological record indicates the continuous existence of the major American climaxes since the beginning of Cenozoic time; but they have not passed through these millions of years without change. The temperate deciduous forest, for example, has expanded and contracted its territory. Under the vicissitudes of changing conditions, selection and segregation have resulted in realignments of species—new associations have come to exist (81, 82, 85, 86, 87). New species have evolved, some old species have become extinct, and former associates have come to occupy separate, and sometimes far-removed, areas. But through it all the essential integrity of the major climaxes has persisted. Subtropical evergreen forests, broad-leafed temperate deciduous forests, deserts, the prairie, etc., appear to have existed continuously since the beginning of the Cenozoic, if not longer.

During the Tertiary the paleontological record indicates many changes, such as the withdrawal southward of a once wide-ranging subtropical forest type (the Goshen flora), the contraction of a once far-flung temperate redwood forest (the Bridge Creek flora), the extension northward and subsequent contraction and segregation of a North Mexican element (the Tehachapi flora), the origin and expansion of a continental interior prairie (156), the development of the Colorado and Mohave Deserts (153), and other great changes. It will be the purpose of the following pages to use enough illustrative materials to develop the principles that are involved, not to tell a consecutive or complete story of the vegetational history of North America.

One of the best-developed accounts of the migration and evolution of a vegetation type in the paleoecological literature is that of Mason (453) on the closed-cone pine forest of the California coast. This forest today ranges through 10° of latitude along the coast but is of markedly discontinuous distribution, as indicated by the following account of its occurrence:

Northern forest
 Trinidad Head
  100-mile interval
 Inglenook, Mendocino County, typical southward for 100 miles to

Fort Ross, Sonomo County
  50-mile interval
Inverness Ridge, Tomales Bay region
  75-mile interval
Ano Nuevo Point
  40-mile interval to the

Central forest

Monterey and Carmel Bays
  60-mile interval
San Simeon, and locally southward to
Pecho Hills, San Luis Obispo County, and
La Purisima Ridge, Santa Barbara County
  500-mile interval on the mainland to the

Insular forest (southern)

Point San Quentin, Lower California, and on coastal islands from
Santa Rosa and Santa Cruz Islands to
Guadalupe Island (200 miles off the coast) and
Cedrus Island

Such a discontinuous occurrence obviously marks the association as one in which the isolated stands are relics. It would not be expected that these widely separated stands are floristically and ecologically identical, but they are nevertheless remarkably similar in several respects. In this connection Mason says,

There is a well-marked endemic population that ranges through the forest units without regard for latitude. Certain species or closely related species occur throughout the range of the forest, though not necessarily in all of the localities. The aspect of the forest in the various localities is strikingly similar. All are dominated by species of pine, *Pinus radiata, P. muricata,* or *P. remorata,* which have the same characteristic growth form and bear a close phylogenetic relationship to one another. It is difficult to distinguish them except on close examination and it is presumed that this similarity of aspect suggests a close ecological relationship as well. Associated with the pines are a few other gymnosperms which are always subordinate as regards the control of the forest—*Pseudotsuga taxifolia, Cupressus macrocarpa, C. goveniana, C. pygmaea,* and *C. guadalupensis.* All of these conifers except *Pseudotsuga* show discontinuity in their distribution. . . . The woody under-story vegetation of this forest is composed of two layers, a broad-leafed tree layer and a conspicuous, highly characteristic shrub layer. The broad-leafed trees are predominantly *Quercus agrifolia, Myrica californica,*

Umbellularia californica, with *Arbutus menziesii* less common. The shrub layer is of two types. There is a hard, rigidly branched "chaparral" type dominated by species of *Arctostaphylos* and *Ceanothus* . . . and a softer more mesophytic type including *Corylus, Baccharis,* and *Rubus.* . . . It is the former type of shrub that is most abundant and which gives to the forest its most characteristic aspect. This aspect remains the same, regardless of the species of pine dominating the particular forest unit and regardless of the particular species of trees and shrubs making up the population.

From a study of the relationships of these forests it is evident that they are parts of one and the same flora and vegetation type. An examination of the fossil records of this association also indicates their fundamental unity. In fact, the Pleistocene records from the Tomales and the Carpinteria floras show that the forest was then much more homogeneous than it is today. There have been three factor groups which have influenced the breakup of the forest into the northern, central, and insular divisions in recent Cenozoic history: isolation, immigration, and climatic change. Mason summarizes these factors as follows: (1) Where isolation has been sustained, the independent development and consequent endemic cast have proceeded farthest; where isolation has been temporary because of uplift and the resulting union of islands, the floras have become more generalized. (2) New elements have entered the flora from outside sources because of the elimination of barriers. (3) There have been a segregation and differentiation due to the selective action of climatic change.

Pinus radiata (Monterey pine) and *Cupressus goveniana* (Gowan cypress) were present in the Pleistocene closed-cone forest of Tomales Bay but they are absent from the equivalent forest in that region today. *Cupressus* spp. and *Pseudotsuga* have disappeared from the forest on Santa Cruz Island. The Bishop pine (*Pinus muricata*) today occurs in the northern division of the forest and south to Santa Barbara County and is then absent for several hundred miles, reappearing in Lower California near Point San Quentin. Within this long gap, however, it is known from fossil material at Carpinteria. Such discontinuities of range and disappearances of species from portions of the former range of the forest association can be attributed only to the orderly processes working through the environment, and not to the result of chance destruction, according to Mason.

Other differences among the modern stands of the forest that have contributed to its present heterogeneity, relative to its geological condition, arise from new additions to the flora. For example, Mason believes that *Rhamnus californica* has made its way into the closed-cone forest from the redwood forest since Pleistocene times, for this species would certainly have been represented in the fossil records had it been a member of the Pleistocene forest.

The closed-cone forest today occurs on island and coastal headlands and mountains that were at times islands in the Later Cenozoic. The forest was continuous and relatively homogeneous during times of emergence when there was habitat continuity. Segregation and discontinuity have developed during times of submergence—of coastal island formation. The geological, ecological, and taxonomic sources provide mutually substantiating data. One problem which has as yet no solution is that pertaining to part of the modern coastal discontinuity. There are abundant sites that appear to be identical with those occupied by the closed-cone forest and, furthermore, are contiguous with them, but are unoccupied by them. This apparent modern lack of aggressiveness may be due to biotype impoverishment (109).

The migration of a vegetation type as a result of a climatic change is illustrated by Axelrod (31) in his discussion of the digger pine (*Pinus sabiniana*) forest. This forest type is represented in the Pleistocene Carpinteria flora and in the Pliocene floras of Mount Eden, San Timoteo, Pico, and San Fernando, which range from 25 to 100 miles south of the present limits of the forest type. The modern digger pine forest forms a savanna association with certain oaks and occupies a location about the Great Valley of California, reaching its optimum in the western foothills of the central Sierra Nevada. "There is a gradual southward decrease in rainfall for 400 miles through the forest, varying from 53 inches in the north to 10 inches annually near the southern border of the forest; temperatures vary but little over the same area." From this it would appear that rainfall is not of great significance in delimiting the area of the association; temperatures, however, seem to be important. Axelrod contrasts the temperature conditions of the modern forest (based on data from 37 stations) with the present temperature conditions at the fossil localities which lie south of the present boundary of the forest. See Table 10.

In the modern area of the forest the temperature means vary only two to three degrees. Axelrod says, "From the data . . . it is apparent that there is a significant discrepancy in winter temperatures. Average winter temperatures are 9 degrees higher in areas of former digger pine occurrence than in the forest to the north, and extreme temperatures for the same season exceed those of the digger pine forest by 8 degrees. Of equal significance is the fact that summer temperatures are nearly identical in both areas. It thus appears that winter temperatures are of critical importance in determining the distribution of digger pine. . . . The higher winter temperatures

TABLE 10

Temperatures	Modern Digger Pine Forest: 37 Stations	Areas of Former Occurrence
Average temperatures		
Annual	59.4° F.	63.4° F.
3 summer months	91.4	91.9
3 winter months	35.2	44.2
Temperature extremes		
High	112.0	113.0
Low	11.0	19.0

in southern California evidently exceed the tolerance of digger pine and favor its successional replacement. Thus, this association has apparently been segregated from the southern California flora in response to rising winter temperatures in that area since the Pleistocene."

In connection with this topic, the migration of associations under the compulsion of climatic change, there are two points to be emphasized. In the first place, parallel migrations of two associations may be caused by different factors. The interior California digger pine association, just discussed, has probably retreated northward since the Pleistocene because of a change in winter temperatures. Coastal communities not far distant appear to have undergone a similar northward migration, but because of a change in the moisture factor. Chaney and Mason (138) have shown, for example, the importance of fogs in influencing the retreat northward of the Monterey pine forest. The second point refers to the importance of extremes, rather than means, in influencing migration. With a long-time point of view and with whole associations being considered, mean climatic conditions appear to characterize associations in both time and space. But in view of the functioning of individual organisms and their behavior at the margins of range, extremes appear to be critical—if we remember that migration is a function of the individual. The significance of extremes has been considered recently by MacGinitie (433), Taylor (622), and Mason (454). The conclusion is that it is the occasional extreme season of a year, or the most critical year or years of a climatic cycle, wherein a critical factor (or factor group) is in excess of the tolerance of a plant for the particular factor in the particular physiological or ontogenetic state of the plant, that often governs the distribution of the species. It is true that most seasonal and annual irregularities (departures from the normal) produce changes in vegetation which are only temporary—changes that are within the fabric of the climax and not destructive to it. Such changes are most apparent near the margin of range for the association, and it appears true that the central core of the association as a whole is characterized by mean conditions. But when the climate is undergoing a definite trend, as revealed through thousands of years, an association must migrate to stay within the general conditions it requires. This migration is a matter of the invasion of new territory not before suitable and a retreat from occupied territory no longer suitable, and is accomplished by individuals that are reacting to extremes of factors in their environment. An association that has migrated from one area to another and has maintained its essential character must have remained under the general average conditions for which it is adapted; but the migration has been accomplished by individuals influenced by extreme factors which, on the retreating margin, have reoccurred with sufficient frequency to prevent continued establishment. This, at any rate, is the only way I can resolve what otherwise appears to be a fundamental contradiction.

It sometimes happens that for topographic reasons under a changing climate a once continuous type will become separated into two areas. For example, a northern vegetation may be forced southward and find two highways for migration on parallel mountain systems, such as the Rocky Mountains and the Sierra Nevada, and consequently the two may become isolated from each other because of the climatic barrier of the intervening lowland. Another type of separation is illustrated by the central development of desert conditions within the area once occupied by continuous non-desert types of vegetation. Clements (153) has discussed this development in considerable detail with respect to the modern Mohave and Colorado deserts. Recently Axelrod (32) has added to the paleobotanical history of the Mohave region. Several botanists, Kearney (388) for example, have commented on the similarity between certain areas of chaparral in southern California and in southern Arizona on the two sides of the desert. This similarity could result in recent time only from migration across or around the desert area—this seems unlikely—or from the fact that the vegetational type at an earlier time extended across the area previous to the development of desert conditions. Evidence concerning the latter interpretation consists of several kinds. In the first place, species in the fossil Tehachapi flora (Miocene) are represented by transads of identical or closely related plants in the chaparral formation of both regions: *Amorpha californica, Arctostaphylos pungens, Ceanothus cuneatus, C. greggii, Cercocarpus betuloides, C. ledifolius, Celtis reticulata, Fremontia californica, Mahonia fremontii, Pinus cembroides, Prosopis juliflora, Prunus fasciculata, Quercus chrysolepis, Q. dumosa, Q. palmeri, Rhamnus californica, Rhus integrifolia,* and *Salix lasiolepis.* In addition, certain species from the Tehachapi flora are represented by equivalent species in Arizona that are absent from California, such as *Robinia neo-mexicana* and *Rhus virens.* Furthermore, Axelrod (32) adds, "When these equivalent species of the Tehachapi flora are considered together with the phylads on both sides of the desert . . . the evidence for the former continuity of the middle Tertiary vegetation across the present desert region seems conclusive." Examples of such paired species on the two sides of the desert include *Quercus engelmanni* and *Q. oblongifolia, Platanus racemosa* and *P. wrightii,* and *Populus fremontii* and *P. wislizenii.*

It may be well at this point to look more specifically at the meanings of the two new terms that have just been employed: *phylad* and *transad.* Clements has introduced both of these useful terms in his paper on the origin of the desert climax and climate (153). He says, "Transads are dominants or subdominants that still exist on both sides of the desert and consequently must have extended through it . . . but may not occur within its borders today." With respect to the other term he says, "In addition to single species that cross the desert, there is a much larger number of phylads, that is, phyletic lines of closely related species." [8] Although these definitions are put in terms of a particular situation, their general implications and usefulness are evident. Phylads, of course, are a special kind of transad.

As a result of his study of the Mohave-Colorado desert situation, Clements (153) mentions 150 transads of the desert. Some of these species, "desert transads," still occur as relics in favorable spots in the desert region; others are the "true transads" and are absent from the desert. The kinds of plants making up this group are listed below, together with the number of species in each group.

1. Grass transads with relics in the desert	17	species
2. Grass transads no longer in the desert	16	"
3. Forb transads, perennial	about 30	"
4. Forb transads, annual	about 40	"
5. Shrub transads of the desert scrub climax	14	"
6. Shrub transads of the sagebrush climax	15	"
7. Shrub transads of the chaparral climax	10	"
8. Tree transads	8	"

[8] This definition implies the probability of evolution coincident with migration and the changing environmental conditions. It is not possible to prove that the members of the phylad did not antedate the development of new conditions, but it is likely that they arose simultaneously with their development.

Names of actual species involved in this comprehensive list should be sought in Clements' publication.[4] He states that phyletic transads (phylads) are much more numerous than single species transads. In some phylads the related species are so similar that they are considered by some botanists to constitute not more than subspecies or varieties of a single Linneon. Clements gives as an example of such intimate relationship the *Quercus minor-gambeli-garryana* complex. Other phylads which are essentially structural continua include the following: *Juglans nigra-rupestris-californica, Fraxinus americana-*

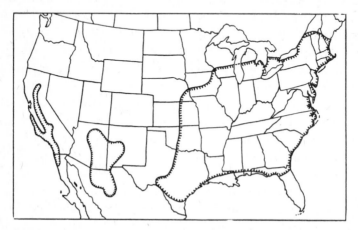

Fig. 5. *Platanus occidentalis* (eastern), *P. Wrightii* (southwestern), and *P. racemosa* (Californian), the American sycamores, constitute a phylad the members of which form a structural continuity and once were connected across the plains and deserts where they are now disjunct.

texensis-velutina, F. coriacea-oregona, and *Platanus occidentalis-wrighti-racemosa.*

All these various lines of evidence (studies of modern floristics and phylogeny, and paleobotany) unite to indicate the former continuity of non-desert types of vegetation across a region which now supports a desert climax. An explanation of the historical dismemberment of the widely spread vegetation by the intrusion of the desert climax is to be sought in the

[4] It is necessary to remark in caution at this point—and this applies in many places throughout this book—that the validity of such conceptions rests to a large degree upon the validity of the basic taxonomic treatment involved. Most ecologists and geographers, unfortunately, are not also taxonomists, and they must depend largely upon available taxonomic dispositions of their materials. For example, this manuscript could never have been completed if I had considered it necessary to go into the taxonomy of every organism mentioned, and I could get on with the geographical considerations only by accepting at their face value the taxonomic matters enunciated by others. Clements, however, is frequently independent in his taxonomic interpretations and sometimes at variance with taxonomists.

climatic deterioration during the Later Cenozoic.[5] This climatic change has forced vegetational migrations, withdrawal of certain elements (grassland) and invasion of others (notably the north Mexican plateau element), and has caused the separation of associations into two distinct geographical groups with a certain amount of differentiation, or segregation. An explanation for the climatic deterioration is found in the desiccation concomitant with mountain making and the interruption of moisture-laden winds.

The invasion of a region by a new climax is always a gradual process and is not everywhere uniformly accomplished at the same time. Relics of preexisting vegetation survive for a considerable time as evidence of what formerly existed. Antecedent vegetation not only is driven out by climatic change but is overcome through competition by the better-adapted invader. Axelrod (31) discusses such a situation in connection with chaparral and the indications of the Mount Eden fossil flora. He says,

This climatic trend [toward a warmer and drier climate] was accompanied by a widespread chaparral expansion, and this scrub formation has apparently supplanted areas of the desert-border unit. . . . The relic occurrence of *Prosopis juliflora* var. *glandulosa, Chilopsis linearis, Yucca mohavensis, Ephedra nevadensis,* and *Coleogyne ramossissima* on dry exposed slopes in the Mount Eden area supports this suggestion, for they are all typical of desert regions. In view of the presence of these species in the Mount Eden area, it seems likely that the living equivalents of the desert-border element could probably exist in exposed situations in that region today, were chaparral not the dominant scrub formation. In interior southern California chaparral is a rapid and aggressive invader into all vegetative types, and, once established, is quickly stabilized. Relic occurrences of former more widespread plant associations may be found throughout the chaparral of southern California. Among these may be mentioned isolated areas of grassland, oak, walnut, big-cone spruce, Coulter pine, digger pine, and, as indicated above, desert-border shrubs. Thus, it seems apparent that the desert-border association was confined to the edge of the present Colorado Desert, 15 miles east of the Mount Eden area, in response to an invasion by a more virile plant formation.

With the passing of time, vegetational changes become increasingly great. Axelrod (31) makes the following statement in contrasting the Miocene with the Pliocene: "Pliocene floras exhibit conspicuous changes within limited areas and show marked floristic differentiation and geographic segregation. In direct contrast to these assemblages, Miocene floras are indicative of

---

[5] This infers not that the desert climate did not exist before but that it enjoyed a northern extension from Mexican areas at this time.

widespread forests and are suggestive of more uniform climatic and topographic conditions in the areas where plants were accumulating. While Pliocene floras are represented by equivalent living vegetation at no great distance from the fossil areas, Miocene assemblages are generally related to regions which are geographically remote from the fossil localities. This is a direct reflection of a greater age, since Miocene floras have been more altered in composition in response to widely changing physical conditions than have Pliocene floras. In addition, it is to be noted that all Pliocene species show closer resemblances to their modern equivalents than do Miocene species."

One of the most remarkable stories of floristic change in paleobotanical literature is that of E. M. Reid (522), summarizing several papers by herself and Clement Reid on the Pliocene seed floras of western Europe. From their studies of the Cromerian, Teglian, and Reuverian floras they had established the following facts: the Pliocene Epoch in western Europe had witnessed the existence and extinction of a flora closely allied to the living floras of the Far East in Asia and of North America, and in whatever part of the northern hemisphere these plants or their nearest relatives, or any Pliocene exotics, were found in lower latitudes, they were nearly always mountain plants. They believed that these plants, once of western Europe also, were driven south by the ever-increasing cold of the Pliocene. For the eastern Asian and North American streams of migration, the way to the south was open and they escaped; but for the western Asian and European southward-retreating stream of migration, from the Atlantic seaboard to the coastal plain of China, the way was everywhere closed in temperate regions by impassable barriers of east-west ranging mountains, seas, and, perhaps, deserts. Successive waves of migrants were driven against these barriers and perished, so that by the end of the Pliocene scarcely a trace of the eastern Asian-North American element was to be found, except in their fossil remains. The Lower Pliocene flora of western Europe, except in eastern Asia and North America, is now found most commonly at a height of 5000 feet or more in the Himalaya and the mountains of western China.

The following explanation (522) makes clear what likely happened: "During a cold period, the warmest flora to survive in any given district must have been that inhabiting the plains or valley-bottoms. As the climate ameliorated, the plains would become too hot for this flora, and, in order to escape destruction by heat, as it had formerly escaped destruction by cold, it must migrate. In a country of great plains, if the changes of climate were rapid, movements to other latitudes might be too slow to counteract the change of climate; but in a mountain country, comparatively small vertical movements would afford the necessary change. Hence, in a warm period following upon a cold—our present condition—we should expect to meet the migrants, when inhabiting more southern lands than those in which their fossil relations occur, not on the plains (where mostly they were exterminated), but upon the mountains."

Subsequent to the study of the three floras mentioned, Mrs. Reid published on the Pliocene floras of Castle Eden and Pont-de-Gail, and it is her summary (522) of the five floras which yielded such remarkable results. In these floras she recognized an exotic element which was, in turn, composed of two elements: an outgoing flora and an incoming flora. In Table 11 are pre-

TABLE 11.—The relationship of five Pliocene seed floras of western Europe
(Data from 522)

I	II	III	IV	V	VI
Cromerian	135	89	5	0.74	Top of Pliocene
Teglian	100	75	40	16	Upper Pliocene
Castle Eden	58	55	64	31	Middle Pliocene
Reuverian	133	46	88	54	Lower Pliocene
Pont-de-Gail	17	35	94	64	Base of Pliocene

I—Name of the flora
II—Number of species compared
III—Percentage of whole flora compared
IV—Percentage of exotic and extinct species
V—Percentage of Chinese-North American species
VI—Age, or supposed age of the strata

sented data concerning these floras, but the surprising results are best shown in the figures in which the exotic Chinese-American element is compared with the total exotic species. The age of certain of the floras was known with fair certainty, but that of others only relatively, although all were known to be Pliocene. When the data were plotted, it was found that the curve verified the dating of the floras which had been less certainly located. Mrs. Reid draws the following deductions: (1) The study of both living and fossil seeds is capable of leading to reliable specific determinations; (2) the results arrived at by using only fossil seed floras are in accord with those derived from stratigraphy and paleozoology; (3) by carrying the curve back into the Miocene it is indicated that about the middle of the Miocene or a little earlier the whole flora of western Europe, north of the east-west mountain ranges, was of the Chinese-American type; (4) the extermination and supplanting of the Chinese-American flora began at exactly the period when stratigraphy and paleozoology indicate that the elevation of the mountain ranges extending from Morocco to Indo-China attained a maximum; (5) by implication from the curve, the rate of change in the western European flora is indicated with respect to the loss of the Chinese-American element and the increase of that part of the flora which survived to form part of the living flora of western Europe. She believes that the incoming element during these changes had two sources: one seems to have been of polar origin, and the other appears to have come from the highlands of central Asia, including Tibet, the Himalaya, and western China.

The space-time concept is important in paleoecology, and, if used judiciously, plant distribution is a guide to age determination. Chaney (130) has developed this concept. In general, strata containing the same or approximately the same fossil biota are usually considered to be of the same age. If larger periods of geological time are employed, such age references are satisfactory but they do not allow for refinement of interpretation. For example, fossil floras of the redwood forest association are known in circumboreal distribution and through 40° of latitude during the Tertiary. If Tertiary time is taken as a unit, it is correct to say that the redwood association had such a wide distribution. An easy inference, however, is that this vast area was covered simultaneously by the association, i.e., that the redwood forest was a far-flung temperate type over most of the northern half of the hemisphere throughout the Tertiary. Another result of a too long time scale is the conclusion that latitudinal climatic and life zones were practically non-existent. Such conclusions are not only unwarranted, they are absolutely false (63). Is it likely that redwood never grew over more than 20° of latitude at any one time, and that there were always latitudinal and altitudinal climatic belts and corresponding biotic formations? Because the redwood association was first known from the Miocene, later discovered temperate associations from Arctic regions were also assigned to the Miocene. During the Tertiary the redwood forest migrated from the high latitudinal positions (Alaska, for example) it held in the Eocene to middle latitudinal positions (such as Washington and Oregon) in the Miocene. Toward the close of the Miocene and during the Pliocene the redwood forest withdrew from continental to coastal regions where it now occupies a few relic stands. Its Tertiary history has been one of southward retreat on one hand and a loss of altitude on the other. This also has been the general history of other types of vegetation that are correlated with temperate conditions in the northern hemisphere.

It would seem correct to assume that conditions similar to the present existed for comparable past communities. For example, species of trees which are coastal in Alaska today are found at progressively higher altitudes as one travels to lower latitudes so that in Pacific Canada or in our northwestern states the species are no longer coastal but strictly montane, and those which enter the Cordilleras are subalpine in the central Rockies of Wyoming and Colorado or in the Sierra Nevada. In Eocene time the redwood association must have had such a distribution insofar as elevations permitted. In Alaska it was a lowland type near the coast, but it must also have extended southward at suitable elevations in the Cordilleras at successively higher altitudes, although there are no fossils to prove it. With the cooling of the climate in the Oligocene and Miocene, the redwood not only migrated southward in the wake of the retreating subtropical forests, but also lowered its altitudinal distribution in the mountains.

When migrational sequences can be established directly through paleo-ecological and other evidence, and indirectly through ecological inference, they become a useful guide to age determination of any fossil flora involved in the migrational history. This, briefly, is the space-time concept of vegetation that is so important in the synchronization of fossil biota.

9.

Certain Aspects of
the History of Cenozoic Vegetation
of Western America

In the preceding chapters there has been a discussion of the principles and working methods of that portion of paleontology known as paleoecology that is of basic importance in an understanding of modern areas. The results of the method have appeared in the account only in fragmentary and non-consecutive form. Although this is no place for a detailed account of the vegetational history of western North America or any other region, a brief recapitulation of the main outlines of part of the story will serve as a summary and an illustration of the type of results that are obtainable from the paleoecological method. Three sketches follow: (1) an account of the forest history of middle latitudes of western America during the Tertiary; (2) the story of the closed-cone pine forests of the California coast from the Pleistocene to Recent; (3) the story of the origin of the southwestern United States desert climax. The first is based on the general papers by Chaney (129, 130, 131, 132, 134); the second, on the papers by Mason (450, 451, 452, 453); and the last on an essay by Clements (153) and papers by Axelrod (31, 32) and Johnston (379).

A partial history of western American Tertiary forests.—Throughout the Tertiary of North America certain general conditions and processes have prevailed. Judging mainly from the vegetation, North America has been a rising continent. The narrow bordering seas have largely withdrawn and mountain ranges have been uplifted, but the continents and the seas have held their same relative positions. The main patterns of air and water circulation have remained similar, but changes on the continent have resulted in climatic deterioration over large areas. The principal trend has been toward lowered temperatures and rainfall, bringing about strong seasonal contrasts in interior localities. Although there is indicated a relatively continuous change in climate, from subtropical to temperate for middle latitudes, there have been some climatic reversals.

This relatively continuous change in climate constitutes a developmental trend that provides a basis for indicating the time when a particular flora passed through a given latitude on its southward migration. The time-space concept provides a critical tool in the recognition of floral sequence, for the members of the clisere, the chain of climates and climaxes, are inexorably bound. Forests rather than continents appear to have been the wanderers, moving under the compulsion of climatic change; and climatic change, in its turn, is a reflection of changes in physiography and, possibly, solar radiation or other non-physiographic factors.

Vegetational types, in their shifts in composition and position, are better indicators of general climatic changes than are species. Furthermore, these shifts frequently indicate the presence of epicontinental seas or their withdrawal, the building of mountains or their planation, as adequately as strictly geological data. Because of their inherent slowness of migration, plants are frequently considered better indicators of change than animals. However, it must be emphasized that the range of animals is closely tied up with the range of vegetation, for the biome (with relations between climate, vegetation, and animals) constitutes an integral whole.

In the past, certain climatic types have been more widespread than at present, and consequently certain vegetational types have been more widespread. The concept of a homogeneous Arctotertiary forest over wide circumpolar latitudes of the northern hemisphere is, however, entirely erroneous. Latitudinal differences in climate and vegetation existed at every stage of Cenozoic history even though the striking, detailed differences of the modern landscape were not developed. Also, where pronounced relief existed, altitudinal belts of vegetation must have existed likewise, even as now. Vegetation of higher latitudes near sea level must have extended southward at higher and higher elevations where mountain chains occurred. Consequently, it is safe to suppose that at middle latitudes, at any time during Cenozoic history, the uplands were clothed by vegetation characteristic of higher latitudes and lower elevations, whether or not such floral elements are known from the fossil record.

In general, the fossil records previous to the Tertiary are derived from deposits along ancient shores. During Tertiary time in western America, however, numerous deposits of fossils were laid down far inland from the sea. Such records are mainly from stream sediments and from pyroclastic materials resulting from the widespread vulcanism of the western American region. Such materials provide an excellent record of terrestrial vegetation, but inevitably such records are principally of the vegetation adjacent to the basins of sedimentation. Upland materials can never be adequately represented in lowland deposits. The reconstruction of upland vegetation must rest mainly on the logic of comparison with contemporary patterns.

During the Cretaceous an embayment extended across the northwest to eastern Oregon. To the east of the Pacific Embayment lay a low plain extending to the great inland sea, which has been called the Mississippi Embayment. In the Late Cretaceous the elevation of the Cordilleras caused the withdrawal of the inland gulf and the westward retreat of the Pacific Embayment, leaving, at the dawn of the Cenozoic, a broad low plain across Oregon, for the Cascade Mountains had not yet been uplifted. This Oregon plain of the Eocene was covered by a rich and luxuriant warm-temperate to subtropical forest. The climate from the Pacific shore to eastern Oregon (and much of the surrounding territory) must have been mild and humid, for the forests were dominated by trees with large, thick, broad, and mostly evergreen leaves. Among the genera were avocado (*Persea*), chumico (*Tetracera*), fig (*Ficus*), palmetto (*Sabalites*), ciricote (*Cordia*), persimmon (*Diospyros*), cinnamon (*Cinnamomum*), and others. Such genera are principally tropical and subtropical in their occurrence, and they belong to families that are almost exclusively of the equatorial belt. Comparable vegetation in the Americas occurs today in the mountains and lower slopes of Guatemala and Costa Rica, in the savannas of Panama, and on the lower slopes of the Andes in Venezuela.

By Miocene time the subtropical forest had retreated from the middle latitudes of Washington and Oregon, leaving behind occasional relics, most abundant near the coast and in southern California. During Pliocene and Recent times the subtropical forest has undergone less change of location. The northern limits have been withdrawn and the forest has moved to lower elevations. From the modern location of the genera of the Eocene forests of Oregon we can interpret the conditions of Oregon of forty to sixty millions years ago. Through this vast span of time, the history of the subtropical forests of western America has been principally one of a southward retreat. Where the subtropical forests of Oregon came from previous to the Tertiary is a more obscure story. To a considerable degree the elements must have migrated northward to middle latitudes during the Cretaceous.

The subtropical forests of Oregon and vicinity were replaced in the

Miocene by a temperate forest, the redwood forest dominated by *Sequoia*. It extended from the coast eastward to the Cordilleras and from shortly above the international boundary well down into California on the south. This magnificent forest contained a mixture of species which we can assign to three elements on the basis of their modern distributions. The principal element consisted of trees and shrubs that still live together in the relic stands of the redwood along the California coast. Among these may be mentioned the following characteristic genera: redwood (*Sequoia*), alder (*Alnus*), pepperwood (*Umbellularia*), tan oak (*Lithocarpus*), dogwood (*Cornus*), hazel (*Corylus*), maple (*Acer*), and Oregon grape (*Odostemon*), together with plants of the modern redwood border forest such as ash (*Fraximus*), live oak (*Quercus*), madrone (*Arbutus*), hackberry (*Celtis*), cherry (*Prunus*), sycamore (*Platanus*), rose (*Rosa*), and willow (*Salix*). Of the common woody genera of the redwood forest today, only *Rhododendron* is missing from the fossil record of Miocene plants. A second element of the Oregon Miocene redwood forests was composed of genera that are missing in western America now, but are common in the temperate forests of the eastern United States and eastern Asia. Some of these genera are basswood (*Tilia*), beech (*Fagus*), chestnut (*Castanea*), hickory (*Carya*), elm (*Ulmus*), and hornbeam (*Carpinus*). Another element consisted of genera now confined to Asia, such as katsura (*Cercidiphyllum*), maidenhair tree (*Ginkgo*), tree of heaven (*Ailanthus*), water chestnut (*Trapa*), and zelkoua (*Zelkova*). These plants are typically temperate, and by matching fossil species with representative modern species it appears that the rainfall had been reduced from about 80 inches or more for Eocene to 40 to 60 inches for Miocene time, and that the temperature was slightly lower. Precipitation was adequate during all seasons, and winter temperatures likely dropped to freezing, a condition the Eocene forest of Oregon could not have endured.

During the Eocene this forest had existed at higher latitudes with a western American center in Alaska. The forest is known, however, from northern Siberia, Spitzbergen, Greenland, and other places in America, Europe, and Asia. The southward migration of the redwood forest to its position at middle latitudes at the beginning of the Miocene occurred under the influence of a changing climate. The temperate and subtropical forests moved together until the former came to occupy in the Miocene the position that the latter had occupied during Eocene time. While the coastal redwood forest was moving southward, the montane redwood forest elements were lowering their altitudinal limits. Although there are no fossils to prove it, while the subtropical forests lived in Oregon, the uplands must have borne at least some of the temperate flora.

During the Miocene, the Cascade Mountains started to rise, working great changes on the climate and vegetation to their east. In fact, the shift in composition of the redwood forest of eastern Oregon provides a good date for the time of elevation of the Cascade range. By Late Miocene the redwood had largely lost its position of dominance in eastern Oregon, and the forest became controlled by plants of the redwood border type, oak, madrone, and others. By Pliocene time and the close of the Tertiary, eastern Oregon had only aspen, cottonwood, box elder, maple, cherry, and willow, all trees which still survive in the region in protected places along streams. During recent time, extremes of temperature and low humidity have marked the eastern portion of Oregon in the rain shadow of the Cascades, and sagebrush, grassland, and occasional junipers grow where once luxuriant forests stood.

West of the Cascades humid forest types persisted. The redwood forest gradually contracted to its present relic areas along the California coast. Over most of the west slope trees of the cool-temperate, coastal Pacific conifer forest have gradually infiltered to form the modern forests of Washington and Oregon at lower elevations: *Picea, Thuja, Chamaecyparis, Pseudotsuga,* and *Abies,* in large part.

The southward-moving temperate redwood forest in Asia lost the redwood itself under the vicissitudes of migration. In western America today the principal area of temperate forests is to be found in the temperate zones of the Cordilleras as far south as Mexico and Guatemala, but with pines taking over the position of dominance. Many of the genera of the redwood forest, during this period of retraction of area, found the increasingly dry summers too difficult, and they became extinct in western America. They survive, however, with representative species, in eastern America and Asia.

The redwood forest relics along the California coast are very similar, except for the losses mentioned, to the redwood forests of Oregon in the Early Miocene, or of Alaska during the Eocene. To walk into these forests today is to walk back forty to sixty million of years in vegetational history to what lived in far-flung places those long ages ago.

The history of *Sequoia* itself goes beyond the Tertiary through the Cretaceous into the Jurassic, further than we can trace the history of the redwood forest type. Species of redwood lived when the dinosaurs trod their heavy way across the earth. But the history of the redwood forest, that distinctive aggregation of genera, is not known beyond the Tertiary. The widespread occurrence of fossils of *Sequoia,* even in Tertiary time, has led to the false impression of a widespread concurrence of redwood forests. In the North America Tertiary, redwood fossils are known through over 40° of latitude, from Ellesmere Island in Arctic America to Colorado. At no one time, how-

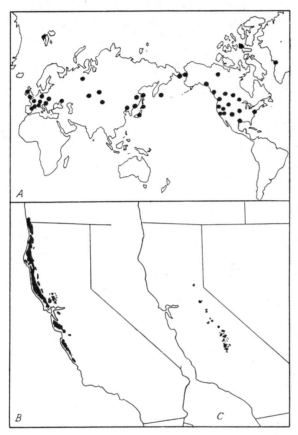

FIG. 6. *A.* Fossil redwood localities (data from Chaney). *B.* The modern relic areas of the coast redwood, *Sequoia sempervirens,* and *C.* the Sierra redwood, *S. gigantea* (*S. Washingtoniana*). Modern areas are redrawn from Munns (472).

ever, was redwood known to have lived over even one-half of this latitude. Eocene fossil localities for redwood are all north of the United States boundary. It was not until Late Oligocene, and principally Miocene, that redwood fossils were found south of Canada. When figs and palms lived in Oregon, the redwood and its associates lived far to the north. It is the telescoping of geological time that gives the appearance of extremely widespread vegetational types. During the ages, the forests have wandered over the earth in response to changing climates. Synchronization depends upon the space-time concept. Floristically similar vegetations are not necessarily contemporaneous by millions of years.

The coastal closed-cone pine forests of California.—The story of the closed-cone pine forests of the California coast has already been mentioned several

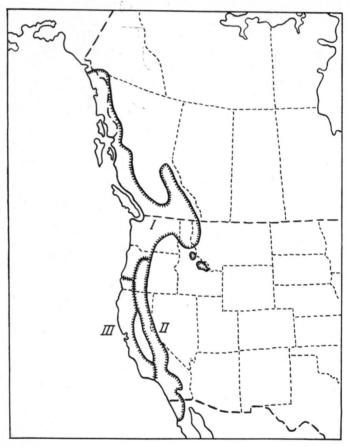

Fig. 7. The three major subdivisions of the Pacific Coast coniferous forest: *I*. The Northern Humid forest; *II*. The Cascade-Sierra forest; *III*. The California Coast Range forest.

times; it presents a most intriguing picture to one interested in the dynamics of plant geography. This story is based upon the researches of Mason, who has employed jointly the methods of taxonomy, phylogeny, paleontology, paleoecology, and plant geography.

For the sake of orientation, let us consider first the Pacific Coast Coniferous forest that extends from southern Alaska southward to northern Lower

California in the Coast Ranges and eastward in the Sierra-Nevada and part of the Rocky Mountains. This coniferous forest contains approximately forty endemic species of Gymnospermae, an assemblage matched nowhere else in the world. A number of the most interesting species are definitely relics of the Tertiary, with roots that go back into the Mesozoic, and are now restricted to ranges that are mere fragments of their former areas. Some of

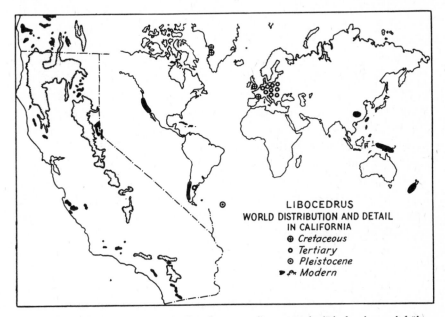

LIBOCEDRUS
WORLD DISTRIBUTION AND DETAIL
IN CALIFORNIA
⊕ *Cretaceous*
○ *Tertiary*
◎ *Pleistocene*
▼ *Modern*

FIG. 8. The world distribution of *Libocedrus,* according to Studt (Die heutige und frühere Verbreitung der Koniferen. Dissertation; Hamburg, 1926), consists of 10 modern species and 5 or more known fossil species from the Cretaceous to the Pleistocene. The detailed occurrence of *Libocedrus decurrens* is taken from Mitchell (U.S.D.A., Bull. 604, 1918). It not only occurs in the Coast Ranges and the Sierra Nevada of California, but extends interruptedly to northern Oregon and into Lower California (not shown). Base map after Goode's copyrighted map No. 101, with permission of the University of Chicago.

these species appear today to be waging losing battles in the struggle for survival, but others are dynamic and spreading. The Pacific Coast Coniferous forest can be subdivided into three floristic regions: the Northern Humid forest, the Sierra-Cascade forest, and the California Coast Range forest. The approximate boundaries of these three floral areas are shown on the accompanying map.

The Northern Humid forest is characterized by several species of general distribution (*Abies grandis, A. nobilis, A. amabilis, Picea sitchensis, Tsuga heterophylla, Larix occidentalis, Thuja plicata,* and *Taxus brevifolia*), together with some that are scattered about timber line (*Pinus albicaulis, Tsuga mertensiana, Larix Lyellii,* and *Chamaecyparis nutkatensis*).

The Sierra-Cascade forest occupies essentially the mountain areas from which it derives its name, but there are outposts in the Coast Range Mountains in southern California and in northern Lower California. Species that characterize this floristic element include some of wide distribution (*Pinus Lambertiana, P. ponderosa, P. Jeffreyi, P. sabiniana, Abies magnifica, Libocedrus decurrens*), together with some of isolated or scattered distribution (*Pinus Balfouriana, Abies magnifica* var. *shastensis, Sequoia gigantea,* and *Cupressus Macnabiana* var. *nevadensis*). In addition to these species, which reach their finest development in the Sierra-Cascade forest, there are certain ones which occur as outposts from the Northern Humid forest and from the Coast Range flora, and which, of course, are non-endemic elements.

The California Coast Range forest occurs from southern Oregon to the Santa Barbara region, with outposts in the channel islands and in Lower California and its islands. Relic species of wide distribution include *Pinus attenuata, Sequoia sempervirens,* and to some extent, *Torreya californica* and *Juniperus californica.* More narrow endemics, of scattered or isolated occurrence, include *Pinus Torreyana, P. Coulteri, P. muricata, P. radiata, P. remorata, Pseudotsuga macrocarpa, Picea Breweriana, Abies venusta, Chamaecyparis Lawsoniana, Cupressus macrocarpa, C. Goveniana, C. Pygmaea, C. Sargentii, C. Macnabiana, C. Forbesii.* There are also outposts from the other two floral regions which occur within this region.

The California Coast Range forest concerns us especially in this discussion because it includes the closed-cone forests which are to be described. Before going on to the closed-cone forests, however, it is worth noting that the California Coast Range forest includes 17 endemic conifer species, only 3 of which are of wide distribution, and outposts of 7 species from the Sierra Cascade and 5 species from the Northern Humid forest, a total of 19 species endemic to the larger Pacific Coast Coniferous forest. The closed-cone forest is dominated solely by relic pines and occupies a markedly discontinuous strip in which the species seldom occur more than a few miles away from the coast. The story of these pines is probably more complete than that of any other group in the region.

The California closed-cone pines consist of four species living today and two that are known only from fossils. The phylogeny, age, and behavior of these species are diagrammed in Fig. 9, in which the descriptive material of Mason has been assembled graphically. *Pinus Masoni* is known only from fossil material of Late Tertiary age (Pliocene). It gave rise to two lines of development that became apparent during the Pleistocene. One line led to *P. linguiformis,* which is known only from Pleistocene fossil material. The other gave rise to *Pinus muricata.* Out of *P. linguiformis* came the modern *P. attenuata,* which is the most widespread of the closed-cone pines, occupying parts of the Sierras as well as the Coast Ranges. *P. muricata* has apparently had a gradual expansion through Pleistocene time and today is

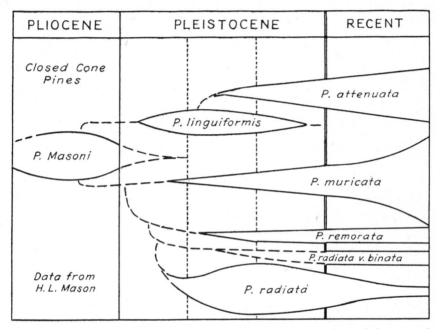

Fig. 9. A diagram combining information concerning the paleontology, phylogeny, and present condition of the closed-cone pines of the California coastal region. This graphic representation was drawn on the basis of various published descriptions by H. L. Mason.

the most widespread of the strictly coastal species and is also morphologically the most variable. Out of *P. muricata,* in early Pleistocene time, came the Santa Cruz Island pine (*P. remorata*) and the insular *P. radiata* var. *binata.* Both of these forms would likely have become extinct had it not been for the protection afforded by their insular habitat. The Monterey pine (*P. radiata*) has been derived from the two-needled form (*P. radiata* var. *binata*). It was more abundant and widespread in Pleistocene time than it is today. The diagram endeavors to express the following relations: (1) phylogeny, through the dendritic connection of forms; (2) geological history, through an indication of the knowledge concerning fossil forms; and (3) modern condition, through the width and trend of the diagrams for each species. For example, with respect to phylogeny, the coastal forms are more closely related to each other than they are to *P. attenuata,* which occurs inland as well as in the Coast Ranges. *Masoni-linquiformis-attenuata* constitutes a direct line of descent. *Masoni-muricata-remorata-radiata* constitutes another phylad, arising early and becoming segregated during Pleistocene time. Geologically, *P. Masoni* is the only form known from the Pliocene and it is considered ancestral, or near to the ancestral type. All the living species, and also the extinct *P. linguiformis,* are known from Pleistocene fossils. In recent times *P. remorata* and *P. radiata* seem to be headed for extinction. This brings us to a consideration of recent behavior. *P. attenuata* appears to be holding its own or expanding slightly, whereas *P. muricata* is variable and apparently in a phylogenetically and ecologically dynamic condition. *P. radiata* had its climax in the Pleistocene and is now on the way out. *P. remorata* and *P. radiata* var. *binata* are probably not extinct only because of their insular areas and the protection that implies.

The closed-cone pine forest ranges today through about 1000 miles from northern California to Cedrus Island off the coast of Lower California. This is not a continuous area—quite the contrary. The northernmost stand is at Trinidad Head, Humboldt County, where *Pinus muricata* occurs as the sole closed-cone species. The forest type is then absent for about 100 miles to Inglenook, Mendocino County. For about 100 miles it occurs frequently in the vicinity of Fort Ross, Sonoma County. It is again absent for 50 miles and reoccurs on Inverness Ridge, Tomales Bay, Marin County. It then skips about 75 miles to Point Ano Nuevo, Santa Cruz County, and near Palo Alto. After another interruption of 40 miles it is well developed at Point Cypress and Point Lobos, south of Monterey Bay, Monterey County. Skipping 60 miles, it reappears at San Simeon, San Luis Obispo County, and continues more or less regularly to the vicinity of Santa Barbara at La Purisima Ridge. The type is represented on the larger of the channel islands, such as Santa Cruz, off the Santa Barbara Coast. After an interruption of about 500 miles, it is present in Baja California, between Ensenada and Point San Quentin, and on Cedrus and Guadalupe Islands, the latter being 200 miles off the coast.

This forest is everywhere close to the coast and is marked by a series of features. Most pronounced, probably, is its conspicuous discontinuity, as described above. It has a uniformity of aspect despite the fact that three different pines are dominant at different places. This uniformity of aspect results from the similar physiognomy of the pines and from the chaparral-like understory that characterizes most of their stands. A third feature of the whole forest is the strong development of endemic species associated with it, especially species of *Cupressus, Arctostaphylos,* and *Ceanothus,* in addition to *Pinus.* The fourth point applying to the whole forest is the fact that all stands are developed on islands or on headlands and ridges which were probably islands during the Pliocene or Pleistocene. Lastly, the modern forest is differentiated into subtypes out of what was historically a much more homogeneous ancestral closed-cone pine forest type.

Today the forest can be divided into three regions: the northern closed-cone forest, the central forest, and the insular forest. The northern forest is characterized by the dominance of *Pinus muricata,* Bishop pine, together with a strong admixture of northern species from the Pacific Coast coniferous forest and from the redwood forest. The limit of the northern type is drawn below Point Ano Nuevo because it is there that the northern species drop out of the picture. The central forest extends southward to the vicinity of Santa Barbara and is characterized by the frequent dominance of *Pinus radiata,* Monterey pine, together with the presence of certain endemic species of *Cupressus, Ceanothus,* and *Arctostaphylos.* The insular forest is characterized in part by the Santa Cruz Island pine, *Pinus remorata,* by *Pinus radiata* var. *binata,* by a return here and there to a dominance by *Pinus muricata,* and by its endemics.

We have mentioned the features which give the whole closed-cone pine forest its unity, and also the fact that the forest today is not a single type but consists of subtypes. If we inquire into its past history, we are led to think that the forest was more homogeneous during Pliocene and Pleistocene times and that certain processes have been active in producing the details observed now. The changing aspects of one and the same flora have been due to the following processes of phytogeographical interest: (1) The closed-cone phylad has itself undergone evolution, particularly during the Early Pleistocene. (2) The flora has been enriched by invasions of species from continuous vegetational types. (3) Under the compulsion of climatic change the forest has undergone migrations, and under migration there have been selection and extinction. (4) Isolation has played an important part in the development of endemism within the forest in the past, and is still operating, especially on those stands that are insular.

Discussion of these phytogeographical processes may be enlarged upon for the sake of clarifying their operation on the closed-cone pine forests. The ancestral species, *Pinus Masoni,* must have enjoyed a fairly extensive distribution along the California coast before the Pliocene development of the California archipelago. This is assumed because it appears that much of the character of the forest has resulted from the insular nature of its stands. The known stands today, as well as the fossil localities, all point to an insular condition, but it is reasonable to assume that the entrance of the

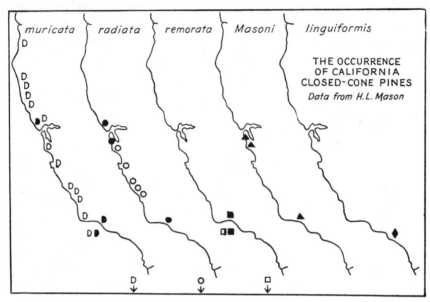

FIG. 10. Diagrams illustrating the approximate modern occurrence and the fossil localities for the closed-cone pines, excluding *Pinus attenuata*. The solid spots indicate fossil occurrences; the hollow ones, modern areas. *Pinus muricata* has its northernmost station near Trinidad Head, Humboldt County. It is more or less continuous from Inglenook, Mendocino County, to Fort Ross, Sonoma County. It reoccurs at Inverness Ridge, Marin County, north of San Francisco, and again at Point Ano Nuevo, Santa Cruz County, and at Monterey Bay, Monterey County, south of San Francisco. Between San Simeon, San Luis Obispo County, and La Purisima Ridge, Santa Barbara County, it is sporadic. On the mainland it is discontinuous to Lower California between Ensenada and Point San Quentin. It occurs on the following islands: Santa Cruz, Guadalupe, and Cedrus. Fossil localities occur at Tomales Bay, Carpinteria, and Santa Cruz. *Pinus radiata* is more or less continuous from just south of San Francisco to Pecos Hills. Fossil localities are known from near San Francisco, at Tomales Bay to the north of its present occurrence, and southward at Carpinteria. *Pinus remorata* is known from Santa Cruz Island and Cedrus Island. It is known from fossils at the former location and at Carpinteria. *Pinus Masoni* is known only from fossils from two beds south of San Francisco and at Ventura, near Carpinteria. *Pinus linguiformis* is known only from Pleistocene fossils from near Los Angeles. Guadalupe Island, 200 miles off the coast of Baja California, has both *Pinus radiata* and *P. muricata*. Cedrus Island, somewhat farther south, has both *Pinus muricata* and *P. remorata*. *Pinus muricata* ranges interruptedly over about 1000 miles of coast. (Data obtained from publications of H. L. Mason.)

ancestral pine to these areas took place before they became islands. *Pinus muricata* is the sole species today that has a wide area.[1] *P. radiata* is known from fossil records to have occurred farther north than it does today, and

[1] *P. attenuata* is excluded from this discussion.

P. remorata has been found in fossil form on the continent at Carpinteria. There is no evidence that either of these species, or the fossil *P. linguiformis,* ever enjoyed a really wide distribution. We can say, then, that part of the diversity of the closed-cone pine forest as known today has resulted from local evolution of new species which have been unable to spread widely.

With respect to the local enrichment of the flora by immigration from contiguous types of vegetation, it is apparent that part of the character of the northern forest is due to the addition of members of the redwood forest and of the Northern Pacific forest. It is apparent, however, that the redwood forest never reached the Pliocene islands on which the closed-cone forest was dominant, and that during Pleistocene time the climate was unfavorable for redwood forest invasion except by selected elements. If this had not been so, it is likely that the closed-cone pine forest would have been swamped out by the redwood forest species. In the central forest there has been an admixture of chaparral species by invasion from the interior and southeast.

Pleistocene climatic changes were not great on the coast, but they were sufficient to cause migrations and extinctions. *Pinus linguiformis* lost out completely, as *P. Masoni* had earlier. The striking discontinuity of the widespread *P. muricata,* and to a lesser extent of *P. radiata, P. radiata* var. *binata,* and *P. remorata,* indicates the selective action of changing environments. Also, *Pinus radiata* and *Cupressus Goveniana* have disappeared from the Tomales Bay region where they occurred during the Pleistocene. *Cupressus* and *Pseudotsuga* have disappeared from Santa Cruz Island and *Pinus remorata* has disappeared from Carpinteria, along with *P. radiata* and *P. muricata.*

The role of isolation can be emphasized by a consideration of the development of endemics. Of the trees and shrubs of the closed-cone pine forests, 58 per cent are endemic to the California Province, and 29 per cent are endemic to the closed-cone pine forests themselves. Certain of the pines are endemic to a single subdivision of the forest, such as *P. remorata* and *P. radiata* var. *binata* to the insular forest. *Cupressus macrocarpa* is endemic to the central forest, being confined to Point Cypress and Point Lobos at Carmel Bay. Within each subdivision there are endemic shrubs, especially of the chaparral understory. It is generally conceded that high endemism can develop only with isolation. In this case it appears that the isolation has largely been a result of the insular character of the areas occupied by the forest, either at present or in the past.

By way of conclusion, let it be emphasized that this story of the closed-cone pine forests, as worked out by Mason and here sketched in the barest outlines, could result only from investigations utilizing simultaneously the methods of and information from paleontology, geography, taxonomy, and ecology. This story is only a small part of the history of the vegetation of western America, but it is significant, particularly because the fossil record

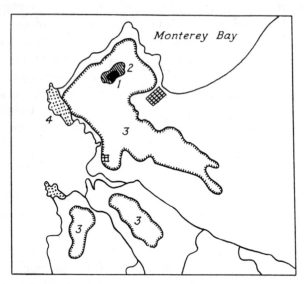

FIG. 11. The occurrence of California conifers in the region south of Monterey Bay. The small black area (*1*) is a stand of Gowan cypress, *Cupressus Goveniana,* surrounded by a stand of Bishop pine, *Pinus muricata* (*2*). The large areas (*3*) are the Monterey pine, *Pinus radiata,* and the stippled areas (*4*) are the only natural stands in the world of Monterey cypress, *Cupressus macrocarpa,* on Point Cypress and Point Lobos.

is unusually complete as is the knowledge of modern distributions of the species.

The origin of the desert climax and climate.—The desert region of the Southwest is dominated by the *Larrea-Franseria* climax. Although the term "desert" is used loosely to apply to a variety of conditions, the desert climax, strictly speaking, is marked by the absence of forest or grassland, by a critical deficiency in precipitation, and by high evaporation resulting from excessive heat frequently accompanied by high winds. According to Clements, an isohyet of five inches marks the disappearance of grass dominants on the climax level and constitutes the best means of limiting the desert climax. The desert, as characterized by *Larrea, Franseria,* and their associates and the conditions mentioned above, is restricted to Death Valley, Mojave, and the Colorado regions, together with a larger area in Mexico.

Farther north, in regions of low rainfall, the desert is replaced by sage-brush and grassland where temperature and evaporation are less critical.

Today the desert is largely encircled by the grassland formation, with mixed prairie on the northeast, the desert plains along the eastern boundary, and bunchgrass prairie on the west. These three grassland associations have been derived from the mixed prairie through the same climatic causes that gave rise to the desert climax from the preexisting grassland. Historically, the desert region was occupied by mixed prairie preceding the origin of the desert climax and climate, the grasses withdrawing and segregating to their present areas and associations as the desert elements invaded.

The evidence for the brief historical account which follows is largely drawn from the following sources: (1) meager fossils, (2) the cliseral rela-tions which exist now between desert, grassland, woodland, and forest, and (3) the presence of relics in the desert area and on its two sides. In this case relics provide Clements with the principal key in reasoning from the present to the past. They are witnesses of past conditions and of changes, including local, segregations (eliminations) and immigrations from the Mexican homeland of the desert species. The relics fall into two principal classes of transads: (1) those which still exist here and there in favored places within the desert area but are really "at home" in the contiguous vegetation on one side of the desert or the other; (2) those which find it impossible to live in the desert area but are found both east and west of it. Both groups, which do not belong to the desert climax and climate, are con-sidered to have extended across the area before the progressive desiccation reached the desert stage. Fossil evidence bearing directly on the problem of the origin of the desert climax is not very abundant, although recent Pliocene and Pleistocene discoveries by Axelrod indicate the Mexican origin of the vegetation and largely substantiate Clements' account.

In the grassland climax three principal grass forms are recognized. The tall grasses are historically the oldest members of the formation (*Andropogon, Imperata,* etc.) and are largely of subtropical origin. The midgrasses (*Stipa, Oryzopsis, Hilaria,* etc.) are of northern origin in North America, and to a lesser extent in Eurasia. The short grasses (*Bouteloua, Buchloë, Aristida,* etc.) are largely desert plains elements today; they came into the grassland climax from their homeland in the mountain plateaus of Mexico and Cen-tral America, probably during the Pliocene, and represent the youngest ele-ment in the formation.

The elevation of the Cordilleras during the Late Cretaceous must have produced a gradual desiccation of the continental interior. At the same time that the Rocky Mountains were being elevated—the process continued into the Tertiary—the inland seas were withdrawing. Gradually the transcontinental forests became of less humid type and finally gave way to grassland in part of the Great Plains region, although forests persisted in the more favorable sites, especially along streams.

During the Early Tertiary (Eocene and Oligocene) a subtropical forest occupied much of the Pacific region in the middle latitudes, but temperate deciduous trees, such as *Fagus, Castanea,* and *Tilia,* must have extended from the Rockies over much of the Pacific slope. The shrinking forests in the Oligocene saw an extension of the area occupied by grassland. Southward of the temperate redwood forest area of the higher elevations in the Upper Eocene and Lower Oligocene, the Southwest probably had a hardwood climax in middle elevations, with such trees as *Carpinus, Ostrya, Morus, Ulmus, Celtis, Asimina, Cornus, Diospyros, Sassafras, Ilex, Maclura, Crataegus, Malus,* and *Rhamnus.* In this rather moist and warm period a tall-grass prairie undoubtedly confronted the deciduous forests. This prairie must have enjoyed a fairly high rainfall (around 35 inches) and have been composed of species such as *Andropogon saccharoides, A. glomeratus, A. Hallii, A. furcatus, Imperata Hookeri, Elionurus barbiculmis, Trachypogon secundus,* and *Tripsacum Lemmoni.*

By Miocene time the subtropical forest had withdrawn from the middle latitudes of the Pacific region and been replaced by the temperate redwood forest from high latitudes and elevations. With cooling and drying, the tall grassland followed the shrinking of the deciduous forests in the southern region. In the central and southern Great Plains region there developed a grass climax that can be classed as modern in type. In the Pacific Northwest the elevation of the Cascade axis caused striking changes in the forests to its east. The temperate redwood forest lost many of its more humid species and much of the upland became dominated by a type comparable to the present oak-madrone redwood border forest. The elevation of this axis must have also allowed some northward expansion of grassland into suitable locations.

By Pliocene time the southwest area (where the deserts are now found) still contained valley forests, with *Juglans, Acer, Platanus, Fraxinus, Salix, Populus, Aesculus, Prunus,* etc., but these forests were not climax. The montane forests included such species as *Pinus Jeffreyi, Pinus ponderosa,* and *Pseudotsuga taxifolia,* which were more widespread than at present. Between the valley forest and the montane forest, there developed a rather widespread grassland of the mixed prairie type. There were some tall-grass relics, but the dominants were a mixture of midgrasses and short grasses. Bordering the montane forest there was likely a certain amount of woodland composed largely of nut pines and junipers, but also containing live oaks and species like *Pinus Sabiniana*. In other places the chaparral and, where drier, the sagebrush climaxes had undoubtedly made their appearance and lay in their proper cliseral positions. The fairly widespread mixed prairie, which had developed in part out of the tall-grass prairie and in part by invasions, had probably largely replaced the tall-grass prairie during the Miocene. In the Pliocene it expanded greatly, taking over places left by the retreating forest. This grassland type contained species such as *Bouteloua gracilis, B. racemosa, B. eripoda, Buchloë dactyloides, Hilaria Jamesi, Poa scabrella, Stipa comata, S. speciosa, S. coronata, S. setigera, S. emineus, S. pennata, Elymus sitanion, Koeleria cristata, Andropogon scoparius,* and *A. furcatus*. With progressive desiccation during the Pliocene the grassland became more and more xeric until dominance was established by species of the desert grassland such as *Hilaria rigida, Oryzopsis hymenoides, Stipa speciosa, Triodia pulchella, Muhlenbergia Porteri, Aristida purpurea, Sporobolus cryptandrus* var. *flexuosus,* etc.

At the close of the Pliocene and the beginning of the Quaternary the present desert had its origin in the rain shadow resulting from the structural differentiation of the Sierra Madre-San Bernardino chain. The *Larrea-Franseria* desert developed through the loss of desert grassland species, through the evolution of endemics, and from the immigration of Mexican Sonoran species from the south where the desert climate and climax have long existed (379), because they are an earth-old feature. Accompanying the development of the desert climax, the grassland withdrew and segregated into the modern contiguous types. Since the glacial and interglacial periods were less clearly marked in the West and Southwest than they were in the North and Northeast, it appears that the grasslands and deserts have shown little change since the middle of the Pleistocene. There have, of course, been extensions and contractions of area. It is likely that each interglacial period saw a culmination of desert much like the present strongly marked Colorado and Mojave deserts.

IO.

Pollen Analysis as a
Paleoecological Research Method

1. *Pollen analysis provides a technique which facilitates a reconstruction of postglacial vegetation and climatic history, supplements the use of macro-fossils, and substantiates the broad outlines derived from a study of con-temporary areas and processes by floristic methods. There is no reason why this method cannot be more extensively employed for intraglacial and Tertiary deposits.*

2. *Pollen analysis has resulted in two principles, discovered from north European observations and substantiated by more extensive European, west Asian, British, American, and New Zealand observations. They are the principle of regional parallelism and the principle of final period reversion or climatic deterioration.*

The use of fossil pollens for the study of vegetational history is a de-velopment of the present century. It began with the investigations of Lagerheim (405, 406) and another Swedish botanist, von Post (505), who was largely responsible for perfecting the working methods and stimulating their wide application. The writings of Erdtman (216, 218, 219, 220) for the first time effectively brought the methods of Lagerheim and von Post to the notice of English-speaking people, according to Godwin (280). By 1927 the literature on pollen analysis numbered about 150 titles (217), and by a little over a decade later (224) nearly 2000 titles had appeared. The growth of this specialized type of micropaleontology spread rapidly over most of Europe and western Asia. The earliest American investigations, so far as I know, were by Auer (28, 29, 30) on Canadian deposits; discussions by Fuller (257) and Erdtman (221) in widely read journals did much to start American students to work. Several midwestern projects were begun at the University of Chicago under the direction of Professor Fuller (664, 356, etc.); Professor Sears (561, 562) started a series of publications by himself and students; and numerous studies by Potzger (509) and his students began to appear from Butler University. The most prolific investigator to come to the field is Hansen (322, 323, and later papers), who has published about twenty papers within five years, mainly on the Pacific Northwest. Wodehouse (711) published a manual of pollen, Godwin (279) and Cain (107) general discussions of methods and limitations, and Erdtman

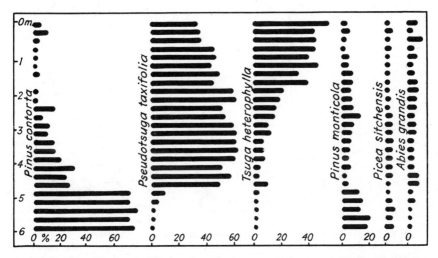

Fig. 12. Pollen profile of Black Diamond bog, Puget Lowland, Washington. Hansen believes that ordinary forest succession has played a more important role in the changes in this region than cliseral migration due to climatic change. Pioneer forests consisted largely of lodgepole and western white pine (*Pinus contorta* and *monticola*), replaced by climax species. Douglas fir (*Pseudotsuga taxifolia*) developed rather abruptly, and western hemlock (*Tsuga heterophylla*) gradually. Data redrawn from (327).

(225) a monographic manual; these provide the best general information available in English.

When it was discovered that peat deposits and other types of sediments contain well-preserved pollen grains, spores, and other microfossils, it was soon realized that their identification and statistical analysis would provide an excellent tool for the investigation of Quaternary vegetational and climatic history. Because the soft peat sediments permit easy drilling and the removal of uncontaminated samples, stratigraphic methods have prevailed from the inception of the science of pollen analysis. Small samples from successive levels are prepared for microscopic analysis.[1] Grains are identified

[1] For various techniques, see (107).

and counted from each sample, permitting a percentage composition de-termination called the pollen spectrum. Successive spectra from a single peat deposit form a pollen profile from which changes in composition can readily be detected. Through the period of time represented by the pro-file, the percentage composition shifts in favor of one or another pollen or pollen group. It was immediately appreciated that these changes in pollen representation reflected changes in the flora surrounding the sedimentary basin and, by analogy, changes in climate or vegetational succession.

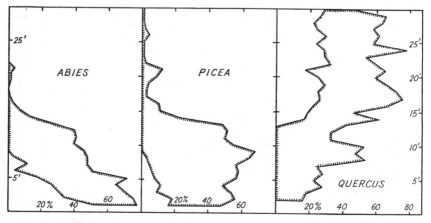

Fig. 13. Graphs showing postglacial pollen percentage trends of three genera in seven peat deposits near the southern limits of the Wisconsin glaciation in Indiana. The curves are synchronized on the basis of deposit bottoms, and only maximum and minimum percentages found in any of the bogs are shown for each foot-level. The graphs were prepared from data published by Potzger and Friesner, *Amer. Midl. Nat.*, 22:351–368, 1939.

The science of pollen analysis rests more or less firmly upon the following facts or assumptions:

1. Most dominant trees of the temperate zone have wind-borne pollen. This fact results in an average pollen rain for a locality which contains a mixture of the grains of the available species.

2. Many of the grains of the pollen rain which fall onto a bog or other receptive surface, or which settle through the water to the bottom of a pond, are preserved under the more or less antiseptic, low-oxidation con-ditions.

3. Year after year, as the peat or other sediments accumulate, stratified pollen deposits occur.

4. Proper methods of sampling and preparation of the samples allow the detection, identification, and counting of the fossil pollen grains, which usually occur in abundance.

5. The structural characteristics of pollen grains are constant for a species, or at least as constant as other morphological characteristics of species. It does not follow, however, that the species of a genus can always be told apart.

6. Consistent general trends in the pollen spectra from a profile, based upon the percentage composition of component pollen grain types, are a fact in most records. As a result, stratigraphic-time-vegetation-climate correlations can be made.

7. When numerous profiles are available over a wide geographical area, regional correlations permit a reconstruction of the general outlines of vegetational migration and succession, together with an approximate dating of the principal events.

8. The phenomenon of regional parallelism of development is that given climatic changes induce equivalent but not identical changes in forest composition in different parts of a country.

Although these points seem simple and straightforward, their application in pollen analysis is a complicated procedure requiring diligent care. Inadequate or misleading data may result in false conclusions because of faulty field technique in obtaining the samples, in preparing them for study, or in identifying the fossils, and because of a variety of assumptions based upon inadequate factual information. I refer to such assumptions, for example, as that the percentage representation of the fossils corresponds to the percentage composition of the vegetation by the represented species, and that a genus—the species being undetermined—can have an accurate climatic significance. What might be called the pitfalls of pollen analysis will be discussed after a brief look at the development of postglacial climatic hypotheses.

Sears (563) points out that ideas concerning postglacial conditions in Europe have undergone a four-period development:

1. The simple assumption that climatic zones, and hence vegetational types, retreated southward before the ice advance and then followed the ice recession northward to a more or less stable postglacial equilibrium.

2. The Blytt-Sernander hypothesis, postulating a series of fluctuating moisture and temperature conditions:

Pre-boreal	cool-humid
Boreal	warm-dry, continental
Atlantic	warm-humid climatic optimum
Subboreal	drier, continental
Sub-Atlantic	return to higher humidity

3. The Andersson hypothesis that temperature changes have been predominant; this denies a second or subboreal dry period and assumes gradual climatic deterioration from the Atlantic down to the present.

4. The von Post hypothesis of three generalized phases:

    a. Increasing warmth—The stage of the approach of the warm period, characterized by the appearance and the first increase of the relatively heat-requiring trees of different kinds.

    b. Maximum warmth—The stage of the culmination of these forest elements.

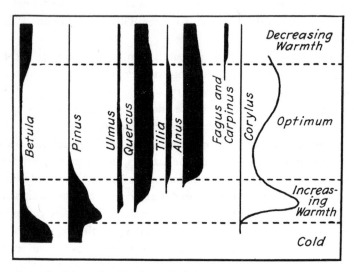

Fig. 14. Generalized form of pollen diagram of the East Anglian type to show the threefold von Post divisibility of the postglacial period, and reversion in the last phase. Figure redrawn from Godwin (280), who found the indicated temperature relationship to hold for all of England and Wales.

    c. Decreasing warmth—The stage of the decrease of the characteristic trees of the warm period and the appearance or return of the predominant forest constituents of the present day.

The Blytt-Sernander concept of postglacial changes was widely accepted, but its terminology was too highly suggestive and, according to von Post (507), affected investigators like a strait jacket. A recent summary by Godwin (280) of pollen investigations in the British Isles corroborates the von Post interpretation of a three-phase temperature change in postglacial time, with a reversion in the last phase. In England and Wales, beginning with a relatively short cold period of *Betula* dominance and *Pinus* co-dominance,[2] the period of increasing warmth is indicated by an increase of *Pinus* (probably all *P. sylvestris*) to a position of dominance at the expense of *Betula,* and a rather abrupt increase of *Ulmus, Tilia,* and particularly *Quercus* and *Alnus.* The second period, that of climatic optimum, endured rather long and was marked by a low composition in the spectra of *Betula* and *Pinus* and a maximum of the quercetum mixtum. The reversion period is marked especially by the return of *Betula* to importance, the consistent presence of small amounts of *Fagus* and *Carpinus,* and a corresponding and gradual decrease of *Quercus* and *Alnus.*

Turning to American results, we may first note the conclusions of Sears (564), based upon a comparison of types of pollen profiles. Finding that regional profile types are a reality, he concluded that the major climatic shift set in with glacial retreat and took place during approximately the time period of the lower third of the profiles, and that since then there has been no comparable change in magnitude. There was, however, a middle period of maximum warmth and, probably, dryness. The third or post-xerothermic period of reversion is shown by an increase of *Picea* at northern stations, and of broad-leafed trees indicating mesophytism (*Acer, Fagus, Nyssa, Liquidambar*) at more southern stations. Recognizing, as did Godwin, that numerous complications are involved in the record, Sears concludes that von Post's simple schedule "is broad enough to allow of future refinement, and does no violence to the facts." Later data of Sears (566), based on a study of five genera from over 100 American deposits, also indicate a recent reversion but suggest a second dry period immediately preceding it. Sears (565) summarizes the difficulties in the way of climatic interpretations of the second and third phases of American profiles, and reserves judgment until the data are more complete. Preston Smith (600) has published a brief discussion of correlations based upon 148 profiles from 123 bog and lake deposits in different parts of eastern North America. His paper is especially useful for its analysis of the early pre-boreal period with its changes of spruce, fir, and pine composition during the oscillations of the late Wisconsin ice age. In a general way, the time following these glacial deposits is divided according to the von Post scheme, but Smith recognizes that numerous complications exist because of the regional differences between seaboard and interior, and northern and southern stations. Such complications are primarily due to the fact that during the past, as now, considerable differences in climate existed to the east and west of the Appalachian mountain chain. The simple temperature relations are

[2] This time may have been glacial, rather than early postglacial.

modified by minor periods of more or less dryness "occurring almost rhythmically." As a matter of fact, Smith stresses the importance of these minor variations for purposes of correlation. They are so consistent that increases and decreases of the same species can be followed through several bogs, allowing cross-dating. These variations, however, cannot be com-

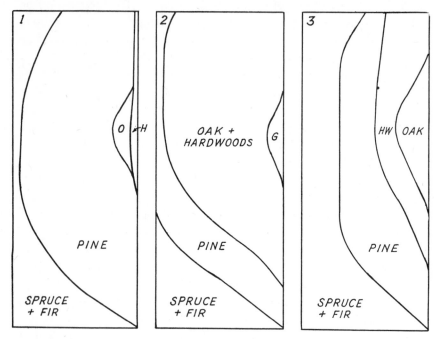

FIG. 15. Schematic diagrams of types of pollen profiles of North America: *1*. Northern lake forest region; *2*. central deciduous forest region; *3*. northeastern oceanic region. H or Hw = hardwoods, O = oak. In Diagram *1* note the early predominance of spruce and fir, the third-period reversion, and the complementing oak period. In addition to the usual interpretation of a middle warm (and probably drier) period, this profile type has been interpreted as due to succession except for the early warming. Diagram *2* shows regional parallelism to Diagram *1*. This profile type likewise is interpreted as indicating a middle warm-dry period, but subsequent reversion is questioned. In Diagram *3* spruce-fir and pine maintain a good showing, with reversion more apparent for spruce-fir and hardwoods, and a complementary oak period indicating the middle warm-dry time. This profile type, however, is also interpreted as being due to succession in the upper part wherein oak is the precursor of mixed hardwoods. Data from (565).

pared with respect to their actual percentages, "but the increases and decreases of the various species are consistent, as well as the appearances and disappearances of previously unconsidered species, especially those near the limits of their ranges." It should be pointed out that when peat samples are taken at infrequent intervals along a profile, these important minor variations may not appear.

We shall now turn our attention to sources of error and difficulty in pollen analysis, caused by the highly deductive basis of much of the pollen work. It cannot be stressed too much that the future of pollen analysis as a science depends greatly upon its striving for a completely inductive basis.

1. **Truncated records and the problem of synchronization.**—Many investigators appear to have assumed that the profiles they have sampled represent the accumulation of sediments during the whole postglacial period, whereas this may not be true for a number of reasons, and their interpretations, as a result, may be highly incomplete or erroneous. The bottoms of various deposits are not necessarily of the same age. The time that pollen accumulation began depends upon the location of the bog relative to the various stages of the ice front during its more or less oscillatory retreat, and to the nature of the basin in which the sedimentation started. For example, a kettle-hole may have been blocked by dead ice for centuries after the ice sheet withdrew from the locality, or the lake may have been large and bog conditions developed only later. The bottom of the record of several profiles runs back into the late Wisconsin period and does not represent solely a postglacial condition (562, 600); in other bogs sedimentation began only after a greater or lesser length of postglacial time had elapsed. Another cause of lack of bottom synchronization is that the profile is truncated because the lower sampling was incomplete. Potzger and Richards (510) concluded that most records in the Middle West indicate too short a *Picea* period because of errors in sampling from the deepest portion of a basin, of failure to sample the early fine-sand deposits, or of failure to penetrate a false sandy bottom to underlying pollen-bearing sediments.

The top of a profile also may be truncated; this is usually attributable to bog senescence. There is no accurate way of determining how long a bog has been "dead" and pollen sedimentation discontinued, except by comparison with nearby "live" bogs. Furthermore, there may be a considerable telescoping of time, as the rate of sedimentation slows up with bog matturity and there is a decrease of surface receptivity for pollen and of conditions favorable to preservation.

Not only may a profile be truncated at either end, but certain intermediate periods may be poorly represented because of climatic conditions causing slow deposition. Preston Smith (600), for example, believes that the pre-boreal in North America (the interval between the *Picea* maximum and the *Pinus* maximum, characterized by a rise in *Abies* and culminating in a final *Betula* maximum) is poorly represented in most profiles because sediments were slowly accumulated during the period.

Synchronization of profiles may be difficult because the profiles were sampled at too infrequent intervals;[3] Preston Smith (600) found that minor variations of increase or decrease of species are of great importance in cross-dating profiles. A good procedure would be to obtain samples at frequent intervals and to analyze a portion of them. When a general outline of the

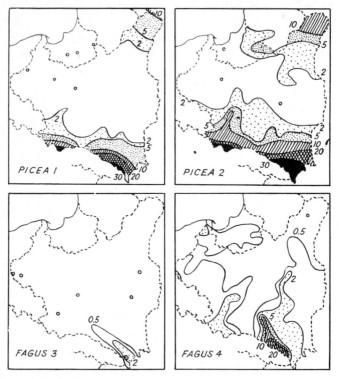

Fig. 16. On the basis of 152 points in Poland where pollen analyses were available, Szafer has drawn isopolls for *Picea* and *Fagus* for each of the five postglacial periods of that region. Diagrams *1* and *3* represent the second period in which there was an increase of warmth-requiring trees. Diagrams *2* and *4* represent the period of culmination of these forest elements. The isopolls connect points of equal average pollen percentage of selected species. Selected maps are redrawn from Szafer, *Bull. Acad. Polonaise Sci. Let., Sér. B., Sci. Nat.* (I), 1935:235–239.

development has been obtained by this means, it will likely appear that certain portions of the record need more frequent analysis.

As it is impossible to make correct correlations on a foot-level basis,[4] they must be made from characteristic features of composition and change in

[3] Several studies are based upon sampling at one-foot or one-meter intervals, whereas it should have been two to several times as frequent. Godwin (281) found it necessary to sample at 5-cm. intervals in portions of certain profiles.

[4] Just as in dendrochronology it is impossible to make correlations on the basis of tree-ring dimensions independently of the pattern of sequence and change.

composition. This requires the disposal of the possibilities of truncation and interval underrepresentation in the record. Also, in connection with the problem of synchronization there is the necessity for recognizing the principles of zonation and regional parallelism (506, 280), that a given climatic change will produce an equivalent but not identical change in forest composition in different parts of a country. The two general rules for postglacial association between climatic evolution and vegetational development (regional parallelism and reversion), based on comprehensive European results, are both applicable to as remote a place as New Zealand, according to Cranwell and von Post (169).

The problems discussed under this topic are important in resolving such controversies among geographers as whether there was, in the vicinity of the ice front, a tundra development, and how wide it was and how long it endured; and the extent, intensity, and duration of the postglacial xerothermic period, or even whether it existed at all.

2. **Bog surface receptivity.**—In his discussion of the boreal hazel forests of the British Isles and Europe, Erdtman (222) emphasized the importance of the receptivity of the bog surface, i.e., whether the bog surface is in a condition to receive and preserve the pollen. In general, the pollen that gets fossilized is that which is quickly carried down from the surface to deeper levels where drying, bacterial activity, and the supply of oxygen are less. Dead bogs in which peat is no longer forming, and living bogs in an inactive state because of cold or exceptional dryness, will not preserve pollen. Thus, in continental regions, the frozen bog surfaces in the late spring may not be receptive to the pollen of early flowering species. *Corylus,* according to Erdtman, is an example of plants that may be present in abundance but may under those conditions leave no trace in the peat record. The extent to which this factor is important in altering pollen records is impossible to determine. That it is not as important as Erdtman thought may be suggested by the fact that superficial moss and liverwort polsters and mats have been shown to contain quantities of pollen despite the fact that they frequently dry out and are well aerated (122). Such accumulations, of course, have not been long in forming, but in Carroll's investigation polsters collected many months after the last rain of spruce, fir, birch, hickory, and other pollens contained quantities of these grains.

3. **Downwash of pollen through the peat.**—Godwin (279) discussed the evidence for the downwash of pollen as a possible source of error and concluded against the likelihood of any serious general errors resulting from that cause. Nevertheless, the possibility of such an error must always be

considered when samples are taken at close intervals and scattered grains are "out of place" in the general sequence of results shown by the profile. Considering the telescoping of time in compact peats, however, it is unlikely that any important distortion of the picture could result from the downwash of grains.

4. **Differential preservation of pollen.**—Certain types of pollen, because of the delicacy of their walls or peculiarities of chemical structure, preserve poorly in peat or mineral sediments. Fortunately, most tree pollen grains preserve well, and apparently equally well. There are some notable exceptions. Godwin (279) reports the general belief that *Myrica* pollen is easily destroyed but that from the work of Jentys-Szafer (378) the absence is apparent rather than real. She shows how to distinguish between *Myrica* and *Corylus,* and believes that the two have been tallied together as *Corylus* by previous workers. Godwin also reports that grains of *Populus, Taxus, Juniperus,* and the Rosaceae are easily destroyed and that the preservation of *Fraxinus, Myrica,* and *Acer* is sometimes questionable. *Larix* grains are frequently absent from bogs where they would be expected, and when they are found in small numbers it is usually along with abundant macroscopic remains of wood, fruit, etc., indicating a great local density of tamarack. Preston Smith (600) says that *Larix* fossilization depends upon the chemical condition of the bogs, and that its presence or absence in a certain bog may have been due to that factor. Erdtman (222) notes that *Populus* may not be well preserved and that as a result of this underrepresentation pollen percentages may be similar for very different situations. As an example he mentions that mountain bogs in the Cordilleran coniferous forest may have spectra similar to those of Canadian muskegs where only scattered conifers occur among abundant aspen (*Populus*). Lewis and Cocke (417) found fresh pollen of *Chamaecyparis* and *Taxodium* to be so delicate that good reference slides were difficult to prepare by a technique which produced excellent results with grains of most species. They concluded that this would account for the small percentage of *Chamaecyparis* found at the surface of the peat of the Dismal Swamp, and its absence at all other levels. Sears and Couch (567) found that *Taxodium* grains were absent from peat in a swamp where cypress had evidently lived for 800 to 1000 years. They also found that *Juniperus* grains break down quickly.

5. **Morphological similarity of unrelated grains.**—One source of error in pollen statistics that has not received the attention it deserves is the convergence in appearance of unrelated pollens. Meinke (461) called attention to the similarity between grains of *Quercus* and *Viola.* When this pollen

type is present in quantity, or in deposits some distance from forest sites, there is little doubt but that the pollen grains are from the anemophilous tree rather than from violets. Meinke also mentioned that *Salix* and *Fraxinus* constitute another pair of similar pollens. Of these Godwin (279) said that grains smaller than 27 microns can safely be considered *Salix*, but that it is perhaps better to count them together as "salicoid pollen" and omit them from the total counts of tree pollen. He also noted similarities of the following groups: *Fagus-Hippophae*, *Corylus-Urtica dioica*, and *Betula-Corylus-Myrica*. Some records of scattered *Fagus* grains that antedate the general appearance of beech in a region may be due to *Hippophae*. *Urtica dioica* is a more or less weedy plant that probably was not very abundant in precultural days so that most *Corylus* records are beyond suspicion in this regard. Although Jentys-Szafer (378) showed that *Corylus* and *Myrica* are distinguishable, and Wodehouse (711) that *Corylus* and *Betula* can be told apart, the fact remains that fossil grains of the "betuloid" type are frequently difficult to identify and have to be counted together unless the improbability of the presence of *Myrica* and *Corylus* can be established.

6. **Determination of species within a genus.**—Although there are good reasons for believing that species characters of pollen grains are as constant as are other small features of species, it is true that the grains of species of a genus are frequently very similar and difficult or impossible to distinguish. It is for this reason that nearly all investigators have apparently been satisfied to stop with generic identification, presenting their statistics for *Abies, Picea, Pinus, Quercus, Carya*, etc. I do not know the extent to which it may prove possible to identify species by the use of refined methods in morphology, chemistry, or size-frequency statistics, but it is certain that every effort should be extended in that direction.

Since the principal objectives of pollen analysis are to discover the nature of past vegetation and climate, the success of the method depends in no small degree upon the identification of species. Larger changes, of course, can be detected by pollen groups, such as the characteristic shift from *Abies-Picea* to *Pinus-Quercus* in many American profiles, or a period of dominance by grass pollens; but most genera contain ecologically different species, and many refinements in postglacial history are impossible on a solely generic basis. For example, *Picea mariana* and *Picea rubens* frequently play different ecological roles; an abundance of *Picea* pollen in one case might mean no more than local, edaphic dominance of *P. mariana* on the bog surface, whereas *P. rubens* might indicate a climatically controlled upland spruce forest. The rather large pine genus contains such ecologically and climatically diverse species as *Pinus Strobus, banksiana, rigida, pungens,* and *taeda* in the East, and among the oaks one needs only to mention *Quercus macrocarpa, palustris, borealis, stellata,* and *virginiana* to suggest that generic identification alone does not tell us much about conditions. I do not mean to suggest that generic identifications are worthless, but that every effort should be made to carry the identification to species where possible.

Sometimes there are morphological characteristics of considerable value. Trela (644) found that *Tilia cordata* and *T. platyphyllos* can easily be distinguished, and Godwin (280) stated that although both species now appear to be native in England, grains of *T. platyphyllos* have never been reported fossil in British peats. Hansen (327) found that *Tsuga mertensiana* has bladders which make it easily distinguishable from *T. heterophylla,* which does not. Although von Post separated *Quercus robur* and *Q. sessiliflora* in Scandinavia, Godwin was unable to do so in the English studies. Nor did his pollen identifications allow the separation of *Betula pubescens* and *B. alba,* although Bertsch (64) concluded from the size distribution of fossil *Betula* pollen from the Federsee diagrams that *B. verrucosa, B. pubescens,* and *B. nana* were represented.

Bertsch used a statistical method to distinguish species of birch and this was taken up by others, including Cranwell (167) in a study of the southern beech, *Nothofagus.* She found that *N. truncata* had a diameter between 23 to 32 microns with about an equal number of grains with 6 and 7 pores; *N. fusca* was 34 to 42 microns with mostly 8-pored grains; and *N. Menziesii* was 40 to 65 microns with mostly 7-pored grains. She extended her morphological-statistical technique to a consideration of all the New Zealand conifers (168), a type of study that should be made in each region where pollen analysis is undertaken.

Among students of American pollens, Erdtman (221) noted that *Picea canadensis* and *P. mariana* can be distinguished, as can *Pinus banksiana* and *P. Murryana.* Hansen has published a long series of papers on pollen analysis of bogs of the Pacific Northwest in which he has recognized the following species: *Pinus contorta, monticola, ponderosa, albicaulis, Murryana,* and *Lambertiana; Abies nobilis, grandis, lasiocarpa, amabilis,* and *concolor; Picea sitchensis* and *Engelmanni;* and *Tsuga heterophylla* and *mertensiana.* In several places (325, 326, 327, 328) he has discussed the problem of species determination on the basis of the size range of grains. He admits that sometimes it is impossible to tell certain species, but in general the method works for his region. In fact, much of Hansen's fine accounts of glacial and postglacial forest history would have been impossible had he been unable to recognize species in the leading coniferous genera. There are two alternatives when grain sizes of two species overlap strongly. It may be possible to tell which species one is likely dealing with because of geographical or ecological differences between the species that for a given situation make one more probable than the other, or the overlapping grains can be discarded from consideration (326). Perhaps a better method than discarding grains of a size represented by the overlap between two species is to prorate such grains between the two species on the basis of the ratio between grains which lie outside of the overlap. This method was used by Carroll (122).

In the eastern United States there are a dozen or more species of pines, and this makes the problem of size-frequency identification much more difficult than in the Northwest or Europe. I have had occasion to look into this problem because of some studies for the Soil Conservation Service. In the Piedmont of South Carolina there have been discovered some buried soils of high organic content. In most layers there were abundant pollens and spores; and in addition to *Picea* and *Abies,* which are climatically far removed from the present conditions around Spartanburg, there was an abundance of small *Pinus* grains that appeared to be the northern *P. banksiana.* This conclusion was reached after the following investigation (110). Measurements of several hundred grains of fossil pine pollens and their size-frequency distribution revealed a strong three-modal curve and the probability that at least three species of pine contributed to the sediments. The next step was to prepare size-frequency curves for each of the modern species of pine in the East in the hope that some of them would match with the modes of the curve for fossil pine. *Pinus banksiana,* the smallest of the species, matched perfectly with the smallest of the modes for fossil grains. Overlap prevented any certainty with respect to what species formed the other two modes. This technique is being carried further because it was realized that one set of grains might not show the whole size variability of a species. Nearly 200 collections were obtained from widely separated places, and enough of them have been analyzed [5] to support the suggestion that *P. banksiana* may have been present in South Carolina at the time of the deposits although its nearest locations today are hundreds of miles northward in northeastern New York and at the southern tip of Lake Michigan. The only other alternatives to the above conclusion are the unlikely ones that the small grains represent some now extinct species, or that fossilization

[5] The work has not been completed although about 15,000 measurements have been made by Louise G. Cain.

shrinks the grains more than does the Erdtman technique of grain preparation. The new data on size frequency, based upon several collections of each species, reveal that many species are variable from collection to collection and that the size-frequency method is of limited applicability. The next smallest grains to *P. banksiana,* however, have a mode that is about 5 microns larger.

Deevey (181) and Potzger and Richards (510) do not believe that *Pinus banksiana* can be distinguished. The latter authors say, "An attempt was made to separate the pollen of *Pinus banksiana* from that of the other two

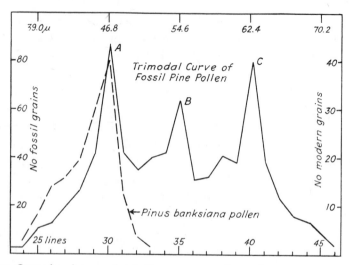

Fɪɢ. 17. Comparison between the size-frequency curve for modern *Pinus banksiana* pollen grains and peak *A* of the trimodal curve for fossil pine pollen grains from buried soils near Spartanburg, South Carolina. The species forming peaks *B* and *C* could not be suggested with certainty by the size-frequency method. Data from (110).

pines, but the recommended diagnostic characteristics did not prove satisfactory in this study when check counts were made by several workers in our laboratory." Be that as it may, I do not claim that species can be identified in all cases or even in many (insofar as I have looked at *Quercus,* for example, it seems pretty hopeless), but only that every effort should be made to do so for the sake of increasing the utility of pollen analysis.

7. **The problem of underrepresentation and overrepresentation.**—Because different tree species produce different quantities of pollen and because the pollen of some species is more easily transported than others, the relations between pollen spectra and forest composition are not direct. That is to say, certain species are undoubtedly overrepresented and others under-

represented. Most workers have not given these problems the attention they deserve and have gone on to their explanations of forest and climatic history as if their pollen spectra were absolutely representative of forest composition in the past.

a. *Pollen production.*—That some species produce great quantities of pollen is a well-known fact; mention of pollen "snow" or "sulphur showers" is not infrequent. Just this spring I walked through a hardwood forest in the Great Smoky Mountains at the time yellow birch pollen was being shed, and my shoes and trousers were soon yellow with pollen that had settled on the herbaceous foliage, and quiet pools were coated with it; at other times I have had a similar experience with *Pinus* and *Alnus* pollen. Cranwell (167) mentions that *Nothofagus* pollen is produced in such quantities that the gently drifting clouds of it are mistaken for smoke. She likewise states that it has been estimated that *Alnus glutinosa* yields about four million grains per catkin, and that during a dust storm in 1934 oak pollen fell at the rate of 3.4 tons per square mile, as measured at Chicago by the exposed slide technique.

It is obvious that those trees which produce relatively enormous quantities of pollen will tend to be overrepresented in pollen statistics and, conversely, other species producing small quantities will tend to be underrepresented. Occasionally one reads such a bald statement as "*Abies* produces more pollen than Picea, so in our diagrams there is a distortion and *Picea* really played a more important role than it appears." When such statements are based upon no more than casual observations and impressions, one can have little confidence in them. In order to obtain a more objective basis for the interpretation of relations between spectra and forest composition, a number of investigators have compared the pollen content of the surface layers of bogs with the tree composition of adjoining forests (1, 223, 343, 535, 536). Hesmer attempts to summarize their conclusions in the following list of genera arranged according to decreasing quantities of pollen production:

Pinus > Corylus > Alnus > Betula > Carpinus > Abies >
Picea > Fagus > Quercus > Tilia

It is usually agreed that those toward the head of the list, especially *Pinus,* are usually overrepresented, and conversely for those at the end of the list. All entomophilous pollens (*Tilia* is the only one on the list) are probably usually underrepresented. But these data are not quantitative and they leave much to be desired. A more direct approach to the problem is discussed under 7c.

b. *The extent of wind dissemination.* — All sufferers from hay fever know that pollen is transported through the air, in some cases in great quantities and over considerable distances. Local eradication of goldenrod (*Solidago*) is an effective control for hay fever from that pollen, but local eradication of ragweed (*Ambrosia*) only reduces the quantity of pollen, for it is apparently blown hundreds of miles. The extent to which pollen can be blown is very important in the science of pollen analysis, but in spite of this little is known that can be directly applied to the interpretation of fossil spectra.

Sears (561) calls attention to pine pollen "sulphur showers," often at considerable distances from pine forests. Erdtman (221) says that conifer pollen can be carried such distances by wind that it might be encountered in recent peats of Greenland. Godwin (279) cites instances of pollen grains collected at sea at distances of 30 to 300 kilometers from land. Malmström (437) presents evidence for the wind transport of spruce pollen from southern Sweden to Degerö Stormyr, a distance of 700 to 1000 kilometers.

Such observations have been supplemented by more direct approaches by a few authors. Firbas (248) has experimental evidence that the winged conifer pollens are not carried as far as certain wingless forms of slighter mass. Dyakowska (204), using a settling flue 12 × 200 cm., studied the rate of fall of various species of pollen under carefully controlled conditions. Her very interesting results give the average rapidity of settling in cm. per sec., the time of maximum pollen rain after the 2-m. fall, and the time spread of the fall. Using the rate of fall and the formula of Schmidt, she calculated the average limit of dissemination in kilometers. All results are compared with average pollen grain size. The list of species which follows is based on decreasing rate of fall and, inversely, increasing theoretical average limit of dissemination: *Abies pectinata, Larix polonica, Picea excelsa, Carpinus betulus, Fagus silvatica, Fraxinus, Pinus cembra, Quercus robur, Pinus silvestris, Populus, Ulmus glabra, Tilia cordata, Pinus montana, Betula (verrucosa* and *pubescens* mixed), *Corylus avellana, Alnus glutinosa, Taxus baccata, Salix caprea.* The calculated average limits of dispersal for certain ones in kilometers are: *Abies*, 0.7; *Larix*, 6.7; *Picea*, 21.6; *Fagus*, 27.7; *Quercus*, 64.9; *Ulmus*, 97.1; *Alnus*, 132.3; *Salix*, 218.1. Numerical irregularities occur which show that other factors than size are important in dissemination. Three species of pine were studied, with the following results: *Pinus cembra*, size 76.2 microns, fall in cm. per sec. 4.46, spread in kilometers 51.2; *Pinus silvestris*, size 59.9 microns, fall in cm. per sec. 3.69, spread in kilometers 74.7; *Pinus montana,* size 66.6 microns, fall in cm. per sec. 3.21, spread in kilometers 98.8. Such studies are useful and more of them should be made, but it is still difficult to interpret the results and to correct spectral percentages for fossil pollens on the basis of such information.

An entirely different approach to the problem is to trap samples of the modern pollen rain and compare the presence of pollen of different species with their nearest source of origin. Lüdi (428) has used this technique at Davos and found a distortion of pollen frequencies from the expectancy based on the forest composition of the region. He concluded that the valley winds caused an overrepresentation of species derived from the direction of the prevailing winds, and that those growing only or mainly on the down-wind side of Davos were strongly underrepresented, as would be expected. In order to study altitudinal wind dissemination, he collected pollen at Davos (1560 m. elev.) and at a station in the precipitous Weissfluhjoch (2670 m. elev.) immediately above Davos. He found a significantly high transport of pollen from Davos to the station over a thousand meters higher in altitude.

One of my students (122) has employed a technique that promises to facilitate the study of problems of overrepresentation and underrepresentation. Polsters and mats of mosses and liverworts of several species were found to contain abundant recoverable pollen that represents one or a few seasons' accumulation of pollen rain. Such accumulations compare favorably in several respects with pollen rain as it falls on a bog surface, and they can be used for examination of over- and underrepresentation with respect to the immediately adjacent forests, and the distance of transport. Comparison of contemporaneous pollen rain with the present forest composition will be discussed in the next section. An examination of the polsters showed that considerable percentages of pollen were composed of species foreign to the spruce-fir forest (where the polsters were collected), and that they must have been carried in from lower altitudes by the wind. Twenty-three per cent of all the pollen recovered from the different polsters was of species foreign to the spruce-fir forest and included *Tsuga canadensis, Liquidambar styraciflua,* and unidentified species of *Tilia, Pinus, Quercus,* and *Carya.* The station where the polsters were obtained was at 5750 feet elevation and, as far as our exploration revealed, the nearest *Tsuga* was 850 feet lower in altitude, the nearest *Tilia* 1450 feet, *Pinus* 1750 feet, *Quercus* 1950 feet, *Carya* 2250 feet, and *Liquidambar* about 3000 feet lower. *Pinus* and *Tsuga* composed the largest contamination, with 9.4 and 7.2 per cent, respectively. Since these studies were made in the Great Smoky Mountains where altitudes change considerably within short horizontal distances, such data are interesting but not as useful as if they had been obtained in flat country such as glacial till plains where most bogs are located. The polster technique can be used even more directly in solving the problem of distance of dissemination and the relation between distance and quantities, by collecting polsters at successively greater distances from a natural boundary for a species. Work is under way on *Abies Fraseri* and *Picea rubens,* which reach their natural boundaries in the Great Smoky Mountains on the slopes of Clingmans Dome, but conclusions have not been reached.

Godwin (280) believes that pollen composition is a good guide to forest composition, and Hesmer (343), after classifying pollen according to the distance of its transport to the place of sedimentation, concluded that trees from the bog to a distance of 500 meters are by far the most strongly represented in the pollen rain and consequently pollen spectra give a good record of local vegetation.

From these data we can draw certain conclusions. Pollen spectra predominantly reflect local forest conditions, but there will be quantities of pollen from greater distances, normally diminishing as the distance increases. Pollen will not arrive equally from all directions, but spectra will be distorted because of prevailing winds. Small percentages of pollens do not necessarily indicate that the species concerned grew in the vicinity of the sedimentary basin, so no importance should be attached to them except when they represent a regional phenomenon. Of course, when the pollen of a certain species shows a consistent percentage trend, such as a rise or a decline, small percentages constitute an integral part of the trend and are significant.

c. *Relations between pollen spectra and forest composition.*—Bertsch (64) compared wood fragments and pollen percentages from Bronze Age hut sites at the time of the beech maximum. Wood of *Populus, Carpinus, Salix, Fraxinus, Malus,* and *Taxus* together composed over 40 per cent of the hundreds of wood samples examined from the sites, but the pollen percentages gave no clue as to their presence in the region. Wood fragment percentages for *Corylus, Alnus,* and *Betula* exceeded the pollen percentages for the same genera. *Fagus* and *Quercus* pollen percentages, however, greatly exceeded those of wood fragments. *Ulmus, Tilia, Pinus, Picea,* and *Abies* pollen were present in the spectra but absent from among the wood fragments. Man likely exerted a selecting influence on the accumulation of wood fragments, but the discrepancies are too great to attribute solely to his selection of wood for use; they must mean something with respect to over- and underrepresentation of pollens.

Lüdi (428) compared the pollen rain at an open station at Davos with the forest composition of the region and found considerable disharmony. *Picea* pollen was about one-sixth of the expected amount, *Pinus* was about seven times as abundant as would be expected from the forest composition, and *Alnus* about four times. These discrepancies were explained by Lüdi as due to the valley winds because when he examined the pollen content of moss polsters within the different forest types he found a far-reaching correspondence with the present forest composition. Rudolph and Firbas (536) collected superficial peat samples in the Riesengebirge between 740 and 950 meters elevation and compared the pollen content with the contemporary forest. They found that the maxima for *Fagus, Picea,* and *Pinus* reflected the actual vertical sequence of forest belts. Nevertheless, the discrepancies of detail made it clear that only a general correspondence existed within such narrow areas and distances.

The work of Carroll (122), referred to earlier, is the most quantitative attempt yet made to correlate forest composition and pollen spectra. Her collections of moss and liverwort polsters were made on Mt. Collins in the heart of the spruce-fir forest of the area. The forest at this station is of simple composition, with *Picea rubens, Abies Fraseri,* and *Betula alleghenien-sis* being sampled on a plot 4500 square meters in area. Table 12 shows the

TABLE 12.—Comparison of the tree composition of the spruce-fir forest on Mt. Collins with the pollen representation in moss and liverwort polsters from the same area (Data from 122)

	Basal Area Percentage	Average Pollen Percentage, Spruce-Fir Species Counted 100% of Pollen	Average Pollen Percentage, All Tree Species Counted 100% of Pollen
Picea rubens, Red spruce	50.3	35.7	27.6
Abies Fraseri, Fraser's fir	34.3	38.6	29.6
Betula alleghaniensis, Yellow birch	15.2	25.6	19.6

percentage composition of these species, together with their pollen percentages in the spectra. These three species composed 77 per cent of all the pollen recovered (the rest being contamination from lower elevations) and were represented porportionally in the various polsters and mats of different species. It is seen that *Picea* is underrepresented about 15 per cent, and *Abies* and *Betula* are overrepresented about five and 10 per cent, respectively, when pollen percentages are compared with basal-area composition of the immediately adjacent forest.

8. **Climatic interpretation of pollen records.**—If, as Godwin (280) says, "Pollen composition is a good guide to forest composition, and there is in fact reason to think the two are closely proportional to one another," and if forest composition is a reflection of climatic conditions, it should be possible to read both vegetational and climatic history simultaneously from the pollen record. There are, however, factors which tend to cause this record to be disproportionate with respect to forest composition—with under- or overrepresentation of certain species—as we have already observed when single pollen profiles or spectra and local forest composition are considered. And, from a consideration of contemporaneous vegetational patterns, we find that a greater or lesser proportion of the communities of an area are not climax and under a predominantly climatic control, but successional or relic (preclimax and postclimax) and under a more immediate edaphic or microclimatic control. For these reasons, then, it is never possible to reason directly from the pollen record of a single profile to the climatic history of its area. As Cranwell and von Post (169) say, "Theoretically, biotic and edaphic factors could cause a succession of different plant communities, power of superseding, etc.—could explain satisfactorily the fact that one element appeared later than another. The pollen diagrams themselves, however, enable us to decide whether such circumstances ought to be considered or not."

Although most American workers have interpreted changes in the profiles they have studied as being due to climatic changes, others have questioned the matter. Wilson (700, 701) believed that the probable belts of forest types in eastern Wisconsin during the interval between the third and fourth substages of the Wisconsin glaciation are not necessarily due to climatic belts but "may also be explained as an illustration of plant succession governed by physiography and speed of migration similar to present conditions in Alaska" (304, 78). Wilson and Galloway (702) made a similar interpretation of a northern Wisconsin bog, and Hansen (324) working in the Puget Sound region, found it difficult to state whether the postglacial succession has been a result of climatic change or merely normal plant succession.

Sears (565) has reviewed the situation and concluded that the broad outlines of postglacial change are due to climatic change, yet many changes in the proportions of existing vegetation may be due to other causes; and he recognizes the cogency of the arguments that have been advanced against postulating all changes in pollen composition as due to climatic change.

In addition to climatic change, changes in the composition of vegetation may be due to (1) topographic changes such as erosion and deposition, emergence, and subsidence; (2) soil development, which is usually a centuries-long process before a mature condition is attained; (3) ordinary biological succession, only gradually reaching the self-perpetuating climax that is supposed to be under intimate climatic regulation; (4) the intrusion of a powerful biotic factor which may cause a reversion of the developmental trends controlled by Points 1 to 3; and (5) catastrophic factors such as fire, vulcanism, or hurricane, which may cause reversion.

Another group of factors which tend to cause difficulty in the climatic interpretation of pollen profiles has already been mentioned in other connections. I refer to the fact that the climatic relations and autecology of living species are only inadequately known, although in a broad way we can assume for many species approximate limits with respect to temperature and moisture, and to the fact that most pollen identifications are only to the genus, and the genus, because of differences among its species, is often only a very broad indicator of climatic conditions.

In spite of all these qualifying factors, frequently inadequately recognized by pollen analysts, there is widespread confidence in the general story of postglacial vegetational and climatic history as worked out from pollen studies. The reasons for this confidence are two. As expressed by Cranwell and von Post (169), "Experience in Europe has established two general laws which express decisively the indisputable and indissoluble association between succession of vegetation and climatic evolution. These laws are (1) that of regional parallelism and (2) that of revertence [506, 507]. Both hold for New Zealand conditions." Godwin (280) found that both phenomena hold for British conditions, and Sears (565) and Preston Smith (600) found that they hold for American conditions. Sears (565) says, "It is the opinion of the writer that certain remarkably consistent, fairly synchronous, and long-time trends appear in the record throughout eastern North America which are difficult to explain on the basis of purely local changes." According to Smith (600), "The above major climatic periods are indicated in the profiles from the entire region. The profiles from the region are also consistent in showing minor variations of climate within these major periods." On the basis of several profiles from an area and studies from several areas of a region, the picture of climatic change emerges, whereas from individual studies it may be impossible to present any climatic interpretation of the record that is beyond serious question. It is, then, the strict regularity with respect to broad outlines (regional parallelism and final period reversion) which distinguishes postglacial evolution as a whole that justifies the application of the climatic-historical theory.

It should be pointed out in conclusion that the principal knowledge as to climatic and vegetational postglacial history derived from pollen studies has merely substantiated conclusions already arrived at by the floristic plant geographers from a study of modern areas and processes. Because of the contemporaneous pattern of major vegetational types and because of the widespread occurrence of relic colonies of both preclimax and postclimax nature, floristic plant geographers (Gleason [276], for example) had already hypothesized the von Post three-phase postglacial history: the early period of warming, the middle period of climatic optimum (including our major xerothermic period), and the final period of reversion.

Evidence for a recent cooling of climate comes not only from pollen analysis but also from a study of timber line. The evidence, however, is not everywhere consistent as to the direction of change. The Arctic timber line in Alaska indicates that the climate is improving (304) and that trees are advancing. In the northern Rocky Mountains conditions seem to have remained static for centuries (305). In northeastern North America, however, there is considerable evidence for climatic reversion and retreating alpine and continental timber lines (237, 306, 445, 517, 518). These conditions would seem to oppose the theory of glacial synchroneity over the northern hemisphere and favor a theory of alternative continental glaciations such as the passage of the center of Wisconsin glaciation from Labrador across Hudson Bay to Keewatin during the epoch, but neither theory is proved (100), and Griggs (306) prefers the one of climatic migration. Shifts of climate such as apparently are now taking place according to evidence from pollen analysis and a study of timber lines do not invalidate the principle of regional parallelism in pollen analysis but do make more difficult the problems of synchronism.

PART THREE

Areography

I I.

Some Terms
and Concepts of Area[1]

Area.—In plant geography, the term area is applied to the *entire* region of distribution or occurrence of any taxonomic unit, or to the region of occurrence of any plant community, whatever its rank. The majority of plant geographical considerations are concerned with *natural* areas which have been attained through natural dispersal mechanisms and the natural agencies of dispersal, except man. Man's activities, either purposeful or inadvertent, have greatly altered the area occupied by certain organisms and communities by extension of the natural area through the production of what may be called the *artificial area*. Natural areas are also reduced by man as a result of his destruction of primeval conditions through agricultural and other practices. Some authors have emphasized what they call the *compact area* from which extend the *Ausstrahlungen* (206, 609, 674). These writers overemphasize the compact area and largely neglect the radiations. Such a practice is almost certain to result in errors for, as Hultén (360) says, "It is just the stations found outside the compact area that are likely to be the most valuable ones, which can give a clue as to how the development has taken place. They are so to speak 'the living fossils' of the species in question."

Topography of area.—According to Wulff (719), the term topography of area is used in connection with the distribution of the organisms within the limits of their area. In other words, topography refers to the local distribution, say of a species, within the area as a whole. *Cephalanthus occidentalis,*

---

[1] The Germans have a convenient term for the science of area, *Arealkunde.* The term *chorology* is already in international usage, but it is less definite and more inclusive than *Arealkunde.* Two possible translations of *Arealkunde* are areology (which has an entirely different usage in astronomy) and spatiology; but perhaps the best term for the science of area is *areography,* that portion of geography which deals especially with area.

for example, has a very wide area in the United States, but within its area its local distribution is confined strictly to ponds, swamps, stream-sides, etc. The same applies, of course, to many other moisture-requiring species such as *Betula nigra, Salix nigra, Taxodium distichum,* and *Quercus bicolor*. To cite another example, an alpine species such as *Rydbergia grandiflora* has a fairly extensive range in the Rocky Mountains, but within this range it has a spotty distribution, occurring on peaks from Montana to Colorado which reach above the timber line.

Shape of area.—The shapes of areas are variable, being influenced by many factors; yet two general statements can be made: (1) an area tends to develop a circular outline as a result of random dissemination of diaspores; (2) this tendency is counteracted by the fact that the principal climatic zones tend to have a greater latitudinal than longitudinal dimension, with the result that areas seldom are circular anad frequently are roughly oval in an east-west direction. It should be understood that these ideal shapes, circular and oval, represent tendencies which seem seldom to be realized. The conditions of topography that modify the shapes of climatic regions (such as altitude, proximity to large bodies of water, etc.) also modify the shapes of major areas, because climatic boundaries constitute the principal boundaries of floristic and vegetational areas. More local features of topography and of microclimate affect the shapes of smaller areas and determine to a large extent the topography of areas. Also, soil conditions provide numerous and varied barriers to dissemination and establishment (the latter being by far the more important) and result in the production of the details of the shape and topography of an area.

Size of area.—The sizes of areas range from the minute, in cases where only a single station is known for a certain species, to the so-called cosmopolitan. There is, however, no such thing as a truly cosmopolitan area, although certain species are known from every major continental mass. In general, it can be said that the wide-ranging species have greater areas than the wide-ranging communities. To express the same phenomenon from the plant sociological point of view, we can say that in all or nearly all cases the species of any plant community range more widely than the community type in which they occur. Small areas are usually referred to as endemic areas, and age is very important in connection with size. Willis' rule of age and area apparently works often, although it frequently is inapplicable. Many north temperate species of small area are relics, surviving in a portion of a once more extensive area, and the same thing is likely true of other regions.

Margin of area.—For a species which has not, in its migration, reached a barrier, the margin of the area will progress in accordance with the migrational capacity of the species and the length of time required to reach reproductive maturity, modified by the vicissitudes of chance encountered in weather and other variables. Sooner or later, however, a migrating species reaches one or more barriers of a temporary or relatively permanent nature. Such barriers can be classified as (1) physical or mechanical (such as a body of water or a mountain range), (2) climatic (as found in a temperature gradient, a moisture gradient, a change in length of day), (3) edaphic (a shift in substratum such as soil structure, soil chemistry, soil water, etc.), and (4) biological, i.e., competition (as found in stabilized closed communities, especially climaxes), parasitism, etc. The first class of barriers is more of a convenience than a distinct type, for mechanical barriers can equally well be classified under climatic and edaphic. Ultimately, it can be said that barriers determine the location and shape of boundaries.

An expanding area tends to have a relatively continuous boundary and a homogeneous topography, whereas a contracting area tends to have a relatively discontinuous boundary and an irregular and broken topography. The degree to which these contrasts are true depends upon a variety of local features of habitat. The principal reasons for the above relationships are the fact that the expanding area has not reached effective barriers and that the contracting area leaves behind relic colonies in local situations where frequently edaphic or microclimatic conditions provide at least temporary compensation for a general climatic unfavorableness. For example, in the central states from Iowa eastward, but especially in Indiana, Ohio, and Pennsylvania, conifers of the Lake forest and Canadian forest (such as *Pinus Strobus, Larix laricina, Picea mariana*) have survived in bogs, canyons, and north-facing bluffs in widely isolated colonies far south of the southern limits of the forest types and completely surrounded by the prevailing deciduous forests. Advance colonies of the deciduous forest do not show any such widely disjunct penetration of the more northern coniferous types. In New England, however, there are stands of the more southern deciduous forest rather widely disjunct northward of the main northern boundary of this forest. It appears likely that these are relic stands from a warmer postglacial time, and that the northern and southern relics are evidence of an oscillatory movement of climates and associations. Such are the broad relations which the above attempts to describe in a general way.

Relic areas.—Relic areas are the areas of epibiotic floras and are usually isolated, contracted, and discontinuous. *Taxonomic relics* may have either narrow areas (*Diphylleia cymosa*) or wide (*Loiseleuria procumbens*), but they are forms which are entirely or nearly without close relatives. They have presumably long survived the extinction of most of their fellows. *Geographical relics,* whether species or communities, are always associated with the idea of *refugia.* A refugium consists of some locality which, for one reason or another, has not been as drastically altered climatically or otherwise as the region as a whole. Widely known refugia are those driftless areas and nunataks within the general zone of Pleistocene glaciation, and the alpine and montane belts more or less far south of the glacial boundary where arctic and boreal plants have survived at least the latest cold period.

Continuous area.—A continuous area is one in which the various stations for a species or a community are not more widely separated than the normal dispersal capacity of the organisms concerned.

Discontinuous or disjunct areas.—Disjunct areas are those in which colonies of plants are more widely separated than the normal dispersal capacity of the propagules of the species. A question of degree is involved in the distinction between continuous and discontinuous areas which may make a decision difficult except in those cases where the disjunction is very wide. Cases of disjunction (together with those of endemism) present some of the most difficult problems in historical biogeography, the solutions of which form much of the core of the science and consist of the main difference between dynamic and static plant geography.

Center of area or center of origin.—We now come to a series of considerations which are also at the heart of dynamic plant geography and evolutionary history. For a taxonomic group, the center of its area (except in a simple geometric sense) refers to the center of its origin, that is, to the territory where dispersal and migration commenced. The determination of the center of origin is not a simple matter in many cases, and a special chapter is devoted to the question.

Center of frequency.—Within the area of a taxonomic group or a community type, the frequency distribution is usually variable. The concept of center of frequency refers to the contained area (or areas) where the individuals of the type (organisms, colonies, or association-individuals or stands) are most abundant. As we shall see later, this center may or may not be also the center of origin.

Center of variation or development.—Within an area as a whole there may be a lesser area which can be designated as the center of variation or development; this is determined as the region where the population is taxonomically most variable. Such a center is not detectable in some species which are inordinately stable. Any wide-ranging, long-known species from which even the most avid taxonomic "splitter" has been unable to establish other species will serve to illustrate the type. We may suggest here *Rhododendron maximum, Kalmia latifolia, Epigaea repens,* and *Monotropa uniflora* of the Ericales. Most wide-ranging species, however, are more or less polymorphous and can be considered to consist of a number of subspecies and varieties which, in some cases, may be elevated to the rank of species. The validity of such specific segregates is not a question to be taken up immediately. A subarea in which such variation and development are pronounced is called the center of variation. It may or may not coincide with the center of origin for the species as a whole. For example, the region of variation-mass may only be the region of a variety of habitats. The concept is perhaps more easily applicable to genera and higher categories than to species.

Unicentric, bicentric, and polycentric areas.—These concepts relate to centers of variation or development, usually of a higher category such as a genus or family, but they are also applicable to origins under the differentiation theory. The use of such terms is sufficiently obvious, as in the disjunct genus *Magnolia,* with its two centers in eastern United States and Asia. The terms likewise apply to discontinuous survival areas.

Center of dispersal.—For young species the center of dispersal is the center of origin. For older species which have suffered from the vicissitudes of changing climates and migration, there may be one or more centers of dispersal which are more or less far removed from the center of origin of the species, genus, etc. Such secondary centers may be *centers of preservation* (refugia) from which the population may expand and acquire new territory following the relief of unfavorable conditions, or they may be secondary evolutionary centers which result from polyploidy, hybridization, and absence of severe competition, or the encountering of a new and variable set of environmental situations which provide a series of "unsaturated" ecological niches.

Equiformal progressive areas.—Hultén (360) says that the chief feature of comparatively recent areas is their concentricity around the place from which they radiated. The plants that have equiformal areas of different size have radiated from the same center, and their center of dispersal can be found if enough such species are compared. Hultén found that the theory of equiformal progressive areas works out practically and that centers of dispersal (for arctic and boreal plants, at least) can be located convincingly.

Vicarious areas.—Wulff (719) defines vicarious areas as mutually exclusive areas belonging to closely related species, differing only in a few specific characters and linked by their ancestral initial form (species). He adds that vicariism is usually spatial and that it can also be altitudinal within a single region. Wulff's definition is much stricter than that employed by some geographers. The term is commonly applied to pairs of species (the two most closely related species of their circle of affinity) which are allopatric (and frequently rather widely disjunct) and often not connected by one or more linking forms. In the common usage of the term, the linking species may have become extinct by being killed out, by having evolved into the vicarious forms, or by never having existed as a distinct form, one of the vicarious pair having given rise to the other.

Monotopic, polytopic, and pantopic origins of areas.—In the early days of plant geography, Grisebach held the hypothesis that under similar physical conditions in different places on the earth similar plants can arise, but von Ihering (374), for example, refutes the belief. According to him, "There are, at remote places, no diphyletic, independently arising species of plants and animals." Most students of evolution believe that a species can only have a single center of origin (the area thus being originally monotopic), but those who accept the differentiation theory believe that a taxonomic entity can be differentiated independently in two or more regions (bitopic and polytopic centers of origin). It has also been suggested that one species may possibly undergo a mass evolution into another species as a result of the subjection of all the individuals of one species to identical selection or their suffering identical mutations. Such a theory appears to have absolutely no genetic status and is upheld only by neo-Lamarckians who believe that the environment calls forth the specific differences. For example, assume that in three altitudinal belts of a mountain—alpine, subalpine, and montane—there are three closely related or vicarious species. Whether the original population of the group was of one belt or another does not matter; as the population spread from one belt into the others the action of the new environment caused new adaptations which characterize the new species. Under this theory the appearance of a new type is not a matter of selection of pre-adapted genotypes, because it is claimed that reciprocal transplants result in the conversion of one species into another through modifications alone. If the so-called species which are converted by transplant experiments have no genetic basis but are respectively merely modifications or ecads, it is true that the transplanting of clonal material from one zone to another will result in a change of form coincident with the new conditions of life. If, however, the respective forms have a genetic basis and really are species, there seems to be no conclusive proof that the conditions of a new environment can produce a conversion from one species to another in clonal material. In brief, the evidence for a pantopic origin of a species is inconclusive, to say the least.[2]

[2] For a comparison of these points of view, see the claims of conversions by Clements in recent *Yearbooks* of the Carnegie Institute of Washington, and the recent publication on experimental evolution by Clausen, Keck, and Hiesey (146), also published by the Carnegie Institute of Washington.

12.

Dispersal and Migration

1. *All organisms are disseminated to some degree during each life cycle, but dissemination is essentially a chance phenomenon in the sense that there is no cooperation between diaspores and agents to assure the movement of a diaspore to a place that is suitable for its germination and establishment.*[1]

2. *Disseminative capacity in certain cases includes active and passive structures which facilitate transport by certain agencies, but areas are ultimately determined by barriers rather than by dissemination. Barriers consist not only of obstructions to dissemination, but especially of conditions unfavorable for the germination and establishment of the diaspores and a successful life for the mature organism. That is to say, dissemination is only a necessary antecedent, and migration is accomplished only upon establishment at a new station.*

3. *That long-distance dispersal has resulted in migration and accounted for discontinuous areas seems rarely to have been the case. This conclusion is based primarily upon existing distributions which largely show symmetrical replicate patterns mostly unrelated to chance and other elements of dissemination, and upon evolutionary phenomena, such as endemism in general and the occurrence of local races in particular.*

4. *In general, the effects of recent dispersal provide little in the way of clues to areal patterns except those of the continuous type, as illustrated by progressive equiformal areas.*

5. *The flora of any region usually contains historical nonendemic elements which frequently can be identified as to source, migratory tract, and time of invasion. A study of areal patterns is one of the most important*

---

[1] This statement needs elaboration. Man as an agent (and several other animals such as ants, birds, squirrels) may transport diaspores to places suitable for their germination and development—and such activities may be purposeful or instinctive on the part of the animal—but all teleological implications for the plant must be strictly avoided.

tools of floristic geography, and the commonplace and continuous types of distribution are equally as important as the unusual and discontinuous in the solution of problems in historical geography.

The science of area has a static phase which consists of the accumulation of records of distribution and their comparative and statistical study, and a dynamic phase which consists of inquiry into the present and past processes that affect areas. As van Steenis (608) has pointed out, the plant geographer meets with three inevitable handicaps: (1) the usual absence of any exact data regarding the age of species; (2) the usual absence of any proof regarding dispersal, that is, whether long-distance dispersal is possible and takes place in reality; (3) an uncertainty whether an area is expanding or retracting. Incidentally, any critical coordination between the geographical data and paleoecology, geology, or paleoclimatology must be regarded as a piece of luck. It should be emphasized that the first task of floristic geography—and one that has always been its forte—is the accumulation of distributional data and their organization on maps. This static aspect of plant or animal geography leads naturally to interpretations of vegetational dynamics, but it is important that hypotheses grow from observations rather than that hypotheses form the framework of the science. Perhaps the most that plant geography can accomplish is the accumulation of coincidences between the occurrence of areas and the possible causes of areas.

It is first necessary to distinguish between dissemination and migration. Migration has taken place when an area has become expanded; that is to say, migration has occurred only when plants of the type under consideration have become established in a new locality. This is understood to consist of dissemination followed by germination and a successful life on the part of the plant in the new territory.

In order to understand the distribution of plants it is of basic importance to know how they are dispersed. An inquiry into dispersal can take two forms: (1) a morphological study of diaspores,[2] the units of dissemination, and the activity of their agents, and (2) the observation of dispersal itself. It is obvious that an inquiry into diaspores and agents of dissemination (wind, water, etc.) provides no proof of dissemination, but can only indicate the possibility or, in certain cases, the probability of dispersal in a

---

[2] We find a variety of terms for what is dispersed: disseminule (149), *Verbreitungseinheit* (663), diaspore (569, 608, 469), propagule, etc. The most widely accepted term is *diaspore*, which is defined as any one of the complexes (spore, seed, flower, inflorescence, etc.) separable from the mother plant and assuring the dissemination of the structures (spores or seeds) responsible for the next generation.

certain fashion over certain distances. On the other hand, direct observations of dispersal are almost non-existent. The difficulties here are obvious and almost insurmountable. For example, the observations that a certain bird eats a certain fruit and that the seed is still capable of germination after having passed through the bird's alimentary tract are preliminary and circumstantial contributions to the problem of dispersal. One must still ask: How long is the diaspore retained? What are the movements of the bird during the period of ingestion? Do these birds migrate? Do they migrate on a full or an empty stomach? What are the chances that the viable seed will be eliminated in an ecological situation favorable for the ecesis of the diaspore? It is clear that those geographers who believe in long-distance dispersal of diaspores endozooically (as between South America and Africa) must assume the burden of proof in the face of great improbabilities. In fact, many modern plant geographers (see 719) emphatically deny the probability of long-distance dispersal leading to migration; they simply dismiss it from their problems. In the subsequent consideration of diaspores and agents of dissemination, it should always be borne in mind that the apparent suitability of a certain structure for dispersal by a certain agent provides no proof of the distance of effectiveness; the data concern only possibilities.

The following notes are taken from a recent study of diaspores and dissemination by Molinier and Muller (469). Following Sernander (569) in part, they distinguish six types of diaspores: (1) The diaspore is composed only of the embryo (*Rhizophora conjugata*) or is a bulbet of adventitious nature as in certain cases of vivipary (*Poa bulbosa, Allium* spp.), and on the fronds of ferns (*Cystopteris bulbifera*), leaves of Crassulaceae, etc. (2) The diaspore is composed of isolated seeds (*Silene*) or of several seeds stuck together (*Helleborus foetidus*). (3) The diaspore is formed of the whole fruit (*Quercus*), of a part of the fruit (Umbelliferae), or by a group of fruits. (4) The diaspore carries, in addition to the fruit, certain floral organs or united bracts, where not in an involucre (*Trifolium, Corylus,* many Gramineae). (5) The diaspore is formed of all or a large part of the inflorescence (*Aegilops, Urtica pilulifera*). (6) The diaspore consists of the entire mother plant (*Plantago cretica*) or most of the aerial part with fruits and inflorescences (*Eryngium campestre* and many "tumbleweeds"). Many plants have more than one of these types of diaspores.

Under the heading of agents of dissemination we can consider two types of phenomena: certain acts of the mother plant, such as mechanical expulsion, and the action of dynamic external agents, such as wind, water, animals, man, and gravity, in carrying diaspores.

Molinier and Muller consider as "adaptations" all structures and qualities which favor the elevation and transport of diaspores.[3] The following are the main points of their classification: (1) Morphological and anatomical adaptations include such features as the plumed pappus of many Compositae, the samaras of *Fraxinus* and *Acer,* the mechanical tensions developed in sporangia of ferns, fruits of many Cruciferae, Geraniaceae, etc. (2) Color and odor adaptations, as in pollination, serve by attracting the attention of birds and other animal agents of dissemination. Color adaptations appear in such fruits as those of *Taxus baccata, Ilex aquifolium, I. opaca, Cornus, Lonicera;* odor adaptations, in *Fragaria, Rubus,* etc. (3) The action of agents of dissemination is a function not only of the form and size of diaspores but also of their position on the mother plant, which can be more or less favorable for certain agents. Here Molinier and Muller list easily detachable diaspores (as in *Taraxacum*), diaspores widely distributed on the mother plant (*Sorbus* type), and diaspores released from fruits, etc., by the agents (*Silene* type, etc.). (4) Cases of a temporal concordance between the maturity of the diaspore and the period favorable for the activity of the agent of dissemination exist. Here are classified the usual types and also those diaspores freed during winter and disseminated by wind over snow, ice, etc. Seed cases are essentially of two types in respect to hygroscopic activity: hygrochases, which open in humid air and close in dry air (*Prunella, Astragalus*), and xerochases, which open in dry air and close in humid (*Silene, Centaurea*).

Molinier and Muller also give a classification of species after their mode of dissemination, in each class of which the efficiency of dissemination is supposed to depend not only upon the activity of the agent but also upon the adaptive value of the diaspore.

I. *Anemochores* are wind disseminated and are the most numerous type. Several subtypes exist, with all intermediate degrees.

A. Types which soar easily: (1) Because of the tenuity and slight mass of the diaspores, as in many orchids. Plants of *Orchis* may produce as many as 200,000 seeds; *Maxillaria* is said to produce 1,500,000 seeds. (2) Because of low density of seeds or air pouches in the tissues, as in Orchidaceae, Orobanchaceae, Pirolaceae, Ericaceae. One seed of *Cephalanthera pallens* averages 0.002 mg., of *Orobanche ionantha* 0.01 mg., of *Pirola uniflora* 0.002 mg., of *Rhododendron ferrugineum* 0.025 mg. (3) Because of the

[3] In the following notes or in any other discussions of "adaptations" the reader should be warned of teleological interpretations. *Apparent* suitability for a certain function is never *proof* of that function. Also, there is no proof that structures have *arisen* because they serve a purpose; there is proof only that they may survive or be selected because they are useful.

presence of plumes, generally formed by the calyx or pappus, in which there is an excessive surface in relation to weight (*Taraxacum officinale, Lactuca, Asclepias, Centaurea, Tragopogon,* etc.).

B. The heavy soaring type, with diaspores which are carried shorter distances and in which the ratio of weight to surface is greater (*Clematis, Fraxinus, Acer*). Seeds of maple are known to be carried 6 miles.

C. The rolling type, with diaspores too heavy to be lifted into the air but which may be rolled considerable distances in steppe country, along shores, or over snow and ice by the action of wind (*Phlomis, Eryngium*).

D. The catapult type, with projecting floral stems from which the wind may throw the diaspores (as far as 15 meters in *Papaver somniferum*).

II. *Hydrochores* are those species in which water is the primary agent of dissemination or in which water or humidity effects discharge of the diaspores.

A. Types with floating diaspores. The seeds or other structures float because of low density, air sacs within the tissues, the adherence of air to the rough surface, or a waxy covering which reduces wetting (*Carex, Nymphaea, Lythrum*).

B. Rain-disseminated types, in which the action of drops effects the release of diaspores through shock or other action (*Marchantia, Sedum, Salvia, Scutellaria,* etc.).

III. *Zoochores* include those species that are normally disseminated by animals.

A. Epizoochores are carried on the body of the animal; they include several types, such as thorny diaspores (*Tribulus*), hispid diaspores (*Tordylium*), hooked diaspores (*Circaea, Medicago, Cynoglossum*), glandular viscid diaspores (*Saxifraga, Linum*).

B. Endozoochores are taken accidentally and are not damaged by digestion (germination may even be facilitated in *Fragaria, Oxycoccus, Juniperus*). The efficacy of this method of dissemination is a result of the frequency with which seeds are taken, the speed of digestion and displacement, and the spatial activity of the animals. Materials may remain in the digestive tracts of mammals for 24 hours and in some birds a very short time (15 to 30 minutes for blackbirds).

C. Zoochores with oil are mainly sought by ants. The ant *Formica rufa,* for example, has been observed to transport diaspores as much as 65 meters; Sernander (1906) estimated that over 36,000 seeds were carried by this species to one colony. Seeds of *Anemone, Helleborus,* etc., are involved.

IV. *Anthropochores* include all those which are regularly disseminated by man, either inadvertently or on purpose; they are usually considered weeds in the former case. Man's activities supplement several of the preceding agents.

V. *Autochores* include all those forms in which the mother plant by its behavior or structure is the main agent of dissemination. Here Molinier and Muller list transporters (*Linaria, Veronica*), physiological projectors (*Ecballium, Cardamine, Impatiens, Oxalis*), mechanical projectors (*Vicia, Lathyrus*), and creeping diaspores (the hygroscopic organs of *Hordeum, Aegilops, Avena,* etc.). The authors state that mechanical projection produces the following results: in *Geranium rotundifolium* (seed weight 0.0002 gram), a distance of 1.8 m.; *Viola canina* (weight not given), 4.6 m.; *Lupinus digitatus* (0.08 g.), 7.0 m.; *Hura crepitans* (0.7 g.), 14.0 m.; *Bauhinia purpurea* (2.5 g.), 15.0 m.

VI. *Barachores* are those species whose diaspores have no special structures and which are largely disseminated by their own weight (*Rhizophora conjugata, Hicoria, Juglans*). The gravity method, of course, is more or less applicable to all diaspores.

It is a self-evident fact that plants are disseminated. It is equally evident that dissemination is a chance phenomenon in the sense that there is no possibility of cooperation between diaspores and agents to assure the movement of a diaspore to a place which is suitable for its germination and establishment. Most diaspores obviously meet a fate that is opposed to their biological *raison d'être.* The presence or absence of plants of a particular species in a particular place is to a certain extent a matter of chance, the suitability of the habitat and the availability of the species being assumed. With all these considerations and handicaps, it is nonetheless frequently useful in plant geography to inquire into the nature of dispersal, if for no other reason than to eliminate the probability of long-distance dispersal in the case of disjunct populations.

The study by van Steenis of temperate mountain plants in the tropics of Malaysia (608) is a case in point. He came to the conclusion that the effect of the probable present dispersal can give no clue to an explanation of the present distribution of mountain plants in Malaysia. Some of the bases for his conclusion are of interest. If it were true, he says, that wind dispersal was of universal importance for dust seeds and the spores of cryptogams, it would be difficult to explain endemic genera and species except by ecology. And numerous cases of absence from suitable ecological situations well within the supposed range of dissemination of such small diaspores argue, generally, against long-distance dispersal by wind even of this most likely type. Related to this point is the fact that Bryophytes, Pteriodophytes, and Phanerogams with minute seeds (Orchidaceae) show the same migratory tracts as do plants with other types of diaspores, and show most of the same disjunctions. That is to say, the same phenomena of distribution are represented in groups with most divergent methods of dispersal.

The failure of correlation between dispersal mechanisms and distributions is illustrated by van Steenis in several ways (any botanist could easily provide parallels). A case or two will serve our purpose. The Valerianaceae is represented in Malaysia by two genera, each with one species: *Valeriana Hardwickii* and *Triplostegia repens*. The former belongs to a large, widely distributed genus and produces plumed fruits in abundance. Unexpectedly, this species is not found outside Java. *Triplostegia repens,* on the other hand, is a Papuan endemic entirely without special means of dispersal, and is 3000 km. distant from its congeners of China. Also, the absence of relatives makes the hypothesis of a polytopic origin of *Triplostegia repens* unreasonable. If these two plants had areas in Malaysia which accorded with their "apparent" dispersal capacities, the species of *Valeriana* would be widespread and *Triplostegia* would probably be confined to the eastern Malaysian border or the Philippines. The absence of a pronounced effect of wind dispersal is also seen in the flora of the Lesser Soenda Islands. On the basis of prevailing winds and their position, it would be expected that they would have a strong Australian temperate mountain element, but their strong affinities are in another direction and with the remote Himalaya.

An analysis of endozoic dispersal showed that it is significant within short distances, perhaps of the order of "tens of kilometers," whereas long-distance dispersal seems to be entirely excluded for temperate mountain plants in Malaysia. For one thing, the development of local subspecies for many birds indicates their limited movements. Too, migrating birds are frequently insectivorous, and what seed-eating birds there are generally destroy the diaspores. In other cases the seeds are rapidly passed by the birds, and flights do not extend over long distances in a straight line in a short time.

While exozoic dispersal meets with many of the same objections as endozoic dispersal, there always remains an occasional possibility of long-distance dispersal. Still too much must not be taken for granted; the case of two sedge genera proves instructive. *Uncinia,* with a typical distribution of "the antarctic type," is provided with a beautiful "crochet-needle-shaped awn (the rachilla) . . . admirably adapted for exozoic dispersal." Van Steenis asks, "How is it that *Uncinia,* so well provided with means of dispersal, shows the same area of distribution as other groups which have no [special] means of dispersal, such as *Ranunculus* (*lappaceus*-group), *Oreobolus, Oreomyrrhis, Plantago, Abrotanella, Libertia,* etc.?" On the other hand, *Carex,* with no particular means of dispersal, is distributed all over the world. The conclusion seems paradoxical; as van Steenis exclaims,

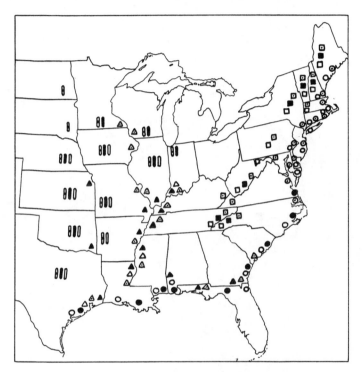

FIG. 18. Migratory tracts and distribution types illustrated by grasses. ATLANTIC AND GULF COASTAL PLAIN: *Ammophila breviligulata* (disk and dot), *Distichlis spicata* (hollow disk), *Uniola paniculata* (black disk). APPALACHIAN MOUNTAIN SYSTEM: *Danthonia compressa* (square and dot), *Cinna latifolia* (hollow square), *Agrostis borealis* (black square). MISSISSIPPI VALLEY AND GULF: *Leersia lenticularis* (triangle and dot). *Leptochloa panicoides* (hollow triangle), *Paspalum repens* (black triangle). PRAIRIE AND FOREST BORDER: *Panicum perlongum* (rod and dot). *Eragrostis trichodes* (hollow rod), *Paspalum stramineum* (black rod). Distribution by states, after Hitchcock (1935).

"Distribution and dispersal adaptability in *Carex* and *Uncinia* are inversely proportional!"

The phenomenon of local races (subspecific endemics) is entirely opposed to the idea of long-distance dispersal, for such variation depends upon isolation which would not exist if long-distance dispersal were generally effective.

Migration, on the other hand, is usually not a random matter. For any region there can usually be recognized only a few main migratory tracts. Van Steenis (608) defines a *migratory tract* as an outlying tongue of the area

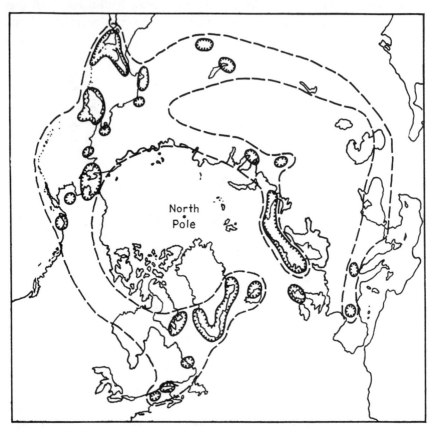

Fig. 19. *Phyllodoce coerulea* is an example of the rather large group of arctic-montane species, many of which spread from the mountains of northeastern Asia to America before the maximum glaciation and were later deprived of a large part of their area and spreading capacity, according to Hultén. This species, with its large gaps between areas in both the arctic and the montane occurrences, would be impossible to explain unless other species with a similar history and similar areas occupied intermediate positions here and there, and unmistakably indicated the routes of migration of *Phyllodoce coerulea*. The map is redrawn from Hultén (360), and is based on an outline after Goode's copyrighted map No. 201PN, with permission of the University of Chicago.

of a genus, extending from the center of distribution. For example, in Malaysia he recognizes three migratory tracts: a Sumatran tract from the Himalaya and West China, a Formosa-Luzon tract also from Asia, and an Australia-New Guinea tract from the southeast. Ecologically, we can consider that a migratory tract (highway) consists of a continuity of habitats which are within the ecological amplitude of the migrating species. Thus the salt marshes, beaches, sand dunes, and brackish to fresh ponds of shorewise regions provide a continuity of suitable habitats for many different kinds of littoral species. The Gulf and Atlantic coastal plains of the United States, along which many migratory tracts are situated, serve as an example.

The migration of a single species can seldom be determined alone; it is usually necessary to consider the areas of all the species of a genus (or other higher category) to obtain reliable evidence. The evidence for migratory tracts is essentially of two types. If the area of each species of a genus is plotted, there will usually be found a center of development where the species are most numerous. The genus will usually show a progressive decrease in polymorphy (number of taxonomic forms) from its center toward its periphery. *Isoflors,* lines delimiting regions with equal numbers of species (within the circle of affinity), can be drawn for the generic area as a whole, and the pattern of the isoflors indicates the migratory tracts for the genus.[4]

Second, when all the specific or generic maps of a flora (or a representative sample) are compared, it will be found that a large number of them coincide in respect to areal pattern. This coincidence points to *waves of migration* which have involved species widely different in respect to taxonomic position, ecological requirements, and methods of dispersal. It is true, however, that a certain wave of migration may have involved plants only or mostly of a certain ecological nature because of the peculiar climatic or ecological nature of the highway. For example, the so-called prairie peninsula (northern Illinois, Indiana, Ohio, and southern Michigan) constituted a highway for an eastern migration of prairie plants—evidence of their migratory tract lies there today—because of the peculiar edaphic, topographic, and climatic conditions which prevailed during a xerothermic postglacial period.

A migratory tract may be active, migration occurring along it at the present time, or it may be passive, the conditions for migration no longer existing.

Closely related to migration (the concepts of migratory tracts and waves) is the concept of *phytogeographic element,* which Braun-Blanquet (88) defines as follows: "A phytogeographic element is the floristic and phytosociologic expression of a territory of limited extent; it includes the taxonomic units and the phytogeographic groups characteristic of a given region." This is a return to the original use of the term in a purely geographical sense. Engler (215) and others have variously spoken of

[4] See chapters 13 and 14.

"Tertiary elements," "ruderal elements," etc., but Braun-Blanquet disapproves of them as destroying the meaning of the term. As an illustration of the use of the concept of element in floristic geography, we may note his treatment of the Massif Central de France (88). He recognizes the following elements: (1) The Eurosiberian-Boreoamerican element, which consists of three subelements: (a) the Medio-European autochthonous subelement, which traces back to the Tertiary in the same region; (b) the Atlantic subelement, which immigrated especially during the warm-humid interglacial and which has produced microendemics in the Massif Central; (c) the Boreo Arctic subelement, which dates to the last glacial period and extends to the Pyrenees and is preserved mainly in peat bogs in regions of abundant rain and fog. (2) The Mediterranean element, which consists of postglacial colonies that have a tendency to extend their areas in some regions, and a few Tertiary relics of Mediterranean-Montagnarde stock, living in regions of low to middle elevation. (3) The Aralo-Caspian element, which consists of a small number of species that reached the Iberian peninsula during the Tertiary epoch and were almost entirely eliminated during the Pleistocene.

It is apparent that such a treatment of elements in the flora of a territory is based upon their being typical of certain well-defined phytogeographical areas elsewhere (that is, of other regions, domains, sectors, districts, etc., such as the Mediterranean region).

It would seem pedantic to insist on a strict usage of the term element. I favor the conclusion of Wulff rather than of Braun-Blanquet, and suggest the usage of element in its widest variety of applications, as is now characteristic in the literature, various elements being distinguished by definite adjectives.

It is likewise sometimes useful to consider floristic affinities in a related but somewhat looser manner. Species in a certain territory can be classified as *intraneous* or *extraneous* according as their occurrence in that territory is well within the area of the form or near the periphery of its area, respectively. For example, the phytogeographical elements recognized by Braun-Blanquet for the Massif Central de France are all *extraneous*. In contrast, species characteristic of the Massif Central are intraneous to it, but if found elsewhere they would be classified as extraneous in the other place.

It is an easy and suggestive procedure in the preliminary study of the floristics of a territory to divide the plants into intraneous and extraneous groups, and further to subdivide them on the basis of extent and direction of area. This process is illustrated by a classification of the woody plants

TABLE 13.—The floristic-geographical affinities of the woody plants of the Great Smoky Mountains (Data from 104)

| | Percentages of the Various Types in | |
	Total Flora	Spruce-Fir Formation
Intraneous distributions	69	41
Eastern North America	36	9
Southeastern North America	11	..
Southern Appalachians	10	5
Endemic	12	27
Extraneous	31	59
Northern areas	25.6	59
Northeastern United States	15	14
Southeastern Canada	7	27
Canadian transcontinental	3.6	18
Not northern		
Southern, Piedmont, etc.	3.6	..
Southwestern	1.6	..

of the Great Smoky Mountains of eastern Tennessee made by Cain (104). Considering the whole woody flora, 69 per cent of the species were found to be intraneous and 31 per cent were extraneous; but for the woody plants of the spruce-fir belt (essentially above 5000 feet elevation) the extraneous plants of northern affinity rose to 59 per cent, whereas they were only 25.6 per cent for the region as a whole. The excellent monograph by Hultén (360) on arctic and boreal areas illustrates the point of these paragraphs.

This suggests the usefulness of an analysis of the distribution of elements *within* a territory, a type of study well illustrated by an investigation by Waclaw (666), who reached some rather broad generalizations from a minute analysis of local distribution. In the canyons of the Dniestr River he found localized species of two types for both of which the canyons had apparently served as a natural route of migration. Plants of Type Ia possess in the canyons their most remote stations to the east (*Aconitum maldavicum, Aposeris foetida, Helleborus purpurascens, Scolopendrium vulgare, Waldsteinia geoides*), and those of Type Ib their most remote stations to the west (*Asparagus tenuifolius, Centaurea orientalis, Inula germanica, Linaria genistaefolia, Rhus cotinus*). These canyons are known to have been formed mostly during the Pleistocene interglacial preceding the Würm glaciation. Plants of Type Ia grow in the canyons in the shady, humid forests and have large areas elsewhere in the lower forests of the Carpathians. It is likely that they immigrated into the canyons of Podolie during postglacial cool-moist subarctic and Atlantic times. Plants of Type Ib occupy drier, more open sites along the canyons and have extensive areas elsewhere in the steppes. It is likely that they entered the canyons during postglacial xerothermic

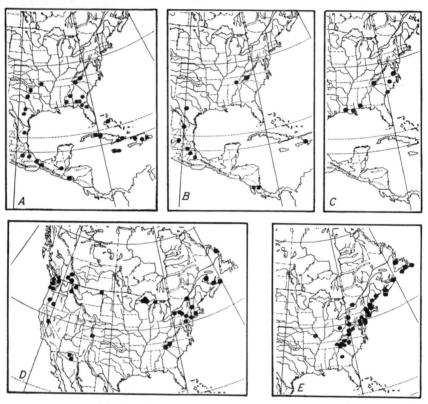

Fig. 20. Maps of some extraneous phytogeographical elements among the bryophytes of the Great Smoky Mountains of Tennessee and North Carolina. *A. Marchantia domingensis,* a species of tropical affinity occurring in the United States coastal plain. *B. Homalothecium Bonplandii,* also a tropical species, but absent from the United States coastal plain. *C. Cryphaea nervosa,* which has its greatest frequency in the southern coastal plain. *D. Dicrodontium pellucidum,* which has its greatest frequency in the northern coniferous forests and extends southward in the mountains. *E. Mnium hornum,* which has its greatest frequency in the northern coastal plain. Spot maps from Sharp (572) are based on herbarium records.

time. In both these subtypes the canyons have been a natural route of migration because of the close sequence of suitable habitat types.

Plants of Type II are found not in the canyons but on the uplands between canyons, and there is no demonstrable relation between the canyons and the distribution (migration) of these plants. They are supposed to have reached the region before the formation of the canyons, and their stations are relic. This type includes three elements: the Carpathian-alpine element (*Anemone narcissiflora, Crocus Heuffelianus*), the Siberian element, disjunctive in Europe (*Avena desertorum, Thalictrum petaloideum*),

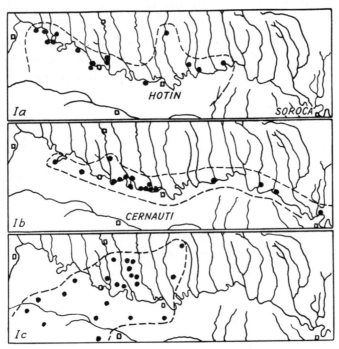

FIG. 21. Examples of local occurrence types significant with respect to general distribution and vegetational history. Type *Ia*, represented by *Scolopendrium vulgare*, occurs only in the canyons where it has its easternmost stations. Type *Ib*, illustrated by *Linaria genistaefolia*, also occurs only in the canyons where it has its most western stations. Type *Ic*, *Crocus Heuffelianus*, occurs only on the uplands between the canyons. It is an ancient Carpathian element of the Podolian flora which obtained its distribution in the area before the formation of the canyons. Squares on the maps indicate towns, some of which are named. Data redrawn from Wacław (666), who maps several species of each type.

and an Asiatic element of the Caucasus and Turkestan (*Evonymus nana*). The conditions of the canyons are not favorable for plants of Type II. In this study by Wacław we have a good example of the far-reaching suggestions which can result from a close study of local distributional patterns coupled with floristic affinities and geology.

Migrations are presumably going on contemporaneously more or less everywhere, as is dispersal, but the process is so slow that, excepting certain weeds (*Galinsoga parviflora*), parasites (*Endothea parasitica*), and insect pests (European corn borer), man is scarcely ever able to observe them. Most conclusions concerning migration must, then, be based upon present evidences of historical processes which are obtainable from modern floristic patterns (species and community areas), taxonomic affinities, ecological requirements, and, in certain cases, fossils. During the last few decades the study of microfossils in peat deposits, especially pollen grains, has produced a wealth of information concerning Quaternary migrations, and the method is being extended to other sediments and more remote epochs.

13.

Center of Area

1. *Most taxonomic entities and many communities have a center of origin, center of variation, and center of frequency, and, with the vicissitudes of change through a long history, may have one or more centers of survival and secondary centers of development.*

2. *From contemporary phenomena—distributional patterns and phylogenetic relationships—it is frequently possible to ascertain with fair certainty the center of origin of a species, genus, community, etc., but abundant fossil remains and a knowledge of geological and climatic history are valuable adjuncts and sometimes necessary for the solution of such problems.*

3. *Because of the chances of climatic change and consequent migrations some species have become extinct and others have been decimated, leaving behind relic colonies in refugia. Such refugia must be areas of compensation, for no species can survive under conditions beyond its range of tolerance. Centers of preservation may subsequently, but not necessarily, become centers of dissemination upon the return of conditions more favorable for the relics. This appears to depend upon the extent of the biotype depauperization which the relic population underwent during its isolation.*

The concept of center of area is not simple, for it includes center of origin, center of development (variation), and center of frequency; nor is it one which is easy to deal with in plant geography, but it is important to the dynamic point of view. In addition, center of survival and center of dispersal may be different from any of the above centers and require independent analysis.

Considering first centers of survival or preservation, we can readily see that they must be numerous for species, species groups, floras, and even communities. Such refugia, as they are frequently called, exist widely over the surface of the earth because all regions have undergone more or less change with the passage of time. Since any change of environmental factors must affect the usually delicate balance between organisms and their habitats, migrations are the universal result of large environmental change (and, also, to a certain extent, extinction and evolution). Changes of climate result in migrations of climaxes,[1] the members of the climax series or clisere moving together in a common direction as determined by the direction of the climatic shift. But no general climatic change ever affects a large area uniformly and no mass migration ever occurs without irregularities. The result is that relic colonies of species, fragments of floras, or communities are left behind in regions of compensation where they constitute centers of preservation. Such centers of survival, with a climatic reversal, may become centers of dispersal—strongholds from which the original area may be more or less completely regained or a new area attained, as conditions permit. Such large cliseral movements were initiated with each glacial advance and recession, and centers of survival exist both northward and southward of the boundary of the glacial advance. Northward are refugia of the interglacial organisms and southward are refugia of the northern types which were forced southward.[2] A few examples will be sufficient to illustrate the concepts of centers of survival and dispersal.

Adams (3, 4, 5) discusses the centers of survival in the United States south of the terminal moraine which provided biotic preserves during the Ice Age and centers of dispersal thereafter, from which successive belts or waves have moved far into the glacial till plain. In the eastern United States the greatest southward extension of northern types probably occurred along the Appalachian mountain system; in the high Southern Appalachians was the principal eastern center of preservation during glacial times. The extent of the southward retreat of northern forms is not well known, nor has the extent of the temperature depression been ascertained, but it appears that contrary to earlier opinions there was in the United States no broad tundra belt south of the ice terminus. Pleistocene records are sufficient, however, to indicate that certain northern forms ranged a considerable distance southward. The extent to which climaxes moved southward is not known. There are published records for such arctic types as the walrus in Virginia and South Carolina, the musk ox and reindeer in West Virginia and Kentucky (336, 334), the now most northern of American pines, *Pinus banksiana*, in South Carolina (110), and tamarack, *Larix*

[1] Such mass movements do not necessarily consist of more than the coincidental movement of individuals.

[2] See Hultén (360) for the most important consideration of these centers and areas for arctic and boreal plants.

laricina, white spruce, *Picea glauca,* and northern white cedar, *Thuja oc-cidentalis* in Louisiana (90). Today the Southern Appalachians contain a biota marked by two features: a high degree of endemism and a large percentage of species, many of them disjunct, now characteristic of north-eastern states and southeastern Canada (104, 572). Much of the whole region south of the terminal moraine provided territory for preservation, but

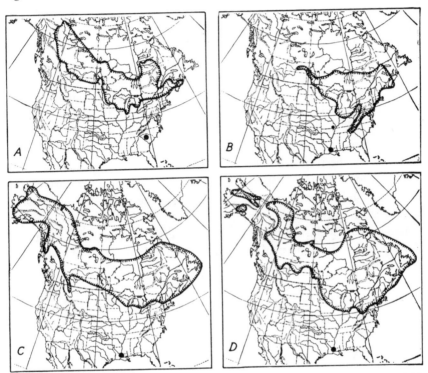

FIG. 22. *A.* The approximate area of *Pinus banksiana* and its fossil occurrence, probably Pleistocene, in buried soils of South Carolina (110). The approximate areas of three other northern conifers: *B. Thuja occidentalis, C. Picea glauca,* and *D. Larix laricina,* and their fossil occurrence in Pleistocene deposits of Louisiana (90). The modern areas of these species are mapped after Munns (472).

the centers were generally the regions of varied topography and conditions (the Southern Appalachians and Cumberlands, the Ozark plateau, etc.). From the southeastern center plants migrated northward by several major routes such as the Mississippi River system, the Cumberland and Appalachian uplands, and the coastal plain. Some species have returned northward over hundreds of miles, others have moved lesser distances, and still others, especially in northern refugia, apparently constitute an epibiotic flora with little capacity for regaining lost territory (276, 242, 360). The Southern Appalachians (because of the optimum conditions prevailing there) have also been a center of preservation of the Tertiary mixed [3] forests and a center from which migration and differentiation have taken place. This idea is developed by Braun (79, 81, 82, 85, 86) in a series of papers on the undifferentiated deciduous forest climax and the association-segregate hypothesis, and is discussed by Cain (111) from a paleobotanical and areographic point of view.

Centers of preservation during the ice ages also existed in more northern regions. They can be roughly divided into two types: certain small, supposedly ice-free areas, frequently called nunataks, where a decimated flora survived at least the last ice advance which completely surrounded it; and larger high-latitude areas (in Alaska and the Aleutians, for example) which lay to one side of the centers of ice accumulation and movement. From the latter centers have been derived the great mass of the modern arctic flora (360). From the former has been derived very little (so far as is now detectable), for it appears that such refugia have been centers of survival, but only to a limited extent centers of dispersal. These are the plants which compose an epibiotic flora (447) which Fernald (242) considers to be senescent but which may better be described otherwise (109). Hultén (360) has brought a great mass of data to bear on this problem of glacial survivors. Fernald's nunatak hypothesis (whether it ultimately will be proved wholly or partly incorrect, or substantiated) has been exceedingly important in stimulating plant geographers and glacial geologists to reexamine their data and assumptions. The Scandinavian plant geographers have been widely in favor of the nunatak hypothesis for an explanation of certain relic elements in their flora,[4] and several papers are supported by many details concerning the distribution and ecology of relics. Dahl (171) has a mass of evidence for much of the mountain flora of Fennoscandia having survived the last glaciation. Nordhagen (491), Holmboe (350), Nannfeldt (484), and others have recently brought forward evidence that the Scandinavian mountain flora, formerly thought to have immigrated from the south and east, persisted during the last glaciation at ice-free refugia on the Norwegian coast. Similar refugia have been described for Iceland and Greenland (for example, by Gelting, 268).

Such northern centers of survival (or secondary centers in mountains) appear now to be occupied largely by relic plants with very limited power of spreading, but there is no reason why only such unaggressive plants

[3] Sometimes called the Arctotertiary forest.
[4] See (229) for a review of recent Scandinavian investigations.

should have survived on the nunataks. There were certainly other species in the refugia, especially in the larger ones, which extended their ranges more or less far during postglacial time and which now compose a part of the surrounding characteristic vegetation and consequently cannot be detected, except by Hultén's method (360) of studying areas and arranging them in progressively equiformal series.

The lowland survival in refugia of species which are now mainly arctic

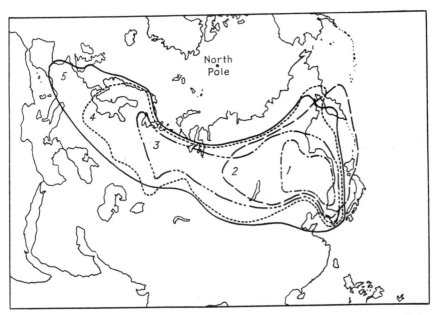

FIG. 23. Equiformal progressive areas represented by a selected series of species: *1. Smilacina dahurica; 2. Stellaria radians; 3. Rubus humilifolius; 4. Carex globularis; 5. Majanthemum bifolium.* These are members of a large group of chiefly lowland plants possessing more or less extensive areas and comparatively northern limits but not really arctic, that coincide approximately with the coniferous belt. According to Hultén (360, from whom the map is copied), there can hardly be any doubt that the plants in question spread from a center in eastern Siberia and Amur westward toward Europe. The outline is based on Goode's copyrighted map No. 201PN, with the permission of the University of Chicago.

and alpine or montane is closely related to the problem of determining whether certain lowland forms are truly glacial relics. Turesson (646), who has brought the technique of genecology to the problem in Scandinavia, comes to the conclusion that several species once considered to have been members of the first postglacial northward wave of migration (and consequently relics when found at lowland stations) have on fuller investigation been shown to be more recent immigrants and "pseudo relics." In this group he places such plants as *Polygonum viviparum, Arctostaphylos alpina*, and *Pedicularis sceptrum-carolinum*, which have an alpine stamp but occur in lowlands, and such plants of the conifer forests as *Pyrola*, *Linnaea borealis*, and *Goodyera repens*, and peat bog plants such as *Betula nana* and *Rubus chamaemorus*. Turesson studied certain of the remaining supposed relics (*Poa alpina*, *Viscaria alpina*, *Pinguicula alpina*) and came to the conclusion that the lowland forms were not true relics. For example, *Poa alpina* consists of three ecotypes: *alpinus*, *subalpinus*, and the lowland type, *pediacus*. It is generally assumed that the rarity and scattered oc-

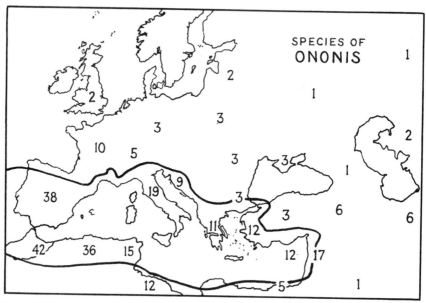

Fig. 24. The numbers of species of the genus *Ononis* in the different countries, and the center of variation of the genus. Map redrawn from (616).

currence of the lowland ecotype indicate its relic nature, i.e., that the colonies have survived from the early postglacial time when a "tundra" climate existed at low elevations in Sweden. Turesson says that this is not so and gives the following reasons: (1) The first immigrating *Poa alpina* populations must have been subjected to such strong selection and elimination of biotypes not hardy enough that the subsequently thinned and impoverished populations would not have been able to differentiate a lowland ecotype. (2) The lowland ecotype (*pediacus*) is not a mere seclusion type (resulting from chance geographical isolation) because of its ecological characteristics in regard to earliness and water requirement. The lowland ecotype, therefore, is considered to represent a later spotty immigration. Turesson believes that *Poa alpina* had differentiated its ecotypes before the last glaciation, and that during that period and its attendant migrations many species were deprived of their lowland ecotypes (*Dryas octopetala, Loiseleuria procumbens, Oxyria digyna*), but that some (such as *Poa alpina*) retained them in fragmentary form.

The cen⸱ ⸱ of origin for a species is difficult to determine unless the species is polymorphic and partakes of many of the characteristics of a species group, such as a section or a genus. The most certain determinations of centers of origin relate to the higher categories except, of course, for young species with very limited area and certain forms of hybrid origin. The spatial and temporal relations of a species are best viewed in the light of its close relatives and their areas. The simplest application of this method consists of determining the number of species of a genus in the various regions of the generic area for which floristic data are available. This can be illustrated by the maps of the genus *Ononis* by Širjaev (582) and Szymkiewicz (616), and by the latter's data in Table 14. From such data one finds the center of development of a genus—at least where the greatest concentration of species is—that is usually considered to be the center of

TABLE 14.—Generic sizes in various countries as a means of indicating centers of development and possibly of origin of genera (Data from 616)

	Spain	Italy	Greece	Littoral Asia	Ana-tolia	Arme-nia	Per-sia	Turke-stan	Al-tai	Russian Orient	Ja-pan
Genista	47	34	13	8	6	5	0	1	0	0	0
Trifolium	54	98	64	53	25	45	15	14	7	2	1
Silene	58	65	86	62	35	65	41	49	14	10	10
Alyssum	13	16	20	29	26	27	14	11	3	0	0
Gypsophila	3	3	7	11	19	23	16	19	7	3	0
Artemisia	20	17	5	5	10	20	23	68	30	30	17
Saussurea	0	0	0	0	0	1	2	41	23	24	19

origin of the genus. In the case of *Ononis* there is a clear concentration of species in the western Mediterranean, with a generally regular reduction of species number eastward and toward the periphery of the generic area.[5] This method, useful as it is in suggestiveness, is subject to errors and false conclusions. In the first place it is difficult to delimit for comparison

[5] As shown on the map, *Ononis* has its center of species concentration in the Iberian Peninsula. Szymkiewicz also discovered from his statistical-floristic studies that 70 genera have their centers in the same region: for example, *Holcus* with 9 species, *Koeleria* with 10, *Rhamnus* with 12, *Trisetum* with 14, *Arenaria* with 29, *Narcissus* with 33, *Armeria* with 37, *Genista* with 47, etc. Such data are useful in the floristic characterization of a region. It should be noted that many of these are critical genera and that floristic statistics depend greatly upon whose monographic revision is employed.

taxonomic treatments and areas which are really comparable. When the data show a geographical regularity, as in those cited, there is some basis for accepting their comparability. A more serious possible error results from the fact that a genus may show a center of development (where there are many species) that is far removed from the center of origin of the genus. A young genus in expanding its area may encounter a region of varied habitats in which the process of speciation (adaptive radiation) may occur extensively, and this center of development may be more or less far removed from the center of origin of the genus.

Objects treated statistically should be equivalent. This is not always true when the species of one flora are compared with those of another because of the differences in interpretation by the writers of manuals and catalogues. Also, species have different values for plant geography because of their different characters. With this in mind, Szymkiewicz (620) devised three categories of species which would reflect their geographical significance: (1) species which are endemic or subendemic, that is, species which are confined, or nearly so, to a single natural area; (2) species which extend from the given area to another that is phytogeographically equivalent; (3) species partly in the phytogeographical domain considered, but extending extensively in other domains. For the determination of generic centers and for floristic characterization of regions, Szymkiewicz considers that the species of Group 1 are the most important and those of Group 3 the least important. These numbers designate the columns of Tables 15 and 16.

In the analysis of *Carex* according to the three species groups it is seen that the eastern Asian center (Orias), as indicated by the total number of species, is strongly represented by endemics (259 species) and that wide species are relatively few. In the other regions of high species number, the Atlantic and Pacific American regions have about half their species in the endemic class, whereas the European and Mediterranean regions have only about one-fourth or fewer endemic species. North America would seem to represent a secondary center in the genus *Carex,* related to the east-Asian center.

Inequalities among species, and taxonomic treatments can be overcome to a considerable extent by a geographical treatment of sections. This is illustrated by the tables on *Draba* and *Saxifraga* published by Szymkiewicz (620). The same three groups are employed for the classification of sections as for species in the case of *Carex. Saxifraga* illustrates a point which was mentioned earlier. The statistics concerning species numbers indicate that the generic center, and the center of origin, lie in eastern Asia (with 143

TABLE 15.—Species of *Carex* of various regions classified according to whether they are endemic (Group 1) or more widely distributed (Data from 620)

Regions	Number of Species in Species Groups			Total Species
	1	2	3	
Europe, excluding the Mediterranean	27	101	14	142
Siberia	8	70	16	94
Mediterranean: Spain and Morocco to Turkestan	31	67	22	120
Orias: eastern Asia of the Himalaya to China and Japan	259	70	18	347
Pacific North America including the Rocky Mountains	61	77	5	143
Atlantic North America including the Great Plains	86	75	6	167
Extra-tropical Mexico	9	11	3	23
Andes: Venezuela to the extreme south	41	16	3	60
Neotropics: American tropics outside of the Andes	12	5	2	19
African tropics	28	6	0	34
Indo-Malaysia	51	17	8	76
South Africa	5	9	0	14
Australia and New Zealand	40	11	1	52

TABLE 16.—A comparison of the number of species and sections of *Saxifraga* in various regions according to whether they are endemic (Group 1) or more widely distributed (Data from 620)

Regions	Number of Sections in Categories				Number of Species in Categories			
	1	2	3	Total	1	2	3	Total
Europe	7	2	3	12	59	22	3	84
Siberia	0	0	6	6	8	13	1	22
Mediterranean	2	0	8	10	35	13	6	54
Orias	3	0	5	8	138	3	2	143
Pacific North America	1	2	4	7	17	17	1	35
Atlantic North America	0	1	6	7	7	13	1	21
Mexico	0	0	1	1	1	1	0	2
Andes	0	0	1	1	4	0	0	4
Tropical Africa	0	0	1	1	1	0	0	1

species, of which 138 are endemic), but the analysis based on sections reveals their center to be in Europe. From this it is reasonable to conclude that the generic center of origin was more likely in Europe than in Asia and that the present Asiatic center of development is a secondary one. This conclusion is obviously based on the reasonable assumption that the sections are older than the species. The data for *Draba* do not reveal any single center of species development, but there is a definite concentration of sections in the Andes, with a secondary center in the Mediterranean. To find the center of origin for the genus *Draba* would require a different approach combining comparative morphology, cytogenetics, and historical data.

TABLE 17.—A comparison of the number of species and sections of *Draba* in various regions according to whether they are endemic (Group 1) or more widely distributed (Data from 620)

Regions	Number of Sections in Categories				Number of Species in Categories			
	1	2	3	Total	1	2	3	Total
Europe	0	1	3	4	29	16	1	46
Siberia	0	1	3	4	6	13	3	22
Mediterranean	3	1	2	6	40	16	5	61
Orias	1	2	1	4	47	8	2	57
Pacific North America	0	5	2	7	40	16	0	56
Atlantic North America	0	1	5	6	4	14	3	21
Mexico	0	1	1	2	4	1	1	6
Andes	5	2	3	10	48	0	0	48
Neotropics	1	0	0	1	4	1	0	5

In general it is apparent that Szymkiewicz's method of handling generic statistics for the indication of centers is one that provides more significant data than the old method of merely listing species numbers. He believes that the center of a genus is in the region presenting the greatest number of variations and that those of sectional rank are more significant than those of lower categories. Incidentally, the resemblance between the Andean and the Mediterranean regions shown by *Draba* is also seen in other genera (such as *Sisymbrium*), and that between the Mediterranean and Pacific North America is paralleled in *Eryngium, Centaurea, Astragalus, Trifolium,* and *Lupinus,* according to Szymkiewicz.

In an excellent study of truly temperate genera (stenotherm genera) in tropical Malaysia, van Steenis (608) has developed several interesting facts concerning the characteristics of genera at their centers and periphery, and the means of determining centers. Taking one example in detail, we find that the genus *Primula* is represented in Malaysia only by *P. prolifera,* occurring between 2050 and 3000 meters altitude in the temperate belt. This genus comprises over 500 species which are mainly distributed in the northern hemisphere, the greatest concentration being in the Himalaya and especially West China. *P. prolifera* belongs to the section Candelabra, with 25 species in southeastern Asia. Although the genus is also represented in Europe and North America, it has crossed the neotropics as well as the paleotropics, with a variety of *P. farinosa* in Chile and Tierra del Fuego. Van Steenis rules out the possibility that the southern hemispheric forms originated there by polytopic development, because no relatives are present in these regions from which they could have been derived. He concludes that the center of the genus is in the Himalaya and the mountains of western China and that the crossing of the tropics was a later development through migration of genetic stocks originated in the northern hemisphere —for *P. prolifera,* in the Asiatic section Candelabra.

Van Steenis suggests, where a genus is sufficiently large, that the mapping of its species will show the number of species in each portion of the generic area and allow the drawing of *isoflors,* lines connecting equal numbers of species of a genus. From the generic center to the borders the number of species regularly decreases (except when two or more centers are discernible), and the isoflors look somewhat like contour lines. In the ideal expansion of a genus the isoflors would be roughly concentric, but such a condition seldom occurs in nature because of barriers and highways hindering and facilitating migration respectively, and because of historical changes of conditions. These phenomena result in the isoflors assuming a pattern which delimits the migratory tracts and strongly suggests the directions and lines of past migrations. Following such migratory tracts backward, we find a convergence upon the generic center. Substantiation of the migratory tracts and center for a genus is usually provided by the fact that several genera, entirely unrelated taxonomically, have the same pattern and relationships. That is to say, it would seldom occur that a single genus would have had a unique history.[6] The coincidence of several generic centers and migratory tracts lends mutual strength to the interpretation of the history of any one. Such a development of centers and radiating migratory tracts is an almost universal phenomenon among genera (and other cate-

[6] In this connection, see Hultén's study of progressive equiformal areas (360).

gories) and the conclusion that there has been a single center of origin appears in most cases to be inevitable. Further evidence, and perhaps the strongest, is supplied by comparative morphology, which forms the principal basis for most systematics. That is, monographic studies which reveal that the migratory tracts are also lines of evolutionary development and specific relationship give to the distributional pattern its ultimate support as an indication of generic centers.

In describing certain characteristics of genera at their centers and periphery, van Steenis says that not only are there usually more species at the generic center, but their polymorphy tends to be greater, hybrids are more numerous, and many species are reticulately related (through hybridization and sometimes amphidiploidy). There seems to be a decrease of genetic polymorphy, perhaps even a decrease of genetic potential through biotype depauperization, from the center toward the periphery of a generic area. One interesting phenomenon, and significant too, is the fact that the lower isoflors, near the generic border, usually contain some of the most widely distributed species which in general also occur at the center. Such species may be the oldest (this sometimes can be detected by their primitive characters) or merely those with the best facilities for dispersal or the widest ecological amplitude, or with all these attributes; but the fact of their occurrence in the lowest isoflors can hardly be otherwise understood than by assuming their former migration from regions within the higher isoflors. When a single species has a fairly wide area it is usually divided into regional facies which may be recognized as subspecies or varieties. This tendency to produce taxonomically (and genetically) distinctive populations is exaggerated at the periphery of the area and toward the apex of a migratory tract. This is understood as an isolation and selection phenomenon, and the more remote the peripheral population the more different it is likely to be from that of the center. Another related aspect, however, is that the internal variability of a population frequently tends to be less for peripheral populations.[7] That is, there is a tendency toward a steady decrease of intraspecific variability from the center toward the apex of a migratory tract which results from partial isolation and the attendant inability of the remote populations to have access to the whole supply of genes (201).

Most plant species are made up of a mosaic or replacement pattern of populations which comprises numerous genetical combinations in various portions of the area. This is not a new observation, although lately with the

[7] That this is not always so is seen, for example, in the bird genus *Junco*. Miller (465) found that coefficients of variability are not low in insular forms.

development of genetics it has been better understood. Hooker, aware of this polymorphy, approached the supposition that the wider a species ranges the more different are the combinations at the periphery. If the central populations become extinct or for any reason there is isolation of peripheral populations, he thought that new species might arise. Kerner also thought that there was a possibility of the origin of species along the border of a polymorphic population through the development of limited interbreeding and the establishment of purer and more constant populations.

We must consider also that there are not only horizontal margins of specific and generic areas but also upper and lower altitudinal limits which constitute other margins. On one mountain a species will usually show rather definite altitudinal limits, whereas on other mountains these limits will likely differ more or less. Such differences in altitudinal limits result from latitudinal position, general climate, peculiar differences in local topography, the Massenerhebung effect, local climate, soil, climaxes, disturbances, and the genetic characteristics of the local populations. Other things being equal, differences in altitudinal limits, as in horizontal limits, may result from the ecological characteristics of the local populations which have a genetic basis and which, in turn, result either from accidental isolation phenomena (seclusion types) or from natural selection. According to the Sewell Wright effect, in the isolation of small populations, as on mountains, there may be a rapid fixation of a random and even non-adaptive genetic combination.

R. Wettstein (686, 687), in his introduction of the important geographic-morphologic method in plant systematics, provided monographers with a valuable new approach to the natural organization of taxonomic groups. In his work on *Gentiana,* section Endotrichia, however, he propounded what may be called a neo-Lamarckian hypothesis of the development of species over the whole area of the ancestral population under habitat influence, and denied the concept of a center of origin with subsequent migration for each specific type. Wettstein's maps show the species of *Gentiana* replacing one another in different geographical areas, and with altitudinal separation where more than one species occurs in a region. This distributional pattern looks like those of the subspecies of birds and mammals as interpreted in modern animal taxonomy—and it may be that Wettstein's species are not deserving of more than subspecific rank. That point aside, Samuelsson (542) has leveled a severe criticism at Wettstein's conclusion concerning the origin of the units. He shows that when the individual taxonomic units are mapped by placing a spot for each known locality rather than merely drawing a boundary line for the entire area of the form, it is clear that there is a *center of density* for each species and that the frequency of stations drops from the center toward the periphery of its area. If Wettstein's assumption of the development of a species over its whole area were true, the density of a species should ordinarily be more or less uniform over the whole area. For ecological reasons there may be occasional exceptions to this expectation. Samuelsson's point maps, however, show that there is a more or less pronounced center of frequency of stations for each species and a rather regular decrease in frequency from the center toward the periphery. This distributional pattern would seem to indicate strongly that each species had a center of origin and subsequent migration. But the pattern does not constitute proof.[8]

Samuelsson goes on to show that among Scandinavian plants there are numerous young species which have not yet attained their natural boundaries and that in each case there is a frequency center from which there are progressively fewer and fewer outlying stations. Many of his data are from apomictic species of *Hieracium,* but the same phenomenon is apparent in panmictic species. Not only are the stations for a form more frequent at its center of area, but the number of individuals per station averages greater and a wider variety of habitat types are occupied than toward the margin of area.

When a genus shows two or more centers of speciation, or at least of species concentration, no regular arrangement of isoflors is possible. For example, *Lobelia* has three centers (southeastern Asia, on high African volcanoes, and in Hawaii), *Magnolia* has two centers (eastern Asia and eastern United States), and *Empetrum* has two centers (in boreal regions and in southern South America). In all such cases two types of explanation would seem to be possible: (1) the widely disjunct centers are of polytopic origin; (2) there was an ancient center from which wide migration occurred and the area was later broken through geological and climatic changes. The latter explanation is the one usually accepted, and it is sometimes substantiated by fossil records for the intervening areas which may even indicate the original center. I cannot agree with von Ihering (374) when he says that the developmental center of a plant family (or other group) can only exceptionally be disclosed from modern relations, but he is certainly right in his conclusion that inferences are safer when "sufficient fossil proofs occur." An assumption of a polytopic origin for a group with two or more

[8] See Chapter 14.

centers would seem warranted only when all other hypotheses fail and when a connecting ancestral population can be demonstrated.

In closing, this chapter cannot omit reference to the conclusions of one of the leading entomological taxonomists, A. C. Kinsey (392, 393), who has specialized on gall wasps. He makes the categorical statement that there are no points of origin or centers of distribution of the units in a higher category such as a section or genus. He says that this is "for the simple reason that the lower categories are not derived by radiate evolution from a single species which represented the higher category. The usual endeavor of students of distribution and phylogeny, to find the place of origin of each group studied, is strongly warped by the conception of evolution following the pattern of a tree. But if, as we now find, the lower categories are only arbitrarily delimited series of species connected linearly on a continuous phylogenetic chain, the several sections may originate at different points remote from the first species in the line. . . . In the phylogenetic map . . . we have marked the points within each line at which the most conspicuous mutations have occurred in the *Philonix-Asraspis* groups in *Cynips,* and the points at which the lines have split to give rise to the most divergent types. These are the 'points of origin' for some of the artificial conventions which might, with some hope of agreement, be recognized as distinct complexes or subgenera in our group. It will be seen that these points are scattered everywhere from Southern Mexico to the Northern United States." Kinsey's concept appears to be that during the migration of a population each genetic character may have entered the phylogenetic stock at a different point from any of the others.[9] Consequently, according to him, higher categories are not groups of similar species or groups of species with a common origin, but merely a group of units (Kinsey's species are approximately the equivalent of the subspecies of most zoologists) connected in a phylogenetic *chain.*

It seems to me that Kinsey has misunderstood the usual concept of the geographer and taxonomist when referring to a generic center. It is not meant that the center is the place where the original species gave rise to *all* the species which, in turn, were the sectional or subgeneric stocks. Reference to a monograph on any ·medium or large plant genus usually shows that there are whole sections which have no immediate connection with the supposed center of the genus. It is true, however, that on the average more

---

[9] Kinsey admits, however, that coincident change of more than one character may result from linked characters, manifold effects of single genes on several characters, and on more than one character controlled by multiple factors, morphological and physiological compensations, etc.

sections will be represented at or near the center of the genus than at any single area away from the center. Evolution does not have to occur only along a lineal system, geographically; it can take place also at the central region or in any other part of the area. Because the center of origin is one to which the genus is obviously adapted, or the forms would never have emerged by surviving natural selection, and because it has long been occupied—major climatic changes and enforced migrations aside—it would naturally be one of extensive variation. Evolution consists in part of the addition of new characters through mutation which we can reasonably assume has some sort of time correlation. We are forced to assume that some sort of isolation is necessary for speciation to occur, for otherwise the new mutations would be merely added to and increase the specific variability. But that this isolation is always one that is attained geographically would scarcely be warranted as a conclusion, for there is little evidence in distributional patterns to suggest that sympatric species can result only from the immigration into a region of additional species which have attained reproductive isolation elsewhere. What I mean by the above remarks can be repeated this way. Nearly all genera of plants—and there is little reason to believe that animals differ significantly in this respect—show one and sometimes two or more regions where there is a concentration of species —the usual so-called center or secondary centers. The phylogenetic relationships and the distributional patterns (natural migratory routes) indicate not that such regions of high species frequency are the result of numerous immigrations of reproductively isolated species from somewhere else—i.e., a congregation from diverse sources—but quite the reverse. These mass centers seem more likely to be areas in which speciation has been going on for a relatively longer period of time. Kinsey is correct in saying that the characters which distinguish a category, whether subspecific, specific, sectional, or generic, are added on to a considerable degree at various points along a lineal migratory path; but his data, and even more so for plant data, do not appear to me to warrant the conclusion that the relation is exclusively that of a chain or that the chains do not in turn trace back toward an original center.

We might say that three different figures may represent phylogenetic relations in different parts of the living world. Some stocks seem fairly to be related in a radiate or dendritic manner; some are linear or chain-like; and in others the relations are reticulate or anastomose.

14.

Criteria for the
Indication of Center of Origin[1]

Forty years ago Charles C. Adams published a pioneer series of papers on postglacial dispersal of biota in North America (4, 5, 6, etc.), outstanding in their conception of process in biogeography.[2] In one of these papers (4) he listed ten criteria for the determination of centers of origin, and they were later reiterated (7) with further comments. Insofar as I know, these criteria have never been critically analyzed, although the concept of center of origin has been attacked by Kinsey (393). Rather, they have been largely accepted without question, despite the lack of substantiating data in some cases, and have been variously and somewhat loosely employed. It is time for an appraisal; hence it is the purpose of this chapter to review these criteria in the light of more recent contributions to the science of plant geography. Findings in the field of genetics in particular, and in the study of wild populations supply reasons why certain of the criteria cannot be tacitly accepted.

The literature of plant and animal geography, taxonomy, and evolution is replete with statements concerning the center of origin of certain species, species groups, genera, etc. For example, Babcock and Stebbins (38) say, "The distribution of the genus *Youngia* taken as a whole is entirely consistent with the conception that it is a natural group which had its origin

---

[1] This chapter was presented as an invitational lecture on the program of the Seventy-fifth Anniversary Celebration of the Torrey Botanical Club, New York City, June, 1942.

[2] It is interesting to note that Adams' inspiration undoubtedly came in part from Engler, who in turn was one of the few scientists to appreciate fully the significance of the early essays by Asa Gray on plant geography. That the science of plant geography has never developed far in the United States may be evinced by the fact that it was all of twenty years after Adams' papers that the first real contribution of the type they suggested was published in American botany, the valuable and frequently quoted paper on the vegetational history of the Middle West by Gleason (276).

in southeastern Asia and that evolution has been accompanied by extension of the geographic range to its natural limit on the south and east and slightly beyond the great mountain barrier to the north and west." Here is seen an example of the almost universal tacit assumption that phyletic stocks have a center of origin; yet there are cases in which this is not so.

Some species do not necessarily have a center of origin in the sense of a restricted geographical spot where they arose. Thus Gleason (276) states:

It is probably true . . . that many of our species have not had a single point or even a limited area of origin. The bulk of later Tertiary plants, so far as paleontological evidence indicates, are of genera still existing, and many of the comparatively few known species of Pleistocene plants are either identical or closely similar to existing species. . . . The distinction between such Pleistocene plants and their modern representatives may depend largely on a break in the record, on a period from which fossils are lacking. Probably if a complete series of specimens were at hand, showing comprehensively the maples of the eastern United States, for example, from the Pliocene to the present time, it would be seen that some of the earlier forms are absolutely continuous with our present species and that the slight morphological distinctions between them are only the result of continuous slow variations throughout the centuries. According to this view, many modern species had no localized origin and are not the offshoot of any parent, but represent the mass development of a species, which, under our present taxonomic ideas, came to a stop at the beginning of a break in our geological record of it and reappeared as a new species at the beginning of our next experience with it.

A different situation is emphasized by Kinsey (393), as was seen in the preceding chapter. Denying both the usefulness and the truth of the concept of center of origin, he demonstrates through his taxonomic work with the gall wasps that species differ by many genic factors that have been added gradually to the population as it has migrated.

Two other situations can be mentioned in which, in the strictest sense, there is no single center of origin. Chromosome (genom) doubling may happen many times in many places in a diploid population. The resulting autotetraploids, which may be good species by any criterion, do not necessarily have a center of origin other than the area of the entire progenitor diploid population. Baldwin's map (44, 45) showing the chromosome races of *Galax aphylla* is of interest in this connection.

It is becoming increasingly apparent that many plant species are of hybrid origin. Sometimes a swarm of diploid hybrids, segregates, and backcrosses have attained a sufficiently distinct character and area that their population has been given specific status. At other times polyploid complexes develop.

Stebbins (605) says, "Dissolution of genetic barriers and exchange of genes between genetic systems that are completely isolated from each other in the diploid condition are made possible by the synthesis of polyploid complexes through allopolyploidy between three, four or more species, following the introduction of genes from all the species concerned."[3] For example, according to Camp's studies (117), *Vaccinium corymbosum* is a tetraploid hybrid complex that has no center of origin in the usual sense. Three interfertile tetraploid species of distinct origin whose ranges were separated in pre-Pleistocene time were commingled as a result of Pleistocene migrations. The result was the development of a great hybrid swarm that found abundant suitable habitats in the open areas left by the retreating ice. Genic materials contributed by the three species are found in all combinations, and with different frequencies in different parts of the area of *V. corymbosum,* accounting for the polymorphic nature of the species. It is impossible to assign a center of origin to *V. corymbosum.* One contributing tetraploid was originally Ozarkian (*V. arkansanum*), one was in the Appalachian upland (*V. simulatum*), and one was eastern coastal plain (*V. australe*).

With the realization that under certain circumstances there may not be a center of origin, let us go on to a consideration of the criteria proposed four decades ago by Adams. I wish it understood that this evaluation of them is in no way a specific attack on Adams' paper, which was breaking new ground at that time, but rather a criticism of the present-day employment of these rules without evaluating them in the light of more modern knowledge and without recognizing their limitations. As a matter of fact, by 1909 Adams was careful to point out that he understood the criteria to be only "convenient classes of evidence to which we may turn for suggestions and proof as to the origin and dispersal of organisms. . . . In some cases a criterion may have great weight, while in another taxonomic or ecologic group it may have no value. . . . It should be clearly emphasized that it is the convergence of evidence from many criteria which must be the final test in the determination of origins. . . ."

Criterion 1. Location of greatest differentiation of a type.—With reference to this criterion of center of origin, Adams (4) says, "It is a very fundamental law that most forms of life are confined to restricted areas and only a small number have extensive distribution. Thus, from the center of origin there is a constant decrease, or attenuation in the number of forms which have been able to depart far from the original home."

This criterion is legitimate and applicable if we make two assumptions.

[3] See also (37) and (289).

In the first place, the basic assumption underlying the whole thesis is that there is a center of origin for a phyletic stock. This has already been discussed in the preceding pages. The other assumption is that there is a time relationship in evolution, that polymorphism increases with time; and that there is an age-and-area relationship, that with age the population of a species or other group tends to increase and occupy a wider area.[4] If we can accept these assumptions, it is clear that there will tend to be more polymorphism in the region of origin of a phyletic stock than away from this center. In such a region there will be more forms (biotypes, subspecies, species, sections, etc.) because of the longer time in which evolution has been occurring in the steadily increasing numbers of different kinds. With time, some of the forms originating in the central region will attain wider areas. They in turn may give rise to new forms away from the center, but in the nature of the relationship the original area will tend to exceed any derived peripheral area in the number of kinds represented. With respect to this point, Payson (499) says, "There is much evidence for believing that *Lesquerella* originated at some point in Central Texas and from this point as a center has spread over the large area that it now occupies. . . . From purely theoretical standpoints also, the greatest number of species might be expected to occur in the vicinity of the point of origin, since there the genus would have existed for the longest period of time." Mason writes in a recent publication on *Ceanothus* (455): "The occurrence of many isolated local species along the coast as against a few widespread species of the interior would indicate that the direction of the *Ceanothus* migration was from the coast to the interior." Employing this criterion alone frequently produces strong circumstantial evidence for the location of a center. Some examples follow.

A good example of the use of this criterion, which also is admirably supported by phylogenetic and geological data, is the study of *Gaylussacia* by Camp (116). According to him, "It becomes apparent that the genus arose in South America for there, today, we find it as a series of interlocked species-groups still differentiating out of a common plexus, only three of which have given representative members to North America." The work of Szymkiewicz (620) indicates a concentration of Mediterranean species of various genera, especially endemic species, in western Mediterranean regions. One example of this type will be sufficient. Sirjaev (582), who has carefully mapped the distribution of the members of the Mediterranean genus *Ononis,* makes the following statement concerning center of origin:

[4] In this connection, see Willis (696, 698) and the numerous expert criticisms of his hypothesis.

"Das Entstehungszentrum der Gattung (*Ononis*) war wahrscheinlich auf der Iberischen Halbinsel und im nordwestl. Mediterranen Afrika, wo jetz noch alle Subsektionen und viele endemische und fast alle älteren Arten sich konzentrieren, während im ostlichen Teile des Mediterraneums keine eigene Subsektion und nur drei endemische Arten anzutreffen sind. . . . Die Migration aus dem Entstehungszentrum fand in verschiedenen Epochen auf verschiedenen Wegen statt." The investigations of van Steenis (608) on isoflors (lines connecting regions of equal numbers of species in a genus) are another example in which a strong indication of center of origin is obtained. Perhaps the most intensive as well as extensive studies of plants and their centers ever made are those by Vavilov (659) and his colleagues. The following quotation is pertinent:

Cultivated species as well as their closely allied wild relatives in their evolution, during the course of their distribution from the primary centers of species-formation, have been differentiated into definite ecological and geographical groups. . . . We found it necessary to elaborate a new, a more detailed morphological and physiological system [of taxonomy] based on a study of the evolution of plants from their primary regions, which are usually characterized by the presence of a great diversity of botanical varieties. . . . Primary regions are at present characterized, as a rule, by the presence of many different species (in the sense of Linnaeus). They reveal practically the entire systems of genera. Transcaucasia, for instance, as regards wheat, rye, flax, peas, lentils, vetch, and chickpeas, is characterized by a great diversity of Linnaean species and closely related wild genera. Here, among wild and cultivated species of wheat, we have found all the basic chromosome sets ($2n = 14, 28, 42$). Here we have found many endemic species. Here we may trace all the links between wild and cultivated types. Here, in the great diversity of conditions specific for mountainous Caucasus as regards humidity, temperature, etc., we have established the presence of a great ecological and physiological diversity of species and varieties of wheat. The same applies to all the other cultivated crops.

It is necessary, however, to recognize that this criterion cannot be accepted as universal, for it only describes a tendency that, under certain conditions, is counteracted by the operation of other factors, as is age-and-area. A few of these conditions will be described.

The development of many species requires either that the forms be allopatric and have geographical isolation or, if sympatric, that they have ecological or some form of genetic (internal) isolation. Regions in which there are many closely related species are usually regions of habitat diversity. Note in the above quotation from Vavilov that the centers of variation are also centers of habitat variety, as in the Caucasus. It is entirely possible, then, that a phyletic stock that has had its origin elsewhere may, through migration, encounter a region in which there are numerous available ecological niches that are unsaturated, that is, in which competition pressure is low. Such a region may provide a variety of habitats with at least partial isolation. Under these conditions a phyletic stock may show a "burst" of evolutionary radiation. It is apparent that such a region of polymorphism is not necessarily indicative of the original center of origin or of dispersal, but is a fortuitously derived center of differentiation. It may constitute a secondary center of evolution and of dispersal. Returning to Vavilov, we find an example. "Secondary regions, as well as primary regions, may be of great importance for practical plant breeding. Chinese wheats and barley, which no doubt are of secondary origin but which have been elaborated during the last four or five thousand years, show many original characters. . . ." Two more examples of this general type can be taken from Fernald's criticism of age-and-area (243). He uses Schonland's conclusions (555) concerning *Erica,* which has nearly 1000 species in South Africa. There is not a single known fact that indicates that the genus arose in South Africa where there are the most endemics and the greatest diversity (species and sections). Willis had concluded that the number of endemics in any genus would rise gradually to a maximum at or near the point where the genus entered a land area (such as New Zealand), or where it had its center of origin. Of this corollary of age-and-area, Schonland (555) says, "Applying this prediction to the genus *Erica* in South Africa, this point would be a part of Southwest Cape Colony west of George, where not only a large number of endemics are massed, but where, moreover, the greatest diversity owing to formation of subgenera and derived genera is to be found; but I fear no contradiction when I assert that it is certainly not the place where the genus *Erica* entered South Africa, or where it originated."

Further evidence as to the care required in arriving at conclusions concerning geographical problems is illustrated by *Senecio.* J. Small (in 696) localizes the evolution of the Composites through *Senecio* in the northern Andes in Upper Cretaceous time, because of the present great expansion of that large genus in the Andean region. *Senecio* in the mountains of tropical America is in the most active stage of maturity, according to Greenman (299), not because it originated there but because it is in a region geologically young and diversified. Small's and Willis' conclusion regarding *Senecio* rests on what Fernald (243) gleefully calls a "colossal geological error," because the present great elevation of the Andes, where *Senecio* now has its magnificent development, did not occur until the close of the Tertiary (Pliocene) and the beginning of the Pleistocene. From Schuchert's recent historical geology (559), however, it appears that the Cordillera Occidental and the still more western and low Cordillera de Choco of northern South America are more ancient elevated land masses than the central and eastern Andes on which pre-Cenozoic plant developments might well have occurred, and from which much of the modern Andean flora must have been derived.

Another condition causing exception to the tendency described by the criterion of center of origin where the greatest differentiation of a type exists is that resulting from polyploid complexes. Stebbins (605) says, "Such a complex consists of two or more diploids, which are well isolated from each other genetically, which therefore are, when taken by themselves, quite distinct from each other. Among the polyploids there may be autopolyploids which are nearly or quite impossible to distinguish from one or the other of the diploids except by counting the chromosomes; there are allopolyploids that are exactly intermediate between two diploids, and there are all sorts of secondarily derived polyploids. Thus among the polyploids the gaps between species no longer exist or are at least very much smaller and harder to recognize. Polyploidy, therefore, tends to break down genetic barriers and to permit exchange of genes between genetic systems that in the diploid condition are completely isolated from each other." Examples of such complexes include *Crepis, Zauschneria, Rosa, Rubus,* and sections of *Potentilla, Antennaria,* and *Taraxacum;* dysploidy may increase the intricacy of the complex. Goodspeed and Bradley (289) note Kostoff's conclusion (403) that "amphidiploids from F_1 hybrids in which meiosis is asyndetic may give rise to monomorphic species, but when gametic sets of the F_1 are relatively equivalent and allosyndesis occurs, and if a series of segregated forms can survive, a polymorphic species is produced. Since inconstant amphidiploidy may originate a series of adaptable forms, they frequently afford more suitable material for natural selection than the highly constant amphidiploids."

In every case, according to Stebbins, the majority of the basic diploids are relatively restricted in area, whereas most of the widespread types are polyploid. According to him, "The center of distribution of the diploid species of a polyploid complex is naturally the center of variation of the complex as a whole. . . . The position of this center with reference to the complex as a whole varies in different genera; the relative distribution of the diploids and polyploids is not always the result of the same climatic and ecological agents; . . . the diploids tend to occupy the older, more stable habitats. This makes the study of polyploid complexes very important from the standpoint of plant geography." Such centers of variation as are due to hybridization and polyploidy may develop at the center of origin of a genus, but that is not necessarily the case. The American species of *Crepis* have such a center in the Pacific Northwest, but the stock immigrated from the Asiatic center of the genus (39).

A third type of exception to the criterion consists of such phylogenetic stocks as have developed a center of variation at the center of origin in the orthodox manner, but which have suffered a decimation of the group at the center as the result of physiographic and climatic changes. Through emigration and extinction the variety of types may be reduced in one region so that a secondary center comes to contain more variety. The center of origin for certain species may now be under the sea, in once poorly drained regions that have subsequently become dissected, or in a territory where now an entirely different climax prevails for which the species have no adaptation.

Hultén (360) has also come to the conclusion that "it must . . . be unsafe to assume that a plant originates in the place where it has its most numerous relatives. In most cases such a consideration will perhaps be correct, but in others it must be misleading." He illustrates this point by reference to old, widespread arctic-montane species, and says, "It is natural therefore that in different parts of the area of a Linnaean species considerably differentiated races should be found. The area has repeatedly been split up, during the glacials under the influence of a cold climate in the north and a pluvial one in the south, and during the interglacials under the influence of drought and heat. Each of these agencies must have caused a selection of biotypes in its particular direction. . . . The idea is current that a district in which a plant shows much variation or has many closely related species must be its original home. According to the above point of view, this would only mean that the plant has been present within the district for a comparatively long time and has developed in different directions under the pressure of varying conditions there. . . . I therefore think that the study of areas will prove valuable and should go hand in hand with the study of variation and taxonomy when we study the questions about the origin and migration of biota. The similarity or dissimilarity of two types alone will hardly be able to settle discussions concerning relationship between them." Hultén arrives at this latter conclusion because of the complication resulting from "parallel selection" of biotypes by separated but climatically similar regions.

We have seen that the location of greatest differentiation of a type may be at the center of origin of the group, and also that the criterion cannot be uncritically applied for a number of reasons.

Criterion 2.   Location of dominance or greatest abundance of individuals.
—In connection with this criterion it is first necessary to note that dominance
and abundance are concerned with different phenomena. Dominance is a
matter of the control of a community through reaction and coaction. Abundance is a matter only of numbers of individuals. It is true that certain
forms may exert dominance through mere numbers—this is possibly more
frequent among plants than animals—but often it is true that less abundant
forms are dominants by virtue of their life form or strong actions.

Species that are dominants in a certain community (there are usually not
many such species relative to the floristic composition of the community as
a whole) range more widely than the area of the community. For example,
beech, sugar maple, hemlock, and yellow birch all range more widely than
the northern hardwood climax association in which they are codominants.
It seems to me that dominance for a species can have no meaning except
in terms of community dynamics. If, however, we consider a genus, there
may be some instances in which the regions where certain species are community dominants or codominants are also the regions where there is a large
concentration of species of the genus. This appears to be true for *Quercus*
and *Hicoria* in eastern North America. They have areas of species concentration and of species dominance—in the oak-hickory climax association—
in the Ozark and Cumberland regions. Even here, however, a different interpretation is likely. These are ancient land areas in which evolution has
long been going on, and the numbers of species and their dominance may
be unrelated phenomena, and unrelated to the center of origin.

The center of greatest abundance of individuals, the center of frequency,
has a special meaning only in connection with the distribution of the members of a population, a subspecies, a species, etc. The assumption that the
center of abundance is also the center of origin for the type has to be based,
it seems to me, on an hypothesis that the species arose in the habitat where
it is best capable of abundant reproduction and establishment. This is a
gratuitous assumption. A species arises because of a number of interacting
factors and, with respect to environment, it is required only that the habitat
does not exceed in rigor the tolerance of the individuals of the population.
It is reasonable that, with migration from the center of origin, a species
population may encounter more favorable conditions than those that prevailed where it arose. In such a case it would be able to increase its relative
frequency in the new locality where conditions are more favorable.

Hultén (360) makes the following remarks concerning the "mass center"
hypothesis: "Christ and other authors considered that a plant is likely to

have originated in a district where its most numerous individuals are now found. Heer already opposed this view. It is natural that if a plant at the border of its perhaps wide original area should find favourable conditions and multiply freely, so that numerous individuals are developed, such a phenomenon will afford no indication of the earlier history of the species." Such cases are apparently found in certain weedy species of *Tradescantia* that have obtained wide areas and relatively high abundance in the eastern grassland and agricultural areas (24). Also, as with Criterion 1, we can conceive of climatic deterioration causing a reduction in the number of individuals at the center of origin.

Shreve (577) has pointed out that shrubs of the Sonoran desert with hard wood, sparse branching, and determinate growth (*Cassia, Mimosa, Acacia, Croton, Karwinskia, Caesalpinia, Lysiloma, Bauhinia, Acalypha,* etc.) belong to genera which are well represented in the thorn forest, with respect to both numbers of species and abundance of individuals. Furthermore, distributional data indicate that this type has spread from the thorn forest into the desert. However, Shreve (575) has clearly shown for *Larrea tridentata* and *Franseria dumosa* what is probably a widespread relationship—that variations in plant size and abundance, and degree of dominance are correlated with environmental conditions.

It is of interest to inquire further into certain characteristics of the distribution of individual plants. Gleason (277), who has studied this matter statistically for species within an association, says, "Environmental differences are not of sufficient magnitude to affect the distribution of the species; . . . the distribution of species is primarily a matter of chance depending on the accidents of dispersal; and . . . the number of individuals of a species, other things being equal, is an index to its adaptation to the environment." But what, we may ask, is the behavior of the species outside its native association or at the margin of its range?

When the area of a population of a new species or subspecies is expanding from its center of origin and when natural barriers have not yet established a boundary, there will naturally be a centrifugal decrease of density. The annual dissemination of diaspores will tend to extend the periphery of area. Further dissemination within the area as a whole will, however, tend toward consolidation of the area and an increase of abundance progressively toward the center. This would seem to be an inevitable result of numbers and random dispersal, and to provide a case in which the criterion is true.

Let us assume that a species population has extended its area to its maximum, having met barriers of one sort or another on all sides. Under these conditions, what will be the abundance relations between the center and the periphery of area? It would seem that there would be a tendency for a greater density of individuals to exist away from the periphery of area because of a central harmony between ecological requirements and ecological conditions. Everywhere outside of this central "typical" climatic region to which the species is adapted there will be, for it, a progressive deterioration of the climatic type. Climatic fluctuations will become more pronounced and frequent, and, relative to the tolerance of the species, extremes will be more often met and will be more often limiting to the distribution and density of individuals of the kind. That is, in marginal regions where the climatic type begins to grade into another climatic type, there will be fewer and fewer suitable spots for the species. Of necessity, if this is true, the density of the species will tend to decrease toward the periphery. Some interesting data concerning the behavior of species at the margin of range have been published by Griggs (303). From a consideration of 123 species that reach the edge of their area in the Sugar Grove district of southern Ohio, he found that 47 per cent were common in many stations, 7 per cent common in a few stations, 3 per cent abundant in one station, and 9 per cent rather common. This is a total of 66 per cent of the species that are abundant in their respective stations at the edge of their ranges. Only 20 per cent of the Sugar Grove flora that occurs at its margin of range is scarce or rare there. According to him, "It is clear from these lists that in this region the species in which the individuals become scarcer and scarcer until it fails altogether is exceptional." Griggs also discovered that failure of the reproductive function was an unimportant factor in the termination of ranges in the region; he says, "Even a considerable falling off in seed production would not necessarily affect the abundance of the species." Certain species are approximately continuous up to the margins of their range, but others are increasingly discontinuous until they are characteristically disjunct, and sometimes widely so, in the peripheral portion of their areas.

In the light of these data, it would seem that the criterion of species dominance and density is by no means an infallible guide to center of origin. Dominance and density are frequently highly irrelevant in this respect.

Criterion 3. Location of synthetic or closely related forms.—The statement of this criterion is obscure. From the context and through correspondence I find that by "synthetic" is meant generalized or primitive forms of a phyletic group. With this half of the criterion we can have no quarrel this far; the most primitive form or forms of a group certainly arose somewhere, and wherever that was, there is the center of origin of the group. But to ascertain that center, after a group has had a long history, is another matter.

It is frequently claimed that the center of origin for a group is where the earliest fossil forms were found, whether or not the group is represented there today. For example, it has been claimed that the shell family Pleuroceridae had a western origin because its earliest record is from the Laramie formation (Colorado, etc.). Adams (8), however, concluded that the family, and especially *Io,* had a southeastern origin centering in eastern Tennessee despite the absence of substantiating fossils.

There are two diametrically opposed views. The most widely accepted view is that the most primitive members of a group are still to be found at or near the center of origin of the group. This is frequently true, to a certain extent at least, because most of our temperate genera date back to the Cretaceous or early Tertiary and their primitive forms are usually found concentrated in the old land areas. In the United States, for example, such ancient land masses with primitive species (276) include the Southern Appalachian center, the Cumberland and Ozark center, the prairie center of Nebraska, Kansas, and eastern Colorado, the southwestern desert center, etc.

In a study of *Lesquerella,* Payson (499) concluded that the center of origin of the genus was in the old land area of central Texas where "not only are these species primitive, but in no other locality may be found anything like an equal display of what have been considered ancestral characteristics for purely morphological reasons. . . . The periphery in general is bounded by highly specialized members of the genus."

The opposite view concerning the location of primitive species of a group is that the primitive forms are to be found at the periphery of area because they have been crowded from the center by the younger and more aggressive members of the group. We can recall also that, according to the tendency described by age-and-area, the oldest species should have become the widest species in area, on the average. The employment of such a criterion as this depends in part upon the validity of taxonomic criteria for the indication of primitiveness. Many of these criteria (as enunciated for botanists by Bessey and others) deserve critical analysis.

The most skillful proponent of the view that primitive forms are peripheral is Matthew (457). The following quotations from *Climate and Evolution* (pages 10, 11, 31, 32) reveal his hypothesis, which is extensively documented by vertebrate paleontology and phylogenetics:

Whatever agencies may be assigned as the cause of evolution of a race, it should be at first most progressive at its point of original dispersal, and it will continue this progress at that point in response to whatever stimulus originally caused it and spread out in successive waves of migration, each wave a stage higher than the previous one. At any one time, therefore, the most advanced stages should be nearest the center of dispersal (original), the most conservative stages farthest from it. . . . To assume that the present habitat of the most generalized members of a group, or the region where it is now most abundant, is the center from which its migrations took place in former times appears to me wholly illogical and, if applied to the higher animals as it has been to fishes and invertebrates, it would lead to results absolutely at variance with the known facts of the geological record. . . . Whether the evolution of a race be regarded as conditioned wholly by the external environment or as partly or chiefly dependent upon (unknown) intrinsic factors, it is admitted by everyone that it did not appear and progress simultaneously and *aequo pede* over the whole surface of the earth, or even over the whole area of a great continent. The successive steps in the progress must appear first in some comparatively limited region, and from that region the new forms must spread out, displacing the old and driving them before them into more distant regions. Whatever be the causes of evolution, we must expect them to act with maximum force in some one region; and so long as the evolution is progressing steadily in one direction, we should expect them to continue to act with maximum force in that region. This point will be the center of dispersal of the race. At any period, the most advanced and progressive species of the race will be those inhabiting that region; the most primitive and unprogressive species will be those remote from this center.

Cytogenetics, however, is providing a body of information for several groups that points undeniably toward the forms that are primitive in a group. One example of this type will be sufficient. Anderson (16) says, "In those species which have both diploid and tetraploid races we . . . know that the tetraploids must have originated from the diploids." Tetraploid *Tradescantia occidentalis* ranges throughout the Great Plains and the eastern Rocky Mountains and has a small diploid area in central and eastern Texas. Tetraploid *T. canaliculata* occupies a wide area in the Mississippi Valley and is diploid in the same territory in Texas. *T. hirsutiflora* and *T. ozarkana* also exhibit the same tendency. The combination of cytology with geological history and taxonomy suggests strongly that the Edwards Plateau area of central Texas was the immediate center from which the American tradescantias have developed in comparatively recent times.

With respect to the other point of the criterion, it can be said that closely related forms can come to be located almost anywhere within the generic area. The nearest relative of any form, however, will tend to be near by, at least at first, because of the filial relationship between them. According to Kinsey (393), the picture of evolution is that of a simple or infrequently branching chain. In this chain each species is a derivative of a previously existing species, usually without extermination of the parental species. In this connection the question of isolation is again of great importance. For the Cynipidae, Kinsey rephrases Jordan's Law as follows: "Closely related species are to be expected in adjacent geographic areas on the same or on closely related hosts, or in single geographic areas on distinct but related hosts." For sympatric species, we may add to host isolation the various types of ecological or physiological isolation which are more generally applicable.

When one looks at a large family, for example, it is apparent that it is not everywhere equally well developed or rich. A certain tribe composing, say, 10 per cent of the family may constitute 30 or 40 per cent or more of the family in one region. This phenomenon is likely to be true for the other tribes. Such regions of differentiation are probably regions of speciation or origin, except where, for historical reasons, they are known to be regions of preservation.

I cannot see, however, that closeness of relationship among species can ever be employed as a criterion to indicate the geographical center of origin of a group without the aid of other facts. We can only say that primitive and closely related forms may or may not occur at the center of origin.

Criterion 4. Location of maximum size of individuals.—In a discussion of the evolution of species through climatic conditions, Allen (12) reiterates a series of "laws" stated by him in 1883: (1) The maximum physical development of the individual is attained where the conditions of environment are most favorable to the life of the species; (2) the largest species of a group (genus, subfamily, or family) are found where the group to which they severally belong reaches its highest development, or where it has what may be termed its center of distribution. In other words, species of a given group attain their maximum size where the conditions of existence for the group in question are the most favorable, just as the largest representatives of a species are found where conditions are most favorable for the existence of the species. These conclusions were reached from the observation that "in the northern hemisphere, in nearly all types of both birds and mammals of obviously northern origin, there is a gradual decrease in the general size from the north southward in the representatives of a conspecific group. . . ." Later on, Allen says, "The variation in size from north southward is as gradual and continuous as the transition in climatic conditions."

It seems to me that within these quotations themselves, employed by Adams and others, the "cat is out of the bag." In the first place, size is a specific character that may not be related to environment. Size differences may be due to biotype selection across a climatic gradient, or to phenotypic expression. Allen's statements concerning size and favorableness of environment are generally correct, but there is no necessary relationship between size and center of origin or center of distribution. It would seem that geographical trends in adaptive characters are usually nothing more than the *clines* of Huxley.

Allen's statements were questioned by Cockerell (158), who said, "I found in that genus (*Hymenoxys chrysanthemoides*) a case which seemed to me to exactly agree with those postulated by Dr. Allen, except that the large form was southern, the small one northern." It is a common observation among botanists that plants on oceanic islands, such as the Azores, Canaries, and the Galápagos, are frequently of larger stature than their relatives on the mainlands from which they were derived. This larger size of herbs, shrubs, and trees would seem to be related to the long growing season, rather than to any hypothetical indication of their island origin.

I have tried to find among plants an authentic case either in favor of the criterion or opposed to it in which the data are adequate, but have failed to do so. *Prosopis,* for example, attains its largest size (height of about 50 ft.) in the Rio Grande valley, where the genus occurs there near its periphery. The genus, however, is taxonomically complicated (52) and has had a long and obscure history as indicated by its split range, for it occurs in the South American deserts as well as in Mexico and our Southwest. The effect of environmental conditions on this group is important and confusing with respect to size. Shreve (576) says, "It is only in the most favorable situations that the mesquite is found as a tree. In less favorable ones it is merely a shrub."

The Southern Appalachians are becoming famous for their large trees as the region is better known. The largest single known specimens of *Picea rubens, Tsuga canadensis, Aesculus octandra, Tulipastrum acuminatum,* and several others are localized in the Great Smoky Mountains, but there is no evidence to indicate their origin in that region.

One situation in which the tendency is opposite to the criterion has been shown by cytology. Autotetraploids, and sometimes allotetraploids, are larger than their progenitor diploids. Furthermore, they have a strong tendency to extend the range of the group and to occupy peripheral positions relative to the diploids (22, 39).

There may be times when the conditions of the criterion hold true, but it is certainly not a safe criterion for the indication of the center of origin or dispersal. As a matter of fact, in a later paper (7) Adams admitted that "this entire subject needs critical study before its value and limitations can be fully understood."

Criterion 5. Location of greatest productiveness and its relative stability in crops.—From his comments, it appears that Adams considers productiveness to be closely related to size and numbers, and essentially a matter of growth and reproduction. According to him, Hyde (372) concluded that crop production, whether it averages high or low, will tend to be more uniform from year to year in the region where the crop is indigenous, and that the variability from year to year increases with departure from that center. In the first place, note that Hyde indicates that crop production is not necessarily high at the region of center, or where the crop is indigenous, but only that it is uniform from year to year. This does not fit well with Criteria 2 and 4. Furthermore, it appears that the term "indigenous" is employed not in its strict meaning of being "native" but in the more general meaning of being "at home," in the sense of being well adapted. It is well known that crop production shows the greatest stability from year to year in climatic areas to which the crop is best adapted (396). When such crops as cotton, corn, or wheat are grown in climatically marginal or submarginal areas, good production is obtained in favorable years and poor production in unfavorable years. The frequence and severity of the bad seasons and poor crop production increase with the extent of the climatic shift from the particular environment to which the particular crop is best adapted. This phenomenon appears to have nothing to do with the center of origin of the crop (658, 659) but is explained by weather and the operation of limiting factors (622).

Criterion 6. Continuity and convergence of lines of dispersal.—When species of a genus or higher category are distributed along natural highways of migration and when these highways converge on a certain area, the distributional pattern suggests that the region of convergence of these routes is the center of origin and dispersal. This suggestion is even stronger when, as is usually the case, unrelated organisms show the same pattern. There is no a priori reason, considering dispersal lines alone, why migrations need have been divergent from the apparent center rather than convergent on it. It is usually not difficult, however, to secure evidence (see Criterion 8) as to which direction the migrations took. Such evidence is largely obtained from comparative morphology and relationships. Sometimes paleontological evidence indicates the direction of migration. In other cases cytogenetic analysis of the related forms reveals without doubt the direction which the movement has taken. As expressed, and by itself, the criterion is not valid.

Migratory tracts are merely lines (however broad) of frequent, suitable habitats; they are not necessarily one-way routes.

In connection with the possibility of divergent and convergent lines of migration, an important question can be raised. Is it possible for migrations to occur simultaneously and in opposite directions within a natural region? The answer is yes, if we are concerned with organisms adapted to the natural area under consideration and if it can be shown that the organisms had their origin on opposite sides of it. In this case the migrations are taking place within the natural area and presumably could go in any direction until the natural boundaries are reached. If, however, two organisms are adapted to two different although contiguous natural areas, it is inconceivable that they can migrate in a convergent or divergent manner and transgress their natural climatic barriers. Any single region examined floristically will be found to contain extraneous forms (276, 303, 104, etc.), that is, species that in the given region occur to one side of their general area. There will be northern species near their southern limits in the region; southern species near the northern limits of their area, etc. In no case is it valid to assume, without further evidence other than distributional patterns, that southward and northward migrations took place simultaneously. In other words, convergence of lines of migration cannot be taken as an infallible indication of center of origin. When climax species are forced to migrate under the exigencies of a general climatic change, the result is a cliseral movement, and members of contiguous climaxes migrate simultaneously in the same direction. The direction depends on the nature of the climatic change. For example, during Pleistocene time climaxes in regions of climatic deterioration due to advancing glaciation migrated southward (speaking generally for continental North America). With glacial regression there was a reversal of climatic trend and a reversal of cliseral movement. Such oscillating movements have resulted in a modern intermingling of northern and southern elements in many regions, largely because relics were left behind in regions of environmental compensation. Any resulting convergence of lines of dispersal, however, is no indication of center of origin. Furthermore, the divergent migratory lines, in this case at least, do not represent simultaneous movements. Because of the inviolate harmony between climate and climax,[5] northward and southward movements must have occurred at different times. Pioneer species that play a role in early stages of succession may have a greater freedom of movement than climax species. If they do, it is because

[5] I am here using climax in the broadest conception as the major life-form landscape feature synonymous with biome or formation.

their climatic tolerances are generally broader than those of climax species, or the young habitats provide compensating conditions that allow the species to occupy more than one natural area.

Criterion 7. Location of least dependence upon a restricted habitat.—The use of this criterion for the indication of center of origin depends upon a species being more polymorphic at the center of origin (Criterion 1) or upon more primitive forms having wider tolerances. Both of these conditions may not be true. A wide species contains a large number of biotypes, perhaps many thousands (645, 651, 201). Progressively from the center of origin, and especially along narrow migratory tracts extending from the main area, there is a biotype depauperization. This can result from partial isolation due to distance alone. A remote portion of a population does not in practice, even if in theory, have access to the entire stock of genes of the species as a whole. When a species is divided into geographical subspecies and ecotypes, these conditions probably apply to them also, but less obviously.

On the basis of the Law of Tolerance (287), it is concluded that each individual organism can live only within the inherent limits of its tolerance for the environment, and that the tolerance of a species is the sum of the tolerances of the component individuals of the species population. Now it seems to me that this summation of Good's can have no real meaning for an individual. No individual can contain (inherit) all the genic variability of the population, although in a panmictic population any individual might theoretically contain any possible combination of genes. In many cases it is an observed fact that morphological polymorphism decreases away from the center of area of a species or subspecies. Although it is more difficult to demonstrate, it is reasonable to assume that individual members of a species differ as much physiologically as they do morphologically. In fact, it seems entirely likely that adaptation and ecological amplitude reside more in unseen features than in the characters of the type usually employed in systematic studies. Both, of course, ultimately result from the genic constitution of the individuals, and may be linked. In this connection Hiesey, Clausen, and Keck (345) say, "Within populations, hereditary variants occur, some of which may possess physiological qualities that give them the potential capacity to survive in different kinds of places. Other variations seem to have no significance for survival, representing random differences that are not incompatible with the main requirements of existence in their population." Just as individuals vary within a population, so may populations show a statistical difference which may or may not be adaptive and favor survival. It would seem to follow, then, that when polymorphism is greater near the center of area than at its periphery, it is entirely likely that there will be less dependence upon a restricted habitat at the center of area. This should not lead to the assumption that any one individual has a wider tolerance and a lesser dependence upon a restricted habitat because it happens to live near the center of area. If it means anything, it means that the various individuals collectively have a total breadth of tolerance that is wider near the center than near the periphery of area.

If primitive members of a group have a wider tolerance than more advanced ones, and if they are more likely to be found near the center of origin, there should be a lesser dependence upon a restricted habitat at the center. The wide ecological tolerance that primitive species are supposed to have is sometimes based on the paleontological evidence of large areas which species of modern genera are known to have had in Cretaceous or Tertiary times. This is frequently a spurious argument because many of these species are known not to have had these wide areas synchronously, and little is known of ecological subdivisions of the species. So far as I know, there are no physiological studies which indicate that primitive species have unusually broad tolerances. Circumstantial evidence, on the contrary, indicates that relic species are frequently markedly restricted in area and habitat type.

This problem has received at least one excellent consideration in paleobotanical literature. After pointing out that certain fossil floras of later Tertiary age contain mixtures of plants from widely different habitats, Axelrod (33) suggests that the explanation may be due not alone to overlap of floras (in ecotonal regions or from migratory mingling) or to the fact that Miocene and Pliocene vegetation was "generalized" and modern forests were derived by "climatic segregation only in the late Cenozoic," but to the ancient existence of ecospecies. For example, *Sequoia Langsdorfii* (close to *S. sempervirens*) was variously associated with species of boreal, warm-temperate, and temperate type. Other modern endemics, now of restricted type but once of wider association, include *Lyonothamnus, Ginkgo, Glyptostrobus pensilis, Picea Breweriana,* and *Quercus tomentosa,* according to Axelrod. He says, "It seems highly probable that many Miocene and Pliocene species related to living endemics may represent extinct ecotypes of more widely distributed Tertiary ecospecies." Probable as this concept is, it still does not show that primitive species are of wide ecological tolerance and recent ones of narrow amplitude. The late Cenozoic was a time of climatic breakup and, for many species, biotype depauperization, with only "senile" relic endemics remaining; but, as Axelrod supposes, the wide area and diversified conditions under which certain Tertiary species lived were due to the biotype (ecotype) richness of the species. That richness represents the mature condition of a species history. In youth and old age a species is likely endemic and poor in biotypes, first because of its youth and last because of elimination through the vicissitudes of change. Each species presumably tends to go through such a cycle; and, as there is no proved synchrony, at any time there should exist side by side, young, mature, and old species.

As with the preceding criteria, we find ourselves confronted by many "if's." The above arguments concerning the region of least dependence upon a restricted habitat are applicable in the determination of center of origin only when this center is also the center of variability and when it has not been disturbed and reduced in biotype richness.

The idea that a species is usually ubiquitous in the center of its range, occurring in all kinds of places and restricted to only the most favorable sites at its areal limits, according to Griggs (303), is probably attributable to Blytt and has been favored by Cowles (162). This contains the assumption that the favorable climate in the central portion of the species range somehow overcomes the edaphic factors. I remember Cowles, when lecturing on the dunes of Lake Michigan's shores, saying of the cactus, "It sits on the southern and western slopes, looking toward its home." There is, of course, a large element of truth in this generalization, as is shown by the usual disposition of preclimax and postclimax communities in any region. But what of the relationship as a rule? In the Sugar Grove district Griggs (303) found several exceptions; he says, "Even so few examples as these (*Castanea dentata, Rhododendron maximum, Aralia spinosa*) are sufficient to destroy the utility of the theory; . . . it cannot be generalized. . . . If climate were the principal factor restricting the spread of plants . . . the ranges should be fringed with outliers occupying habitats where the climatic conditions are locally favorable." He found this to be obviously true in only seven of 25 examples studied carefully.

Let us turn again to the often-cited polyploids. According to Anderson (16), "The diploid species are of limited distribution and even in those areas where they do occur are usually restricted to one particular habitat. By contrast, the tetraploid species and races have wide distributions and most of them have the ability to flourish under a variety of situations." Allopolyploids especially may combine the tolerances of their diploid progenitors.

In amplifying his discussion of this criterion, Adams (7) selects what seems to me to be a particularly vulnerable example. He says, "Outlying colonies tend to have a limited or restricted range. At the same time such colonies are peculiarly liable to become extinct, as they are usually near the limit of favorable conditions. . . . This is true of the 'boreal islands' in swamps within the glaciated portion of the continent. For example, members of the tamarack bog association, toward their southern limit, have very restricted or local range; but to the north, *the bog forest conditions, as it were, spread from the bogs proper and become of extensive geographic range,* as the water beetles invade the damp mosses. . . . These restricted, attenuated, or isolated colonies, dependent upon special conditions, are clearly indicative that they are pioneers or relics, which point toward the region where the range is spread out and becomes of geographic extent." I have italicized a portion of the above quotation to emphasize the fact that the areal pattern is apparently wholly dependent upon the pattern of occurrence of suitable conditions. This is an ecological matter that denotes nothing concerning origin. Adams goes on to say that the isolated colonies are *either* pioneer or relic, thus destroying his own thesis, in my opinion.

Criterion 8.   Continuity and directness of individual variations or modifications radiating from the center of origin along highways of dispersal.—This criterion, related to No. 6 (continuity and convergence of lines of dispersal), is often reliable. With respect to changes in character frequency, as shown by the mass-collection techniques (234), we can conclude only that there can be a gene flow in any direction through a population. Any attenuation of the frequency of a certain gene is presumably direct evidence of the center of origin of that gene in the region of highest frequency. One of the most interesting cases of this sort concerns the distribution of the recessive melanistic mutation in *Cricetus cricetus,* the hamster. Timofeeff-Ressovsky (627) has described this situation in his recent discussion of mutations and geographical variation. An exact study of the hamster was possible because it has long been important in the fur markets of Russia. In 1771 Lepekhin observed that in the region between the rivers Kama and Belaia in northern Bashkiria the hamster population contained a high percentage of melanistic animals. In 1934 Kirikov brought together the subsequent information concerning the hamster. Timofeeff-Ressovsky says, "In the course of the last 150 years this mutation has spread from its original center of high concentration along the northern border of the species-area. . . . Populations with rather high concentration of this gene are spread westward as far as the river Dnieper." Apparently the melanistic form is adaptive in the wood-steppe ecotone along the northern portion of the species area. This is one of the few cases in which it is definitely shown that mutations participate in the origination of geographical races.

When introgressive hybridization (20) is demonstrable and when a series of chromosome changes, such as a polyploid series, can be shown along highways radiating from a center, it would seem that the indication of center of origin is incontrovertible. When several characters show a parallel and direct continuity of gradation of frequency or of modification, it is likely that there has been active migration of the population from a center. This is sometimes recognizable by chains of subspecies, pairs of species, etc. For instance, Gleason (276) says that postglacial forest migration in the eastern United States was sometimes accompanied by specific evolution, and he gives as an example *Agalinis paupercula* of the glaciated region which was probably derived from *A. purpurea*. Sometimes the parent species, "existing through the glacial period south of the ice margin, has followed both routes of migration and in so doing has become segregated into a pair of closely related forms, one of which took the southern route leading northward from the Ohio Valley and the other the northern route along the Alleghenies and the Great Lakes. Thus *Trillium nivale, T. declinatum,* and *Cynoglossum virginianum* of the southern portion of the Middle West are paired with *T. undulatum, T. cernuum,* and *C. boreale* respectively of the northern portion." Payson's work on *Lesquerella* (499) provides a good example based on comparative morphology. He says, "In a graphic representation of the subsectional groups they may be shown by lines radiating from a common center. Such a diagram could be superimposed upon a map and in nearly every case the species at the base of each line of development would be nearer the Texas region [center of origin] than species derived from it."

Of this criterion Adams (7) said later, "This is perhaps not of universal application but carries much weight under certain conditions. For example, continuity of variations, as dwarfing or increasing size, have a certain definiteness which clearly points in a limited number of directions when correlated with highways of dispersal." Once again it can be said that this criterion alone is of no significance. A geographical series of size expressions may be due to environmental conditions reflected in growth responses (phenotypic changes in a genotype) or it may be due to selection operating through a region of gradually changing environment. One thinks immediately of clines (371).

When morphological, phylogenetic, and geographical data are used to support one another, the validity of the conclusions regarding direction of migration depends upon the validity of the morphological criteria employed. The validity of morphological differences as an indication of evolutionary progression (such as described by Bessey 65) needs critical evaluation in some cases.

Criterion 9.   Direction indicated by geographical affinities.—This criterion is frequently valid for organisms located at stations removed from the major area they occupy. As mentioned earlier, in any region there are usually numerous extraneous species representing two or more different floristic elements and recording as many different migrations in the vegetational history of the region. In this connection Grinnell and Swarth (307) say, "We cannot expect to derive universal laws for the behavior of species, to be applicable uniformly in any region . . . where two faunas meet. Perhaps the only general rule that can be laid down is that there is no exact concordance in the distributional behavior of all the animals of a region. . . . Upon reflection it is difficult to conceive of precisely the same set of delimiting factors operating upon any two species alike." For extraneous spe-

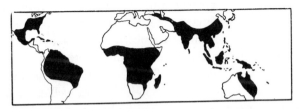

F<small>IG</small>. 25. Probable direction of migration of *Diospyros* to the United States as indicated by geographical affinity which is all with the south. Map from Fernald (245).

cies, it is frequently a fairly safe assumption that they were derived from the areas where they have their principal distribution. For example, a northern species found southward of its principal northern area was likely derived from the northern area. If a genus or family is largely characteristic of a single formation or climatic type and has one or a few species of a different type, it is likely that the latter migrated and evolved from the generic center. Bromeliads have migrated away from the humid tropics and entered the deserts of southern Mexico, and, conversely, cacti have migrated out of the desert region and established themselves as epiphytes in the tropical forest, according to Gleason (276). No one suspects certain rather large tropical groups as having a temperate origin because of a few temperate representatives, as in *Diospyros, Tripsacum,* and *Phoradendron;* quite the contrary. The point is well illustrated by a quotation from Merrill (463): "When a genus is described from material collected in a certain place and is known only from that region for many years, we more or less automatically accept it as a group characteristic of that region. If a representative of it is later found in another area, we are apt to consider it as an extraneous entity there. *Eucalyptus* is such a genus. It is tremendously developed in Australia, has a few species in New Guinea, and one which extends to the Bismarck

Archipelago, the Moluccas, Celebes, and the Philippines. We are justified in accepting it as an Australian element in the other regions. The same is true of the few phyllodinous species of *Acacia* outside of Australia. The one species of *Ulmus* in northern Sumatra two species of *Pinus* in Sumatra and the Philippines, and one species each of *Taxus* and *Gleditschia* in the Philippines and Celebes may be unquestionably accepted as Asiatic (continental) types in Malaysia." But this is not always the case, as Merrill continues to demonstrate. Returning to our own region we can cite a typical example.

Typical Atlantic and Gulf coastal plain species have long been known from the Appalachian and Cumberland uplands (267, 387). Sometimes these inland plants are rare, and stations are of small area and widely disjunct from the coastal plain where the species are now common. Fernald (245) has correctly hypothesized the origin of some of these species on the old lands that are now part of the Cumberland plateau, and Braun (83, 84) has found them most abundantly in the undissected portions of the now elevated peneplain. Fernald says, "With the Tertiary uplift of the Appalachian region and its final conversion into a vast well-drained mesophytic area . . . the Cretaceous xerophytes and hydrophytes which had previously occupied the ground gradually moved out to the newly available and for them more congenial Coastal Plain and similar habitats to the west and northwest." In such a case as this, the principal area is a derived one and is no indication of the center of origin. It really is a question not of coastal plain plants in the Appalachian and Cumberland uplands, but of upland plants in the coastal plain, if we view the relationship historically.

Not all coastal plain species in the interior have had this history. In his monographic study of the Scrophulariaceae, Pennell (502) has detected some forms that have migrated from the coastal plain into the Piedmont and the Blue Ridge provinces. Similarly, many coastal plain plants around the Great Lakes, but not all, were apparently derived from the Atlantic coast during early postglacial time.

When there is a close biological association between diverse types (as in parasitism or various symbiotic relations) and when the evolutionary and geographical history of one is known, it is probable that the other offers a parallel. In amplification Adams (7) says, "This criterion can be illustrated by reference to the Ajax Butterfly (*I. Ajax*). The sole food plant of the Ajax larva is the Pawpaw, a shrub clearly of tropical origin. The allies of *Ajax* are also tropical; thus the associated biogeographic (plant and animal) affinities clearly point to the tropics."

The direction of dispersal and the center of origin are many times indicated by geographical affinities, but the criterion cannot be used alone; the principal area and biographic type may be derived and the minor area relic.

Criterion 10. Direction indicated by the annual migration routes in birds. —Applied to plants, this criterion would be restricted to species whose diaspores are bird-disseminated, either epizooically or endozooically. If the migration takes place both northward and southward over the same route, as for some species employing the Mississippi Valley and others using the Appalachian uplands, the direction of plant movement is not necessarily indicated. In cases where the northward and southward migration paths are not coincident, the direction of movement is indicated.

Criterion 11. Direction indicated by seasonal appearance.—Although Adams was aware of this criterion at the time of publishing his first list (4), he did not include it until later (7). In the northern hemisphere, vernal activity suggests boreal origin. He also thought that there is an altitudinal as well as a latitudinal relationship, i.e., that mountain forms spreading downward should belong to the vernal aspect, and lowland forms spreading upward should belong to the aestival aspect.

It is undoubtedly true that such relationships between origin and aspect occur. It does not seem to me, however, that this criterion expresses any inherent indication of origin. The described relationship could exist, for example, for a form or series of forms occupying montane, subalpine, and alpine belts (or the corresponding latitudinal zones), with the center of origin in either terminal belt or the middle. The limitations to the spread of a form are found in the action of the whole environment upon the physiology of the form, with such factors as temperature, light intensity, and photoperiod operating. Therefore, it would seem as easy and sound to conceive of a vernal form of the south spreading northward with a change to aestival aspect, as the reverse. In my opinion, this fact illustrates perfectly the pitfalls of deductive reasoning and generalization.

Criterion 12. Increase in the number of dominant genes toward the centers of origin.—This criterion could have been proposed only after the development of genetics and is appended to Adam's older criteria because of its apparent validity. It can, I think, be attributed solely to Vavilov (657), who said, "The direct study of the centres of the origin of cultivated plants . . . has revealed not only a great diversity of forms but also a prevailing accumulation of dominant forms characterized by dominant genes in the centres. A considerable number of plants investigated show this regularity. . . . The secondary centres of the origin of forms are, on the contrary, characterized by a diversity of chiefly recessive characters."

Several cases are discussed by Vavilov, but only one will be mentioned here by way of illustration. The center of origin of cultivated rye and of the genus *Secale* to which it belongs lies in eastern Asia Minor and Transcaucasia. Here are all the species of rye and the whole diversity of characters of the varieties; but also here are concentrated the dominant characters of red ears, brown ears, black ears, and marked pubescence of flowering glumes. In the secondary centers are such recessive characters as liguleless leaves, yellow ears, and glabrous glumes. The recessive or dominant nature of at least some of the characters on which Vavilov bases the criterion has been shown by studies of segregation of heterozygous forms.

Cultivated plant types in their progress from their principal genetic centers seem to exhibit a "falling out" of the dominant genes and "proportionally to the spread of isolation proceeds the accumulation of recessive forms."

Criterion 13. Center indicated by the concentricity of progressive equiformal areas.—This criterion, developed by Hultén (360) primarily concerns centers of dispersal for arctic and boreal biota from refugia; but it also concerns centers of origin when evolution as well as migration has occurred. Hultén's thesis is as follows: From a refugium, each species tends to spread in all available directions, but because of different tolerances and capacities for dissemination it could not be expected that all plants would spread to the same extent or with the same rapidity. The result is a tendency toward the development of approximately circular areas of different size around the center; but in nature the theoretically circular form of areas is seldom attained because of various barriers. There still remains, however, the chief feature of areas: those plants that radiate from the same center have progressive equiformal areas of different size.

This criterion is obviously related to No. 6, stated by Adams. As developed by Hultén, however, there is a clean-cut scientific basis and the conclusion is reached through strictly inductive reasoning.

Conclusion.—Only one conclusion seems possible, and it carries implications far beyond the scope of the present discussion of criteria of center of origin. The sciences of geobotany (plant geography, plant ecology, plant sociology) and geozoology carry a heavy burden of hypothesis and assumption which has resulted from an overemployment of deductive reasoning. What is most needed in these fields is a complete return to inductive reasoning (519), with assumptions reduced to a minimum and hypotheses based upon demonstrable facts and proposed only when necessary (360). In many instances the assumptions arising from deductive reasoning have so thoroughly permeated the science of geography and have so long been a part of its warp and woof that students of the field can distinguish fact from fiction only with difficulty.

15.

Endemics and Endemism

1. The concept of endemism includes two types of organisms whose areas are confined to single regions: endemics, sen. str., which are relatively youthful species, and epibiotics, which are relatively old relic species. The concepts are applicable to groups other than species.

2. Youthful endemics may or may not have attained their complete areas by having migrated to their natural barriers. Epibiotics may, but frequently do not, contain the biotype richness which will allow or has allowed them an expansion of area, following their historical contraction of area.

3. The percentage and kinds of endemics in a flora are significant with respect to the history of the flora. A high degree of endemism is usually correlated with age and isolation of an area, and with the diversification of its habitats, as these factors influence both evolution (the formation of new endemics) and survival (the production of relic endemics).

Although the term *endemic* is very useful, it is exceedingly difficult to give it an exact definition. Its general meaning in biology is that of being confined to, or indigenous in, a certain region, as an endemic plant or animal. In this sense its antonym is exotic. The concept, as introduced by A. de Candolle, referred to a taxonomic unit or its area, particularly a species, limited to one natural region or habitat. This strict usage of the term is preserved by Szymkiewicz (620), who divides the species of a region, for the sake of floristic studies, into (1) endemics, species which are confined to a single natural area, or nearly so; (2) species which extend from a given domain to another that is phytogeographically equivalent; and (3) species partly in the given domain but extending extensively or throughout other domains. Although plants which are limited in their distribution and not to be found elsewhere in the world have constantly been referred to in geographical writings as endemics, it has long been known that plants of restricted range are of more than one type. Since the term *endemic* signifies that the organisms live with their own people, Ridley (526) proposed the term *epibiotic*, signifying survivors, to distinguish those endemics which are relics of a lost flora. If we follow him, the term endemic would be reserved for those organisms which are related to, or evolved from, other plants in the same area. If circumstances permit, there would appear to be no a priori reason why endemics could not spread and become ordinary, widespread, successful plants. He considers them to be "newborn" species. In contrast, epibiotics are relics of an earlier flora which has nearly disappeared from the region as a result of climatic change or other environmental vicissitudes. He says, "Epibiotics are at the end of their species life. They do not spread. Remains of a flora which has long disappeared, they persist only in an isolated spot which, for some reason, has not been overwhelmed by the later invading flora, but are unable to spread any further. They are usually unprovided with sufficient means of dispersal to cross the barrier of the modern flora, and to reach another suitable spot for their growth." (Such questions as those raised here are discussed elsewhere, as under the topics of "senescence" and the "differentiation theory of evolution.")

It should be noticed for endemics, whether relic or young species, that there is no means of an exact delimitation of the area. What constitutes "one natural region or habitat"? Sometimes the use of the term is precise, as in the case of islands, isolated mountains, nunataks, or strongly marked habitats such as serpentine soils or a particular salt desert; but generally speaking, the limits of an area in which a species is endemic are more or less arbitrarily defined. Nevertheless, the term is very useful, and an understanding of the phenomenon of endemism is exceedingly important in biogeography. It should also be noted that the term is commonly used in reference to political and larger geographical units. A plant may be said to be endemic to a certain state, to a country, or to a continent. Below are a few examples of the use of this term.

In a study of the families of flowering plants, Irmscher (375) found that certain ones are confined to the large continental masses, i.e., endemic to them. He divided the world into four large continental masses and found that 33 families are endemic to the Americas, 16 to Europe-Africa, 7 to Asia, and 5 to Australia-Polynesia. Szymkiewicz (616), in a characterization of the Mediterranean portion of the Iberian Peninsula (Portugal and all but northern Spain), lists 22 genera which are endemic there, including members of 11 families. Pax (497) notes that in *Acer* the Section Negundo is endemic in North America and, so far as fossil records show, always has been, but that the Section Rubra, now endemic in North America, was also Eurasian in the Tertiary. Skottsberg (589) says that in the Hawaiian Islands the palm genus *Pritchardia* contains over 30 species and that no species is found on more than one island out of the eight. Oahu alone has 9 species, each found in a separate valley and consisting of a single or a few groves.

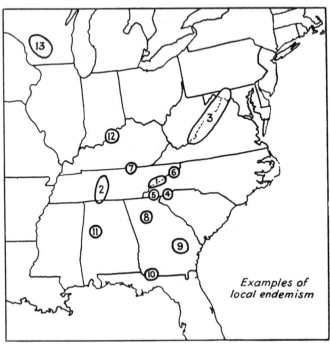

Examples of local endemism

FIG. 26. Illustrations of local endemism in the eastern United States. *1*. Great Smoky Mountains: *Calamagrostis Cainii* Hitchc., *Senecio Rugelia* Gray, *Glyceria nubigena* W. A. And., *Rubus carolinianus* Rydb; *2*. Cedar barrens of Middle Tennessee: *Lesquerella Lescurii* (Gray) Wat., *Petalostemon Gattingeri* Heller, *Lobelia Gattingeri* Gray, *Psoralea subacaulis* T. & G.; *3*. Shale barrens: *Allium oxyphilum* Wherry, *Eriogonum Alleni* Wats., *Oenothera argillicola* Macken., *Pseudotaenidia montana* Macken., *Solidago Harrisii* Steele, *Aster schustosus* Steele, *Convolvulus Purshianus* Wherry; *4*. *Shortia galacifolia* T. & G.; *5*. *Lindernia saxicola* Curtis; *6*. *Buckleya distichophylla* (Nutt.) Torr.; *7*. *Conradina verticillata* Jennison; *8*. *Amphianthus pusillus* Torr.; *9*. *Penstemon dissectus* Elliott; *10*. *Torreya taxifolia* Arn.; *11*. *Neviusia alabamensis* Gray; *12*. *Penstemon Deamii* Pennell; *13*. *Penstemon wisconsinensis* Pennell.

Turning to the eastern United States for a few examples, we can note that, among hundreds of others, *Conradina verticillata* has been found only in the Cumberland Mountains in the vicinity of Rugby, Tennessee; *Calamagrostis Cainii* is known only from Mt. LeConte in the Great Smoky Mountains; *Shortia galacifolia,* lost for nearly a century, has been rediscovered in two localities in the foothill valleys of South Carolina; and *Cirsium*

Pitcheri is known only from the sand dunes near Lakes Michigan, Huron, and Superior.

The number of endemics (and percentages of them) vary widely from region to region, depending largely on the history and condition of the region. The condition of endemism in islands is probably best known. Skottsberg (590) says that the flowering plant species of the Hawaiian

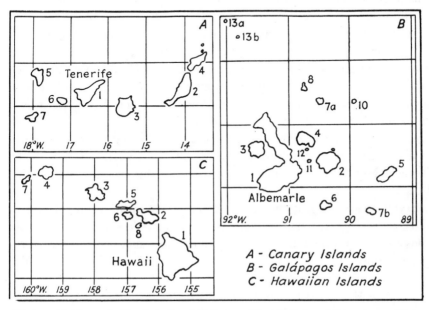

FIG. 27. Position and size of certain archipelagoes. The islands are numbered according to area. *A. Canary Islands: 1.* Tenerife, *2.* Fuereventura, *3.* Gran Canaria, *4.* Lanzarote, *5.* Palma, *6.* Gomera, *7.* Hierro. *B. Galápagos Islands: 1.* Albemarle, *2.* Indefatigable, *3.* Narborough, *4.* James, *5.* Chatham, *6.* Charles, *7a.* Bindloe, *7b.* Hood, *8.* Abingdon, *9.* Barrington, *10.* Tower, *11.* Duncan, *12.* Jervis, *13a.* Culpepper, *13b.* Wenmon. *C. Hawaiian Islands: 1.* Hawaii, *2.* Maui, *3.* Oahu, *4.* Kauai, *5.* Molokai, *6.* Lanai, *7.* Niihau, *8.* Kahoolawe.

Islands are nearly 90 per cent endemic, that the genera are about 20 per cent endemic, and that many of them altogether lack close relatives. Wulff (719) gives the following list of endemic percentages for certain islands: Canary Islands, 45; Corsica, 58; Madagascar, 66; New Zealand, 72; Hawaii, 82;[1] St. Helena, 85.

Mountains on land are in many respects like islands in the seas, because their isolation is relatively complete for reasons of climates and climaxes. According to Willis (695, 698), in the flora of Ceylon there are over 100

---

[1] Note the discrepancy between estimates for the Hawaiian Islands.

species confined to one or a few hilltops rising from a plateau. There are 10 such endemics in *Strobilanthes,* 8 in *Eugenia,* 6 in *Hedyotis,* and 4 in *Anaphalis;* all together there are 19 genera with more than one such endemic. In continental regions and large islands it appears that as a general thing endemism tends to be high in mountainous areas. Many endemics among flowering plants are known from the higher Southern Appalachians, and this phenomenon extends to the lower plants. Bryophytes can be selected for illustration because they are rather thoroughly known. Sharp (572) reports 15 species—each of a separate genus—known only from this area.

Endemism appears also to be higher, on the average, in older land masses than in younger. For example, the lands of the northern hemisphere which were covered by the Pleistocene ice sheets seem to be conspicuously low in endemics. Raup (518) says, "In general, the flora of boreal America is poor in local endemics, and such as occur are not evenly distributed." Endemics are commonest in the western mountains and the Bering Sea region. There is a considerable number about the Gulf of St. Lawrence, a few in the Torngat Mountains of Labrador, and a few in the region of Lake Athabaska.

Szymkiewicz (621) has made a study of endemism in oceanic islands of three archipelagoes: Canaries, Galápagos, Hawaii. We shall give this paper considerable space because it leads to some important conclusions concerning endemism. In Table 18 the islands are arranged according to size. The

TABLE 18.—Endemism in the Canary Islands (Data from 621)

	Area in Sq. Km.	Max. Altitude in M.	Endemic: One Island	Endemic: The Archipelago
(1) Tenerife	1946	3715	81	233
(2) Fuereventura	1722	860	11	51
(3) Gran Canaria	1376	1898	57	155
(4) Lanzarote	741	684	8	43
(5) Palma	726	2356	16	111
(6) Gomera	378	1340	17	102
(7) Hierro	278	1512	7	76

numbers of species endemic to each island are listed, together with the number on that island endemic to the archipelago. Both types of endemics (to one island, or to the archipelago) diminish regularly with the decreasing size of the islands except for Fuereventura and Lanzarote. These exceptions to the otherwise striking correlation are easily explained by the low elevation of these islands. Szymkiewicz states that the number of species evolved in a region (such as one of these islands) would seem to depend upon the age of the land area, i.e., the duration of evolution there, and the nature of the terrain, for evolution apparently depends upon its general climatic favorableness and the variety of conditions offered. In the Canaries the climate is dry at low elevations and humid in the high altitudes. As a result, it is more favorable and varied in the higher islands, and, on the average, more space and variety are offered on the larger islands. The factor of age is not analyzable for this archipelago, or at least the data are not available.

In considering the Galápagos the data are again arranged in Table 19

TABLE 19.—Endemism in the Galápagos Islands (Data from 621)

	Maximum Altitude in M.	Endemic: One Island	Endemic: The Archipelago
(1) Albemarle	5000	9	91
(2) Indefatigable	2300	11	75
(3) Narborough	5000	2	27
(4) James	2850	12	73
(5) Chatham	2100	14	84
(6) Charles	1500	18	91
(7b) Hood	640	0	19
(7a) Bindloe	800	0	15
(8) Abingdon	1950	1	38
(9) Barrington	900	0	19
(10) Tower	210	0	9
(11) Duncan	1300	3	26
(12) Jervis	1050	1	14
(13b) Wenman	830	1	4
(13a) Culpepper	550	0	3

according to size of the islands, and again there is the same tendency toward impoverishment of the flora with decrease of area. Also, the influence of climate is very clear, as in the Canaries, for the low islands (Hood, Bindloe, Tower) have fewer endemics than the islands of higher elevation. But the situation is more complicated by the intervention of the age factor. Among the first six islands, Narborough shows evidence of most recent volcanic activity (which parallels its impoverished flora), and James, Chatham, and especially Charles seem to be most ancient (which again parallels the floral condition). Thus, one has some basis for understanding the poverty of Narborough in comparison with Charles, although Narborough has about three times the area and is about three times as high as the latter.

TABLE 20.—Endemism in the Hawaiian Islands (Data from 621)

	Area in Sq. Mi.	Max. Altitude, Ft.	Endemic: One Island	Endemic: The Archipelago
(1) Hawaii	4210	13,823	43	163
(2) Maui	728	10,032	63	232
(3) Oahu	600	4,030	72	198
(4) Kauai	547	5,250	61	153
(5) Molokai	261	4,958	14	156
(6) Lanai	139	3,480	7	111
(7) Niihau	97	1,304	0	51
(8) Kahoolawe	69	1,472	1	44

In the Hawaiian Islands, the difference of age between the islands again plays a part. The age of these islands decreases from west to east, the most recent being Hawaii, where volcanic activity is intense. The most ancient is Kauai. One comprehends easily why Hawaii has fewer endemics than Maui, Oahu, and Kauai, although its surface is larger.

There would appear to be no reason why the conclusions which Szymkiewicz reached concerning the evolution of endemics on islands cannot be applied to other isolated areas. For example, endemism in a mountain system on a continental mass should be in proportion to its relative age (i.e., the time it has been available for occupancy by flowering plants), its area, its general favorableness for vegetation, and its variety of conditions. These are factors (especially that of age) which were emphasized by Fernald (245) in his study of specific segregations in eastern American floras.

In a study of temperate mountain plants in Malaysia, van Steenis (608) states that there is a very low proportion of endemics above the rank of subspecies. His line of reasoning is somewhat as follows: Widely distributed species are usually polymorphic. Migrating diaspores can, of course, carry to any new locality only a small portion of the specific polymorphy. Under the influence of isolation (which is characteristic of mountain forms) and of the inbreeding of members of small populations (which is always present in the early stages of immigration) the peculiar local combination of characteristics *quickly* becomes stabilized, perhaps in a few generations. Such isolates have frequently been called species, but van Steenis considers that they are usually only subspecific and that "the local subspecies will remain the same forever and there is no chance that it can furnish a source of new species. The idea that in such a case [where a single species only is represented by a local population outside of the generic areas as such] the formation of subspecies is the first step towards the origin of new species by isolation must be abandoned." Temperate mountain forms in Malaysia, then, consist of disjunct populations of conspecific relations with southeastern Asia (or elsewhere) for the most part, and the endemism in this group is principally that of subspecies. Endemism aside, the statement of van Steenis quoted above is preemptive, ignoring the possibility of mutations which could raise the endemic to specific rank (including the appearance of reproductive isolation of intrinsic nature) and the possibility of the future immigration of another member of the genus.

It may not be too great a digression at this point to consider in more detail some of the features of island life. Charles Darwin (177) was one of the first botanists to become interested in the problems of insular floras. In the *Origin of Species* he brought out three basic facts: (1) the species of all kinds which inhabit oceanic islands are few in number compared with those in a comparable continental area; (2) despite the relatively scanty flora, the proportion of endemics is often extremely large; (3) the proportional numbers of the different orders are very different from continental areas, certain whole taxonomic groups, which might be expected, failing to occur. In 1867 Hooker delivered his renowned lecture on "Insular Floras" before the British Association, and about the same time Wallace (670) made the first clear distinction between oceanic islands and continental islands, the latter being remnants of former continental masses or bridges. Wallace's famous book, *Island Life* (671), appeared in 1880, and Hemsley (339), in the publications of the Challenger Expedition, summarized the status of knowledge of various insular floras at that time. He divided islands into three groups: (1) those containing a large endemic element, including endemic genera the nearest affinities of which are not to be found in any one continent—St. Helena, Juan Fernandez, Sandwich (Hawaii), Galápagos, Seychelles groups; (2) those containing a small endemic element confined mostly to species the derivation of which is easily traced—Bermuda, Azores, Ascension, the Antarctic islands (including southern Indian Ocean groups); (3) those containing no endemic elements—Cocos Keeling and the low coral groups in various parts of the Pacific.

Ridley (527) lists the following islands as continental in type: Canaries (50 miles from Cape Juby, Africa), Madeira (450 miles from Morocco), Cape Verde (50 miles from Senegal), Chatham Islands (360 miles from New Zealand), Auckland Isles and Campbell Island (180 miles from New Zealand), Norfolk Island (400 miles from New Zealand), Andamans and Nicobar Islands (formerly connected with Burma and perhaps North Sumatra), Galápagos Islands (600 miles from South America). Such islands, he says, are really "detached" fragments of continents or larger islands. They possess larger numbers of species than comparable oceanic islands and have more species to the genus, on the average. They contain plants and animals of types for which transoceanic dispersal seems impossible or highly improbable (such as snakes and land tortoises on the Galápagos, and freshwater fish on Kerguelen). Ridley lists the following as oceanic islands: Azores (900 miles from Portugal and 550 miles from Madeira), Ascension (700 miles from St. Helena), St. Helena (1140 miles from South Africa), S. Trinidad and Martin Vaz (600 miles from Brazil), Bermudas (580 miles from North Carolina). Oceanic islands are supposed never to have had any land connections from which even a part of their flora could have been derived. Their total number of species is small and the genera average only a few species each. The more remote such islands are, the smaller their flora. Sea-borne plants reach the remote islands in greater abundance than wind- and bird-borne plants.

We have seen that the more remote islands are—especially oceanic islands —the poorer their biota, and that mountains on the continents are similar to islands in respect to their high percentage of endemism. It is interesting, however, to find a rather unexpected relationship for isolated mountains. Grinnell and Swarth (307) have studied the boreal fauna of a series of California mountains that differ in boreal area and in distance from the main Sierra Nevada massif terminating with Mt. Whitney, and have concluded as follows: "The smaller the disconnected area of a given zone (or distributional area of any other rank), the fewer the. types which are persistent therein." These authors had anticipated that the boreal fauna would be represented by diminishing numbers of species the farther the disjunct area was from the main boreal mass. That this was not entirely the case is shown by the following results, in which the mountains are listed according to boreal area:

1. Mt. Whitney (Sierra Nevada)—51 species of boreal mammals and birds

2. San Bernardino Mts. (325 miles distant)—35 species of boreal mammals and birds
3. San Jacinto Mts. (350 miles distant)—27 species of boreal mammals and birds
4. San Gabriel Mts. (260 miles distant)—18. species of boreal mammals and birds
5. Mt. Pinos (160 miles distant)—19 species of boreal mammals and birds
6. Santa Rosa Mts. (365 miles distant)—12 species of boreal mammals and birds
7. Cuyamaca Mts. (410 miles distant)—8 species of boreal mammals and birds

An explanation is not difficult to find. The larger areas on the average must provide more suitable ecological niches during times of stress than do smaller areas. The plants of these mountains appear to provide a parallel with the mammals and birds (320).

Several islands have been considered oceanic by some geographers and continental by others. In this group can be mentioned Tristan de Cunha (5520 miles from Montevideo and 1760 miles from South Africa), the Kerguelen Group (960 miles from South Africa and 4510 miles from the Falklands), and South Georgia (900 miles from the Falklands), which some consider to be remnants of a large extent of Antarctic land and flora now lost except for certain relics on the islands. This idea was proposed by Hooker, and Ridley gives some evidence for it. The Polynesian Islands, Easter Island, Pitcairn, Rappa, and Hawaii, among others, have usually been considered oceanic islands, but some geographers believe that they also are remnants of continental masses.

One of the first problems to arise concerning island biota and whether a certain island has always been an island is dispersal. Guppy (309), for example, believed that ordinary dispersal agencies are adequate for an understanding of the stocking of the Pacific islands, and Setchell (570) saw no necessity for postulating any fundamental changes in the permanence of the Pacific Ocean as such, and the purely volcanic origin of the islands. An interesting point here is that Setchell attributes the apparent cessation of transoceanic migrations of certain elements, which so puzzled Hemsley, Guppy, and others, to the recent ecological view of the impenetrability of closed plant associations. That is, he believes that dispersal over great distances of sea or land is possible, and that much of the absence may be ex-

plained by the difficulty of germination and establishment by the disseminules under the competition of preexisting closed vegetation. Brown (91,
92) also believes in over-water transportation in the origin of the Hawaiian
flora and says that only 31 immigrants of 16 families during a Jurassic-
Comanchic-Cretacic period would be sufficient to account for 322 modern
species and practically all the endemic genera of Dicotyledons. He places
their source as the American Isthmus region. On the other hand, Campbell
(118, 119, 120) found it necessary to suppose a more or less direct land connection between Hawaii and the southwest. Skottsberg (590) says that if
islands were oceanic *ab initio,* their flora or its ancestors must have traveled
across the ocean. But if this were true everywhere, the plant distribution of
the world would be quite different from what it is. The endemism on the
individual islands of an archipelago, such as Hawaii, does not speak well
for the power of long-distance dispersal. If there were over-water dispersal
there should also be recent arrivals (limited littoral and bird-carried species
excepted), but the high endemism argues against this. In Hawaii, for example, according to Skottsberg, an analysis of the floristic elements shows that
the islands have had a long and complicated history, and that there were
probably land bridges of different sorts and at different times.

The geneticists have widely accepted the possibility of relatively rapid
evolution under conditions of complete or partial isolation, especially of
small populations, which may be, and in fact is, very often of a non-
adaptive type. This phenomenon has become generally known as the Sewell
Wright effect because of Wright's demonstration of it mathematically.
Supporting field data are more often derived from island forms than otherwise, because the conditions of complete and partial isolation and of small
populations are more often fulfilled under island conditions. The previously
mentioned Hawaiian palm genus *Pritchardia,* with over 30 local endemics
in the archipelago, appears to be an illustration of the phenomenon. Of them
Skottsberg (589) says, "The *Pritchardia* assemblage suggests an ancestral
population which has split up in a number of segregates under isolation. It is
generally believed that the archipelago once formed one continuous land."
He mentions several genera in which local endemism occurs in Hawaii,
suggesting segregation of different gene combinations: *Schiedea, Geranium,
Hibiscus, Palea, Fagara, Acacia, Sanicula, Cheirodendron, Tetraplasandra,
Lobordia, Cyrtandra, Plantago, Wikstroemia, Euphorbia, Pipturus, Astelia,
Dianella, Panicum, Deschampsia,* etc.

One view of endemics, and perhaps the most generally accepted for the
majority of them, is that endemics are relics or epibiotics, the last survivors

of a once prevailing flora. This theory is widely held by geographers in the north temperate, where indubitably there are numerous relics. Probably the most abundant local species of this type are Tertiary and intra-Pleistocene relic[2] members of floras which were decimated by the ice sheets and the attendant climatic changes.

Sequoia sempervirens, of the coastal valleys of California and Oregon, and *Sequoia gigantea,* of the Sierra Nevada, are both endemic to their respective regions and relics of a widespread Cretaceous and Tertiary genus. Likewise, *Taxodium distichum, Sassafras variifolium, Liriodendron tulipifera,* and *Liquidambar styraciflua* are all relics and endemics, although their areas are comparatively large in the eastern United States. Also, *Iris setosa* var. *canadensis* is an endemic, since the varietal form is confined to the Gulf of St. Lawrence region, and it is a relic of the wide-ranging pre-Wisconsin *Iris setosa* specific population.

A study of relics in general provides valuable clues to vegetational history, as ably pointed out by Clements (152), because each major climatic shift produces a corresponding clisere and leaves behind relics of each climax type in formerly occupied regions. (Clements [150] defined the clisere as a series of climax formations or zones which follow each other in a particular region in consequence of a distinct climatic change. The great mass movements of contiguous climaxes under the impulsion of climatic change leave behind them, within the general region of a climate, relic communities and species populations of the pre-existing climax and climate.) Such relics, however, are seldom endemic to their relic station, but are merely disjunct from the major area of the community or species population which has moved on elsewhere. This is the status, for example, of the numerous members of the Canadian spruce-fir formation which are found disjunct south of the glacial limits and especially in the Southern Appalachian Mountains, and of prairie plants in Ohio and elsewhere eastward which are relics of the xerothermic period of postglacial time (163, 273).

The literature on relics of this sort is exceedingly abundant, and all botanists are familiar with examples, so there is no need to labor the point. But the converse interpretation has never received much attention from ecologists and geographers of north temperate regions. Numerous endemics of our eastern states, for instance, are obviously young species, but their biological status has scarcely been noticed. Such local endemics as *Vernonia*

---

[2] Perhaps, at this point, it may be useful to point out a rather common fallacious use of the term endemic. A relic that survives in two or more places more or less widely disjunct cannot be called an endemic. A relic of a former transcontinental population surviving in the Gulf of St. Lawrence and the Rocky Mountain regions cannot be said to be endemic to the two regions.

Lettermannii, of the southern Ozarks, are obviously newly formed species, and not relics.

The strongest proponent of the idea that most endemic species are young species, rather than old relics, is Willis, whose studies were brought together first in *Age and Area* (696) and more recently in *The Course of Evolution* (698). His discovery of the hollow-curve distribution of species numbers in genera and of a similar relationship in area sizes (within a circle of affinity) would seem strongly to indicate that many local species are young. Also, his attack on the natural-selection theory and the idea of adaptation by small degrees (of morphological change) provides other evidence for his conclusion that most endemics—in fact, nearly all endemics in tropical regions—are young species. The following paragraphs review his understanding of endemism, for, as he says, "it is a crucial feature upon whose proper explanation largely hangs much of the whole structure of evolution and of geographical distribution."

One problem of endemics is brought sharply to the fore by an example given by Willis. On the summit of Ritigala Mountain, lying isolated in the flat "dry" zone of Ceylon, occur two species of *Coleus. C. elongatus* is endemic and morphologically quite distinct. It is accompanied in the same habitat by *C. barbatus,* a close relative which is widely distributed in tropical Asia and Africa. From the width of its area, *C. barbatus* is the most "successful" of all the species of *Coleus,* but on Ritigala it grows in the same way upon open rocky places as the narrowly endemic and most "unsuccessful" *C. elongatus.*

Willis says that natural selectionists are hard put to find an explanation for such a situation. They usually bring to bear on the problem one or the other of two mutually contradictory hypotheses. Some say that *Coleus elongatus* is a local adaptation and a success on Ritigala. If this is the case, why does it occupy the same habitat with *C. barbatus?* And how did it gradually attain its specific status in so small an area associated with *C. barbatus?* The reverse hypothesis is that *C. elongatus* is a relic of a previous vegetation and condition. That is to say, it is, in effect, a misfit and a failure. But again, why does it continue to grow together with *C. barbatus?*

Structurally, *C. elongatus* differs from its near relative in several characters, none of which would seem to provide any basis for selection. According to Willis, the final refuge of the selectionist is usually that the structural peculiarities must have been useful at some other time or in some other place. There is no evidence, however, that the conditions at Ritigala have changed since the Tertiary. Neither of the two supplementary hypotheses of the selectionists offers any acceptable explanation of the endemic *Coleus elongatus*. Willis' theory for this case—and hundreds of similar ones could be mentioned—is that *C. elongatus* was derived from *C. barbatus* on Ritigala by a few large mutations (he usually supposes one mutation) which included also the establishment of reproductive isolation. The derived species carries the same internal adaptations as the parent species (and so can live along with it), and the morphological differences are without adaptational significance. This is in line with the theory of evolution by differentiation which is discussed elsewhere.

A common type of distribution is that of a polymorphic, widely ranging species accompanied by a few to several endemic species confined, or essentially so, to different relatively small portions of its area. *Anemone rivularis*, widely distributed in all the mountains of India and Ceylon, is cited by Willis as such a species; it is accompanied, especially in the northern mountains, by no less than 12 related, narrowly endemic species. Such distributional patterns are easily explained by the hypothesis of age-and-area and evolution by divergent mutations or differentiation. Willis describes the relationship as follows: "There are very many factors that may affect dispersal, but if one suppose factor a to produce an effect in distribution in a long time x that may be represented by 1, one may reasonably expect that in time $2x$ it will produce an effect 2. If the effects of all the factors be added up, the total effect in time x may be represented by m, and in time $2x$ by $2m$." He was not unaware that this mathematically expressed relationship would be liable to exceptions and deviations in nature. Knowing that there would be great individual differences among species, he makes the proviso that comparison (for determination of age) must be made only between allied forms, which are most likely to behave similarly under similar conditions. In addition, he suggests that comparisons should be made on the basis of averages when species are taken by tens, to allow for averaging the effects of the many factors that might take part in distribution. But bearing these reservations or precautions in mind, one can say, according to Willis, that in most cases a large area of distribution usually indicates considerable age and a small area of distribution (endemic) indicates relative youth. Another set of factors which tend to prevent the complete realization of this tendency in distribution is barriers. Barriers may be of many kinds; simple physical barriers, changes from one climatic type to another, and peculiarities of soil and other ecological conditions may impede the spread of a population of a certain type. One of the most important things that would necessarily follow from age-and-area, and that lay behind much of the criticism of the theory, was the replacement of the long-accepted idea that endemics in general are either relic forms or local adaptations by the supposition that when they occurred in very small areas they were mostly young beginners as species and had not yet had time to occupy large areas.

There are, of course, many instances that support Willis' hypothesis of local differentiation of species out of the stock of a widespread species. The genus *Elytrigia*, which is part of *Agropyron*, presents an interesting case.

Fig. 28. The species of *Elytrigia* (*Agropyron*) according to Nevski. (Map after 360.) The species are: *1. strigosa, 2. stenophylla, 3. propinqua, 4. Gmelinii, 5. amgunensis, 6. Jacutorum,* and *7. reflexiaristata.* This group of closely related species, which probably should not be accorded more than subspecific rank, illustrates the development of an old arctic-montane species which was split into seven isolated areas by Pleistocene effects. The northern branch of the probable area of the old species (indicated by broken line) was less well preserved than the southern branch, having a relic only in the Urals. Outline is after Goode's copyrighted map No. 201PN, with permission of the University of Chicago.

Hultén (360) published a map[3] of its seven isolated endemic and closely related species with the comment that arctic-montane plants must be a group of old species and that it is to be expected that in different parts of their areas considerably differentiated races should be found. This, he thinks, is the case of *Elytrigia*.[4] "The area has repeatedly been split up, during the glacials under the influence of a cold climate in the north and a pluvial one in the south, and during the interglacials under the influence of drought and heat. Each of these agencies must have caused a selection of biotypes in its

[3] Taken from Nenski in *Acta Inst. Bot. Acad. Sci. U.S.S.R.,* Ser. I, Fasc. 2, 1936, p. 74.

[4] *Elytrigia* is better treated as a Linnaean species and, as Hultén points out, these small species of *Elytrigia* are probably better classified as subspecies so that their close relationship is more apparent.

particular direction." This case parallels that of *Anemone* mentioned by Willis, except that the wide-ranging parental species no longer exists and that the derived endemics which remain are not new species, except in a geological sense. It seems to me that the fact which stands out clearest is that endemics may be either new or old species, and that it is frequently rather difficult to know which.

Another significant result of Willis' studies is the demonstration that over the world as a whole the majority of endemics are members of large and successful genera. For example, the mountain endemics in Ceylon, which number over 100, belong to genera which average 14 species per genus in Ceylon and 216 species per genus in the world as a whole. These figures are to be compared with the average of 2.7 species per genus for the entire Ceylon flora and about 13 species per genus for the flowering plants of the world. In northern regions endemics frequently belong to small genera and even very small families (and are mostly relics) such as *Podophyllum peltatum* and *Jeffersonia diphylla* of the Podophyllaceae, and *Galax aphylla* and *Shortia galacifolia* of the Diapensiacae. In warmer countries, it seems, one rarely finds endemics of this kind, for they occur chiefly in the large and successful genera such as *Ranunculus, Poa, Eugenia, Vernonia, Senecio, Coleus,* etc. The same phenomenon, however, is also common in temperate regions in many of the same genera, and in others such as *Erigeron, Aster, Solidago, Veronica, Hieracium, Allium, Artemisia, Viola, Festuca,* etc.

In conclusion, we can say that the relic nature of an endemic should never be accepted without some form of positive evidence.

16.

Species Senescence

The concept that certain species are senescent is fairly widespread in taxonomy and plant geography. Populations of this sort are variously referred to as old, weak, conservative, and unaggressive, whereas other species are characterized as young, strong, competitive, dominating, and vigorous. One immediately noticeable aspect of this concept is the use of adjectives which are ordinarily applied to individual organisms, but which, in the concept of species senescence, are applied to whole populations that are taxonomic entities. The other conspicuous attribute of the concept is the reference to the behavior of the population.

An example of this usage is found in a discussion by Fernald (242) of Cordilleran elements found in the flora of the Gulf of St. Lawrence region. He says, "This failure of the plants of the unglaciated spots to extend their ranges into closely adjacent areas which, upon the melting of the Labrador sheet, became open territory ready for invasion is interpreted as a further evidence of the antiquity of these plants; at the close of the Pleistocene they were already too old and conservative to pioneer, although they were able to linger as localized relics in their special undisturbed crannies and pockets." Further on we read, "The younger Arctic flora which made its way south, in eastern America . . . coincidentally with the advance of the comparatively recent Labradorean or Wisconsin . . . ice-sheets, has not yet had time to set off local species . . . and far from being the rarest and most retiring of relics on our mountains, these young plants, like the young of the human species, are aggressive, dominating and from mere youthful vigor are inclined to crowd out the survivors of a more ancient cycle whose territory they had only recently invaded."

From Marie-Victorin's interesting discussion of botanical problems of eastern Canada (447) we can extract several comments related to the question of species senescence. According to him, "We are naturally led to infer that our living allogenous flora is a survivor of an important and now largely extinct flora cornered on the nunataks during the Wisconsin glaciation. . . . We picture that since the forty-odd thousand years, perhaps, they wage a losing fight against the sturdy northern conifers and their natural ecological associates. . . . It seems too much for them. Through ecological pressure or through the failure of some essential biological processes, intrinsic or extrinsic, the old species have already surrendered most of the ground. They are mostly local, some of them extremely so, being confined to one mountaintop, or to one secluded cove by the seashore." A few paragraphs later he writes, "A last word on a point that might have been raised at the very outset of this study. Is the conception of a senescent preglacial or interglacial flora, vanishing gradually from northeastern America, defensible on biological grounds? Are there really such things as senescent floras in the present world?

"I have often caught smiles on the lips of my biological auditors, when indulging in such anthropomorphic utterances. But I do not think we can escape a more or less guarded admission of this kind, if we look at the facts without bias . . . Such extremely local plants as *Cirsium minganense* certainly seem to be doomed. If we could prove with reasonable certainty that these forms are disappearing because they are 'old' to the point that some essential mechanism has ceased to be functional, we could open a new chapter of auto-ecology, and grasp at least one of the negative aspects of evolution."

If we now look at the characteristics of the species which have been called senescent, we find that they all have certain attributes in common. With respect to area, they are circumscribed. Their populations, although sometimes locally abundant, are not large in the usual sense, and their total area is conspicuously small. From the historical aspect, senescent species are considered to be relic. They constitute the remnants of a once more widely spread flora, and when several such species occur together they are referred to as constituting an epibiotic (527) or an allogenous (447) flora. Senescent species are also usually endemic. We must, however, examine briefly the concept of endemism. If, for example, one of the so-called senescent species, say on the Gaspé, occurs also in the Rocky Mountains, it cannot be referred to as endemic to the Gaspé. If, also, the Gaspé population behaves as if it were senescent and the western population does not, the concept of senescence would seem to fail. Furthermore, endemics of a circumscribed nature can be of two types, new species or relic species. If the eastern disjunct population is taxonomically separable, as a variety or a closely related species, from the western population, it would be correct to refer to the former as an endemic. There are other characteristics of senescent species. They are relatively constant or invariable morphologically, whether given specific or subspecific rank. Ecologically they are of narrow amplitude; that is, they are confined to certain usually rather extreme habitats. Their competitive ability, except in their peculiar niches, is apparently very low. They cannot make any headway toward penetrating the prevailing habitats that are dominated by the typical vegetation of the region. The senescent species constitute an anomalous element in the flora of a given region.

It is easy to understand how it came about that species with the previously described characteristics should be referred to as senescent, as weak, conservative, and unaggressive. If we accept the implication that species can "run down" or "wear out," the implication of the concept of senescence, with its teleology, is perhaps satisfactory. But this is no explanation of the cause of the characteristics of species which are described as senescent, and it has the danger of being superficially credible. We may, then, well inquire into the possibility of known causes of such characteristics of circumscribed relics. I have already published a note on the possible causes of apparent senescence (109) and the argument can now be expanded.

In the first place, an endemic species of strikingly limited area may be a young species recently evolved which has not yet had time to extend its range to natural boundaries as determined by intrinsic and extrinsic barriers. According to Willis' age-and-area hypothesis (696), most endemic species are considered to be youthful. It is a truism of biology that populations tend to expand their areas in ever-increasing concentric circles, other things being equal. This ideal is seldom realized, for other things are seldom equal. Nevertheless, it is not possible to accept the fact, for an endemic species of narrow range, either that it is young or that it is old from a knowledge of its area alone. It is necessary to inquire into its other characteristics.

A second line of inquiry involves the correlation between the ecological amplitude of the species and the distribution of suitable habitat types. Ordinary ecological field observations have led to the conclusion that species differ with respect to the breadth of their tolerances to changing environmental factors; that is, species have different ecological amplitudes. The ecological amplitude of a species, of course, is the result of the characteristics of the individuals of the different biotypes which are classified together as a species (or other taxonomic category). There appear to be two possibilities here. The different biotypes of a species may have rather narrow tolerances, but, differing among themselves, they collectively may have more or less broad (at least broader) amplitudes; or the individuals of the biotypes may have relatively broad amplitudes and collectively have more or less broad tolerances. As a result of the operation of the above limits, plants of some species are found in nature only in certain strongly characterized habitats, but the plants of another species are found in several related habitat types or even in apparently widely diversified situations. It is true, however, that widely ranging species can usually be recognized as consisting of geographical races or of ecotypes, groups of biotypes within the species which are more closely adapted to certain types of environment within the range of the species as a whole (146). The above inherent physiological situation being true, it follows that certain species populations may be narrowly limited in the field because of the relatively small area in which there is a suitable combination of habitat factors. Certain narrow endemics, then, cannot express their potential spread because of climatic and edaphic limits which circumscribe their range. Such species may even be locally abundant. A classification of such populations as senescent would be false. On this point there is an interesting statement by Marie-Victorin (447): "We are now in possession of a first series of experimental data in connection with the 'health' of the allogenous units of eastern North America. A certain number of the rarest relics and endemics have been grown for the last two years in the Montreal Botanical Garden. Quite unexpectedly they are very successful. . . . What of it? Since there does not seem to be anything fundamentally senescent or lethal in the plants themselves, we are naturally led to concede the senescent condition to the allogenous flora as a whole."[1] Before assuming that an endemic species is of limited area because it is senescent, the question of the extent of area of suitable habitat should be investigated. For illustration, suppose that a certain species is adapted for life in soils of circumneutral to basic reaction. In a region where such soils sporadically occur as small islands in an area of prevailingly acid soils, the basophile species will have many of the attributes assigned to senescent species, whereas the prevailing flora will be composed of apparently youthful and aggressive species, but their youthfulness is an illusion resulting from their tolerance of acidic soils. If the same species happen to occur in a region in which acidic soils occur as islands in prevailingly circumneutral to basic soils, their roles will be reversed and the allegedly senescent species will seem to have youthful characteristics.

[1] There must be many biologists who would not like to be included in Marie-Victorin's editorial "we" and be carried along to his conclusion that a flora—the endemics of his allogenous flora in this case—can have a senescent condition when the individual plants of the species which compose it do not have this nature.

The phenomenon of competition has some bearing on the area which a species attains. Salisbury (541) has noticed that species of open areas are frequently more variable than species of closed communities. He suggests that the increase in variability which a wild species frequently shows when brought under cultivation and protected from competition is due not to a change in inherent conditions but to the preservation of forms which, in nature, would be rapidly eliminated. At any rate, it is a common observation

Fig. 29. Generalized areas of *Pinus Jeffreyi* (broken line) and *P. ponderosa* (solid line). Outlines based on detailed maps by Munns (472).

that certain species for which a region is both climatically and edaphically suitable do not have a chance because they cannot compete with the better-adapted species which form the closed communities of the region. Invasion of climax or near-climax vegetation seems to be especially difficult because of the intense competition provided by the members of the closed communities.

There are differences among species with respect to their characteristics which suggest that one type is old and the other is young, so we must inquire into their nature. For example, two closely related pines of the mountains of the western United States are strikingly different in their areas. *Pinus Jeffreyi* is essentially Sierran and endemic to California, but *P. pon-*

derosa is widespread in nearly all of the western mountains. Mirov (466), who has investigated these pines, says, "Jeffrey pine is extremely stable, that is, it does not vary much, thus exhibiting the characteristics of racial senility. On the other hand, the extreme variability of *P. ponderosa* may be considered an indication of its relatively younger age." He has published biochemical evidence favoring the conclusion that *Pinus Jeffreyi* is more ancient than *P. ponderosa*. The data consist of iodine numbers which are a measure of the abundance of unsaturated fatty hydrocarbons. The theory,

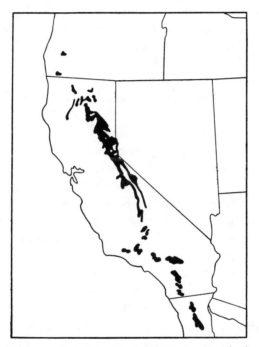

Fig. 30. Topography of area of *Pinus Jeffreyi* in the Sierra Nevada and related mountains. Map redrawn from Munns (472).

based on fairly wide investigation, holds that the more ancient phylogenetic stocks have higher percentages of saturated compounds and lower iodine numbers. Mirov found that *P. Jeffreyi* has the iodine number 134, whereas *P. ponderosa* has the number 151. This fits in with a general series of gymnospermous numbers and is one indication that the Jeffrey pine is the older of the two. We have here a case of two closely related species in which independent evidence suggests that the more restricted and more constant species is the older. There is no evidence, however, as to why the older species should have the observed characteristics.

Mason (453) has made a careful study of the closed-cone pine forests of the California coast as to both modern distribution and their Pleistocene and late Tertiary occurrence.[2] He has decided that these relic forests are growing in their widely disjunct regions in essentially the same places as those they have lived in since the forest type was characteristic of the coastal region as a whole; that is, *Pinus radiata, P. muricata,* and *P. remorata* are relic species occupying relic areas. Mason takes into account Fernald's hypothesis of species senescence (242) and says that, considering only the pines, their discontinuity, limited distribution, and failure to occupy adjoining favorable habitats might be attributed to senility. There is an important difficulty, however, for associated with these highly endemic and conservative pines are such highly plastic genera as *Arctostaphylos* and *Ceanothus.* These genera, in part at least, have had the same history as the closed-cone pines and have endemic species in the areas of the pines, but their evolutionary potentiality is vastly different. While the relic and disjunct pines have remained relatively constant, the shrubby *Arctostaphylos* and *Ceanothus* stocks have undergone extensive speciation. Age alone will not account for the differences; neither will isolation.

Another good contrast is provided by *Shortia (Sherwoodia) galacifolia* and *Galax aphylla.* These genera are both monotypic in the Southern Appalachian Mountains and are probably equally old. *Shortia* consists of four species, the one in the Southern Appalachians which is extremely localized, and three in Asia. These species of *Shortia* and *Galax* are the only two members of their family in the Southern Appalachians and are probably relics of the Arctotertiary forest of circumboreal distribution. *Shortia* is so rare that it was lost (not known in the field) for nearly a century after its original discovery (500) and it shows no tendency to spread into apparently suitable adjacent territory, although it grows well in gardens. *Galax,* on the other hand, although endemic too, is aggressive in several forest types, notably in pine, oak, and chestnut forests. Why does one of these species seem to be senescent and the other one young and aggressive? The facts of distribution of the family indicate the ancient origin of these stocks. Neither genus in the Southern Appalachians has shown any tendency to speciate, but one spreads and the other does not. Is one species senescent for some intrinsic reason other than age? Ross (534) says, "Although *Shortia galacifolia* has considerable potential value for horticulture, it is known to a comparatively few people in its own habitat. This is because of the fact

[2] See Chapter 9 for a more complete account of these pines.

that it is restricted to a very few small areas in the Appalachian Mountains.[3] No reason for this extreme localization has yet been given; it is not due to soil conditions, nor to exposure or other such influences; similar soil and exposures exist over very large areas; also, the plant is very adaptable when transplanted, flourishing in a variety of garden soils." Ross did find, however, that the seeds have none of the well-known agencies for dispersal; in fact, they germinate within the ovary and the stem decays and falls over. Even should such seedlings make contact with the soil, they would have to undergo competition with the mature plants, for *Shortia* grows in a dense mat, spreading by underground runners. *Galax,* on the other hand, produces an abundance of small, easily transported seeds. What appear superficially to be senescence in *Shortia* and youthful vigor in *Galax* are attributes of their seed production and germination behavior. Another aspect of the contrast between the pair of species is reported by Baldwin (43, 44, 45), who found that *Galax aphylla* is diploid and tetraploid and *Shortia galacifolia* is diploid. He suggests that the range of *Galax,* in open woods of the Piedmont and Appalachian plateau from Georgia and Alabama to West Virginia and with extensions into the coastal plain of North Carolina and Virginia, results from greater ecological adaptability as a consequence of polyploidy than is characteristic for the more limited diploid, *Shortia.* This suggested difference between diploids and tetraploids is not constant.

Let us make one more comparison of the ranges and behaviors of certain species, all of which are known to be of ancient phylogenetic stocks. If we consider species of genera which are known from fossil records to have had the so-called Arctotertiary distribution (temperate North America, Europe, and Asia) or which appear logically to have had such an occurrence because the species today are confined to eastern Asia and eastern North America, we can compare their areas and behavior with confidence that age, per se, is not an explanation. The following species of this type are arranged according to increasingly greater areas: *Shortia galacifolia* (Blue Ridge of South and North Carolina), *Hugeria erythrocarpa* (Appalachian plateau of Georgia and Tennessee to Virginia and West Virginia), *Pyrularia pubera* (Appalachian plateau from Georgia to Pennsylvania), *Liriodendron tulipifera* (northern Florida to Arkansas, Michigan, Ontario, and Massachusetts), *Epigaea repens* (Florida to Mississippi, Saskatchewan,

---

[3] House (357) lists it for McDowell County, North Carolina—a station now exterminated—Keowee River headwaters, Jocassee Valley, and along the Whitewater and Toxaway creeks, South Carolina.

and Newfoundland), and *Cephalanthus occidentalis* (Florida to Texas, California, Ontario, and New Brunswick, and in the West Indies and Mexico). These species are all of the old Tertiary forests; a few million years one way or another does not matter much in comparing them. Are some of them old in the sense of being senescent, i.e., physiologically old, while others are less so? *Liriodendron tulipifera* of today can be matched against Tertiary fossils (in fact, modern variations of the one species can be matched against several fossil "species"); yet it is one of the most aggressive old-field species of the southern states, and together with *Sassafras*, another Tertiary relic, gives good competition to other woody old-field plants of *Rhus, Pinus, Cornus, Crataegus, Halesia, Viburnum*, etc.

Finding evidence for a difference in behavior between species of approximately the same age—such a difference that one can be characterized as "youthful" and another as "senile"— let us inquire into the genetic nature of species for a possible answer. Elsewhere in this book there is a rather extended consideration of the effects of polyploidy on areas and aggressiveness, but here we may mention a case or two. Manton (442, 444) found that within the species *Biscutella laevigata* there are two chromosome races. She says that the diploids are virtually endemic forms of relic nature and high antiquity relative to the surrounding vegetation. "The most cursory ecological comparison reveals a difference in vigour of colonisation between a more ancient (diploid) and a more recent (tetraploid) type that is not unexpected, though the underlying causes are not thereby specifiable. . . . Whereas a colony of diploids gives a strong impression of uniformity, the subspecies or variety being in consequence easily recognized, in a population of tetraploids many differences in details such as leaf shape, hairiness, etc., occur between adjacent plants." The diploids are thought to be relics of a temperate flora which survived the last glacial period in the protected and isolated spots of certain north German river valleys where they are even now found. The tetraploid forms are wide-ranging in the central European mountains in glaciated areas. The tetraploids are derived from the diploids, of course, so there is an age difference; but does that signify that age is the cause of the difference in behavior? Or that polyploidy, either, is sufficient explanation? The importance of the decimation of the original diploid range and of the subsequent isolation of the partial populations needs consideration from a genetic point of view. This topic will be returned to shortly.

Iris setosa var. *canadensis* is one of the Cordilleran relics of the Gulf of St. Lawrence region which is of limited area and high morphological con-

stancy. *Iris setosa,* however, is of wider area and has considerable morphological variation in central Alaska. Anderson (15) says that the results on
the two sets of irises are just what a geneticist might predict. "This conservatism of *Iris setosa* var. *canadensis* is distinctive of most of the glacial
endemics (or near endemics) of the region around the Gulf of St. Lawrence.
. . . The invariability can not be a direct effect of time, for the highly
variable irises of Alaska are quite as aged. It is more probably, as Professor

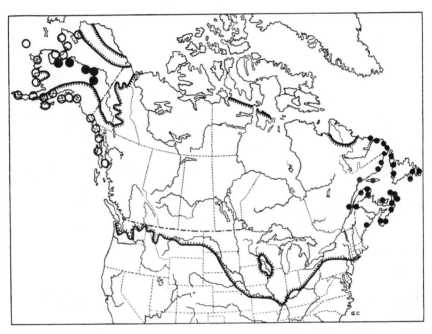

Fig. 31. The discontinuous range of *Iris setosa* (open circles) and its varieties *canadensis*
(small black disks) and *interior* (larger black disks). This transcontinental discontinuity undoubtedly is a result of Pleistocene glaciations, the maximum extent of which is approximately
shown by the hatched lines. The map is redrawn from (15).

Fernald suggested, an innate conservatism; a conservatism founded
genetically upon the fact that these irises are descendants of a small and
highly selected stock. Hard times removed from the region all the luxuriant
types which may once have existed there. When the ice age was over the
immediate area was repeopled from the few plucky survivors. Their
descendants, *Iris setosa* var. *canadensis,* bear the scars of the glacial period,
so to speak, in their conservatism; an innate invariability which, on the one
hand, gives them greater uniformity, and, on the other, prevents their
adapting themselves readily to other environments."

To follow up this line of genetic reasoning, we may remember that species which are wide-ranging in nature are composed of a large number of biotypes. The tendency in a continuous population is toward panmixy and stabilization of certain gene frequencies. This tendency is not realized, however, in wide-ranging populations because different conditions result in certain biotype groups being selected in each of the different ecological regions of the range; that is, a large species population becomes subdivided into ecotypes or geographical subspecies. Furthermore, distance becomes a barrier to panmixy, and wide-ranging species tend to differentiate at various places around the periphery of their range. Both ecological and spatial differentiation appear in part to be a matter of the elimination of biotypes and a consequent genetic simplification of local populations. Instead of having access to the entire stock of genes of the species, local populations are considerably more simple, constant, and specialized. With complete isolation such populations have only a limited capacity for variation due to sexual reproduction and, as long as they remain isolated, can become variable, plastic, aggressive, etc., only with the introduction of mutations of some sort.

Since the Cordilleran relics in the Gulf of St. Lawrence region have their closest relationships with populations of the western part of North America and since they are mostly confined to small special habitats in the east, it can be assumed that the eastern populations have descended from decimated portions of the populations which were transcontinental or nearly so preceding glaciation (or at least the last glaciation). The eastern populations, prior to glaciation, were probably already genetically differentiated from the species as a whole. In the severely glaciated east, the peripheral populations were further reduced by increased selection pressure resulting from the changing climate, but the western populations had more avenues of escape and wider suitable territories for survival in a genetically richer condition. The condition of the eastern species, varieties, etc., is due not to age but to combined genetic and geographical factors. It would seem a misnomer to refer to them as senescent. They are populations of narrow ecological amplitude because of their limited genetic constitution, and their genetic homogeneity is a result of their vegetational history.

It is of especial interest to find that Hultén (360), in a study of the distributional features of all species known to him as growing in northeastern Asia and northwestern America, arrives at a conclusion which substantiates Anderson's study (15) of *Iris setosa* and its forms. The conclusion is that there is a general tendency for the populations of a formerly widespread species which become isolated in the various refugia to be more plastic the larger the area of the refugium. Of the populations trapped in small areas Hultén says, "It might be pointed out that the fact that rare plants are often thronged together at the top of the mountains . . . gets its natural explanation from the discussions in this paper. During the hot, dry periods many plants could only persist on the moist alpine summits. They then turned to rigid species there, and lost the capacity of widening their area later." Elsewhere he says, "New systematic units were proposed to have arisen through climatic selection in the biotype material of isolated parts of former split-up areas, once occupied by ancestral types. The Linnaean species were established during the great interglacial or earlier. Smaller segregations arose later." Thus we see again that the Cordilleran and other relics on the small refugia (whether nunataks or not) in the Gulf of St. Lawrence region are "senescent" not because of age but because of biotype depauperization.

Perhaps the explanations so far are really too simple. There are cases in which isolation and genetic simplification are difficult to apply as factors in apparent senility. *Pinus* is certainly old as a genus, but we have such contrasts between closely related pine species as in *P. Jeffreyi* and *P. ponderosa,* and *P. radiata* and *P. attenuata,* in which isolation does not seem to be operative, yet one is relatively conservative and the other is relatively variable and aggressive. *Sequoia* was once widespread and there were several species, but today two relic species alone remain in circumscribed areas of California. *Sassafras, Liriodendron,* and *Liquidambar* are millions of years old in the southeastern United States, but each now consists of a single though wide-ranging species.

The relation of polyploid sequences to senescence has been mentioned by Stebbins (605). According to him:

Since the polyploid members of a complex are more numerous and widespread than the diploids, one would naturally expect that as a polyploid complex becomes older and as conditions cease to be favorable for the type of plant represented by that particular complex, its diploid members would be the first to go. An old or senescent polyploid complex, therefore, is one that consists only of polyploids. With increasing age the polyploids also begin to die out, so that in the last stages of its existence a polyploid complex is simple once more, and is a monotypic or dytypic genus without any close relatives. Examples of such vestigial complexes, that is, of isolated monotypic or small genera with high chromosome numbers, are scattered throughout the plant kingdom. Perhaps the most striking ones are the two living genera, *Psilotum* and *Tmesipteris,* which are the only survivors of the most ancient order of vascular plants, the Psilotales. Both of these genera are frequently considered to be monotypic; their species have more than a hundred chromosomes in their sporophytic cells. They may represent the remnants of polyploid complexes which flourished hundreds of millions of years ago in the Paleozoic era. We know from fossil evidence that this order formed a dominant part of the earth's vegetation at that time. Other vestigial polyploid complexes are probably the redwood, *Sequoia sempervirens,* and the genera *Lyonothamnus* and *Fremontia,* familiar relic species of our California flora which have high chromosome numbers. The evidence from the plant kingdom as a whole, therefore, suggests that polyploidy has been most important in developing large, complex and widespread genera; but that in respect to the major lines of evolution, it has been more important in preserving relics of old genera and families than in producing new ones.

The basic chromosome number of several genera is so large that the only a priori explanation is that they are relic polyploids, although today they may pass as diploid. Be that as it may, a relatively high polyploid condition may be one form of "senescence" in species, because it results in a certain amount of invariability through the masking of mutations. The polyploid sequence in a genus does not seem to be reversible; hence it is a development which may, in some cases at least, tend toward ultimate extinction for the phylogenetic stock.

There are certain evidences for a cyclic nature in evolutionary phenomena which will be mentioned in closing this discussion of senescence. Huxley (371) says, "Even where gene mutation is the sole or main source of heritable variation, its rate may differ in different types. We know comparatively little as yet on this subject, but it is quite possible that in certain species the rate may vary cyclically to quite a considerable extent, and also as between different species. The rate of mutation may then be a limiting factor of evolutionary change, condemning some forms to stability, stagnation, or eventually extinction." DeBeer (180) also has concluded that there is a shift in mutation type as evolution progresses. He writes, "It is concluded that, as evolution proceeds, paedomorphosis is succeeded by gerontomorphosis which actualizes the further evolutionary potentialities opened up by paedomorphosis and exhausts them. The group then lingers or becomes extinct unless a new bout of paedomorphosis supervenes. It is interesting to note that palaeontologists . . . have on independent grounds come to the same conclusion regarding the occurrence of alternate bouts of 'large' and 'small' evolution."

Good (286) proposed a theory of generic cycles in which there are the following stages of phylogenetic history: A genus starts out juvenile, monotypic, and endemic; through speciation and migration it becomes mature, polytypic, and continuous over a more or less wide area; and, finally through extinction and radical evolution (specialization?) it becomes senile and discontinuous and may end up by being relic-monotypic. Note that there are two types of endemism, at the beginning and at the end of this cycle. With respect to this theory, propounded by a taxonomist and geographer, Senn (568), who is a cytogeneticist, says, "These concepts may be associated with the concepts of chromosome evolution to provide criteria for estimating the relative phylogenetic positions of closely related genera. Genera, as we have seen above, may arise as a result of aneuploidy, polyploidy, or accumulation of genic changes. In groups in which the basic number is well established, endemic genera with the basic number could be distinguished at once as relic endemics, whereas genera with derived polyploidy or aneuploidy would be juvenile endemics. Thus a genetic basis for distinguishing 'old' and 'new' endemics may be established."

There is much speculation regarding some of the concepts that have been discussed, but I do not believe that there is a sound basis for the conclusion that certain species are physiologically old, or young, for that matter. Other explanations can be found that have a factual basis and do not have an anthropomorphic taint.

17.

Discontinuous Distributions

1. *Within their areas, species populations do not have absolutely continuous distributions: the so-called continuous areas are only relatively so, the disjunctions being less than the normal dispersal capacities of the diaspores of the species.*

2. *Environmental discontinuities—consisting of regions of particular topographic, climatic, edaphic, or biotic characteristics separated from each other by regions of different character—constitute a sufficient basis for floristic discontinuities, since organisms are limited by their tolerances to the particular kinds of habitats to which they are adapted; but these facts provide no explanation of how similar organisms, such as members of the same species, got into separated regions.*

3. *Minor discontinuities of areas probably frequently result from recent migrations, but major disjunctions seem almost exclusively to have resulted from historical causes which have produced the disjunctions, in a once more nearly continuous area, through destruction or divergent migrations caused by climatic or some other changes.*

The phenomenon of disjunction is one widely encountered in studies of plant distribution, and it presents some of the most interesting problems of geography. It is treated in one connection or another at many points in this book, but in this chapter our desire is to make some general observations and to treat a few types and cases in some detail. It is obvious that an absolutely continuous distribution of a species is a condition seldom if ever encountered, but there are numerous instances of distributions in which the known individuals of the population are not more widely separated than the normal dispersal capacity of the type. All such forms are considered to have a continuous distribution; on the other hand, the term *disjunction* is applied to those in which two or more populations are more widely separated. The cases of wide disjunction are the ones with which we are concerned.

Perhaps the first question that arises is whether the distribution of plants takes place gradually, by small steps, the plants embracing more and more territory until stopped by some barrier, or whether a species can attain a wide and disjunct distribution by occasional, accidental, wide dispersals. The answer to this question determines whether the geographer proceeds with his inquiries into disjunctions or stops with the assumption that long-distance dispersal can happen and that no other problem in disjunction exists except to make this assumption reasonable. Some geographers believe in the efficacy of long-distance dispersal for many types of organisms, but the weight of evidence in most cases seems to be strongly opposed to such an assumption.[1]

We may proceed, as most geographers do, with a consideration of disjunction types and their problems on the basis of the assumption that the area of a species is gradually attained by ordinary migration. On this basis, discontinuities are usually understood to have resulted from the extinction of the members of the species in the area now constituting the disjunction as a result of climatic or other changes. That is to say, the explanation of most, if not all, disjunctions is to be found in historical rather than in contemporaneous biological causes.

Irrespective of the questions of history per se, existing floristic discontinuities can be observed to result from four types of environmental discontinuity: topographic, climatic, edaphic, and biological. By *topographic discontinuity* is understood the separation of islands from each other or from a mainland, the separation of highlands by lowlands, etc. By analysis, this type can be reduced to a subdivision of climatic or edaphic discontinuities. *Climatic discontinuity* is seen in cases of the interjection of a monsoon climate (with a pronounced dry season) in a rain-forest region, rain forest surrounding temperate montane conditions, steppe climate between two tropophytic forests, etc. *Edaphic discontinuity* is based upon substratum differences, more or less local, arising from the nature of the parent rock, from topographic position, from climate, from the clothing vegetation, from age, or from any combination of these factors affecting soil. *Biological discontinuity* results, for parasites, simultaneously with the development of discontinuity by the host. It also includes separations due to climaxes and other communities.

The coincidence of floristic discontinuity and any one of these environmental discontinuities may depend upon the dispersal capacity of the organisms concerned or be an historical phenomenon—the discontinuity of en-

---

[1] See Chapters 15, 19, etc.

vironment once having been less—or it may result solely from the ecological limitations of the organisms, the environmental discontinuity being easily bridged by dispersal. Which explanation is best depends upon the individual case and a consideration of all the circumstances.

In arriving at any conclusion concerning this problem, the following questions must face the investigator: What faculties for spreading are shown by the plants, including their biological adaptations that facilitate transport? What are the characteristics of the natural agencies of transport, such as birds, winds, water currents, etc.? What capacities for germination, establishment, and competition do the disseminules and seedlings have, especially for the places where their dissemination is likely to lodge them? Dissemination is only a necessary antecedent; migration is not accomplished until ecesis has been performed and the immigrant has made itself "at home" in the new place, habitat, and community. Wulff (718, 719) comes to the conclusion that the importance of accidental transportation for the distribution of plants is not sufficiently proved. As a case in point, he mentions the history of the colonization of the island of Krakatao, which has become a *Pradesbeispiel* of the importance of accidental factors. He concludes, as does Backer (40), that these data are of little scientific importance for the question under discussion. For riparian and aquatic vegetation there are many cases of long-distance dispersal, but for the majority of ecological types the improbability is so great that it usually can be dismissed. In fact, one would expect the distribution of plants over the whole world to be entirely different were long-distance dispersal really generally effective. Setchell (571) does not so much impugn the power of plants to accomplish long-distance dispersal as stress the general inability of disseminules to consummate migration through establishment after their dispersal. Most surfaces of the world are already occupied by closed associations and a disseminule must not only by chance reach a suitable environment but also be able to compete successfully in an established community. This, he decides, is in nearly all cases (except for certain littoral and aquatic plants) entirely impossible.

With these limitations to the probability of long-distance dispersal in mind, we may now turn our attention to major discontinuities, the chief classes of which have been listed by Wulff (719) as follows:

1. Arctic-alpine (*Salix herbacea*)
2. North Atlantic (*Eriocaulon septangulare*)
3. Asturian: Ireland and the Mediterranean (*Dabeocia polifolia*)
4. North Pacific: East Asian and American, either western or eastern North America (*Liriodendron tulipifera* and *L. chinense*)

5. North-South America (*Sarracenia*)
6. Europe-Asia (*Cimicifuga foetida*)
7. Mediterranean (*Rhododendron ponticum*)
8. Tropical:
    a. Asia-Africa (*Olea, Pandanus*)
    b. Africa-Madagascar (*Viola abyssinica*)
    c. Asia-Madagascar (*Nepenthes, Wormia*)
    d. Africa-America (*Symphonia*, Vochysiaceae)
    e. India-Malay (*Araucaria, Engelhardtia*)
9. Gondwana: Africa, Madagascar, (India), Australia (*Adansonia*)
10. South Pacific (*Jovellana, Nothofagus*)
11. South Atlantic (*Paullinia, Telanthera*)
12. East-West Australia (*Eucalyptus* spp.)
13. Antarctic (*Nothofagus*)

Another author would probably modify such a list of major disjunctions, usually by the addition of types not mentioned above. It will be noticed that many of these disjunctions are transoceanic and that some are transcontinental. Each type is subject to elaboration by the citation of subtypes.

Irmscher (375), making an extensive inquiry into the distribution of families and genera of flowering plants in connection with the Wegnerian continental-drift hypothesis, came to the conclusion that the data can be regarded as a support for the theory. Our present concern, however, is not with this conclusion but with the data on major discontinuities. Irmscher considers the land surface of the earth to consist of four continental masses: (1) North and South America, (2) Europe and Africa, (3) Asia, (4) Australia-Polynesia. In the following data the continental masses are referred to by the preceding numbers. He found that the 289 families of flowering plants have the following distributions:

Endemic on single continental masses:

(1) The Americas	33	families
(2) Europe and Africa	16	"
(3) Asia	7	"
(4) Australia-Polynesia	5	"

On two continental masses:

On (1) and (2)	9	"
On (2) and (3)	10	"
On (3) and (4)	4	"
On (1) and (3)	7	"
On (1) and (4)	2	"

On three continental masses:

On (1), (2), and (3)	27 families
On (2), (3), and (4)	9 "
On (1), (3), and (4)	9 "
On (1), (2), and (4)	1 "
On all four continental masses:	150 "

Of these 289 families, 228 have one or more major disjunctions, occurring on two or more of the continental masses. For the sake of illustration and reference, some of the categories will be listed in detail.

Families with a major disjunction between (1) and (2): Hydnoraceae, Humiriaceae, Winteranaceae, Turneraceae, Caricaceae, Loasaceae, Mayacaceae, Rapateaceae, Velloziaceae.

Families with a major disjunction between (2) and (3): Cynocrambaceae, Moringaceae, Salvodoraceae, Dipterocarpaceae, Tamaricaceae, Ancistroclodaceae, Punicaceae, Cynomoriaceae, Globulariaceae, Dipsacaceae.

Families with a major disjunction between (3) and (4): Himantandraceae, Eupomatiaceae, Stackhousiaceae, Philydraceae.

Families with a major disjunction between (1) and (3): Saururaceae, Lardizabalaceae, Platanaceae, Trigoniaceae, Sabiaceae, Nyssaceae, Phrymaceae.

Families with a major disjunction between (1) and (4): Batidaceae, Eucryphiaceae.

Families with major disjunctions between (1), (2), and (3): Salicaceae, Myricaceae, Juglandaceae, Betulaceae, Rafflesiaceae, Berberidaceae, Resedaceae, Podostemonaceae, Crassulaceae, Hamamelidaceae, Empetraceae, Buxaceae, Staphyleaceae, Aceraceae, Hippocastanaceae, Balsaminaceae, Cistaceae, Datiscaceae, Cactaceae, Elaeagnaceae, Clethraceae, Pirolaceae, Diapensiaceae, Styracaceae, Polemoniaceae, Adoxaceae, Valerianaceae.

Families with major disjunctions between (2), (3), and (4): Casuarinaceae, Nepenthaceae, Pittosporaceae, Sonneratiaceae, Alangiaceae, Pedaliaceae, Pandanaceae, Aponogetonaceae, Flagellariaceae.

Families with major disjunctions between (1), (3), and (4): Chloranthaceae, Magnoliaceae, Calycanthaceae, Elaeocarpaceae, Epacridaceae, Symplocaceae, Stylidiaceae, Centrolepidaceae, Stemonaceae.

The family with the major disjunctions between (1), (2), and (4): Haemodoraceae.

The 61 families which are endemic to a single continental mass and the 150 families which occur on all four continental masses will not be listed here. Irmscher also considers 'generic discontinuities between the continental masses. Of these, only the genera of the disjunction between (1) and (3), in northern extra-tropical regions, will be listed, and they later in connection with a consideration of Asiatic-American disjunctions. A few genera are selected for somewhat fuller consideration.

Menodora has the major disjunction between (1) and (2) and in addition has a wide disjunction in the Americas. Steyermark (611) revised the

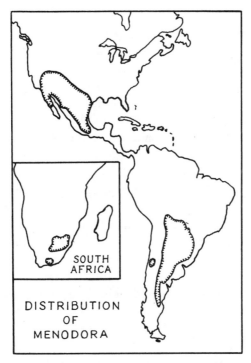

FIG. 32. Approximate discontinuous areas of the genus *Menodora*. Redrawn from (611).

genus and decided that there are 9 species and 11 varieties confined to southwestern United States and Mexico, 6 species and 1 variety confined to central and southern South America, and 2 species and 1 variety in South Africa. Most of the species have restricted ranges, but a few of them are fairly wide within any one of the three widely separated areas.

His most startling conclusion is that *Menodora heterophylla,* which occurs widely in the North American region, has a variety in South Africa. This is the only species complex of the genus which seems to bridge a major disjunction within the genus. One would expect that the anomaly of a type of *M. heterophylla* occurring in South Africa might be a case of convergent evolution (within the genus, of course), rather than a matter of direct genetic affinity by descent. This question aside, however, the distribution of this well-defined genus calls for some sort of historico-geological explanation, for the genus would seem completely incapable of having attained its whole area under existing conditions. The principal reason for this conclusion rests on the fact that the entire genus inhabits a definite belt between 18 and 45 degrees north and south latitude, and a similar set of environmental conditions exists in the three regions as to rainfall, temperature, and topography. In general, the genus is limited to arid or semi-arid rocky plateaus or stony slopes in mountainous regions where there is a characteristic xerophytic vegetation. Steyermark finds the center of variation of the genus to be in Mexico and concludes that it had a more continuous geographical continuity at least before the end of the Cretaceous period. He believes the most logical explanation to result from the postulation of a land bridge across the south Atlantic which was obliterated by the Upper Cretaceous. The geologists allow such a land connection (Devonian through Jurassic to Upper Cretaceous), but the theory of continental displacement should also be considered as a possibility along with the "lost continent" idea.

Of considerable interest in connection with the discontinuous distribution of *Menodora* is the fact that a genus of bees, *Hesperapis,* has a parallel distribution, according to Cockerell (159, 160). It occurs in the arid districts of southwestern United States and under similar conditions in South Africa. The agreement between plants and animals in their centers, migratory tracts, discontinuities, etc., affords mutual support for the historical explanations which meet the needs of either case.

A recent paper by Johnston (379) provides some interesting data concerning North and South American disjuncts of the xerophytic regions which are separated by some 3500 miles of wet tropics. These two desert areas have communities that resemble each other superficially but are so completely different floristically as to indicate a different history. Nevertheless, there are certain species of the two regions which are taxonomically identical and several pairs which could have come only from common ancestors. The following three shrubs are identical in the two regions: *Larrea divaricata* (called creosote bush in northern Mexico and southwestern United States and Jarilla, where it is characteristic from Patagonia to Salta) is the most famous shrub of the group. Johnston considers this clearly to be a South American type because of its several congeners in the Argentine deserts where its family, the Zygophyllaceae, has one of its principal centers of development. In the Capparidaceae there are two well-marked monotypes. *Atamisquea emarginata* is widely spread in South America and occurs again in a more limited area about the Gulf of California in northwestern Mexico. In both regions it is an associate of *Larrea*. *Koeberlina spinosa*, on the other hand, is widely spread in the deserts of northern Mexico and southwestern United States and occupies a more limited area of the dry chaco of Bolivia. Paired forms are known in *Ephedra, Acacia, Caesalpinia, Condalia,* and *Prosopis*. As an example Johnston cites *Prosopis cinerascens*, of northern Mexico, and *P. strombulifera*, of Argentina. After adding other evidences of relationship between these two widely separated desert regions, he concludes that "we are concerned with a very old American desert flora formerly shared by both continents. In South America it is now relatively well preserved but in North America it lingers in a few recognizable remnants." He believes that desert climates and vegetation are an old earth-feature: [2] "The world must have always had its deserts, at least those just outside the tropics." The connection between these two desert regions he believes antedates the development of the characteristic northern shrubs, for they are absent from the southern region; it probably dates from early Tertiary time. One reason for believing that the connection was very ancient is those genera which not only are disjunct between North and South American xerophytic regions but also have affinities across the Atlantic. In addition to *Menodora,* described above, there is *Hoffmanseggia* of northern Mexico, South America, and South Africa; *Fagonia* of the Mediterranean, northern Mexico, and Chile; *Thamnosma* of northern Mexico, Socotra, Somaliland, and South Africa, all well-marked xerophytic genera.

There are many disjunctions in north temperate regions, especially among woody plants, which are reduced in extent by the discovery of fossil remains in intermediate stations. We shall cite a few examples from among the many. Pax (497) has published a series of distributional maps for sections of *Acer* which give fossil records as well as modern areas. The genus is circumboreal and the sections make contact across Eurasia. The genus consists of about 150 species and has its center in West China (Yunnan, Szechwan, Hupeh), where there are 30 species and also the related genus *Dipteronia*. The genus apparently had its origin in the Cretaceous, and during early Tertiary time extended much farther northward than now, to Greenland, Iceland, and Spitzbergen. Fossils indicate that the section

[2] Compare with Clements (153).

Rubra was once circumboreal, although it is now endemic to eastern North America. The section Saccharina occurred also in **Europe during** the

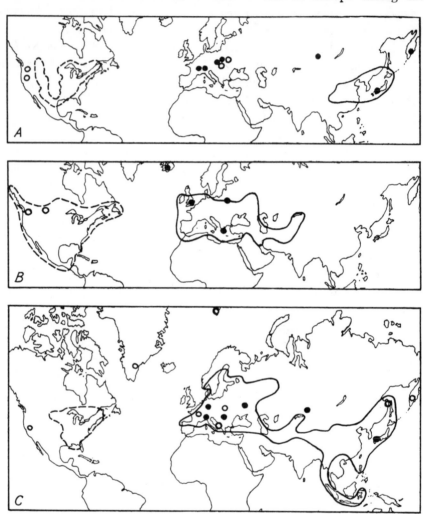

FIG. 33. Modern and fossil areas of the sections of the maple genus *Acer*. American sections are shown by broken lines, and their Tertiary stations by hollow circles. Eurasion sections are shown by solid lines, and their Tertiary stations by black disks. *A*. Sections *Saccharina* and *Palmata*. *B*. Sections *Negundo* and *Campestria*. *C*. Sections *Rubra* and *Platanoidea*. Maps redrawn from (497).

Tertiary and is now confined to the Rocky Mountains and eastern North America. Other sections are on essentially the same land masses today as in the Tertiary, so far as available fossils indicate.

A study of the Hippocastanaceae by Pax (498) shows the genus *Aesculus* to consist of a series of completely disjunct species, with only a few exceptions, and fossil stations for some of the intermediate regions. The greatest aggregation of *Aesculus* species occurs in the eastern United States; the other genus of the family, *Billia,* has two species in Central America.

A third example is taken from Cretzoiu (170) and deals with the present and interglacial distribution of *Rhododendron ponticum* in the Mediterranean region. This species has a discontinuous distribution and consists of three taxonomically distinct units. *R. ponticum* var. *baeticum* occurs in the Spanish mountains near Algerciras and Tarifa and in Portugal in

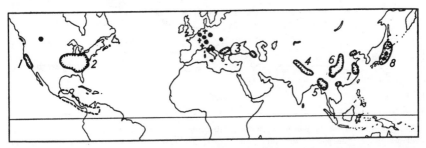

Fig. 34. Discontinuous relic areas in the family Hippocastanaceae. Of the two genera, *Billia* (broken line) is confined to Central America and nearby areas and *Aesculus* is widely spread through north temperate regions. The modern areas include: *1. californica, 2. glabra, octandra, parviflora, pavia,* etc., *3. Hippocastanum, 4. indica, 5. punduana, 6. Wilsonii, 7. chinensis, 8. dissimilis* and *turbinata.* Black disks indicate Tertiary stations, and crosses Pleistocene stations for *Aesculus.* Data from (498).

Algarbia and the Sierra de Monchique. Variety *brachycarpum* occurs at the eastern end of the Mediterranean in the Lebanon Mountains of Syria, extending to the subalpine region. Variety *ponticum* is found from southeastern Bulgaria and European Turkey (Strandscha Mountains), through northern Asia Minor, to the western Caucasus, vertically distributed from 100 to 2500 meters and the upper tree limit. Interglacial fossil stations are known from the present areas and also from five intermediate localities. *Rhododendron ponticum,* with its disrupted range today, is best considered as a relic of a species once widely distributed in central European mountains.

We shall now turn our attention to a somewhat detailed consideration of two sets of disjunct species of special interest in the eastern United States: Asiatic-American disjuncts and coastal plain-interior disjuncts.

Asiatic-American disjuncts.—An interest in the floristic affinities between eastern (and western) North America and eastern Asia (especially Japan) is commonly considered to have commenced with the pioneer work of Asa

Gray in the field of plant geography. Fernald, however, has pointed out that floristic entities common to the two regions had much earlier attracted the attention of Linnaeus. The publication in 1845, by Zuccarini and Siebold, of the *Florae Japonicae Familiae Naturales* led to Gray's first comparison of the floras of the two regions (292). Gray's interest persisted through a long time, as shown by a fine series of papers terminating in

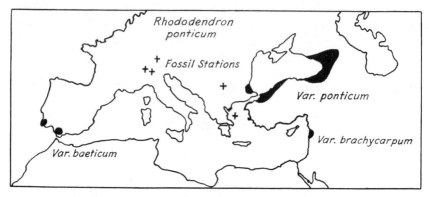

FIG. 35. *Rhododendron ponticum*, with its disrupted range, is a Tertiary relic. Interglacial fossil records of the species are indicated by +. Today var. *baeticum* occurs in Spain in mountains near Algeciras and Tarifa, and in Portugal in Algarbia and the Sierra de Monchique; var. *ponticum* occurs in southeastern Bulgaria and European Turkey, northern Asia Minor to the western Caucasus, with a vertical distribution from 100 to 2500 meters and timber line; var. *brachycarpum* is found in the Libanon Mountains of Syria, extending to the subalpine region. Map redrawn from (170).

1878. Commenting on the first section of the above Japanese flora, Gray wrote as follows:

It is interesting to remark how many of our characteristic genera are reproduced in Japan, not to speak of striking analogous forms. Thus the flora of Japan has not only Wistaria,[3] Lespedeza, Sieversia, Chimonanthus (in place of our Calycanthus), Philadelphus, several species of Rhus closely resembling our own, and two peculiar genera of Juglandaceae, but also a Pachysandra, some Berchmias, a Staphylea, and a peculiar genus of the tribe (Euscaphis), besides, not only a dozen Maples (Acer) but also a Negundo, a Stuartia, two Tilias, a Phytolacca, an Opuntia (surely not indigenous?), a Sicyos referred to our own S. angulata, two Drosera, a Nelumbium, a Nuphar and two species of Nymphea, Gynandropsis, a real Dicentra (Dielytra) and an allied new genus, with several species of Corydalis, a Trollius, our own Coptis and two new ones like the western C. asplenifolia, an Isophrum, two species of Aquilegia, one of them near A. cana-

[3] Italics were not used by Gray.

densis, a Cimicifuga, a Trautveterria, an Illicium, some Magnolia, Kadsura and Sphaerostemma in place of Schizandra, a Mitellopsis, two species of Astilbe (Hoteia), many Hydrangea as well as peculiar hydrangeaceous forms, a Hamamelis with two other characteristic genera of the family, some true Dogwoods (Cornus) as well as Benthamia, the analogue of our Cornus florida, some true Vines (Vitis), and two species of Ampelopsis, three species of Panax, and four of Aralia, one of which is near our A. nudicaulis: and among the Umbelliferae are Hydrocotyl, Sanicula, Sium, Angelica, but what is most remarkable, Cryptotaenia, Archemora, and Osmorrhiza. Further cases of generic conformity abound in the remaining divisions of the vegetable kingdom; thus, for example, Diervilla, Mitchella, Maclura, Liquidamber, Torreya, and Sassafras are represented in the flora of Japan.

In 1855, Charles Wright, botanist of the U. S. North Pacific Exploring Expedition, collected a number of temperate Japanese plants which were examined and reported on by Gray (293). In a table Gray listed nearly 600 species from Japan and gave their known occurrences in central and northern Asia, Europe, and western and eastern North America. The following data are taken from the above report by Gray:

I. Species occurring in Japan and eastern North American (21 species): *Asplenium Pennsylvanica,*[4] *Caulophyllum thalictroides, Diphylleia cymosa, Brasenia peltata, Vitis Labrusca, Prunus Virginiana, Penthorum sedoides, Archemora rigida, Aralia quinquefolia, Viburnum plicatum* (*lantanoides*)*, Asarum Canadense, Liparis lillifolia, Pogonia ophioglossoides, Trillium erectum, Smilacina trifolia, Polygonatum giganteum, Cyperus Iria, Carex rostrata, Onoclea sensibilis, Osmunda cinnamomea, Lycopodium lucidulum.*

II. Species occurring in Japan and in western North America (15 species); *Coptis asplenifolia, Geranium erianthum, Thermopsis fabacea, Photinia arbutifolia, Pyrus rivularis, Ribes laciflorum, Cymopteris littoralis, Echinopanax horridus, Achillea Siberica, Boschiakia glabra, Polygonum Bistorta, Iris setosa, Juncus xiphioides, Carex macrocephala, Festuca pauciflora.*

III. Species occurring in Japan and both western and eastern North America (19 species): *Trautvetteria palmata, Rhus Toxicodendron, Spiraea betulaefolia, Heracleum lanatum, Archangelica Gmelini, Osmorrhiza longistylis, Cornus Canadensis, Artemisia borealis, Vaccinium macrocarpon, Menziesia ferruginea, Pleurogyne rotata, Rumex persicarioides, Streptopus roseus, Veratum viride, Carex stipata, Sporobolus elongatus, Agrostis scabra,*

---

[4] I have not attempted to modernize these names.

Adiantum pedatum, Lycopodium dendroideum. Gray also listed about 60 genera which are represented in Japan and vicinity and in eastern America but are unknown in Europe and western America.

Here may be interpolated a list from Irmscher (375) of 108 genera of flowering plants which have the Asiatic-American distribution, being absent from Europe. They are genera of the northern extra-tropical region. Those marked by an asterisk appear also on Gray's list already referred to. *Achlys, Actinodaphne, Astilbe*, Amphicarpa, Apios, Aesculus* sec. *Pavia, Ampelopsis*, Ascyrum, Amsonia*, Adenocaulon, Arundinaria*, Aletris, Buckleya, Boykinia, Boschigia, Castanopsis, Chamaerhodos, Cladrastis, Coelopleurum, Clethra*, Chiogenes, Chionanthus, Capsicum, Chamaesarocha, Castilleja, Campsis, Catalpa*, Cacalia*, Croomia, Clintonia, Diphylleia*, Dicentra, Deutzia, Decumaria, Diarrhena, Disporum, Echinopanax, Epigaea, Fothergilla, Gymnocladus, Gordonia*, Hydrastis, Hortensia, Hamamelis*, Halenia, Illicium*, Itea*, Jeffersonia, Liriodendron, Lindera, Leptarrhena, Liquidambar*, Lysichiton, Morus, Mirabilis, Merckia, Mahonia, Menispermum*, Magnolia*, Mitella, Menziesia, Mitchella*, Muehlenbergia, Notophoebe, Nyssa, Nephrophyllidium, Nothoscordum, Oxygraphis, Osmorrhiza, Pasania, Pyrola, Podophyllum, Persea, Phoeba, Parrya, Penthorum*, Philadelphus, Plantanus, Physocarpus, Photinia, Pachysandra*, Parthenocissus, Panax, Phellopterus, Phlox, Pentstemon, Phryma, Romanzoffia, Saururus*, Schizandra, Stylophorum, Smelowskia, Sorbaria, Sageretia, Stewartia (Stuartia), Shortia, Stylocline, Symplocarpus, Smilacina, Trautvetteria, Tiarella, Thermopsis, Triosteum, Trillium, Tipularia, Wistaria, Xanthoxylon*, Zizania.*

The genera Gray listed as being represented in Japan and vicinity and eastern America but not in Europe and western America, which are not in Irmscher's list, are as follows: *Archemora, Ardisia, Asarum* sec. *Heterotropa, Arisaema, Arethusa, Aletris, Brasenia, Berchemia, Benzoin, Boehmeria, Cocculus, Caulophyllum, Cissus, Cryptotaenia, Callicarpa, Cedronella, Coprosmanthus, Chamaelirium, Desmodium, Dioscorea, Hydrangea, Juglans, Lespedeza, Leucothoe, Leptandra, Laportea, Microptelea, Maclura, Onoclea, Oldenlandia, Pieris, Phytolacca, Pilea, Pogonia, Rhynchosia, Sapindus, Symplocos, Sassafras, Tecoma.*

Other students added later to the details of the floristic affinity between eastern Asia and eastern North America, and some revision of the earlier work of course has resulted from closer taxonomic studies and from wider knowledge of distribution. In an introduction to Wilson's *A Naturalist in Western China,* Sargent (545) summarizes the statistics which reveal the similarities and differences between the Chinese and the Atlantic American ligneous floras. He says that 92 families and 155 genera are common to the two regions, but that it was impossible at that time to give exact figures for the species. A comparison of these ligneous floras, based largely on Chung's catalogue, has recently been made by Hu (358). His data reveal that the Chinese woody flora is richer even than Sargent thought. Chung's catalogue listed 778 genera containing woody plants, and Hu added 181 more, making the impressive total of 959 genera. This is more than three times the number (313) of eastern North American genera with woody members. Hu's figure for genera common to the two regions is 156, an addition of only one to Sargent's list. The greater richness of the Chinese flora is due to its greater topographic, climatic, and ecologic diversity, together with the absence of glaciation and the presence of numerous tropical elements.

The fact remains, however, that the floristic relationship between these two widely separated regions is a striking phenomenon, considering the fact that so many plants common to the two regions are absent from western Asia, Europe, and western North America. The most striking relationship is revealed by the identical species which are common only to these regions and by the vicarious pairs or sets of congeneric species. As Gray said, "There has been a peculiar intermingling of the Eastern American and Eastern Asian floras, which demands explanation."

Gray's original explanation of these data is remarkable for its use of the historical point of view at a time when many leading biologists were still proponents of special creation. Gray wrote:

All the facts known to us in the Tertiary and post-Tertiary, even to the limiting line of the drift, conspire to show that the difference between the two continents Asia and America as to temperature was very nearly the same then as now, and that the isothermal lines of the northern hemisphere curved in the directions they do now.[5] A climate such as these facts would demonstrate for the fluvial epoch would again commingle the temperate floras of the two continents at Bering's Straits, and earlier—probably through more land than now—by way of the Aleutian and Kurile Islands. I cannot imagine a state of circumstances under which the Siberian elephant could migrate, and temperate plants could not. . . . Under the light which these geological considerations throw upon the question, I cannot resist the conclusion, that the extant vegetable kingdom has a long and eventful history, and that the explanation of apparent anomalies in the geographi-

[5] In this connection, see a recent paper by Chaney (135), which suggests palaeoclimatic data from fossil isoflors in agreement with this early opinion of Gray.

cal distribution of species may be found in the various and prolonged climatic or other physical vicissitudes to which they have been subject in earlier times;—that the occurrence of certain species, formerly supposed to be peculiar to North America, in a remote or antipodal region affords of itself no presumption that they were originated there. . . .

Something of the prevailing ideas which Gray's reasoning was eventually to overcome may be noted by referring to the minutes of the American Academy (294). Agassiz argued that the present distribution of organisms was linked with that of earlier periods in such a manner as to exclude the assumption of extensive migrations or a shifting of floras and faunas from one area to another. He viewed the similarity of the biota of northeastern America and northeastern Asia as a primitive adaptation of organic types to similar corresponding physical features which have remained respectively unchanged since the introduction of these organisms upon the earth. He did not believe in a single center of origin for conspecific individuals widely separated. Bigelow (then president of the Academy) argued for an original biota with a circumpolar distribution and homogeneity in the various zones, and said that subsequently the dying-out of types in various places resulted in similarities between far-flung lands. He too was opposed to the idea that plants could undergo any considerable migration. Gray rejoined by saying that he believed the apparently early homogeneity of temperate floras and faunas was a resulting or derived condition and not an original one. Still less, therefore, could he agree with Agassiz in regarding the present distribution, with all its dislocations, as a primitive state. He reiterated his opinion that plants and animals had migrated under the compulsion of climatic changes and along highways of favorable climates at various times during their history. He had no idea that recent migrations had anything to do in accounting for the present existence of the same species in such widely separated stations. For him long-distance dispersal was not to be considered as an explanation. More than a decade later Gray (295) reaffirmed his views, and in 1878 gave his classical lecture on forest geography and archeology which terminated nearly four decades of his interest in the relationships of the Asiatic and American disjuncts.

It was Hooker (355), however, who pointed out the topographic reasons for the impoverishment of the European flora with respect to the once circumboreal forms of the Tertiary. He said that it was the mountain masses of western America and the east-west trending mountains and bodies of water in Europe and western Asia that resulted in the extinction of many of the forms from these regions during the forced southward migrations of the Pleistocene. In eastern America and eastern Asia the forms had an opportunity for a southward retreat during the ice advance which was denied many species in the intervening areas.

Fernald (240, 241, 243, 244, 245) has many times concerned himself with the ancient mesophytic forests of Cretaceous and Tertiary times which are today represented by disjunct areas, especially in Atlantic America and eastern Asia. The laurel which Sir Joseph Hooker handed Asa Gray for his essay on the "Flora of Japan"—"the first entirely satisfactory contribution of its kind to the science of botanical geography known to me"—can certainly be passed on to Professor Fernald by saying that his contributions to floristic and historical plant geography are unsurpassed in recent American botany. We shall use only one quotation from Fernald (245) with respect to the survivors of the ancient Tertiary vegetation. " . . . Not only are these Appalachian genera of today the genera of a thousand ages; their species are also ancient and usually sharply differentiated. . . . When the hundreds of species of Appalachian angiospermous genera are compared with their Old World representatives the general conclusion is apparent; that in nearly all groups the species of the Western Hemisphere are completely segregated from those of the Eastern; that we have stable or essentially stable specific entities." Thus we bring to a close our consideration of Asiatic-American affinities and disjunctions, a phytogeographical problem which has attracted the best geographical investigators and resulted in many of the fundamental concepts of the science. Recent taxonomic studies sometimes result in the segregation of what were once considered conspecific entities (frequently only the description of a variety for one of the disjunct areas) or the union under one species of what were once considered vicarious species, but the essential picture and its interpretation have not been materially altered since the days of Gray and Hooker.

Eastern American coastal plain-interior disjuncts.—Under this topic we shall deal with two types of interior disjunction from the modern Atlantic coastal plain flora: that which occurs mainly about the Great Lakes, and that which is found principally in the southern uplands of the Appalachian and Cumberland regions. Coastal plain plants, and even maritime plants in a narrower sense, have long been known at isolated places far removed from the Atlantic and Gulf coastal plain. As early as 1807 Pursh noted the interior occurrence of plants of a seaside nature. Torrey (643) was apparently the first to make the suggestion that the distribution of such plants as *Cakile maritima, Hudsonia tomentosa, Lathyrus maritimus*, and *Euphorbia polygonifolia* was connected with oceanic submergence. Subsequently Drummond (197) and Hitchcock (347) argued for the western extension of Atlantic coastal plain plants along oceanic arms and the "salty" Great Lakes. Svenson (615), after reviewing the geological evidence of later decades, concluded that while the post-Pleistocene marine sea covered large areas in the St. Lawrence and Champlain valleys, the saline waters were absent from the Great Lakes west of Lake Ontario, from

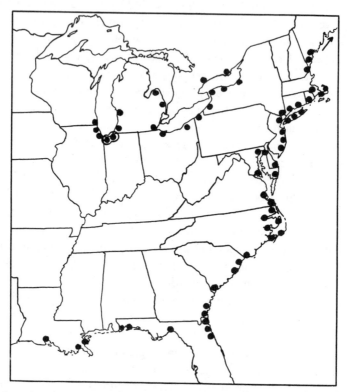

FIG. 36. *Euphorbia polygonifolia* is a strand plant typical of the Atlantic coastal plain which also occurs generally about the Great Lakes, a type of distribution that is suggestive of its route of inland migration. Data from Peattie (501) and McLaughlin (436). *Eleocharis melanocarpa* has a coastal distribution similar to the above, but is disjunct to the southern tip of Lake Michigan (large circles), as are many other species. Data from Peattie (501) and Svenson (615).

most of the Hudson valley, from southern New England and the Connecticut valley, and from the St. John valley in New Brunswick. In general, the occurrence of coastal plain plants inland is not a result of oceanic submergence areas.

Harper (331) was impressed by the similarities between the soils and the habitats (sand hills and plains, shores, marshes and ponds) of the coastal plain and the area of glacial drift. The westward inland extension must have occurred in early post-Pleistocene times. Harper leads one to believe that these plants are present or absent inwardly from the coastal plain because of the presence or absence of suitable habitats. No suggestions are given as to the paths of migration and the reasons for the floristic disjunctions (sometimes a thousand miles), which are greater than the habitat disjunctions.

Several years later Peattie (501) attacked the problem of the disjunction of the Atlantic coastal plain plants about the Great Lakes. The fact that some species, like *Euphorbia polygonifolia,* are known to extend westward along the shores of the Great Lakes to southern Lake Michigan by small disjunctions between numerous stations, and that other species make wider and wider "jumps" until those like *Eleocharis melanocarpa* are not found between southern Lake Michigan and the Atlantic coastal plain caused him to look for a single explanation for all types. He says, "Those cases of one thousand miles inland disjunction are not to be explained by that stock method—a most overworked and uncritical method—of dispersal by birds, nor yet by winds." He turns to the assumption of a once more nearly continuous line of suitable habitats along the region of the present Great Lakes. The answer he finds in the evidence for suitable paths of migration along lines no longer available—lines of glacial lake shores and connections, which connections, and hence migrations, were severed soon after Algonquian time. Peattie is certain that the path into the Lake Michigan region led from the northern Atlantic coastal plain and not from the Gulf. The plants typical of the northward extension along the Mississippi from the Gulf reach their limits, usually, in southern Indiana, Ohio, and Illinois (*Taxodium distichum*), whereas the plants with which he dealt have none of that distribution.

Svenson (615) took up the problems of the interior distribution of maritime plants in eastern North America, particularly in connection with the post-Pleistocene marine submergence. In this regard it should be remembered that many of the so-called maritime plants are not halophytes, and that it is frequently impossible to draw a sharp distinction between brackish and salt swamps on one hand and brackish and fresh swamps on the other. Also, many plants of the shores and dunes which occur inland are related not to salt but to the nature of the sandy soil and to the excessive light and heat of such sites. Except for salt springs, the maritime plants of interior distribution are non-halophytes or, at the most, indifferent halophytes. According to Svenson, none of this latter group is confined to the area of Champlain submergence. He thinks it probable that several plants of boreal tendencies (*Triglochin maritima, T. palustris, Ranunculus Cymbalaria*), which have by far their greatest distribution in western America, migrated eastward.

McLaughlin (436) entered extensively into the problems of the interior occurrence of Atlantic coastal plain plants, especially in the sand-barrens region of northwestern Wisconsin. He adds data to those already cited and concludes that the region about the Indiana dunes of Lake Michigan may receive coastal plain species from three sources: (1) the Mississippi basin, (2) the Mohawk-Hudson outlet and the eastern Great Lakes, and (3) the Ottawa connective. According to him, "It appears that this region, as a gathering point, was a center from which the greater number of coastal plain species migrated northward along the Wisconsin shore of Lake Michigan, west and southwestward down the Fox River valley, down the sandy bottoms of the Wisconsin River to the Mississippi, and thence northward to the region of the northwestern Wisconsin sand barrens." This sand-barrens region is largely a feature of the old Barrens Lake of early post-glacial times, and several plants of the coastal plain now occur closely correlated with its features and limits; but McLaughlin concludes that the influence of this old lake on distribution was local. In this connection it should be remembered, as Gleason (276) competently argues, that the migration of plants is intimately linked with a continuity of suitable habitats.

In order to understand the differences among the ideas of the above investigators and to make an evaluation of their evidences, or of the inland occurrence of any species, we must bear in mind that the term coastal plain flora does not refer to a homogeneous group of species. It includes plants of a dozen different types of sites and of many more communities, as well as plants within a physiographic area which extends as a more or less narrow band from warm temperate to cool temperate climates. Furthermore, the coastal plain flora includes plants both characteristic or merely common there, and, in both categories, plants of a greater or lesser shorewise distribution from Mexico to Nova Scotia. Because several elements make up this flora, it would be expected that inland coastal plain disjuncts would have to be explained in several ways, according to their particular history. Fernald (247) has emphasized the necessity of avoiding confusion by distinguishing between coastal plain plants, *sen. str.*, and plants which live in many other places as well as on the coastal plain.

Of equal interest to the coastal plain disjuncts about the Great Lakes region are those which occur on the highlands of the Southern Appalachians and Cumberlands. The species in question are common, typical, or characteristic of the coastal plain but on the highlands are usually local to rare. There is a great disparity in age between these two regions. Fernald (245) remarks as follows on the age of the upland region: "Never, since it was first occupied by Angiosperms, has the Appalachian Upland of the United States been invaded by seas; and except for the northern extension, it lies wholly south of the limits of the Pleistocene glaciation. During the Cretaceous, while this southern half of the Appalachian region was covered by land vegetation, the lower marginal country, east, south and far to the west and northwest, was submerged under Cretaceous seas. In the Tertiary, likewise, much of the low-lying coastal plain was again covered by shallow seas; and, furthermore, the outer margin of the coastal plain is often of very modern or Quaternary origin." In the upland area, that has been exposed since the Paleozoic, land plants have had an opportunity to develop and spread since the advent of Angiosperms.

In the above-mentioned study, Fernald concerns himself with the austral and tropical element which is so typical of the Atlantic coastal plain and which also has disjunct occurrences on the interior uplands. He says that here and there along the ancient tableland crests of the Appalachian system, wherever the primitive, open xerophytic or hydrophytic conditions prevail, we are learning to expect rare and highly localized members of the so-called coastal plain flora. In favorable points on the now uplifted ancient peneplain are found relics of a coastal plain element of tropical or austral nature from such genera as *Schizaea, Lygodium, Stenophyllus, Eriocaulon, Xyris, Lobelia, Stomoisia (Utricularia)*. Mingled with these typical lowland plants are sometimes found similar relic colonies of endemic American genera of the Atlantic coastal plain: *Orontium, Xerophyllum, Helonias, Calopogon, Cleistes, Sarracenia, Hudsonia, Rhexia, Leiophyllum, Bartonia,* etc. Fernald points out that this relationship of the coastal plain and the Appalachian highlands has long been known to such botanists as Canby, Gray, Sargent, Kearney, and Small.

An explanation of this disjunction which many earlier botanists held is expressed by Harper (331), as follows: "The few which now occupy isolated stations in the Piedmont region and southern mountains may of course have been there before the Pleistocene, but it seems more likely that they have migrated there in comparatively recent times from the regions where they are now much more abundant." On the other hand, Kearney

(386, 387), Harshberger (332), and Small (593) were of the opinion that the upland plants are a remnant of a very ancient flora, that their first home was on the ancient peneplain that later elevated to form the present mountains and plateaus, and that they migrated to the later-formed coastal plain where they are now so abundant. The rejuvenated erosion resulting from the uplift has all but eliminated suitable habitats for this ancient floristic

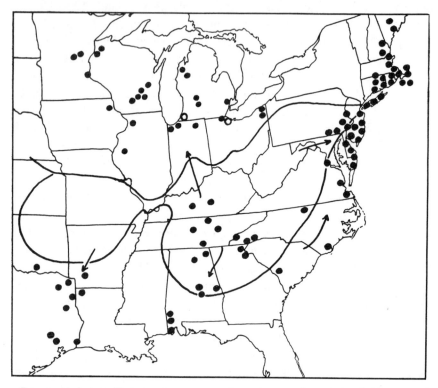

Fig. 37. *Xyris torta* illustrates the movement from the now elevated Appalachian area of lands exposed since the Paleozoic out to the coastal plain and the Great Lakes regions. The hollow circles indicate two Great Lakes stations for *Woodwardia areolata;* this is also a coastal plain species which, like *Xyris torta* and a number of other species, found its way to the coastal plain and the Lakes from the old uplands. Data from Fernald (245) and McLaughlin (436).

element on the tableland, and this accounts for their present scattered and local distribution there.

The detailed and excellent discussion by Fernald (245) summarizes the best thought on the subject today: "The Coastal Plain flora of Atlantic North America is distinguished by the abundance of tropical groups represented. Although these plants now chiefly occupy open silicious, peaty and aquatic habitats in comparatively youthful regions of eastern North America, it is probable that they or their progenitors formerly existed on the area of the ancient Appalachian Upland, especially in the Cretaceous, when that primitive region of the continent was base-leveled and reduced essentially to sea level and at the time when the tropical groups of today were widespread in the North. Then, with the Tertiary uplift of the Appalachian region and its final conversion into a vast well-drained mesophytic area available to the groups which now constitute the climax forests of the Appalachian Upland, the Cretaceous xerophytes and hydrophytes which had previously occupied the ground gradually moved out to the newly available and for them more congenial Coastal Plain."

Somewhat later Braun (83), in a discussion of some relationships of the flora of the Kentucky Cumberland upland, enumerated several typical coastal plain plants of disjunct occurrence inland. These plants are usually local, and sometimes rare, but several are also known to occur together at a single station. They are to be sought especially in the less dissected portions of the Cumberland Plateau near its western margin. Some of these species are *Andropogon glomeratus, Erianthus alopecuroides, Aristida affinia, Uniola laxa, Panicum longifolium, Pogonia divaricata, Schwalbea australis, Cyperus retrofractus, Itea virginica, Viola primulifolia, Bartonia paniculata, Ascyrum stans, Gratiola pilosa, Lobelia Nuttallii, Orontium aquaticum, Diodia teres, Gymnopogon ambiguus, Quercus phellos.* Ten of these coastal plains plants have been found together in one small area of swamp woods and wet meadow in Laurel County, Kentucky (84).

Braun's explanation of these inland coastal plain occurrences is in agreement with Fernald's (245). She points out that what had been called the "Cretaceous Peneplain" is now referred to as the Schooley or Cumberland Peneplain and is not dated as older than Miocene. It was the home of plants of swamp and poorly drained sandy flats, which retreated before the changing environment that resulted from the erosion of the uplifted peneplain, and migrated onto the emerging coastal plain, and in part perhaps northward where the uplift was less. But these shifts of area left behind telltale relics on the plateau where now their known distribution coincides with the undissected remnants of the old Schooley Peneplain. Of all the species mentioned, only two (*Itea virginica and Quercus phellos*) are found in the Mississippi embayment region and might have migrated onto the Cumberland plateau. The preponderance of the evidence is for the reverse—the derivation of the coastal plain flora from the upland regions, at least for most of the species with wide disjunctions. There have undoubtedly been some relatively recent migrations from the coastal plain inward. In his excellent monograph of the Scrophulariaceae (502), Pennell says, "It is natural to expect it [the direction of migration] to have passed from the older inland formations to the younger ones of the Coastal Plain. . . . But when we analyze the relationships of the various species involved, we cannot hold to the simple hypothesis that advance has been uniformly coastward. It has proceeded in both directions."

It is seen, then, that geographical affinity and age of land mass are insufficient data for determining the relative age of different discontinuous areas. It is necessary to combine such information with probable evolutionary relationships. The probable direction of migration of one group may be reversed by other species.

18.

Vicarious Forms and Areas

Vicarious species are closely related allopatric species which have descended from a common ancestral population and attained at least spatial isolation. The concept of vicariism is equally applicable to other than the specific category, i.e., to subspecies, sections, etc., and to communities.

Vicarious species and other groups are particularly instructive in the study of dynamic plant geography. The subject is one which has received considerable attention in the past, although the term vicarious has not always been used in the same sense. It will be well to see just how various authors have considered it.

Although the term was used earlier, Drude's definition of it (196) will serve for the early concept: "The different newly originated forms, in their different places of origin and showing their formation from a common beginning, are vicarious forms." They were also referred to as corresponding or representative forms. In a series of papers on evolution, Jordan (381, 382, 383) developed what has become known as Jordan's Law: "Given any species in any region, the nearest related species is not likely to be found in the same region, nor in a remote region, but in a neighboring district separated from the first by a barrier of some sort or at least by a belt of country the breadth of which gives the effect of a barrier." This he called the Law of Geminate Species. Such pairs of species (twin or geminate species) are no different from those known as vicarious species, for Jordan says that in all matters of adaptation to environment, presumably as a result of selection, they may be absolutely identical, as also in habits unless confronted by some novel condition. They differ only in minor characters, undoubtedly of later origin than their common characters. Drude's definition was essentially repeated by Diels (187), who wrote, "Vicarious sets are essentially similar, showing differential units of degree only, and in their distribution having mutual exclusion of area. One is accustomed to consider them as derivatives from a common type." Vierhapper (660) introduced *vicariad* to supplement vicarious, and distinguished between true and false vicariads. The units in the regions or formations concerned which have arisen from a common stock constitute *true vicariads*. *False vicariads* are units which do not bear this close genetic relationship. He recommends the term *substitution species* to refer to the occurrence of essentially similar units in different regions or formations when it is not desired to distinguish between true and false vicariads or knowledge will not permit it. False vicariads frequently are members of different sections of a genus, rather than the same section. Vierhapper gives examples in *Erigeron, Trifolium, Soldanella,* and *Dianthus,* in which disjunct species have developed similar life forms but are not immediately descended from the same stock. This is the phenomenon usually referred to as convergent evolution. Setchell (570), taking a broader concept, defines vicarious species as two species only slightly discontinuous morphologically but widely discontinuous geographically. Wulff (719), however, adheres to the stricter and more classical concept of mutually exclusive yet closely related species linked by their ancestral initial form or species.

The concept of vicarious pairs or sets is applicable to groups of various taxonomic rank (subspecies, species, sections, genera, etc.) and to communities under climatic and edaphic control. Also, the separation between vicariads is of several types. Vierhapper (660) mentions the following types: those with (1) horizontal or geographical separation, (2) altitudinal separation, (3) habitat or ecological separation. To these three types Wettstein (688) has added those with (4) seasonal separation, or seasonal dimorphism. From the application of the concept of vicariism, we find in the literature that the stricter definitions frequently are not followed and that the idea is taken in its widest sense. A series of examples follows.

Systematic vicariads.—For higher categories, Vierhapper suggests that the Cactaceae of America and the Aizoaceae of the Old World, especially South Africa, represent vicarious families. The subfamily Taxoideae of the Taxocupressaceae may also be considered to consist of vicarious tribes, the Podocarpeae in the tropics and extra-tropical regions of the southern hemisphere, and the Taxeae and Cephalotaxeae in extra-tropical regions of the northern hemisphere.

Most of the studies of vicariads have referred to species, and many such studies have suggested the historical causes of their development of the vicarious state. Engler (214, 215) pointed out that a number of vicarious species owe their origin to the Pleistocene segregation of Tertiary floras. He mentioned as examples of this phenomenon *Ostrya carpinifolia* in southern Europe and *O. virginiana* in North America, *Cercis siliquastrum* in the Mediterranean region, *C. chinensis* in China and Japan, and *C. canadensis, C. occidentalis,* and *C. reniformis* in North America.

Fernald (244) listed a few of the fairly large number of closely related species of eastern North America and Europe. Nine pairs of these European-American vicariads are, respectively: *Juniperus Sabina* and *J. virginiana, Vallisneria spiralis* and *V. americana, Anemone nemorosa* and *A. quinquefolia, Oxalis Acetosella* and *O. montana, Hepatica triloba* and *H. americana, Luzula pilosa* and *L. saltuensis, Maianthemum bifolium* and *M. canadense, Ranunculus Flammula* and *R. lexicaulis,* and *Scrophularia nodosa* and *S. marilandica*. Fernald points out that these species occur in North America in the general regions of *Liriodendron, Comptonia, Carya, Magnolia, Sassafras, Hamamelis,* etc., whose ancestry is clearly traced back to Mesozoic or early Cenozoic forests of the northern hemisphere. Such woody types have mostly vanished from western Asia and Europe, but most of them have relics in eastern United States and eastern Asia. With them occur many herbaceous plants which likewise, it would seem reasonable, should be interpreted as descendants of the Cretaceous and early Tertiary circumboreal flora: *Polygonum virginianum, P. arifolium, P. scandens, Symplocarpus foetidus, Podophyllum, Caulophyllum, Panax, Shortia, Phryma,* etc. According to Fernald, there would seem to be no reason for not treating the European-American vicariads in the same way. Although many of the circumboreal Tertiary plants were eliminated from Europe, many of these vicariads apparently survived the maximum glaciation of the Pleistocene in southern or eastern Europe.

Eastern American ferns have undergone numerous nomenclatural changes, largely because of the close similarity of many forms, but among them we can detect several vicarious units of subspecific rank, according to relationships accepted by Wherry (690). Such cases include *Osmunda regalis* var. *spectabilis* of the Americas and a variety in Europe, *Osmunda claytoniana* var. *vera* of eastern North America and another variety in eastern Asia, *Osmunda cinnamomea* var. *typica* of eastern Canada and the United States and another variety which is tropical, *Pteridium latiusculum* var. *verum* of the United States and other varieties elsewhere, *Cystopteris fragilis* var. *mackayi* of eastern United States and the circumboreal var. *genuina* and the Southern Appalachian var. *protrusa, Onoclea sensibilis* var. *genuina* which is widespread in eastern North America and has a close relative in Asia, *Phyllitis scolopendrium* var. *americanum* and a European variety, and

Polystichum braunii var. *purshii* of America and a European variety. It is of interest to note that the subspecific status of these eastern American ferns has in most cases only recently been recognized. Gray reduced *Osmunda spectabilis* to *O. regalis* var. *spectabilis* in 1857 and Lawson recognized *Cystopteris fragilis* var. *mackayi* in 1889; but all of the remaining varieties were established by Fernald, Wherry, or Weatherby between 1928 and 1937. In each case they appear to represent vicariads in the usually accepted sense of the term.

To give some examples very different from the foregoing, we quote Setchell's illustrations (571) of vicarious species with widely discontinuous distributions selected from the maps of "sea grasses" by Ostenfeld (494). These pairs occur in the Indo-Pacific and the Caribbean regions, respectively: *Thalassia Hemprichii* and *T. testudinum, Halophila decipiens* and *H. Baillonis* of the Hydrocharitaceae; and *Diplanthera uninervis* and *D. Wrightii, Cymodocea isoetifolia* and *C. manatorum* of the Potamogetonaceae.

Vierhapper (660) refers to section Thylacites of the genus *Gentiana* as a natural group which occurs in six vicariads in middle and southern European mountains. They are in part disjunct and in part meet at their borders; they include *Gentiana latifolia, G. alpina, G. vulgaris, G. divaricata, G. angustifolia* and *G. occidentalis.* There is some basis for considering them subspecies of the collective species *G. acaulis,* but such a taxonomic disposal of them would not affect their status as vicariads. The genus *Elytrigia* is a similar case (360). In such cases as these mountain vicariads, where the morphological distinctions are weak and the areas not much separated, it is likely that their age is not great, being Pleistocene or even post-Pleistocene. Such vicariads as those of eastern Asia and eastern North America are patently much older.

As examples of altitudinal vicariads, Vierhapper lists several pairs at lower and higher altitudes, respectively: *Rumex acetosa* and *R. arifolius, Dianthus deltoides* and *D. myrtinervis, Myosotis silvatica* and *M. alpestris, Solidago virga aurea* and *S. alpestris, Centaurea pseudophrygia* and *C. plumosa, Senecio Fuchsii* and *S. cacaliaster, Phleum pratense* and *P. alpinum, Avenastrum pratense* and *A. versicolor, Trisetum flavescens* and *T. alpestre, Trifolium pratense* and *T. nivale,* and *Soldanella major* and *S. hungarica.*

Turrill (654) states that in Crete and Greece there is a considerable number of altitudinal vicariads and that the high-mountain species are frequently limited to a small area, sometimes a single mountain. The mountain species are closely related in morphological characters to species occupying the lowlands, and their ranges may be more or less in contact. He says that the low-elevation species usually have wider occurrences than the mountain species "and the inference of *in situ* development of the high-mountain species seems justified on a number of lines of evidence."

Van Steenis (608) emphasizes the fact that "each species has an altitudinal distribution of its own, which may show differences in different localities, islands, mountains, etc., but ought not be compared with the altitudinal range of another species of the same genus." It usually does no good and may be confusing in geographical comparisons to consider collectively the altitudinal limits of a genus or higher category because it often happens that in one such group there are marked lowland as well as mountain species. Such cases may even be cases of altitudinal vicariism; van Steenis cites two Malaysian examples. *Bulbophyllum tenellum* occurs in mountain forests between 1000 and 1500 meters in altitude, and its vicariad, *B. xylocarpi,* is found only at sea level in mangrove forests. He also mentions *Sopubia* as a typical case of altitudinal vicariism. *S. stricta* is the lowland member of the pair, occurring in southeastern Asia and Madoera Island northeast of Java, and *S. trifida* is a mountain plant widely distributed from tropical Africa to Asia through Malaysia to Australia. In Malaysia its altitudinal range is between 950 and 1800 meters.

Ecological vicariads.—Under local habitat vicariads there are numerous cases of ecological pairs of species affiliated with different associations or facies of vegetation of a region. In such instances the character of the soil may be of great importance in accounting for the separation of the species. According to Turrill (654), the relatively local extent of soil distribution is the reason why a considerable number of ecological pairs of species can be found in any region. He gives the following examples. In salt and fresh marshes, respectively: *Triglochin maritima* and *T. palustris,* and *Scirpus Tabernaemontani* and *S. lacustris.* In soils with high and low available water, respectively: *Alopecurus geniculatus* and *A. pratensis, Stachys palustris* and *S. sylvatica, Lotus major* and *L. corniculatus,* and *Geum rivale* and *G. urbanum.* In calcium-rich and -poor soils, calcicoles and calcifuges, respectively: *Gentiana Clusii* and *G. Kochiana, Primula auricula* and *P. viscosa, Anemone alpina* and *A. sulphurea, Lithospermum purpureocaeruleum* and *L. diffusum, Rhododendron hirsutum* and *R. ferrugineum,* and *Achillea strata* and *A. moschata.* Vierhapper (660) had already listed some of the above pairs and in addition included for the Alps the following vicariads of acidic and limestone rocks, respectively: *Dianthus glacialis* and *D. alpinus, Soldanella pusilla* and *S. minima, Juncus trifida* and *J. monanthos, Poa laxa* and *P. minor,* and *Silene acaulis* f. *norica* and f. *longiscapa.*

The existence of closely related pairs of species with divergent soil acidity preferences was recognized by Kerner (389). Warming (676) summarized the earlier work of this nature in which plants were usually classified rather broadly as calciphile or calciphobe. Many botanists—among them Atkins (27) and Fernald (236, 238, 239)—have studied the distribution of related plants and soil reaction, and Wherry has published a long series of papers on the subject of soil reaction, systematics, and plant distribution. In one paper (689) he compared a series of closely related species pairs (or groups) with their typical soil reactions. With respect to the data shown in Table 21, Wherry says, "The general trend of the relations in the rather consid-

TABLE 21.—A selection from Wherry's data (689) of closely related species of divergent soil reaction and, in many cases, of geographical separation also, whose members can be considered vicariads

	Characteristic pH Range of a Species				
	4–5	5–6	6–6.9	7.0	7.1–8
Woodsia ilvensis		x			
W. glabella and W. alpina					x
Smilacina trifolia	x				
S. racemosa			x		
S. stellata					x
Iris verna	x				
I. cristata			x		
I. lacustris					x
Pinguicula elatior	x				
P. vulgaris					x
Asplenium montanum	x				
A. ruta-muraria					x
Uniola laxa		x			
U. latifolia				x	
Malaxis unifolia	x				
M. monophyllos					x
Listera reniformis	x				
L. convallarioides				x	
Vincetoxicum carolinense		x			
V. obliquum				x	
Lobelia Canbyi	x				
L. Kalmii					x

The following species groups are without conspicuous morphological distinctions and presumably are closely related:

	1	2	3	4	5
Ophioglossum arenarium	x				
O. vulgatum			x		
O. Engelmanni					x
Arisaema Stewardsonii	x				
A. pusillum		x			
A. triphyllum			x		
Spiranthes cernua	x				
S. odorata		x			
S. ochroleuca			x		
Gentiana saponaria	x				
G. clusa			x		
G. Andrewsii				x	
Silene caroliniana		x			
S. Wherryi					x
Pyrola incarnata	x				
P. asarifolia				x	
Pyrola convoluta	x				
P. chlorantha			x		
Agalinis tenuifolia		x			
A. Besseyana				x	
Campanula aparinoides		x			
C. uliginosa				x	

erable number of plant groups here considered is so definite that it seems safe to state that in eastern North America, as a general rule, in any group of closely related plants those of southern distribution are likely to favor more acid soils than northern and western ones." In 22 sets there is geographical separation of related species, and in 20 of these the acid-soil species are southern and the less acid-soil species are northern. By northern is meant in general that the plants are found mainly above the glacial boundary. The members of all these groups can be considered ecological vicariads.

I wish to insert a word of warning at this point. There has been little or no investigation that has been able to prove single ecological factor limits for certain species. If we take as an example these pairs and triplets of species which have contrasting soil-reaction "preferences," we must realize that because a certain species has been found mainly or exclusively in soils of a certain pH range, this does not prove that its ecological amplitude with respect to pH is shown by the observed limits. An alternative explanation for the general correlations illustrated by Wherry's data is possible. These data may show only reaction differences between the soils of different geographical regions and not pH "preferences" on the part of the plants. The plants designated as northern are essentially those of the young soils of the glacial till plain; the southern plants are mainly of the coastal plain and the Appalachian uplands. The latter are generally acid soils because of their relatively great age, strong leaching, and other factors, whereas the more northern soils of the till plain are more often circumneutral or weakly acid. Perhaps these pairs of species with geographical separation have developed their vicarious nature because of differences in pH preference, but it is entirely unproved; one might as well use another factor and call them "temperature pairs."

Johnston (380) has called attention to the long-neglected phenomenon of gypsophily—plants growing in pure soils or those rich in hydrous calcium sulphate—and has enumerated a number of species that tolerate or require gypseous soils. He finds that all the known species of *Dicranocarpus* (*D. parviflorus*) and *Sartwellia* (*S. humilis, S. mexicana, S. puberula, S.* sp.) are gypsophilous, and that such genera as *Nerisyrenia* (*N. gracilis, N. Castillonii*), *Drymaria* (*D. lyropetala, D. elata*), and *Nama* (*N. canescens, N. hispidum* var. *gypsicola, N. Purpusii, N. Stewartii*) have groups of species characterized by gypsophily. One of his observations, however, is of special interest for the present topic. He says, "Another example of the dramatic way in which species refuse to transgress gypsum boundaries is found in the behavior of two species of *Fouquieria* growing north of Mohovano, Coahuila. One of these species is frequent on gypsum flats while the other replaces it on the surrounding non-gypseous soils. In a few cases I observed the shrubs growing near one another with interlocking branches, but *F. Shrevei* was always rooted in gypsum and *F. splendens,* beyond an abrupt gypsum boundary, always rooted in non-gypseous soil."

Vicarious associations.—Several different forest types of Asia, Europe, and North America offer examples of true vicariads; the members of the circumboreal spruce-fir formation are a conspicuous example. Sissingh (583) has made a phytosociological comparison of the forest associations of Holland and their vicarious associations in North America. Du Rietz (199), who made a comparative study of the vegetation of the Alps and of Scandinavia, accepted Vierhapper's definition of vicarious species. With respect to associations, he says, the concept would mean that vicarious associations consist of numerous similar species and especially numerous vicarious species, so that they are obviously directly comparable associations that have developed from a common stock. To select an extreme example, Du Rietz considers the *Empetrum nigrum* heath of the northern hemisphere and the *Empetrum rubrum* heath of the southern hemisphere to be true vicarious associations despite their exceedingly wide disjunction.

Two common associations of the Alps are the *Loiseleuria-Cetraria islandica*-Assoc. and the *Loiseleuria-Cladonia rangiferina-silvatica*-Assoc. According to Du Rietz, these two associations are represented by false vicarious associations in Scandinavia, for although the same species exist there, they do not combine in the same way. It may be that the associations are not vital in Scandinavia, but he thinks there are other explanations. For example, in Scandinavia *Cetraria islandica* undergoes severer competition with other fruticose lichens than in the Alps and consequently cannot develop dominance in heaths in which it occurs with *Loiseleuria*. Also, in the case of the *Loiseleuria-Cladonia*-Assoc. the associations are similar in Scandinavia but the species combinations are different because the species are locally better separated by their tolerances (to altitudinal factors, to snow duration, etc.) than they are in the Alps.

Another type of false vicarious associations develops where the inferior forest layers (synusiae) are all different floristically in the various regions, and the apparent vicariad nature of the associations results only from the uniformity of the dominant species. Such cases occur in the northern and alpine forests dominated by *Larix* and in others dominated by *Pinus cembra,* according to Du Rietz. These synusial changes in the composition of phytocoenoses produce what have been called twin associations. This has been reviewed by Cain (108). Still another type of false vicarious association is that in which the members are ecologically comparable (in life form) but floristically different. Gams (264) illustrates this phenomenon by the *Carex curvula* heath of Alps and the *Carex rigida* heath of Scandinavia.

19.

Polytopy and Polyphylesis

1. *When the occurrence of a species (or members of any other category) consists of two or more discrete areas, the species is said to be polytopic. There are four theories in active use that are invoked by geographers to explain the origin of the areas.*

2. *There is a belief that polytopic forms are genetically and immediately related, and that the intervening area, however great, has been completely bridged at one step by long-distance dispersal. Opposing this theory as a general explanation of polytopy are the observed facts of endemism, of replicated distributional patterns involving all types of organisms and refuting any random character of major biotic patterns, and the high improbability, for most organisms, of long-distance dispersal successfully culminating in establishment.*

3. *Therefore, the most widely accepted hypothesis is that the polytopic forms are genetically and immediately related, and that the intervening area has been bridged in the past by a continuous series of populations, although not necessarily at any one time. For this there is a preponderance of evidence, including fossil records of occurrence in intervening areas, and the highly probable theory of climatic migrations.*

4. *A third view holds that the polytopic forms have had an independent origin, in the several existing areas, from an unlike ancestral population that now bridges or has bridged the gaps between the areas. The members of the polytopic areas are similar because of parallel descent. This hypothesis invokes the differentiation theory and frequently some form of orthogenesis, and eliminates from immediate consideration in geography most of the problems presented by the first two hypotheses. There is some direct evidence and a considerable quantity of inferential evidence for the hypothesis, but it should not be employed except when other explanations fail or are inadequate.*

274

5. *Some few believe that the polytopic forms have had an independent, autochthonous origin from taxonomically and presumably genetically different ancestors, and have become similar through convergent evolution. This is the phenomenon of polyphylesis which, for most students of evolution, genetics, and taxonomy, represents only the result of inadequate knowledge, or the forming of groups (genera, families) for practical convenience, except in cases of hybrid descent such as amphidiploidy. When polytopic forms may reasonably be considered polyphyletic, problems of their geography recede in importance because of the remoteness of the relationship of the forms.*

We have already observed that complete continuity among the individuals of a kind is a negligible phenomenon occurring only in vegetative clones, and that in all or nearly all such cases the colonies become fragmented and discontinuous. It is consequently the practice to speak of discontinuous distribution for a form or a group of related forms only when the distance between populations is greater than the usual effective dispersal distance. Wide discontinuities are always of interest to geographers, and they pose some of the most interesting and difficult problems in the history of plants and animals. When two or more areas of a form, a species or subspecies, are entirely separated from each other by some geographical barrier, the form is said to be polytopic. The phenomenon occurs also within the higher categories such as genera and families.

The fact of polytopy brings up the question of its origin. The most frequent assumption made by geographers is that the area that now separates the related forms has been bridged by the forms or their immediate ancestors, either by long-distance dispersal or by slow migration. The other explanation, which has always had some adherents, supposes a polytopic *origin* of the disjunct populations; that is, the two populations do not stand in the direct relationship of parent and descendant. The theory of polytopic origin also has two explanations. According to one the intervening area was in the past or is at present occupied by an ancestral syngameon out of which the modern related populations have had an independent evolution in the two or more regions now occupied by them. The other explanation is polyphylesis which supposes that the two populations (or related forms) which now occupy the discontinuous areas and are classified as taxonomically identical or closely related have not had an immediate common ancestry but have arrived at the condition of identity or similarity through convergent evolution from originally more diverse phylogenetic stocks.

We have seen what the four theories[1] purporting to explain polytopic areas are. Each of these hypotheses poses its own secondary problems and requires particular assumptions which will now be examined.

The first theory assumes at least occasional long-distance dispersal, sometimes over hundreds or even thousands of miles, without loss of viability of the diaspores, under the action of such agents as wind, water, currents, rafts, and bird carriers. The big problems in this connection are to provide suitable carrying agents and, in particular, to demonstrate the fact or probability that the dispersal can result in ecesis, the second requisite for migration. It should be noticed that this hypothesis is based upon an assumption which nullifies the only reasonable definition of discontinuity, i.e., two areas are not discontinuous if their gap can be bridged at one jump.

The second theory requires the demonstration of a series or a continuity of suitable habitat types that act as intermediate stations for the migration. If such a continuity of suitable ecological situations does not now exist between the disjunct areas, it must be hypothesized for some time in the past; the problem is to marshal data concerning geological, physiographic, and climatological history in support of the theory. Further aspects of this theory need emphasis. It must be assumed that the population as a whole had a single center of origin and that consequently one area is parental and the others derivative, or that a once broad area was segmented by the development of regions of unsuitable habitat across the preexisting continuous range, or that the original area was somewhere near the now unoccupied region and that the derived and disjunct areas resulted from migration from the original area in two or more directions. Under these assumptions the members of the disjunct populations are part and parcel of the same genetic stock. It is only a question of explaining the existing fragmentation of the area. In connection with this theory, it can be stated that modern discontinuities are frequently resolved by the discovery, in the intervening areas, of fossils of the species in question. Furthermore, this theory is applicable not only to disjunct occurrences of members of a species or subspecies but to disjunct occurrences of closely related species of a section, genus, etc.

The third theory, that the disjunct populations have had a polytopic origin from a parental syngameon, requires the demonstration of suitable intervening habitats for the parent population, just as the second theory does. But an entirely different kind of problem must also be met—the genetic question of whether there can be differentiated from a syngameon, independently in two regions, two populations that consist of the same

---

[1] The old conception of special creation is excluded.

group of biotypes and that can be truly classified as constituting a single species. Let us call the parental syngameon species A; the question is whether species B can arise independently from species A in separate regions. This is a question of independent parallel evolution from the same phylogenetic stock. Such an evolution could conceivably result from chance, from the operation of similar selective factors of the environment in two regions, or from the expression in two regions of the same internal orthogenetic tendencies of variation.

The fourth theory offers no immediate problems of migration or suitable habitat type but presents fundamental genetic and taxonomic questions. Starting with distinct phylogenetic stocks, is it possible that nearly identical conspecific genotypes can be arrived at through convergent evolution? If species C is the modern disjunct form under consideration, is it conceivable that it had an independent origin from species B in one place and from species A in another? The taxonomists know of or suspect several taxonomic groups, principally among higher categories such as tribes, families, and orders, which are polyphyletic in that they include forms that have no immediate connection by descent. The classification of such genera into one family, or of such families into one order, has resulted from their superficial resemblance. Such mistakes in classification arise from incomplete knowledge, from the employment of superficial rather than fundamental diagnostic criteria, or from practical reasons. If it is assumed that the classification strives to be natural, in that forms placed together are related by descent, such cases are resolved by reclassification into monophyletic groups. Apparent problems in the geography of such disjunct groups disappear when the polyphyletic origin of these groups is recognized. The only other alternative seems to be the assumption that conspecific biotypes can arise independently from diverse genetic sources. The likelihood that a taxonomic group includes forms of polyphyletic origin increases with the rank of the group. The members of a species, for example, have so many genetic properties in common that it seems highly unlikely that the same genotype can have had independent origins.

Now that we have seen what the problems are, there is a further consideration. Frequently, in an approach to geographical problems, one of the first necessities is to dispose of the question whether the group under consideration could have had a polytopic origin. For example, in a discussion of the floras of Japan and North America, Gray (293) found it necessary to settle this question because at that time many biologists, following the philosophies of Schouw (556) and Agassiz, believed that species originated where they are found. Gray said, "I am already disposed . . . to admit that what are termed closely related species may in many cases be lineal descendants from a pristine stock, just as domesticated races are; or, in other words, that the limits of occasional variation in species (if by them we mean primordial forms) are wider than is generally supposed, and that derivative forms when segregated may be as constantly reproduced as their progenitors." He admitted that the discovery of numerous related species in two widely separated places might not of necessity require former continuity, migration, or interchange, but he was certain that the occurrence of identical species populations under such conditions would require the supposition of a former continuity. "Why should it?" he asked. "Evidently because the natural supposition is that individuals of the same kind are descendants from a common stock, or have spread from a common center; and because the progress of investigation, instead of eliminating this preconception from the minds of botanists, has rather confirmed it."

Agassiz maintained substantially that each species originated where it now occurs, and that probably as great a number of individuals occupied as large an area, and generally the same area or the same discontinuous areas, as at the present time. Schouw's hypothesis of the double or multiple origin of species and Agassiz' more comprehensive doctrine of autochthonal origin had already been abandoned by A. deCandolle (121) (who was originally a strenuous supporter of the idea of polytopic origin) because he felt that the hypotheses were no longer necessary in the light of the increased knowledge of geological history and plant distribution. Gray also wrote that "in referring the actual distribution, no less than the origin, of existing species simply to the Divine will, it would remove the whole question out of the field of inductive science." He proposed two answers to the problem of discontinuous areas: (1) the enormous multiplying powers and facilities and means for dissemination which all creatures possess, and (2) the ever-increasing paleontological evidence of the previous occurrence, in areas that are different from modern ones, of species which are even yet extant.

Turning from these early opinions, we find that recent geographers are still working under the necessity of accepting or eliminating the possibility of polyphylesis and polytopic origin from their studies. Good (286) says that *Coriaria* is the only genus occurring in four widely discontinuous regions: the western Mediterranean, continental and insular Asia, Australasia including parts of Polynesia, and Central America and western South America. According to him, the discontinuities of *Coriaria* within each hemisphere can be explained as a result of the great climatic changes of the late Tertiary, and their origin can be considered as geologically comparatively recent. No such conditions are applicable to the discontinuity between the two hemispheres. A different solution must be sought for this problem. He says:

The first question that arises is whether or not the present distribution of the genus represents the outcome of a former continuous range—whether, in fact, the genus is monophyletic or polyphyletic. The more orthodox view holds that it is monophyletic, and that its present range is due to the geographical divergence of the two parts of the genus from a single point or area culminating in their structural differentiation and geographical segregation. The important question becomes by what path and in what direction this geographical divergence pro-

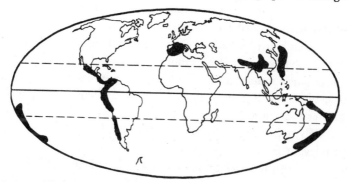

Fig. 38. Despite its wide discontinuities, the genus *Coriaria* is considered to be monophyletic, i.e., to have had a single area or origin and to have gained its present areas through migration and extinction. Map is redrawn from Good (286).

ceeded, and, in particular, where and when the tropics were crossed. But there is another possible view which supposes that the genus is diphyletic or polyphyletic, and that the different parts of it are merely the products of convergent evolutionary trends and have no direct phylogenetic relationship. They must of course have had a common ancestor at one time, but what this was and whether or not it was an Angiosperm at all is unknown. Presumably the only evidence which could prove such a history for the genus would be very complete paleontological records in both hemispheres, and these are lacking. Whether or no this theory is applicable to *Coriaria* in particular, it cannot reasonably be supposed to account for the many other similar cases of discontinuity in the Angiosperms and, moreover, it runs counter to all the more generally accepted ideas of phylogeny and evolution. The point of view is not elaborated further here because its acceptance would automatically remove the necessity for any further discussion and, until there is direct evidence to the contrary, *Coriaria* is assumed to be monophyletic and to have arisen once only at a definite part of the world's surface.

Wulff (718) states categorically that there are no sufficient data for accepting the polytopic hypothesis of the independent origin of a species in more than one place. According to him, "We may continue the study of the origin of the area of a species within the limits of historical evolution, which we would have entirely to discard if we were to adopt the point of view of polytopic species origination, and return to the view of pre-evolution biology."

Turrill (654) considered the question of polytopy in his discussion of the principles of plant geography. He writes as follows: "The question as to whether species originate in more than one place independently (polytopism) or always arise in one locality (monotopism) or at more than one time (polychronism) is difficult to answer. Species that are morphologically and genetically little, if anything, more than varieties, such as *Fraxinus Pallisae* Wilmott, may well have arisen polytopically, as may those which have arisen through hybridization. On the other hand, a large number of species have probably arisen by gradual differentiation following the accumulation of small mutations after some kind of isolation of a population. The chance of parallelism in such isolated populations is proportionately less the greater the number of different characters common to members of the species. Polytopism should only be accepted after a very full investigation of all relevant facts."

Certain aspects of these hypotheses and their attendant problem will now be given fuller consideration. Other phases of the problem are treated elsewhere in these pages in connection with special topics.

Although the question of long-distance dispersal has already been discussed, let us look at it again from the point of view of the first of the theories mentioned at the beginning of this chapter. It has always been an acute question in connection with island biota, separated by hundreds and sometimes thousands of miles of open seas from other islands and continents. Likewise, isolated mountains are in some respects like islands on the continents, the intervening land of a radically different climatic character constituting an impressive barrier. In between these extremes is an infinite variety of kinds of barriers and degrees of discontinuity which the geographer must attempt to explain.

Dispersal to islands is accomplished by the action of several agents of dissemination aided by the adaptations of diaspores of various sorts (469). Ridley (527) says that sea-borne seeds and fruits can reach every island and coast in the world. The first necessity is that such organs be salt-water tolerant and that they float. The absence from an island of at least some plants of this source is a matter not of failure of transport but of the unsuitability of the coast because of either climatic or edaphic reasons, and the consequent inability of the plants to germinate and establish. Next to sea-borne plants, the wind-borne propagules usually constitute the largest group of species on oceanic islands. Wind-borne diaspores, if they cross any great expanse of sea, must obviously be small and light. Ridley says that the wind-borne element of islands is due largely to dust-seeds such as orchids, which are known to have been transported as much as 700 miles through the air, and to the spores of cryptogams. Bird-borne diaspores usually constitute the group of third importance in most oceanic islands. A small fraction of an island biota may have been transported by rafts of vegetation or on cakes of ice (at suitable latitudes). It is tolerably certain, according to Ridley, that some seeds can be transported across wide oceanic spaces by being caught in crevices of logs or in pumice. The presence of land snails was formerly thought to be a sure indication of the continental nature of an island, but their recent discovery on Krakatau shows that they cross the sea, probably on driftwood. The presence of ants, reptiles, shrews, and rats on Christmas Island and elsewhere is taken as a corroboration of the driftwood route of migration to islands. Matthew (457) is a strong supporter of the raft theory of transport of animals. The importance of these methods of transport is illustrated by Table 22, which is adapted from Ridley's data.

Transcontinental discontinuities, which are purported to be explained by long-distance dispersal, do not have the agency of water for transport; hence exceptional activity of winds, bird flight, etc., must be invoked.

The problems of wide disjunction are dramatically presented in the numerous cases of bipolar distribution which have recently been discussed by Du Rietz (203). The bipolar group consists essentially of plants which are boreal and austral, in the broad sense of the terms, and which are completely absent from the tropical zone or at least from tropical lowlands. Following are some of the types of bipolar distribution and some of the plants which have these polytopic areas:

1. Plants with arctic and boreal circumpolar area, with one isolated austral area and no intervening tropical station:

a. With the austral center in New Zealand or Australia, or both regions: *Carex Lachenalii, C. diandra, C. stellulata, Sphagnum centrale, S. palustre, S. papillosum, Climacium dendroides, Grimmia trichophylla, Encalypta vulgaris, Polytrichum gracile, Bryum affine, Brachythecium plumosum, Isopterygium pulchellum, Calliergon cordifolium,* and the genera *Plagiobryum, Seliferia,* and *Eucladium.*

b. With the austral center in southern South America: *Armeria elongata,*

TABLE 22.—A comparison of the floras of three oceanic islands and the types of transport probably responsible for their presence on the islands (Data from 527)

	Krakatau	Christmas	Cocos-Keeling
Distance from Java in miles	25	140	700
Soils	Ash and muds	Impregnated with lime	Sands and limes
Winds	E or SE, monsoon from NW	SE or ESE, monsoon from NE	SE or ESE, rarely NE
Sea currents	From Java	From Java	From Papua and Moluccas
Age	58 years in 1942	Eocene	Unknown
Flowering plants:			
Sea-borne diaspores	60	44	17
Wind-borne diaspores	34	9	0
Bird-borne diaspores			
Berry or drupe	34	36	0
Adhesive	9	15	5
In mud on feet	3	0	0
Doubtful types	4	7	0
Total seed plants	144	111	22
Vascular cryptogams:			
Pteridophytes	48	25	0
Bryophytes	19	18	1
Total cryptogams	67	43	1
Total flora analyzed	211	154	23

Carex magellanica, C. microglochin, Draba magellanica, Koenigia islandica, Vahlodea atropurpurea, Sphagnum plumulosum, Polytrichum strictum, Plagiothecium Toeseanum, Bartramia pomiformis, B. ithyphylla, and the genera *Litorella, Empetrum, Chrysoplenium, Primula,* and *Oncophorus.*

c. With two austral centers corresponding to the two above: *Carex canescens, Sphagnum fimbriatum, Bartramia Halleriana, Calliergonella cuspidata, Andreaea rupestris* (=*A. petrophila*), *Campylium polygamum, Cornicularia aculeata, Lopadium duscoluteum, Alectoria nigricans,* and the related genera *Fagus* (boreal) and *Nothofagus* (austral).

2. Plants with boreal and austral centers, but with one to several populations in temperate belts on high tropical mountains:

a. Bipolar plants with mountain stations which are Asian, East Indian, New Guinean, etc.: *Carex Gaudichaudiana, C. breviculmis,* and the genus *Euphrasia.*

b. Bipolar plants with mountain stations in the tropical Andes and in Central American and Mexican mountains: *Carex Macloviana, C. maritima (= C. incurva), Phleum alpinum, Pleurozium Schreberi, Alectoria ochroleuca, Sphagnum magellanicum, Thamnolia vermicularis, Cetraria islandica, Drepanocladus uncinatus,* and the genera *Saxifraga, Alnus, Gentinella, Leptobryum, Aloina,* and *Conostomum.*

c. Other bipolar groups which have tropical connections but are completely absent from tropical lowlands: *Poa, Festuca, Luzula, Juncus, Ranunculus, Cardamine, Epilobium, Galium,* and the families Juncaceae, Ranunculaceae, Cruciferae, and Umbelliferae.

The problem of bipolar distribution involves all groups from species up through higher categories to families and, in some cases, orders.

According to Du Rietz (203), a number of geographers have been firm believers in the effectiveness of long-distance dispersal, and he mentions in this connection Gibbs, Beccari, Grisebach, Schimper, Schenck, and Heintze. Wallace (671) believed that the various modes of long-distance dispersal across the sea were sufficiently ample to account for the vegetation found on oceanic islands. It was even easier for him to assume that a mountain chain constitutes a highway for continental migration. In bipolar distributions, he attached little importance to the fact that the Andean chain is broken in Panama by the absence of high mountains. We may readily admit that mountain systems do provide a highway for migration, as long as the interval between similar climatic belts on successive mountains (such as the alpine tundra or the subalpine forest belt) is not greater than the normal dispersal maximum for the species involved. If the intervals are greater, long-distance dispersal must be invoked unless the questions of climatic and topographic changes, and the possibility of the evolution of certain types, are brought under consideration. The general opinion held by Engler (214, 215) is not much different from that of Wallace. Engler believed in long-distance dispersal as an explanation of trans-tropical bipolar areas and also as an explanation of the wide disjunctions found in circum-antarctic regions, as between South America, Kerguelen, and Australia and New Zealand. Guppy (309), who is likewise a convinced believer in long-distance dispersal, thought that bipolar distributions, as in *Carex,* resulted from the plants halting temporarily, during their traverse of the intertemperate regions, on the tops of such mountains as there are in the tropics.

In contrast, many geographers have not believed in the effectiveness of long-distance dispersal. Hooker (353), for example, thought it necessary to suppose that there was a more continuous system of mountains than now exists, which provided temperate conditions for the migration of bipolar organisms across the tropics, and that continental bridges were required for the dispersal of plants and animals to now widely disjunct antarctic biotic areas. Stapf (603) found the assumption of a tropical highland migratory route absolutely necessary in an understanding of the highland Malaysian flora. Van Steenis (608) reviewed the literature concerning the origin of the Malaysian mountain floras and, on this basis and his own knowledge of the plants, concluded that there was no evidence that long-distance dispersal played a role in their origin. In fact, he reached the following general conclusion: "On the whole I cannot trace any relation between distribution and what is known as dispersal." Von Ihering (373) considered Wallace's hypothesis of long-distance dispersal a "fantasy"; he could not believe that plants could bridge at one jump the gap over the whole Central American lowlands which lie between the available mountain peaks. After analyzing the dispersal ecology of the Kerguelen flora, Werth (680) concluded that it is impossible that Schimper's theory of long-distance dispersal is correct. Skottsberg (584, 586, 587, 588, 590) has probably done the most to destroy the hypothesis of long-distance dispersal. As a result of his studies of Pacific Island floras from the point of view of dispersal ecology, phylogenetic relationships, and distribution, he concludes that the present circumscribed distribution of endemics argues against the probability that long-distance dispersal occurs now, or has ever occurred as a general phenomenon between isolated islands and continents.

The earlier belief that organisms with light diaspores (especially mosses, liverworts, lichens, and ferns) could not be used for geographical problems because of their wide dissemination has been strongly refuted by the evidence gained from the increased knowledge of areas occupied by certain members of these groups. These plants show the same kinds of areas and disjunctions as do the relatively heavy-seeded flowering plants. Phytogeographical conclusions concerning the flowering plants can be extended to cryptogams, and are supported by cryptogams, as shown by studies of these groups: ferns (142, 708), mosses (341, 342, 376, 572), hepatics (192, 572), and lichens (200, 431, 182).

Against the theory of long-distance dispersal can be marshaled the following objections: (1) Modern distributional patterns and areas show so many specific parallels under both continental and oceanic conditions that, because of the element of chance involved in occasional random long-distance dispersals, this theory cannot reasonably be generally applied. (2) The high degree of endemism which is related to the extent and age of isolation in islands, mountains, etc., is opposed to the concept of long-distance dispersal. (3) The assumption of dispersal is inadequate in itself, because it must be demonstrated that the arriving diaspores can be delivered, so to speak, in a viable condition to a suitable habitat where they must also be able to enter and compete in a closed community. It is at this point that most cases fail to convince one that long-distance dispersal is effective on the whole.

In general, the problem of plant geography is to explain the differences and similarities between the various floras of the world today. As we have remarked, the usual modern theories attribute the present distribution of plants mainly to the continued effects of the changes that have occurred in the environment to which they have been exposed during their history. Such theories may be said to have originated with the "Theory of Climatic Migrations" proposed by Forbes (253, 254). He takes for granted the existence of specific centers of speciation and recognizes that there may be three explanations for polytopic areas, i.e., that isolated areas may be populated in three ways: by special creation, by transport from an original area to derived areas, and by migration from one area to another before isolation. Forbes was convinced that the first possibility was improbable, the second inadequate, and the third of the greatest importance. He says, "The specific identity to any extent of the flora and fauna of one area with those of another depends on both areas forming or having formed part of the same specific center, or on their having derived their animal and vegetable population by transmission, through migration over continuous or closely contiguous land. . . ." According to Good (283), "The migrational hypothesis suggests that the distributional phenomena in plants are due to and are directly caused by those various environmental changes which are generally believed to have taken place, and for which there is abundant evidence." With respect to the question of dispersal, Good emphasizes that dispersal mechanisms and agents are the means by which plants actually accomplish their movements from generation to generation, but that not until such movements are permitted by external factors to result in ecesis can they constitute migration. We may readily admit that environmental changes provide the compulsion for retreat and the opportunity for advance which result in migration and changing areas. With evolution, however, a species may add types that permit an expansion of area without environmental change.

The problems posed by the second of the four theories stated at the beginning of this chapter are those of evidence for land connections and for suitable environmental conditions between areas now disjunct. Also, it is necessary to explain the causes for the extermination which has resulted in the disjunction. We shall use a single example, the Magnolieae, for the purpose of discussing this theory further.

The Magnolieae are treated by Good (283) as consisting of three genera (*Liriodendron* is omitted from consideration): *Magnolia* with 63 species, *Michelia* with 25 species, and *Talauma* with 40 species. According to geographical regions the distribution of species is approximately as follows:

		Total	
Western Hemisphere:			
North America	8		
Central America	7		
West Indies	7		
South America	2	24	
Eastern Hemisphere:			
Malay Archipelago	34		
Malay Peninsula	8		
Himalaya-Burma	20		
China-Indo-China	35		
Japan	7	104	128

Magnolia, the largest genus, is also the most characteristically temperate and provides all the species of North America (8) and Japan (7). The great majority of the Eu-Magnolias are found in the Himalayan-West China region; they become increasingly fewer southward. *Talauma* occurs in both hemispheres and is more characteristically equatorial than *Magnolia*. *Michelia* occurs only in the eastern hemisphere and falls into two main groups of species, one in Himalaya-Burma (9) and the other in China (7). There are no really wide species in the tribe—all of them may be said to be endemic—and no species is common to the two hemispheres, although in *Magnolia* there are very similar species in southeastern United States and southeastern Asia, as first pointed out by Gray (293).

The Magnolieae occur nowhere with a total rainfall of less than about 50 inches annually, and 160 inches is not too much for them; they do not endure a markedly dry season. They are not found where the average temperature is below 50° F., according to Good; if the North American and Japanese species of *Magnolia* are excluded, the minimum is between 65° and 70° F. Certain of the North American magnolias are the only ones which have to endure frost, with a northern limit at the January isotherm of 32° F. When rainfall and temperature are considered together, it is seen that the Magnolieae occupy nearly all the regions of the earth that provide suitable climatic conditions. It is a fair assumption that these plants require such conditions for their perfect development. Our remaining remarks will be limited to *Magnolia.*

The fossil record is based mostly on leaves, and some of the specimens are doubtfully assigned to *Magnolia;* but Fritel (256), Marty (448), Berry (58) and others have made careful comparisons with modern species and other possible genera to which the fossils could be referred and have verified many records. Some of the records include seeds (521) and fruit (448). Despite the frequent difficulties involved in making generic reference, the record is sufficient to prove that during the Cretaceous and at least the early Tertiary the genus *Magnolia* was widely distributed in the northern hemisphere, extending to latitudes much farther north than is now the case.

Berry (58) gives the following summary of the fossil occurrences of *Magnolia,* a record which is long and extensive. Twenty-three species of *Magnolia* have been described from Upper Cretaceous sediments. One group of forms found in western Greenland ranged southward along what is now the Atlantic coastal plain and westward to Texas. A second group of species, including some of the former ones, are fossilized in the Dakota sandstones of Kansas and Nebraska laid down along the shores of the Mississippi Embayment; magnolias of this age are known also from Wyoming, western Canada, Vancouver Island, and Tennessee. In the Old World, Upper Cretaceous magnolias are known from Portugal, Bohemia, and Moravia. From the Eocene there are about 20 species, all different from their Cretaceous ancestors, which have an equally wide distribution. They are known from Greenland, Spitzbergen, Alaska, Sachalin Island, and western Canada on the north, and Oregon, California, Wyoming, Colorado, New Mexico, Mississippi, Tennessee, Louisiana, France, Germany, Bohemia, and Croatia farther south. During the Oligocene, 8 magnolias are known from European deposits in Italy, France, and Russia. Sixteen Miocene species are recorded from North America and Europe, and they were probably also present in Asia. At the close of the Tertiary, during Pliocene time, 11 species of magnolias are known from eastern Asia, Europe (mostly in Mediterranean regions), and North America. The last magnolias of Europe and central and western North America were apparently exterminated by the Pleistocene conditions.

With respect to paleogeography, it appears that the northern hemispheric continents were connected by land bridges in Cretaceous time and during much of the Tertiary. Greenland and Europe were probably connected by way of Iceland until well along in the Pleistocene, and most of the time there was a land connection between eastern Asia and northwestern North America. This circumboreal continental mass was broken in central Asia during the Eocene, but continuity was restored during the Miocene. A second break appeared in the Pliocene, separating Ellesmere Island and Greenland. Details concerning these matters can be found in Arldt (25). It is apparent that land surfaces for migration have been amply available (whether we accept the land-bridge hypothesis or some form of continental displacement) and that the *Magnolia* species of the regions now so widely disjunct are descendants of the same phylogenetic stock. Fossil records of *Magnolia* bridge the gaps between present areas.

It is then probable that the regions not now inhabited by *Magnolia* once had a suitable climate, as attested by the fossils themselves. The wide latitudinal spread of the fossils, especially northward, does not mean that what we may call a "*Magnolia* climate" existed over this vast expanse simultaneously. Latitudinal zones of vegetation have undoubtedly always existed—although they may have been broader or narrower—as is well shown by Chaney (135). *Magnolia* had attained a wide distribution by Cretaceous time; its Tertiary history was apparently largely one of withdrawal southward and of loss of western American and European areas. E. Reid (522) has shown that the withdrawal of the so-called Chinese-American element from Europe began in mid-Miocene time with a cooling climate and was virtually complete before the onset of the Pleistocene. The magnolias have apparently long been disjunct, but there is every reason to believe that the Theory of Climatic Migrations, when coupled with evidence of adequate land bridges, is an adequate explanation in their case. Many plants of the northern hemisphere have probably had a history similar to that of *Magnolia,* and former connections for some of them are substantiated by fossil evidence. It appears possible, also, to extend the explanation given *Magnolia* to forms which have a similar polytopic distribution today, especially herbaceous plants, even when they are unrepresented by fossils from Europe, western North America, etc. (419, 111).

It seems reasonable to conclude that *Magnolia,* and other members of the eastern Asian-eastern American element, have attained their polytopic condition as a result of climatically forced migrations. Although the whole world surface now or once occupied by magnolias was likely never simultaneously inhabited by them, there has been, if we employ the space-time concept, a continuity. No theory but the migrational theory is required for an understanding of their present polytopic condition.

The third hypothesis of the series of four brings us to a problem that is not purely geographical and geological, since it involves questions of genetic and evolutionary possibilities as well. We may refer to this as the differentiation theory (discussed in the next chapter), which supposes that similar taxonomic groups can evolve independently in two or more regions from a parental syngameon. The results of accepting this hypothesis are that two disjunct forms are not considered to be lineally related and that it is unnecessary to explain the dispersal or migration of the modern related forms from one area to another. The latter necessity is not completely obviated, however, for the connecting syngameon must still be recognized among modern related forms bridging the disjunction, or a former continuity or migration must be hypothesized for it. In this respect, the differentiation theory and the migration theory are on common ground. But great difference appears in genetic and evolutionary conceptions. This is the problem of independent and perhaps parallel evolution from the syngameon in two or more regions. Two hypotheses seem to be involved in the differentiation theory.

Similar environmental conditions may result in the selection of similar biotypes from the parental syngameon. We can illustrate this concept by supposing that a temperate species is widely distributed along a mountain system. Certain biotypes of this species are more suited for alpine conditions of life, and on two or more peaks of the system there develop identical alpine ecotypes of the subalpine temperate syngameon. We may suppose, further, that the differentiation of the alpine forms proceeds to the point that taxonomists recognize the alpine plants as constituting species B, fully distinct from ancestral A, and that eventually speciation may become complete, because reproductive isolation develops between A and B. The alpine populations of B may be completely disjunct in that the distances between mountain peaks where they occur are greater than the dispersal capacity of the species.

Geographers who do not accept such an evolutionary hypothesis would propose as an alternative that at sometime in the past, when the climate was cooler, the alpine belts extended downward until they were confluent or the gaps between them were less than the dispersal capacity of the diaspores of the species, and that at that time the population spread from one peak to another. With the subsequent warming of the climate, the discontinuity was developed by the upward retreat of the lower limits of alpine conditions.

Adherents of the differentiation theory must be able to demonstrate the presence of species A, the parent syngameon, or hypothesize its existence at some past time over the whole range of territory through which populations of species B are now scattered. If species A and B differ by several genetic characteristics, as species usually do, it must be assumed that the same or closely similar genotypes can probably be independently selected in widely separated portions of the parental syngameon. In this connection it can be said that this is not impossible genetically; however, the more characters the two or more populations of B have in common, and the more characters by which they differ from A, the greater the improbability of such an evolutionary history.

It is, of course, a common evolutionary belief that species A, under the conditions we have hypothesized, could give rise to alpine species B on one mountain peak, but that on another mountain peak the similar-looking population would be genetically different and constitute a closely related species, C. This is the ordinary theory of divergent evolution. This supposition is especially reasonable when it is remembered that wide species are in most cases subdivided into geographical allopatric subspecies, and further into local populations, each of which differs genetically from all other subdivisions of the species.

An alternative to the idea that similar or identical biotypes are selected in similar environments is that the similar environments cause similar mutations to come into being. Among biologists there are several neo-Lamarckians who adhere to this possibility. It seems possible to reconcile these two views in the case of autopolyploids. If temperature conditions affect plants of A so as to result in the development of tetraploids from diploids, then this change can be conceived of as happening independently to plants in two or more regions, and autopolyploid species B could have developed independent alpine populations in different parts of the range of diploid species A.

The second explanation of evolution under the differentiation theory apparently requires the supposition of orthogenesis rather than parallel selection. This is a theory applied not so much to the development of one species from another polytopically as to the independent and parallel development of species within a genus or of genera within a family. In this case the closely similar species or genera are not linear descendants but constitute parallel series of forms developed from a more ancient syngameon. They attain their similarities because of internal predetermined abilities to mutate only in certain directions with respect to certain structures. The evidence for orthogenesis seems to be circumstantial, but the idea is strongly supported by certain scientists, especially paleontologists.

The concept of polytopic origin, or the differentiation theory, has recently been brought to the fore by Du Rietz's consideration of the problems of bipolar species (203). Our succeeding paragraphs will depend heavily upon his publication. Du Rietz says, "It has grown into a habit among most biologists to assume that a taxonomic unit must have originated within a single region and from there spread over the distribution area it occupies or is known to have occupied. If, however, we do not uncritically accept the orthodox theory of divergent evolution but take also the differentiation theory [309, 693, 694, 696, 698; cf. 201] into serious consideration, it seems to be equally possible for a genus or any taxonomic unit to differentiate or 'crystallize' out of its more polymorphic ancestral syngameon simultaneously over a very large area. . . . The general way of looking upon the genesis of taxonomic units roughly sketched above has brought me to a very sceptical attitude towards the problem whether a certain bipolar population has ever migrated across the tropics in one direction or the other."

Willis (696) wrote, "Chiefly important . . . is the new view of evolution, first proposed by Guppy in 1906, and by the writer in the following year, that evolution did not proceed from individual to variety, from variety to species, from species to genus, and from genus to family, but inversely, the great families and genera appearing at a very early period, and subsequently breaking up into other genera and species." He cites as evidence (697) for the differentiation theory some recent monographs in Engler's *Naturlichen Pflanzen familien* in which there is wide separation geographically of closely related species, genera, subfamilies, and families. For example, in *Cardamine* three closely related species occur in New Zealand and Polynesia, in the Azores, and in Chile, respectively; and in *Euphorbia* there are closely allied species in Venezuela and Cape Colony, in Persia and Africa, in Central Asia and North America, and so on. Willis' view is compactly stated in the following paragraph:

Now with relationship like this, which is so much complicated by the great separations over the surface of the globe, to get an explanation by the method of accumulation of small differences is an extraordinarily difficult matter, and it is much simpler to call in the *linking genera* that cover the enormous gaps than to suppose that the related genera, say in Chile and Siberia for example, once overlapped or nearly overlapped each other, and that destruction took place upon an enormous scale, and through all varieties of conditions and climates. . . . These subfamilies have various genera that cover the whole or much of the range, and it is much simpler to regard them as connecting links—as in fact the *ancestors,* directly or at times indirectly through intermediate genera, of the small scattered though often closely related genera. . . . The only necessary thing is to get rid of the idea that small genera and species of restricted area are necessarily relics. . . . If one supposes a genus to give off new species more or less in proportion to the area that it covers (which again will be more or less in proportion to its age among its peers), it is clear that all the offspring will carry a large proportion of the characters of the parent, and that therefore while offspring arising near together will be most likely closely to resemble one another, there is no reason why a close resemblance should not arise with a wide geographical separation.

In addition to the introduction here of Willis' age-and-area hypothesis (696), which has been widely criticized, there is one point to be noticed. Related genera, for him, are genera that are morphologically similar although they may be distant with respect to descent; i.e., by descent one genus or portion of a genus is related to a linking series of forms, and another genus or portion of a genus is related to the other end of the linking series. Closeness of relationship is usually judged according to descent, rather than to morphological similarity which may result from so-called convergent evolution.

In connection with the genetic aspects of this problem, the hypothesis of the taxonomic importance of polymery which Winge put forward in 1923 is of interest. From one of his later papers (705) we take the following description of the genetic basis for higher categories: (1) Historically, chromosome conjugation, duplication, and translocation tend to result in a gradual, manifold appearance of the same factors in more than one chromosome and possibly in several or all chromosomes. (2) The larger the systematic unit under consideration, the more often must the genes characteristic of that unit be repeated in the chromosomes and probably in all the chromosomes. (3) As two varieties chosen within a species differ only by a few factors, the "variety factors," but are homozygotic and isogenous with regard to all the species-, family-, etc., factors, the effect of "variety factors" alone is observed in segregations. (4) There is, of course, no absolute boundary between "variety factors" and "species factors," for instance, but we must suppose that the factors which are widely represented in all the chromosomes afford a basis for classification into large taxonomic units. Their frequent presence in the chromosomes will entirely prevent segregation into types in which these factors are lacking, and the loss or modification of a few of them could hardly provoke any immediately visible mutation, because many others of the same nature would be left.

This hypothesis of the duplication of the genes responsible for the characters by which higher categories are recognized can be applied directly to the concept of a syngameon from which there could have been a polytopic origin of related species. The problem in the bipolar distribution of species, of closely related species, of genera, etc., is to demonstrate, across the tropical lowlands, a continuity of linking syngameons out of which the bipolar populations could have been differentiated.

The work of Fischer-Piette (249) on the limpets (*Patella*) of the Atlantic coast of Europe and Africa and the islands of the Macronesian group (the Azores, Madeira, the Canaries, Cape Verde Islands) can be considered as describing a group that is undergoing evolution by differentiation. Although a large number of forms have been named, this author considers them to constitute one vast syngameon capable of splitting into several species which are morphologically and ecologically distinct. The stages in the differentiation process are illustrated by *Patella caerulea,* which partially takes shape at the Isle of Wight, and which succeeds in speciating locally on the Biscayan coast and fully in the Mediterranean, where it lives sympatrically with related forms. According to him, "Geographical isolation is not necessary for the formation of new species of patellas; but, of course, the natural tendency of the complex to split into particular types is greatly favored where geographical isolation happens, as it is not held back by the arrival from other localities." Out of the syngameon of wide distribution, *Patella caerulea,* for example, has differentiated or is being differentiated independently in different geographical centers, and these disjunct areas are connected by the syngameon. There is no problem of explaining the dispersal or migration of *P. caerulea* from one of its centers to another. An interesting side point in this connection is the fact that a systematist studying the forms where they occur sympatrically would consider each as a species. Another taxonomist, working in a region where the forms show intermediates, would consider them all as parts of one species. The former would consider the latter a "lumper" and the latter would consider the former a "splitter," and neither alone, working in his respective regions, could gain a true picture of the biology of the complex.

Among plants the genus *Espeletia* seems to present a good case for the polytopic origin of a series of endemic mountain species from a connecting syngameon of lower elevation. Smith and Koch (596) have studied this genus, which is a good illustration of páramo endemism in the northern Andes, and Smith has drawn certain geographical and evolutionary conclusions that are pertinent to our present topic. The genus *Espeletia* (a composite of the Heliantheae, closely related to *Polymnia*) consists of 30 known species, of which all but one are confined to the páramos (alpine meadows above the timber line). There are no widespread species of this genus living today, but each group of páramos is inhabited by one or more species, the range of which is limited by the lower forested mountain levels that separate the mountain peaks and are unsuited for members of this genus. According to Smith and Koch, 18 species are known from the complex of páramos in Venezuela, 5 species are known from the Norte de Santander páramos of Colombia, 4 species from the Cundinamarca, and one species each from the Central Cordillera, the Western Cordillera, and the Sierra Nevada de Santa Marta.

The disjunction of these species is maintained because of their inability to live below the timber line and their poor dispersal capacity. Long-distance dispersal seems very unlikely because of the absence of a pappus on the achenes, and this conclusion is supported by the fact that the species are quite distinct from each other, showing no intermediates or evidences of hybridization. An explanation of this endemism must depend upon the history of the region and the evolution of the genus. Smith states that the formation of the Andes in this region antedates the probable time of the evolution. "The formation of the Andes is thought to have taken place during the Cretaceous period. The origin of composites was doubtless more recent than that, so such a genus as *Espeletia* has probably orginated by the migration of a lowland population up to the páramos." He disposes of mutation, hybridization, and adaptation as probable causes of speciation in this upward migration from a wide population in lower elevations by the following reasoning: "In the case of mutation as the chief cause of variation, it is assumed by Willis and others that parent species migrate and here and there give rise to other species by more or less sudden mutations. Such mutations are considered comparatively young species and are thus limited in range, while the parent species has a considerably larger range. Thus there should be one or a few species spread over the entire range of a genus, while here and there are found endemic species. In *Espeletia* there are no such parent species, as no single species has a wide range beyond one large páramo group." Hybridization is dismissed because of the wide disjunction, poor capacity for dissemination, and the absence of intermediate forms. Smith says, "Here the species are very stable, and one can tell by examination of a plant approximately where it was collected." Adaptation is eliminated because of the striking environmental uniformity from one páramo to another, so that the probability of the species having an ecological-selection basis seems remote.

Smith consequently turns to a different explanation and finds a satisfactory hypothesis in the "law of differentiation by means of the automatic reduction of potential polymorphy" proposed by Hagedoorn (310) and discussed by Du Rietz (201). According to Smith, "If such a theory accounts for the evolution of the present-day species of *Espeletia,* we must assume one (or more) ancestral species spread over the northern Andean region at middle elevation. This species was presumably most polymorphic in the Mérida Andes [where most species are now found], while toward the periphery of its range its polymorphy was less pronounced, due to the more limited possibilities of inter-racial hybridization. By a movement upwards to the páramo zone (probably accompanied by the extinction of middle elevation individuals), this polymorphic species became permanently divided into smaller populations. It would be expected that the peripheral portions of the original population would develop into fewer present-day species than the central portions, where, due to more pronounced polymorphy, a resultant large number of present-day species were evolved."

Espeletia appears to be a good example of a fairly large series of congeneric species which have developed polytopically out of a parental syngameon. I wish now to pick up the speculation where Smith dropped it and see what the possibilities and their necessary assumptions are. If it is assumed that the ancestral syngameon was of the *Espeletia* type, there seem to be two possibilities. The present-day absence of the parental syngameon or connecting species of the *Espeletia* type could be explained by the assumption that at some time in the past climatic conditions were such that the upper tree line was depressed, allowing essential continuity of páramo plants. At this time there developed the *"pro-Espeletia"* as a polymorphic and fairly widely distributed syngameon. With a change in climatic conditions the complex, following the upward trend of the páramo zone, became broken up into isolated populations which have become the alpine endemic species of today. Fragmentation of the population and speciation have gone on simultaneously with the changing climatic conditions. The species, although closely related, are polytopic, and no connecting species are found because the entire phylogenetic stock is limited in its adaptations to páramo conditions. In this connection there is the unfortunate circumstance that at present there is no evidence of climatic changes to the extent which would be necessary.

The second possible hypothesis is that the ancestral *Espeletia* syngameon was adapted to montane or subalpine conditions and that the whole phylogenetic stock underwent parallel evolution which resulted in its adaptation to the alpine conditions of the páramos. If there were a residual temperate portion of the complex, its extinction from intermediate elevations would have to be explained.' If there were no such residue, it would mean that the whole complex of populations had undergone parallel physiological evolution. In either case such changes are scarcely conceivable.

The third possible hypothesis is that the connecting syngameon is not *Espeletia* at all, but some related genus present today in the region at intermediate elevations, but unrecognized as the ancestral stock. Perhaps, as Smith thinks, some closely related genus of the Heliantheae, such as *Polymnia*, may constitute the connecting syngameon. Since the species of *Espeletia* form a compact and well-characterized genus, such a theory would seem to require an assumption of orthogenesis within the ancestral genus or species. We can express the idea symbolically as follows: Out of species A, or a series of species A, B, C, D . . . of the parental genus have evolved species A', B', C', D' . . . of *Espeletia*. The first species or genus had certain orthogenetic tendencies that, under alpine conditions, resulted in the mutations of similar character which define the genus *Espeletia*.

Van Steenis (608) believes that *Symplocos spicata,* a common and rather variable lowland and montane species in Malaya, has independently given rise to the same variety (ecotype) on several mountains. In other species of *Symplocos* and *Ilex* he suspects origin by polytopic development. Briquet (89) believed that *Poa Balfourii* developed at several points in the area of mother species, and he gave the phenomenon the name "polytopic process." *Poa Balfourii* was originally known only in the mountains of Norway and Great Britain but was later found in central Europe and the Alps. Briquet thought that several new stations might be found by critically reexamining *Poa caesia.*

Let us cite one more case in which there may have been polytopic origin in a group of related species. Gleason (272, 274, 275) has done monographic work on the ironweed genus *Vernonia* which is based upon the combined indications of geographical distribution, migration, phylogeny, and comparative morphology. This genus has made its way into the United States from Mexico, and several lines of migration and evolution have spread out from Texas—one, for example, up the central grasslands and another eastward along the coastal plain. Of the numerous species and species groups that represent the most advanced stage in inflorescence evolution, one group of closely related species, the Fasciculatae, will serve to illustrate a characteristic evolutionary pattern. In the Fasciculatae there is one species, *Vernonia fasciculata,* that is present over a much wider area than the remaining forms (*V. tenuifolia, V. marginata, V. Lettermannii,* and *V. fasciculata* var.

nebraskensis and var. *corymbosa*). This phenomenon, of one species of wide and several of narrower area, is repeated in most species groups of *Vernonia*. In the Fasciculatae, Gleason considers that the wide species probably repre-

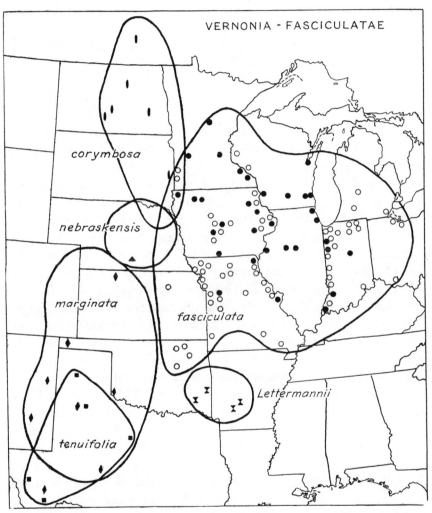

Fig. 39. Species and varieties of *Vernonia* belonging to the group Fasciculatae. These forms have likely differentiated polytopically from a *"profasciculata."* Black symbols represent herbarium material examined; circles are additional records.

sents the ancestral syngameon from which the other taxonomic entities have evolved, or at least is now closest to it. The southernmost form, *V. tenui-folia* of western Texas, and the Ozarkian *V. Lettermannii* are most closely related to the group Texanae out of which the Fasciculatae evolved. This is shown best by the acute involucral bracts. *V. fasciculata* var. *corymbosa* is the northernmost form, extending up the Red River valley to Manitoba, and, with blunt, broad involucral lobes, is the most different from the Texanae. The northern and eastern forms are characteristic of the moist soils of prairies, meadows, and valleys. The western and southwestern forms are found in dry grassland types; *V. Lettermannii* occurs on sandbars in the southwestern Ozarks.

The group of species which now compose the Fasciculatae were once a polymorphic species that migrated from southern Texas up the eastern prairies and eastward, mainly along the prairie peninsula, into the deciduous forest region as far as Ohio. *V. fasciculata* itself, in its broader variation, is probably similar to the early syngameon. Along the western border of the range of the syngameon, which is probably the older area to be occupied, have appeared the species and varieties derived from *V. fasciculata* (or perhaps a *"profasciculata"*). In western Texas arose *V. tenuifolia*, differentiated from *V. fasciculata* by broader, more acute involucral scales and narrower leaves. *V. marginata* arose in the plains grasslands farther west, from southeastern New Mexico and northwestern Texas to southwestern Nebraska and Colorado. *V. Lettermannii* arose in the southwestern Ozarks eastward of the path of the *"profasciculata."* These species are apparently isolated from each other and from *V. fasciculata* geographically and, where their ranges overlap, ecologically. Also on the western border of *V. fasciculata,* but more to the north, there have arisen two varieties in the moister prairies and river valleys, var. *nebraskensis* covering the eastern half of that state, and var. *corymbosa* reaching northward. Variety *nebraskensis* differs from the species in having shorter and narrower leaves with crowded heads, and var. *corymbosa* differs by having blunter and broader involucral scales and broader, thicker leaves. Intermediates between the varieties and the species indicate that these populations have attained only partial reproductive isolation.

The absence of derived forms on the eastern and other portions of the periphery of *V. fasciculata* may be due to the more recent immigration of the species eastward and northeastward, a movement that probably started during the post-Pleistocene. At any rate, the forms of this group illustrate the phenomena of geographical replacement and, apparently, of polytopic origin. With respect to the last point, it should be noticed that the series of western forms, consisting from south to north of *V. tenuifolia, V. marginata, V. Lettermannii,* and *V. fasciculata* var. *nebraskensis* and *corymbosa,* do not constitute a linear series of variations. Rather, they appear to be isolations or partial isolations of *V. fasciculata* (or our hypothetical *"profasciculata"*), including in each case some of its normal variability but accentuating certain characters and combinations of characters that have built up populations now given specific or varietal rank. They thus represent, in part at least, the breakup of *V. fasciculata* and the polytopic origin of a group of forms which are nearest of kin. This conclusion is based on the assumption that *V. fasciculata* appears to be the parental species of the group Fasciculatae, and that a *"profasciculata"* once occupied much of the area and path of migration (in the southern part of the prairies) where now reside the derived species *V. tenuifolia,* etc. There does not appear, however, to be a polytopic origin of taxonomically identical forms in this group.

Generally speaking, evolution in *Vernonia* appears to accompany migration and to consist of gradual changes in characters, both quantitatively and qualitatively, with speciation resulting upon isolation.[2] For example, the Fasciculatae differ from the Texanae from which they were derived in the following ways: They have retained the pitted leaves and the narrow, undifferentiated outer pappus bristles of the Texanae, but they have lost the primitive involucral type and have developed long heads with regularly imbricated, numerous involucral scales. Evolution, then, results in part from the retention of certain primitive characteristics and the loss of others, and in part from the addition of new characteristics. Genetically this can be understood in terms of shifting gene frequencies resulting in new collective genotypes, and in the development of new allelomorphs by mutation. The relationship of a group of close forms is evident in the characters which the populations retain in common (as between the members of the Fasciculatae or between the Fasciculatae and the Texanae). Their differentiation is evident in the characters that are lost or added.

In *Vernonia* we can conceive of the evolutionary pattern as consisting first of the development of wide polymorphic populations along the principal highways of migration (such as the Fasciculatae in the prairies, the Interiores in the western part of the deciduous forest bordering the prairies, the Angustifoliae along the Gulf coast, the Altissimae northward from the Gulf in the moist forests of the Cumberland and related highlands). These early species are the hypothetical *"profasciculata," "proaltissima,"* etc., of which the now widest ranging species of their respective groups (such as *V. fasciculata, V. altissima, V. missurica,* etc.) are the central residual portions. From these early, wide, and polymorphic species that developed along

---

[2] In this connection "speciation" does not mean that physiological isolation has been attained, but that by ordinary plant-taxonomic standards populations of specific difference have developed.

migratory paths with more or less complete isolation have developed the species groups, largely by polytopic differentiation. Whether this is the correct story is largely beyond proof, but it seems to fit well with the present distributional pattern of related forms.

In concluding our discussion of polytopic origin we may say that apparently it can consist of several types. (1) There ·may be cases of the polytopic origin of one species (B) from one other species (A). This seems unlikely except in cases of autopolyploidy and certain "microspecies," but theoretically it is not impossible—although it is highly improbable—for species that differ by many characters. (2) The polytopic origin of two or more species which are next of kin (B, C . . .) from a single species (A) appears to be a common phenomenon. (3) The polytopic origin of related species (B and B') from other related species (A and A') seems likewise to be possible, especially in groups where there is strong evidence for orthogenesis. (4) There may be polytopic origin of congeneric species (A, B, C, D . . . of one section, and A', B', C' . . . X, Y, Z of another section) from a widespread and probably ancient syngameon.

In none of these cases is there the problem of explaining a modern connection between the plants of the polytopic areas, by either dispersal or migration, and the derived species; the problem is merely pushed back in the history of the group to the parental connecting syngameon, species, or complex.

The fourth hypothesis, applicable to questions of polytopic areas, holds that the disjunct taxonomic groups may not be closely related by descent but may be polyphyletic. If it can be shown that the species which constitute the populations of two areas are similar as the result of convergent evolution and have only remote relations phylogenetically, problems of disjunction are pushed back so far into the past that it is futile to deal with them. The question which must be asked is whether there really are polyphyletic taxonomic groups; the answer is that among the higher categories there certainly appear to be many.

Woodward (714) says that paleontologists generally are dissatisfied with the Linnaean system of nomenclature because of its inadequacy in dealing with polyphyletic origins. He gives several examples of well-known polyphyletic groups among animals. A study of the polyp cells and other characters of Graptolites shows that during the successive geological periods there were several distinct lineages which passed from grade to grade in a definite way. The usual classifications cut across lines of descent and resulted in polyphyletic groups, whereas (if the system is to be natural) the Grapto-lites should be placed in related groups which follow the lineages. Amphibia pass into the earliest Reptilia through at least three distinct lineages. It is difficult to explain the early differentiation of modern teleostean fishes except by supposing that they originated independently from several lineages of the Ganoids which preceded them. Birds are also pólyphyletic; among mammals parallel lineages are described from horses, rhinoceroses, camels, elephants, etc. The data which support such general statements as the above, and many others, are of such a nature as to cause paleontologists to believe in the poly-phyletic and polytopic origin of many of the larger groups of animals.

Among plants there are frequent cases of probable polyphyletic higher categories. The lichens are an outstanding example. Clements (148) well disposed of the arguments for the independence of the lichens and showed their separate origin through parasitism (symbiosis) from different fungal groups. The conclusion seems inevitable that the lichens should be distrib-uted through the fungal groups according to their phylogenetic affini-ties. Flowering plant families are frequently probably polyphyletic; the Cyperaceae, Flacourtiaceae, Compositae, and Melastomaceae are good exam-ples. Gleason, who is a specialist in the Melastomaceae, says [3] that this fam-ily has had sources in both the Lythraceae and the Myrtaceae. Only one American tribe of the Melastomaceae is polytopic, however. In the tribe Rhexieae the large genus *Monochaetum,* with about 40 species, extends from Mexico to Peru. Skipping about one thousand miles of Amazonia, a second genus, *Pachyloma,* with two species, occurs in southeastern Brazil. And, with a northward disjunction from *Monochaetum,* a third genus, *Rhexia,* occurs in coastal United States. It is not certain that this tribe is polyphyletic, but it well may be. There are numerous polytopic species in the family, but, according to Gleason, there is little likelihood that any species is polyphyletic. But several of the polytopic species appear to have had a polytopic origin. One of the most strikingly disjunct species is *Miconia Matthaei,* of which apparently identical specimens are known from eastern Peru, Trinidad, and Central America.

The interesting bipolar genus *Empetrum* (studied by Good, 285) is a case of the spurious polyphylesis of a subspecific group. According to Good, *Empetrum* consists of *E. nigrum* (with three forms), which is widely dis-tributed in the boreal regions of the northern hemisphere, and *E. rubrum* (with a subspecies, a variety, and three forms), which occurs principally in the southern part of South America but has outliers on Juan Fernandez,

[3] Personal communication.

the Falkland Islands, and Tristan da Cunha. A red-fruited form, *E. nigrum f. purpureum* (Raf.) Good, of Arctic regions, was classified as *E. rubrum* by Durano. Such a treatment, throwing all red-fruited forms together, would make *E. rubrum* polyphyletic, according to Good's understanding of the evolutionary relationships within the genus. Incidentally, the northern and southern areas of this genus are separated by about 70 degrees of latitude.

We shall now give one example to show that hybrid complexes are polyphyletic and that such a complex may enjoy specific status. Again referring to the Vernonieae studied by Gleason (272, 274, 275), we propose at this point to center our attention around *Vernonia illinoensis*. This species was

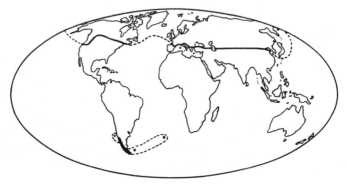

Fig. 40. The bipolar distribution of *Empetrum*. *E. nigrum* is boreal, occurring north of the heavy line. *E. rubrum* is austral, as indicated by the black areas, and includes a disjunction of about 2500 miles from South America to Tristan da Cunha. Map redrawn from Good (285).

described by Gleason in 1906 and dropped by him in 1922 because it was found to give every evidence of being a hybrid complex resulting from the crossing of members of three species: *V. missurica, V. fasciculata,* and *V. altissima.*

Vernonia missurica is essentially a species of the open uplands and requires considerable light. It ranges through eastern and northern Texas to eastern New Mexico, thence northward and eastward through Oklahoma, eastern Kansas, Arkansas, Missouri, and into the prairie peninsula of Illinois, Indiana, northeastern Ohio, and southern Michigan. It is also known in western Kentucky, Tennessee, Alabama, and Mississippi. Natively, it is a species of the prairies and the drier soils of the uplands. *V. fasciculata* is likewise a species of the open places and also requires light, but it prefers wet to moist soils and is found in low prairies and along valleys. Its range is essentially northern Mississippi and Ohio valleys, from northeastern Oklahoma up through eastern Kansas, Nebraska, and South Dakota, thence eastward through southern Minnesota, Wisconsin, and Michigan to Ohio and Kentucky. *V. altissima* is characteristically a species of the moist woodlands and definitely is not a sun-tolerant form. Its range includes much of the Mississippi and Ohio valleys and it extends eastward to include the Cumberland uplands and part of the Appalachians. It reaches its northeastern limit in western New York and its southeastern limit in northwestern Florida. A

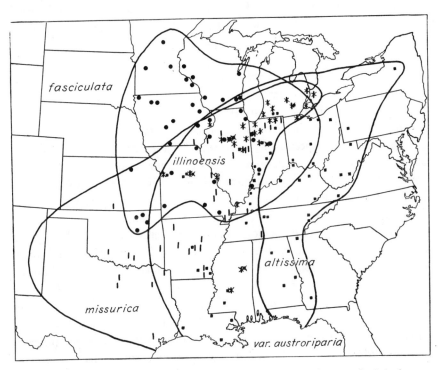

FIG. 41. *Vernonia illinoensis* appears to be a hybrid swarm of crosses and backcrosses among *V. fasciculata, V. missurica,* and *V. altissima*. This supposition, based upon comparative morphology, is substantiated by the occurrence of *V. illinoensis* within the area of overlap of the putative parents.

reference to the map shows these areas and the striking fact that *V. illinoensis* lies in the area from· Missouri through Illinois and Indiana to southwestern Michigan where the three species mentioned overlap.

In view of these ranges it would appear that *V. illinoensis* should be viewed with suspicion and that the supposition of its hybrid nature seems to be sound. The hypothesis can be further supported even without breeding experiments. *V. fasciculata* is glabrous and its leaves are thickly covered on the under surface with small, deep pits which are characteristic of the species group Fasciculatae. *V. altissima* and *V. missurica* do not have these pits; their leaves are hairy beneath, but the hair types are strongly contrasting. *V. missurica* is typically densely covered with long, stout hairs, whereas *V. altissima* bears scattered to dense very short, stout, almost conical hairs. *V. illinoensis* reveals itself to consist of a hybrid swarm of these three species by possessing all possible combinations of these under-leaf characters. For example, eastward where only *V. altissima* and *V. fasciculata* overlap their ranges are plants with pits and short hairs. Westward where *V. missurica* and *V. fasciculata* predominate, combinations of pits and long hairs are most abundant in *"illinoensis."* Elsewhere all combinations occur. These, then, can be used as key characters in the solution of the *"illinoensis"* problem. Other characters common to the three species also show intermediates and combinations in the *"illinoensis* complex."

Further aspects of this interesting situation may be commented on. For example, the three species (*V. missurica, V. fasciculata,* and *V. altissima*) are geographically isolated in parts of their ranges, and in the region of coincidence they have only ecological isolation. Neither geographical nor ecological isolation alone (or together) will serve to keep species apart more than temporarily. With climatic changes and migration, geographical isolation can be overcome. Ecological isolation is locally overcome through the activity of agencies which destroy primeval conditions. Man is, of course, recently the most important of these agents. Under primitive conditions it is likely that there were not many plants of *"illinoensis,"* but with man's occupancy of the Middle West—cutting timber, draining lands, etc.—numerous intermediate sites have been formed which are now occupied by the intermediates or hybrids which compose *"illinoensis."* There are places in Illinois, Indiana, and Michigan where the original species have been completely or almost completely swamped out. Furthermore, it is obvious that the three contributing species have not developed reproductive isolating mechanisms. There is, however, no evidence of introgressive hybridization, i.e., that genes from one species are spreading into the population of another species and contaminating it. The hybrid complex seems to be limited almost to the area where the three species overlap. This deserves closer study.

Another aspect of interest, and one of the principal reasons for introducing the problem of *Vernonia illinoensis* at this point, is the fact that the contributing species are not at all the closest of kin. That is to say, the "species" *V. illinoensis,* such as it is, is a population of polyphyletic origin. *V. fasciculata* is the key species of the group Fasciculatae, which includes also a series of derived species and varieties, and it is derived from the group

Texanae. This complex migrated northward through the prairies and spread eastward through the prairie peninsula (in the form of the present-day *V. fasciculata*). *V. missurica* belongs to another species group, Interiores, which includes *V. interior, V. Baldwini,* and *V. aborigina. V. missurica* is the species of widest range in its group, and is also probably the parental species. Its general path of migration was similar to that of the Fasciculatae, but it apparently lay eastward in the deciduous forest margin, being typically Ozarkian. It, too, came from the Texane. *V. altissima* belongs to the species group Altissimae that includes *V. ovalifolia* and *V. flaccidifolia* of the Gulf coastal plain and lower Piedmont. *V. altissima,* the only wide-ranging member of this group, has spread northward to occupy much of the western slopes of the Appalachians and the Cumberland uplands as well as the Mississippi and Ohio valleys. The Altissimae are derived from the Angustifoliae of the coastal plain, and the latter from the Texanae.

We see, then, that the three wide-ranging species which enter the *"illinoensis"* complex are each derived independently and more or less remotely from the Texanae, that they have migrated along separate routes, that they have long been isolated, in part at least, and that when they have met again (let us say, in postglacial time, since the majority of the hybrids are on the glacial till plain) they reveal that they have failed to develop reproductive isolation. *Vernonia illinoensis* is shown to be a hybrid complex, with individual plants containing various proportions of the characters and germ plasm of the three species, and it is also shown that these three species are far from being the nearest of kin. *V. illinoensis,* as long as it stood on the books, was a species of polyphyletic origin. And, as the result of some quirk of isolation, the population which it represents might even yet give rise to a true species.

Van Steenis (608) believes that a species may have a polytopic origin when hybridization is followed by amphidiploidy. The literature supplies several cases of a polyphyletic origin of a form, but none, to my knowledge, where it is proved to be also polytopic.

In conclusion, we can make a series of statements concerning the four hypotheses employed in explanations of widely disjunct areas. Long-distance dispersal operates for some organisms, and it is especially applicable to littoral species and a portion of the biota of oceanic islands. The hypothesis, however, is much too widely used; in most cases of wide disjunction, a careful investigation shows that the dispersal mechanisms, agents, and ecesic requirements of the species rule out this explanation. All too frequently the assumption of long-distance dispersal is merely a careless and easy way out of a difficult problem and it leads to fanciful and even ridiculous conclusions. The climatic migration hypothesis, which is the usual one employed by geographers to explain wide discontinuities, is apparently valid in the majority of cases and is supported many times by paleontological evidences and by information concerning paleogeography and paleoclimatology. The polytopic differentiation hypothesis appears to be genetically possible, but the polytopic origin of a species seems far less likely than the polytopic origin of related species. It is a reasonable hypothesis when the migration theory is inapplicable or inadequate and when a linking syngameon can be demonstrated. The theory of polyphyletic origin is applicable to higher categories but probably never to species, except in the case of allopolyploids and hybrid swarms which pass into the literature as species. When the polyphyletic nature of a polytopic group can be demonstrated or reasonably inferred, many of the problems of geographical distribution are essentially obviated.

20.

Theory of Differentiation in Relation to the Science of Area

1. *The theory of evolution by differentiation relegates natural selection to a subsidiary position, acting primarily with respect to adaptation, and accounts for the development of a group through a series of relatively large morphological mutations progressing downward from the family to species. As a consequence of this hypothesis, "taxonomic characters" are rarely concerned with adaptation or natural selection.*

2. *The differentiation hypothesis accounts for the "hollow-curve" relationship in the size of genera in species and in areas, and leads to the conclusion that, in general, large and complex families and genera, and those of large area, are old among their confreres.*

3. *According to the differentiation hypothesis, earlier mutations within a family were generally large, and later mutations have been progressively smaller and possibly more frequent, with the passage of time and the increase of complexity of the family. This phenomenon tends to produce relatively poorly differentiated species in the older and larger families.*

4. *An acceptance of the differentiation theory, with respect to polytopic areas, causes a complete change in the problems of geography from considerations of migration, highways, barriers, climatic change, etc., to a consideration of possibilities concerning evolutionary processes.*

5. *The differentiation hypothesis, because of its more deductive nature, should not be invoked unless all inductive methods of plant geography fail or are inadequate.*

Willis has recently published a book (698) which presents the argument for the theory of evolution by differentiation or divergent mutation rather than by selection. The principal ideas of the theory were first published independently by Guppy (309) and Willis (693, 694), although their roots go back to observations made by St. Hilaire (538).

St. Hilaire believed that evolutionary change goes downward from the family toward the species, not from varieties to species, species to genera, genera to tribes, tribes to families. A family begins as a family. When it begins, it is of course at the same time a genus and a species; that is, it is monotypic. Guppy formulated the idea underlying the theory of differentiation, as he called it, as follows: In the early days of the flowering plants the climates of the world were moister and more uniform, and the subsequent history has generally been one of breakup or differentiation into regional and more local climates which are damper or drier, warmer or colder, etc. With the climatic differentiation came plant differentiation, not by gradual adaptations, but by the development of specific or generic differences at single abrupt steps. Willis accepts the concept of evolution by single large mutations, but he is not prepared to admit that they necessarily accompany climatic changes. To Guppy's concept he added the facts brought out in his studies of plant distribution and set forth in *Age and Area* (696). A family begins as a family because of some large mutation; it is not formed by separation from other families by the destruction of intermediates. If the family meets with some disaster (which may be natural selection, some climatic catastrophe, etc.) while it is in the monotypic stage, it disappears. As the family grows, however, single species and genera become less and less important to its survival. Contrary to the natural selection theory, according to Willis, there is no reason why, under the differentiation theory, parental species do not continue to survive along with the derived species. Starting with the original genus of the family (which is also the original species), in time there will appear a new genus as the result of a large mutation. Subsequently new genera will be added by mutations from the original genus and from the derived genera; of course, lesser mutations from the generic stocks result in new species. According to this theory, the family grows, however slowly, roughly according to the law of compound interest and, in general, its size and complexity are a measure of its age. To this must be added the corollary that area is also generally coordinate with age. This refers to the areas of the family, its genera, and its species. As a rule, the oldest units within the family will tend to have the largest areas.[1]

Many exceptions to this generalization are known. For example, Good (284) decided that in the Stylidaceae the large and successful genus *Stylidium* (with 112 species and the widest continuous area) is probably the youngest,

[1] Cf. Chapter 14.

and the most primitive genus of the family, *Oreostylidium,* is monotypic, endemic, and epibiotic in the sense of Ridley. The other genera are also small.

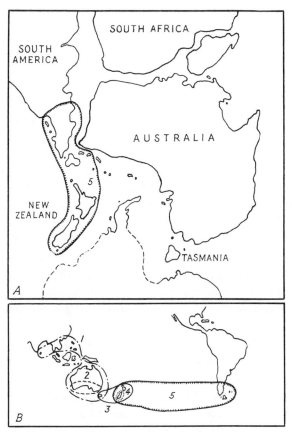

Fig. 42. Distribution of the Stylidiaceae. *A.* The genus *Phyllachne* (*5*) has one species in the Magellan region of South America (*P. uliginosa*) and three species in New Zealand. *B.* Areas of the five genera of the family. *Stylidium* (*1*) appears to be an advanced genus and contains about 112 species, of which all but four are Australian. *Levenhookia* (*2*) is closely related to *Stylidium* and contains six Australian species. *Fostera* (*3*) contains four species, with *F. bellidifolia* in Tasmania and three other species in New Zealand. *Oreostylidium* (*4*) is monotypic with *P. subulatum* in South Island. Data are taken from Good (284). By way of contrast with the ordinary Mercator projection, map *A* is based on a reconstruction by H. B. Baker (privately printed, *The Atlantic Rift Theory,* Ann Arbor, 1932) to illustrate his hypothesis of continental displacement (42).

If a newly formed type of plant can pass through the "sieve of natural selection" and survive the dangers which threaten its early existence, it can spread and differentiate. The factors of persistence, differentiation, and spread lead to the phenomenon of relationship that is graphically described as the hollow curve. It is Willis' emphasis on the hollow-curve relationships in intrafamily differentiation and intrafamily areas that appears to be his most important contribution to the theory of evolution by differentiation and to the science of area, because the hollow curve is readily explained by this theory and not at all (according to Willis) by the theory of evolution by natural selection of minute differences.

The hollow curve of differentiation is an expression of the fact that in families which are large enough there are numerous genera with one species each and progressively fewer genera with more species until the family is capped, so to speak, by a few large genera and one or two very large genera. The hollow-curve relationship also appears in areas because most large families show numerous local or endemic species and progressively fewer species of wide distribution (the same sort of relationship holds for genera). Willis (698) says, "As the original species thus survived *as well as* the offspring, the family must necessarily increase in number in such a way that when plotted by numbers of species, its genera would form the 'hollow' curve." This does not deny that some species may die out; in fact, according to him, "It is quite possible that after a certain lapse of time a species *must* die out, and it is still more possible that it may change into another by some simultaneous mutation." Such possibilities have not been realized with sufficient frequency to destroy the general hollow-curve relationship.

A corollary of the differentiation theory would seem to be that the earlier mutations, in general, were of a larger order than later ones and that mutations have tended to become progressively more frequent but of smaller degree. As Willis puts it, "The general impression that one gains here [in discussing the Cyclanthaceae], as in almost all cases, is that after the big mutation which first gave rise to the family, there followed others which gradually became less and less marked, and which kept more or less closely within the boundaries that were differentiated by the first mutation. . . ." In a later paragraph he describes this corollary in even more detail in discussing differences in generic rank between small and large families. The divergence between two genera in a small family will be about equal to the first divergence in large families; this makes the generic divergence of the former about equal to the tribal or subfamily divergence of the latter. Willis says:

It has long been known as an axiom in taxonomy that genera in a small family are much better separated than genera in a large one, and here is a simple explanation of this. As the family grows in size, new mutations will come in at more and more frequent intervals, but within, or close to, the original divergent mutation. In other words, all the family, sprung from the original genus and its first mutation, will show some at least of the characters shown by these first two genera, which by the hypothesis of divergent mutation will tend to be very divergent. The original family characters will show best in the largest genera, which will be the oldest in the families, and carry the most of the earliest characters. The genera sprung from later mutations will not have, so well marked, many of the characters of the earlier mutations. The generic characters will necessarily become on the whole closer and closer together as the mutations to which they are due are less and less far back in their ancestry. It would seem as if there were a tendency in each family for mutation to become less pronounced as time goes on, so that the appearance of what we call family characters becomes less frequent in proportion to the total of characters that appear. . . . Upon the natural selection theory it has always been a great difficulty to explain why the divergences became greater the higher one went in the key from species upwards. Why should natural selection cause the disappearance of just those forms necessary to make the divergences increase? This is inexplicable by natural selection, working upwards from small differences, but simple to differentiation, working the other way.

It is interesting to note that Blum (73) has arrived at a similar position from a study of thermodynamics. He presents a physico-chemical basis for the assumption that during the history of plant and animal life mutations have progressively become of lesser degree although probably of increasing frequency. We have no means at hand of proving such an hypothesis. It can be admitted, however, that some circumstantial evidence is apparently in favor of mutations diminishing in degree and increasing in frequency.

Two of Willis' earlier suggestions must be viewed with considerable suspicion. The idea that in time a species must die out, that there is a definite limit to the life of species, is treated elsewhere under the topic of senescence and in connection with polyploidy. The remaining suggestion, that a species —a whole population—can enjoy a simultaneous mutation, is apparently in the realm of metaphysics and is incomprehensible on the basis of genetics, at least as I understand it. Attention should also be called to another point in Willis' argument. Small families with well-distinguished genera are frequently relic groups; the clear demarcation between genera today is not a condition that has always prevailed but one which has been attained by extinction of numerous forms. It does not necessarily follow that small families with "neat" genera are young in their evolution and that in time they will become large families, with the generic distinctions blurred by numerous minor intermediate mutations. The evolutionary theorist, like the geographer, must always be on the lookout for decimated phylogenetic stocks because their condition can be understood only in the light of their history.

An essential idea in Darwin's establishment of evolution (177) stemmed from Malthus (438) with the realization that the capacity for multiplication possessed by every kind of organism inevitably results in a struggle for existence, because the young are crowded for space and the materials for sustenance. The Darwinian concept of natural selection, as a picking out of the variations that gave some survival advantage and the working by gradual adaptation, was the principal feature of the evolutionary hypothesis. Willis contends that the natural selection theory attempts to work evolution backwards. He says that there are many weak points in Darwinism and that there has never been any proof "(1) that evolution proceeded essentially by improvement in adaptation, (2) that it was gradual and closely continuous, (3) that the phenomena of the structure of plants reflect the adaptation that has gone on in them, or (4) that groups of plants can compete as units." He presents a list (698) of 32 assumptions on which the theory of natural selection rests. He is convinced that most of them have never been backed by proof and that natural selection—a picking out of gradual improvements in adaptation, chiefly structural—is still a long way from being established, although evolution itself is thoroughly established.

It is Willis' intention not to deny natural selection but rather to relegate it to its proper place. He says:

After fifty years of work, the author has come to the conclusion that *evolution and natural selection work at right angles to one another,* with but slight mutual interference, the latter being quite possibly greater in animals. The evolution *provides* the structurally different forms of life, while natural selection works upon the functional side, and *adapts* them in detail for their places in the local biological economy. There is no obvious reason why selection should not develop small structural variations, though one will not expect specific changes, unless rarely. In general, selection will simply kill out those individuals, whether species or not, that commence anywhere with functional characters that are unsuited to the conditions of the moment, or that simply have ill-luck. Each new species, by mere heredity, will probably have functional characters more or less closely suited to the place in which it arises, but as time goes on, and the number of species increases, chiefly by arrivals from elsewhere, more and more careful adjustment will be needed to fit in each newcomer. It is in this work that natural selection is of the first importance, doing work that nothing else could do with the same efficiency.

Willis is convinced that real adaptation is largely internal and that most structures, especially those used diagnostically by taxonomists, have little if any adaptational significance. I believe that this conclusion is sound and that it is proved by the large areas of many species without any important morphological changes (i.e., taxonomic changes) from one region to another or from one ecological situation to another, and by the large number of morphologically and taxonomically different plants which characteristically constitute a community and occupy a single ecological situation.

There is certainly a great weakness in the usual assumptions made by students of ecological morphology and anatomy, for they frequently are wholly unwarranted. This is no place to enter into a detailed examination of the problems of adaptation or a criticism of physiological plant geography, but a case or two are mentioned to illustrate the point. For a few decades Schimper's theory (553) of the physiological drought of bog and salt-marsh plants went unchallenged because of its apparent reasonableness in the light of numerous xeromorphic plants in these situations. The plants without such xeromorphic structures were largely ignored and no studies of comparative physiology were made. Recently, however, it has beeen shown that many such xeromorphic plants are not xerophytes because they have transpiration rates equal to those of mesophytes and in some cases hydrophytes, and that the theory of physiological drought is entirely unnecessary and unwarranted. The writers in this field include Maximov (459), Montfort (470), Steiner (610), Stocker (612), and Walter (672, 673). Likewise, most studies of comparative ecological anatomy dealing with degrees of xeromorphy of different plants of the same species grown in different situations are entirely without evidence as to the physiological characteristics of the diverse forms. In this connection note such studies as those of the author and his associates (112, 198, 113, 114) on the leaf structure of certain ericads grown in different situations in which the more exposed plants have a more xeromorphic leaf structure. These studies at least have the virtue of *not* assuming that the increased xeromorphy of the plants in the more xerophytic situations results in a more favorable water balance, as is usually the case; one study of the group (105) indicates a much higher transpiration rate for the more xeromorphic forms. Concisely, the point to be made here is that it is never safe to reason from structure to function and that adaptation does not necessarily reside in the obvious; in fact, it seldom does.

For one more example, we can note certain characteristics of the vegetation of the Mediterranean region. Szymkiewicz (616) points out that the boundary of the Mediterranean region is falsely drawn when ecological characteristics prevail over floristic. In the Italian portion of the region Engler shows the littoral parts of the peninsula as Mediterranean, and the central Apennines as belonging to the central European floristic regions. This sort of mapping can be paralleled, as shown by Adamowicz, by mapping the olive (as characteristic of the Mediterranean) and the beech (as characteristic of the central European region). On a floristic basis, employing the limits of 573 species, Adamowicz showed (see Fig. 43) that the entire peninsula except Lombardy belongs to the Mediterranean region. Moreover, Szymkiewicz (616) brought out the startling fact that in the Iberian portion of the Mediterranean region only *Rhamnus,* out of 70 genera which have their

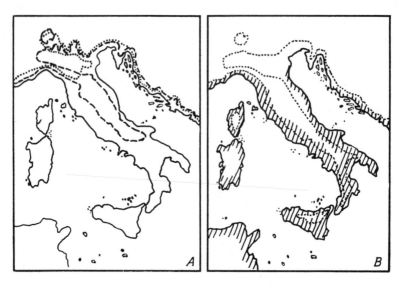

Fig. 43. Maps to illustrate the contrast between the mapping of vegetational areas on an ecological basis and on a floristic basis. Map *A* shows the northern boundary of Mediterranean vegetation after Engler (broken line) and after Adamowicz (dotted line). Map *B* shows the southern limits of beech (dotted line) and the area of olive culture (shaded by diagonal lines). Engler's ecological map shows vegetational boundaries based upon a compromise between these lines, whereas Adamowicz's northern limit for Mediterranean flora is based upon an analysis of hundreds of species without reference to ecological "key" species. Data from (616). Four maps have been combined in two and simplified.

centers there, has the coriaceous leaves that are supposed to be characteristic of the Mediterranean floristic region and type of climate, according to physiological plant geographers. Inasmuch as such illustrations can be found almost ad infinitum, we can more readily accept Willis' point that real adaptation is largely internal and that most structures, especially those used taxonomically, have little adaptational significance.

To return to Willis' arguments, we note that any theory of evolution to be satisfactory must explain the facts of numerical composition of families and of geographical distribution, as well as the facts of morphological relationship. Systematics, comparative morphology, geography, and evolution must handle common materials in a mutually satisfactory and compatible manner. This, of course, is a large order. To present the case for the differentiation theory as opposed to the natural selection theory, Willis examined 34 "test cases" and found that the former hypothesis was superior. A few of the cases will be briefly reviewed here.

Increase in number with evolution.—It is admitted that the variety of plants has increased with time. On the basis of natural selection one would expect a few "super-plants," not a vast and increasing number with related forms about equal in competition. If small genera are usually to be considered relics on the basis of natural selection, why are there so many of them? Of the 12,571 genera of flowering plants, 38.6 per cent are monotypic and 12.9 per cent are bitypic, and all genera together show a hollow-curve distribution. The larger the family, the more accurate is the hollow curve. Why should a successful family have so many "relic" genera with a few species? With natural selection, one has to admit a continual reduction in numbers of species to explain the curves, or admit that the majority of forms are poorly adapted. Why did nature produce so many at first, only to cut them down? Was there no selection, or was it less stringent, during earlier times? According to the differentiation theory, however (699, 720), the manner of production of genera and species necessarily results in the hollow-curve relationship.

The size of the largest genus in a family.—The assumption that the size of the largest genus in a family goes with the size of the family is borne out by the data. Willis' figures are shown in Table 23. These data, according to

TABLE 23.—The relationship between the size of the largest genera and the size of their families (Data from 698)

Number of Genera in Family	Average Number of Species in Largest Genera of Family
1	12
2–3	43
4–8	94
9–20	129
21–40	153
41–70	195
71–100	313
101–250	330
251–over	611

Willis, fully support the differentiation theory and are inexplicable on the basis of natural selection or gradual adaptation.

The origin of large genera.—Upon what grounds of adaptation did *Senecio,* for example, come to have about 3000 species, and other large genera also become enormous? If they owe their success to adaptation it could only have been generic. "There are no characters in the individual species that one can point to as adaptive, and how could an adaptive and generic feature be produced in a genus formed from below upwards by the dying out of intermediates between it and its near relatives? The bulk of the species in these big genera are local in distribution, and it is far simpler to explain the whole matter by differentiation and by age. . . ."

Some morphological puzzles.—Even where a structure is apparently of some physiological advantage, like the tenacles of Droseraceae, natural selection is hard put to explain its gradual formation. There is almost no end to the inexplicable difficulties in structure that can be brought up for the selectionist, such as the windows in the leaves of Aponogetonaceae, the pitcher leaves of many Bromeliaceae, the tetradynamous stamens of Cruciferae, the stinging hairs of Loasaceae, the corona of Asclepiadaceae, etc. Many characters are so puzzling that intermediates are difficult to imagine and no traces of them are found either today or among fossils. Most of them must obviously have beeen formed at one step, and far enough back to provide time for the development of the families which they characterize. Even where intermediate structures seem possible, they usually provide no apparent basis for selection.

It appears that there is some circumstantial evidence for the theory of evolution by divergent mutations or differentiation. Many facts of distribution and morphological discontinuity are explained by the assumption of large mutations in the early history of a family. On the hypothesis that mutations have been growing progressively smaller with the passage of time I have no basis for an opinion, but modern physics (astrophysics, thermodynamics, etc.) provides a parallel and supporting body of data. One point, largely neglected by Willis, seems evident. Although large mutations may have been more frequent in the past, there were also small mutations. In the early Tertiary, for example, many genera existed which were composed of numerous species as closely related to one another as the average species of today. It would seem that mutations of various degrees from small to large may have always occurred. This does not deny that there may be a family mutational "cycle," but only that different families are and have been at any one time in different stages of development. Whether large mutations still occur is completely unknown; in fact, that they ever occurred seems beyond real proof.

With respect to Willis' conclusion that the mutations which he supposes are responsible for evolution cross the "sterility line" between species at one step, there seems to be little genetic evidence. The intrinsic causes of inter-specific reproductive barriers remain essentially unknown. Morphologically very similar forms may be reproductively isolated for intrinsic reasons, whereas allopatric species, sometimes even those of different genera, have failed to develop reproductive barriers of a genetic nature, although they have been isolated from one another for millions of years.

How the differentiation theory operates with respect to problems of area has been discussed in the preceding chapter. If similar or related species can differentiate independently, in two or more separated areas, from a common parental syngameon, many of the problems of plant geography are obviated. I am certainly in no position categorically to deny the hypothesis, but I prefer to look for solutions to such geographical problems as disjunction that do not require the differentiation hypothesis.

PART FOUR

Evolution and Plant Geography

21.

Some Principles Concerning Evolution, Speciation, and Plant Geography

In this chapter are gathered together some conclusions regarding the relations between evolutionary processes and their results and plant geography. The brief comments which accompany the statement of these principles are intended to serve only as an introduction; more extended remarks are to be found elsewhere.

1. *Evolution results from the action of one or more of several processes: mutation and hybridization, selection and isolation.*

Mutations as now understood provide new materials for evolution, adding to the variability of populations; they consist of gene or point changes, changes in chromosomal structure such as inversions, segmental interchange between non-homologous chromosomes, loss or duplication of segments, changes in position of fiber attachments, and chromosomal duplication or loss, including genom doubling, or polyploidy. Hybridization, both intra- and inter-specific, results in new combinations of genetic materials. The processes of mutation and hybridization are primary causes in evolution; they affect individuals and result in new combinations of genetic materials—new genotypes. Hybridization tends to increase variability by breaking down distinctions which exist between populations. A form known as introgressive hybridization (20) results in the infiltration of the germ plasm of one species into another. Hybridization between species that results in a more or less sterile F_1 may be followed by speciation when amphidiploidy occurs.

Evolution includes changes in individuals and populations. Selection and isolation can be thought of as qualifying factors which affect populations but do not provide new genotypes. Selection promotes population differentiation by producing a different allele frequency through differential survival and reproduction of types that differ in their pre-adaptation. By pre-adaptation is meant inherent fortuitous factors which, at the time, place, and conditions of their origin or existence, give an organism characteristics that have a survival value greater than that of similar but not identical organisms.

2. *Speciation is understood as an ultimate step in evolution which results in discontinuity between populations and which, strictly speaking, depends upon reproductive isolation that results from intrinsic factors.*

An increase of genetic variability through mutation, hybridization, and isolation does not necessarily result in speciation. That is to say, speciation does not automatically result from the accumulation of a certain quantity of ordinary differences; probably it is the result of the evolution of definite isolating factors. There is a considerable quantity of data to support Dobzhansky's hypothesis that speciation usually requires isolation, followed by a meeting of populations and hybridization. During isolation the separate subspecific populations have undergone mutations that, when recombined by hybridization, result in reproductive isolation. It is seen, then, that reproductive isolation probably cannot occur at a single step; it requires at least two complementary genes, or gene groups.

3. *Species are not genetically homogeneous populations and consequently they are composed of individuals which vary in morphological and physiological characteristics within definable limits. Species consist of a number of biotypes, sometimes very large, that are usually not uniformly dispersed throughout the species population but are more or less concentrated in local populations, ecotypes, and geographical races.*

Species are not homogeneous and composed of genetically identical individuals because no individual can contain all the hereditary factors of the species, the factors existing in allelic pairs and series. In cross-fertilizing populations panmixy tends toward a sharing and an even dispersal of genetic factors, but this tendency is opposed by selection pressure in different environments of the specific area and by partial isolation, with the result that taxonomically recognizable subspecific groups are frequent. The members of a subspecific group bear a constant internal resemblance that is closer than their resemblance to members of other subspecific groups of the species, and collectively the groups form a population that has a certain unity because of their community of hereditary factors and their discontinuity with other specific populations.

Species that do not have obligate cross-fertilization, such as self-fertilizing plants and apomictic forms with parthenogenesis and vegetative reproduction, are not as variable as those described above. Some populations may have attained reproductive isolation in the absence of conspicuous morphological differentiation, and relic species with biotype depauperization are relatively homogeneous. Relic species, however, are frequently strikingly discontinuous from their near relatives.

4. *The species concept has undergone an historical change from the time when the species were conceived as separate acts of creation, to modern times when the species is understood on a population and cytogenetic basis. All taxonomists, however, have not accepted the genetic definition of a species, nor could they be expected uniformly to put it into practice for a variety of practical reasons. The result of this situation is that the literature contains "species" which are biologically very diverse and which, in ecological and geographical investigations, must be understood and distinguished.*

Following Darwin and Wallace, considerable attention has been given to the phenomena of variation, selection, and geographical isolation, with the result that a large portion of taxonomy consists of species recognized on the basis of morphological discontinuity and geographical separation. When few specimens are available, as in all exploratory work, and few stations are known for a type, it is reasonable and practical to assign morphologically discontinuous types to specific status. With increased knowledge of plants, however, greater variability of individuals is encountered, and frequently types are found actually not to be discrete in nature. Because of the all-pervading belief in evolution and because of neglect to consider isolating factors and the causes of species stability, much taxonomy is practiced as an art. On the other hand, a study of populations and their geographical pattern, together with recognition of the significance of hybridization where such populations meet, has led geneticists to realize the importance of intrinsic or reproductive isolation as the basis of speciation, and to attain a scientific basis for the species concept.

When biological types have only geographical or extrinsic isolation, it is not known whether they also have reproductive isolation unless they are brought together with their close relatives and this fact is experimentally determined. In spite of the ability of such disjunct sympatric forms to cross, when they do not do so in nature and when they have morphological discontinuity it is probably reasonable and useful to regard them as species. But when such forms are sympatric and meet and hybridize at their margins of area, it would be to the interest of botanical taxonomy to assign the populations to a subspecific status as is done in mammology and ornithology.

Even among groups which have reproductive isolation, however, there are biologically different types of species, and it is important in ecology and geography to recognize the nature of their differences.

5. *Taxonomic systems form an immediate and basic tool in geographical studies; hence it is necessary for the geographer to understand the basis of taxonomic arrangements, for his interest is always in natural relationships.*

Most modern classifications attempt to express phylogenetic relations, but evolutionary patterns and a systematization of species and other categories are often incongruous. Taxonomy is necessarily static and two-dimensional, whereas phylogeny, as an expression of evolution, is dynamic and three-dimensional, involving changes in time. Because phylogenetic systems are of necessity largely hypothetical and based upon circumstantial evidence, and because systematic arrangements and nomenclature must result in practical schemes, the two ideals can only approach one another. In addition to the problem of marshaling the circumstantial evidence of relationship, which consists mainly of morphological and geographical facts and only partially of cytogenetic facts, there is the difficulty in expressing phylogenetic relationships because of the phenomena of parallel and convergent evolution, and because of anastomosis, as in hybridization and allopolyploidy.

Certain geographical problems which arise because of endemism, disjunction, etc., may not exist if relationships are not as supposed by taxonomic classifications. On the other hand, new problems may be presented when taxonomic arrangements are revised on a more natural basis. In other words, actual closeness of relationship between species or other groups is of fundamental importance in ecology and geography.

6. *Geographical isolation promotes speciation but it does not of necessity result in speciation.*

Geographical isolation of two closely related populations promotes their evolutionary divergence because their mutations subsequent to isolation cannot be shared, the isolated populations may come under different selection pressures, and there may be a different non-adaptive genic drift. When such isolation is only partial, or when it is temporary and contact of the two populations is restored by migration, it appears that conditions are favorable for speciation, but geographical isolation may endure for long periods of time without the appearance of reproductive isolating mechanisms. Closely related organisms may have an allopatric distribution without being reproductively isolated. When they have a sympatric distribution they must have either ecological or reproductive isolation; only the latter, with its intrinsic basis, cannot be overcome.

7. *The rate of evolution depends upon the independent and conjunctive operation of a number of factors such as mutation rate; the richness of the population in allelomorphic genes; the replication of chromosomes as in euploidy and aneuploidy; population size; the size, availability, and variability of ecological niches; the vicissitudes of environmental change and concomitant migrations, and the nature of selection pressure. There are suppositions of evolutionary cycles or mutation spates, but there are no unqualified evidences of strikingly different mutation rates or of the fact that age of a phylogenetic stock results in evolutionary senility.*

For a variety of reasons different evolutionary stocks appear to have different evolutionary potentialities, but the better evidence seems to indicate that actual evolutionary rates result from extrinsic rather than from intrinsic causes. External conditions to a large extent control the ability of a phylogenetic stock to realize its potentialities. Species populations that consist of a relatively small number of biotypes and have comparatively restricted ecological amplitudes or adaptability may, with changing conditions, enter on a phase of evolution that dooms them to quick extinction or to a state of lingering relic survival. Ecological opportunity, as when a population enters a region which is "unsaturated" and in which there are numerous ecological niches where competition pressure is low, or when plants are brought into culture where competition is eliminated, may result in an apparent quickening of evolution. However, this phenomenon has not been shown to be due to an increased mutation rate. Polyploids may appear to have a reduced rate of change, but it is more likely that this is due not to a lowered mutation rate but to the masking of recessive mutations which must be duplicated in each homologous genosome in order to gain expression. The evidence provided by paleontology of a rapid rate of evolution in certain phylogenetic stocks may be more apparent than real, and be due to incomplete records and the telescoping of time. It is concluded, then, that differences in evolutionary rate and speciation shown by different phylogenetic stocks are due more to the operation of external qualifying factors than to differences in mutation pressure, although it is not denied that mutation rates may differ.

22.

Nature of Species

---->>><<<----

A brief historical account.—It is only to be expected that the species concept has undergone change as the findings of science have accumulated. Although differences in delimiting species are of fundamental importance to plant geography, only a brief history will be attempted.

To Linnaeus[1] the species was inviolable; a species was created as a unit and its constancy was beyond doubt. The two general qualifications of biological classification—variation and discontinuity—were known to him. Normal variations were within the fabric of the species, but such things as horticultural varieties were monstrosities. Linnaeus was no evolutionist; hence specific discontinuities were perfect, and no species ever gave rise to another. Pre-evolutionary taxonomy had only a morphological basis, and questions of plant geography were concerned solely with where plants lived.

Lamarck was one of the first biologists to obtain insight into the variability of species; he did so because he saw the possibility of evolution in the variability which species everywhere revealed. He believed that variation was due to individual adaptation to the environment, and that variations thus acquired could be transferred to progeny and gradually fixed in the heredity of the strain. Evolution resulted from the inheritance of acquired characteristics because changes in the milieu caused new "needs" in the organism and these needs led to individual, profitable variations which "satisfied" them.

Darwin believed in the general occurrence of spontaneous, inherent, and heritable variations which were acted on by natural selection in the evolution of species. In his chain of evidence—variation, heredity, and struggle for existence—the question of the origin of varieties, which were considered to be incipient species, ultimately forced him to a sort of

---

[1] An excellent account of Linnaeus' concepts is to be found in Ramsbottom's Presidential Address before the 150th Session of the Linnaean Society of London (513).

326

Lamarckian hypothesis.[2] Darwin saw species as the traveler-naturalist sees them, with discontinuities imposed by broad geographical barriers; but he was also aware of gradients of character change which occur between related forms in contiguous and unisolated areas. He believed that two or more such intergrading populations could attain the specific level through the continued operation of natural selection.

De Vries held that the important materials for evolution were not the minor random variations considered important in the Darwinian hypothesis, but the mutations, which are generally larger, sudden, inherent, and heritable variations of the genetic constitution. He believed that latent genes could become active, genes could be formed *de novo,* and genes could be lost. He did not believe in the inheritance of acquired characters because the occasional mutations which he observed among the hundreds of thousands of plants in his garden were apparently unrelated to environment and selection. He helped rescue Mendel's findings from oblivion. According to De Vries, recombinations which resulted from hybridization were important in the origin of new forms, but Mendelian recombinations were not enough; mutations were necessary for the production of new materials.

With the further development of cytology and genetics, there has been an extensive accumulation of information concerning variation and discontinuity. Müntzing (481) has organized the material related to these phenomena as follows:

A. The origin of variation:
1. Environmental modifications occur within the inherent limits of the plasticity of organisms. Modifications are not transmitted to progeny and have a significance for taxonomy only when they are confused with heritable variations.
2. Hybridization results in genetic recombinations.
3. Mutations are heritable variations not due to recombination; they include:
   a. Gene mutations.
   b. Alterations of chromosome structure.
   c. Changes of chromosome number.

B. The origin of discontinuity:
1. Natural selection results in the subdivision of species into adapted biotype groups which correspond to the different habitats. Such intraspecific discontinuities are incomplete.

[2] The theory of pangenes.

2. Discontinuities between species are complete or nearly so, and result from reproductive isolating mechanisms:

    a. There are species that are so completely isolated that they cannot cross, or the embryo fails at an early stage.

    b. Other species can cross, but the F_1 generation consists of more or less sterile individuals.

    c. Some populations have completely lost sexual reproduction. Such apomictic populations have a strict constancy because they have only mother inheritance.

From this brief historical account it is seen that the concept of species has progressed from a strictly morphological basis to one that involves the hypothesis of evolutionary relationship, together with the addition of geographical and, finally, cytogenetic criteria of species.

Species definitions.—There are dozens of definitions of species. Some involve fundamental differences of opinion and understanding which result largely from an acceptance or denial of the importance of certain genetic relationships. There is so much divergence among the definitions that many biologists believe that no single definition of species can be devised. This point of view is reflected in statements such as the following: "A species is a name in a book." "Species are judgments." "A species is what a competent taxonomist thinks it is." "A species is a group of individuals which, in the sum total of their attributes, resemble each other to a degree usually accepted as specific, the exact degree being ultimately determined by the more or less arbitrary judgment of taxonomists. . . . Any attempt to define a species more precisely in terms of particular attributes breaks down." From this standpoint, much of the practice of taxonomy is more of an art than a science.

A series of definitions occur in *The New Systematics* (371), and many of them emphasize genetic rather than morphological characteristics. One of the best is by Timofeef-Ressovsky: "A species is a group of individuals that are morphologically and physiologically similar . . . which has reached an almost complete biological isolation from similar neighboring groups of individuals inhabiting the same or adjacent territories." According to this author, a species usually consists of a number of groups of lower taxonomic rank; the simplest real systematic group is composed of "individuals characterized by one or several common heritable characters and having a common area of distribution." Emerson (210) has written what in my opinion is the most satisfactory and most scientific species definition: "A species is an evolved (and probably evolving), genetically distinctive, reproductively isolated, natural population." In the reasoning behind it, this definition does not differ materially from the one proposed earlier by Dobzhansky (190): "A species is that stage of evolutionary process at which the once actually or potentially interbreeding array of forms becomes segregated in two or more separate arrays which are physiologically incapable of interbreeeding."

There is no doubt that these definitions of species describe a real phenomenon in nature, and their validity is attested by the fact that species so defined correspond, in the main, to species as recognized by taxonomists. It is not to be expected, however, that taxonomists will always find it possible to employ such definitions. For one thing, if species are evolving populations, there must be many such populations in nature that have not attained complete reproductive isolation. It is not required, in genetic definitions of species, that closely related species be cross-sterile or even that they be unable to interbreed (as in the unnatural conditions of experimental plots and cages), but only that in nature they interbreed slightly, if at all.

An important question is whether permanent reproductive isolation is required for two populations to be specifically distinct. No one, so far as I know, denies that physiological and genetic barriers to interbreeding constitute a definite line of demarcation between species; but what about geographical and ecological barriers which may be overcome by migration? Must the taxonomist and geneticist test each group of closely related but spatially isolated forms that are now called species to ascertain whether intrinsic reproductive barriers exist? Emerson says that any mechanism which prevents panmixy has similar evolutionary consequences, so he is apparently willing to accept geographical isolation as sufficient when the isolated populations are genetically distinctive. Hybrids of related plants may occur sporadically without confusing either the morphological or the chorological limits of species, and sometimes where two species meet there is a large and complex swarm of hybrids so that the characteristic genetic constellations of the parent species are locally broken down. When species of the types just mentioned maintain their distinctiveness as a general rule, and hybrids are the exception, most botanists prefer to consider them "good" species despite the lack of a complete reproductive barrier, and this practical solution apparently is also acceptable to many geneticists. Emerson uses an interesting analogy when he says that occasional hybrids between two species do not invalidate the species concept any more than mitosis invalidates the cell concept. When there is a ring or chain of forms that have a geographical replacement pattern the members of which interbreed freely at their group borders, many zoologists and a few botanists consider the whole syngameon to be a species and the lesser groups subspecies. An interesting situation is presented when such a chain of geographical subspecies completes a circle and the members of the two end populations live sympatrically without interbreeding, although they are connected along the chain by other subspecies that do interbreed where they meet. According to Dobzhansky, "A situation of this kind appears paradoxical unless one adopts a dynamic species concept: the extreme forms have already developed physiological

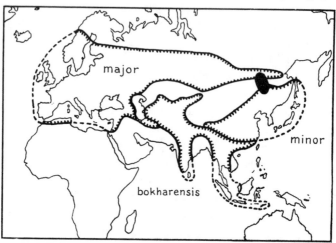

Fig. 44. Subspecies of the titmouse Rassenkreis, *Parus major,* hybridize where they meet except in the east Asian region shown in black where *P. m. major* and *P. m. minor* live sympatrically without crossing. Base map after Goode's copyrighted map No. 101, with permission of the University of Chicago. Data on distribution from Rensch, *Verh. Dtsch. Zool. Ges.,* 1933:19–83.

isolating mechanisms that keep them separate from each other, and yet each of them has not become isolated from certain other forms that unite them."

Treatments of the Linneon.—Most of the species of Linnaeus and of many later taxonomists are broadly delimited and include members which show considerable variation. In many instances, subsequent investigators have drawn specific lines more finely; this resulted in the erection of several species and subspecific units within the original limits of a Linneon. It will be useful at this point to consider some of the different types of treatment accorded the Linneon.

The taxonomist working solely with morphological data must delimit species without any proof of evolution. His species are abstractions that are conceived on the basis of individuals which are more or less representative of the ideal or the "type," and which he regards as natural. Suggestions of evolutionary relationship between morphologically related but differing forms may come, however, from structural series which coincide with the geological sequence or from structural series that today present a geographical sequence. On the basis of morphology alone, a taxonomist could carry classification and naming to an absurd extreme, because within a species the finest recognizable unit is the biotype. When the morphological approach to taxonomy is restrained by geographical considerations, the smallest unit consists of a group of similar biotypes which have a real unity. On the other hand, the same classificatory *reductio ad absurdum* would be reached with the naming of a group of individuals of identical genetic constitution which are unable to produce more than one kind of gamete. Somewhere between the biotype and the genotype, at one extreme, and the specific boundary set by reproductive isolation, at the other, are the intraspecific groups which are reasonably subject to taxonomic classification and are of use in geography.

Lotsy (according to Goddijn, 278) used the term species in the sense of the Linneon and referred to microspecies as Jordanons. All the micro- and macro-species that can exchange genes through crossing he called a syngameon. Syngameons have been variously treated taxonomically as genera, sections, and species. Danser (172) also tried to delimit groups of individuals whose origin could be understood, but he did not attempt to coordinate his conceptions with taxonomic usage. According to him, a comparium consists of all individuals held together by the possibility of hybridization, whether the products are sterile or fertile. The comparium has phylogenetic value even though it may include a section, a genus, or even a portion of a family. The next smaller category is the commiscuum, which includes all individuals that can successfully exchange genes. It is close to the usual species. Danser's final category is the convivium, which is a population differentiated within the commiscuum and isolated by geographical influences.

At about the same time, Turesson (647) established concepts which are somewhat similar to those of Danser, but which stress the natural selection of biotypes on a habitat basis. According to Turesson, a coenospecies is a group of species which may not all be able to exchange genes directly but which have the possibility of doing so indirectly through various hybridizations. Within the larger group, an ecospecies is an amphimictic population with vital and fertile descendants, but with reduced fertility with other members of the coenospecies. It compares with most species of the taxonomist. The smallest group is the ecotype, which can cross freely and successfully with other ecotypes of the ecospecies and which maintains its individuality because of partial isolation and environmental selection of adapted biotypes. It seems to me that an ecotype is a convivium caused by ecological factors. Turesson found it necessary to erect a fourth category to include those forms which do not have sexual reproduction, the agamospecies.

Du Rietz (201) has provided definitions for intraspecific categories and the species, and they have become more popular with taxonomists than the earlier terminology used by Danser and Turesson, although the latter is gaining an increasing following.[3] Du Rietz says, "A species is a population

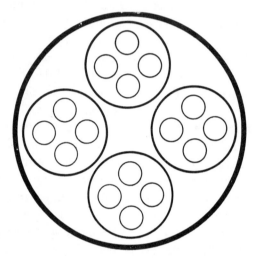

Fig. 45. The large circle represents a coenospecies in the sense of Turesson; it is composed of ecospecies and ecotypes, the smallest circles. Each ecotype may consist of one or more forms and a rather large number of biotypes, not illustrated by the diagram. See the text for definitions and genetic relationships.

consisting either of one strictly asexual and vital biotype, or of a group of practically indistinguishable, strictly asexual and vital biotypes, or of many sexually propagating biotypes forming a syngameon separated from all others by more or less complex sexual isolation or by comparatively small transitional populations." A biotype is a population that consists of individuals of identical constitution; a species may consist of many thousands of them. A variety is a population consisting of individuals of one or more biotypes and forming a more or less distinct local facies of a species. A subspecies is a population of several biotypes that forms a more or less distinct regional facies of a species. The species is the smallest possible natural population; it is separated from other such populations by distinct discontinuity

[3] Cf. (146).

in a series of biotypes, for this is not true of the variety and the subspecies.

Babcock (34) says that subspecific groups usually develop within species on the basis of structural and fertility differences which are greater between themselves and other facies of the species than within themselves.

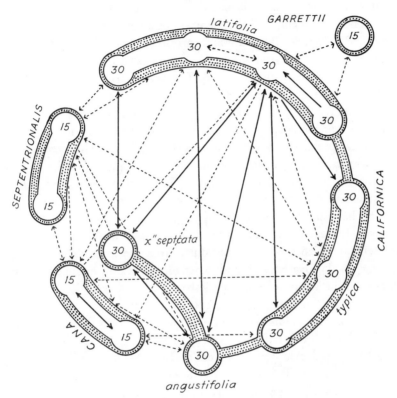

Fig. 46. A diagram of the genus *Zauschneria*, which is shown to compose a coenospecies. The four species of the genus (*cana, septentrionalis, Garrettii,* and *californica*), each included within a dotted zone, are ecospecies. Within the ecospecies *Z. californica*, three subspecies in the sense of ecotypes are recognized (*typica, latifolia,* and *angustifolia*). A series of 13 different forms was selected for polygonic crossing (circles within the ecospecies or ecotypes) and the direction the pollen was carried is indicated by the arrows. The solid lines indicate crosses resulting in fertile hybrids and vigorous F$_2$ plants. The dotted lines indicate less successful crosses of various types. The figures are gametophytic chromosome numbers. The diagram is simplified from (146), which should be referred to for details.

When these systems are collated, it is seen that they all agree in accepting the phenomenon of rather complete reproductive isolation for the demarcation of species. Within the species there is a certain commonalty because of descent from common ancestors and because of the direct or indirect exchanges of genes between members of the subspecific groups. Subspecific categories are usually two and are erected on the basis of habitat or area and the freedom or extent of genetic exchange. Apparently it would seldom serve any useful purpose in geography to name or deal with any smaller unit than the subsubspecies, whether it be called variety, ecotype, or convivium. On the other hand, it would seldom seem practical to extend the species concept to include all the members of a coenospecies or comparium.

The kinds of species in nature and in systematics.—The first necessity in geographical and ecological studies is a knowledge of the biological units with which the problem is concerned. The organisms of the community or area belong to what species or subspecific group, or, perhaps, to what genus, tribe, or family? In their simplest form, these questions are answered by reference to suitable taxonomic manuals. Since, however, the nature of many geographical problems, or even the existence of a problem, depends upon the identity of the organisms, it is necessary to inquire into the biological nature of the species described and named in the manuals and monographs. Unfortunately, the taxonomist's array of species erected for the various forms he has studied does not always coincide with the species in nature. Moreover, the extent to which the species of two taxonomists differ and the extent to which they both may differ from the facts of nature depend not alone on the relative skill and experience of the taxonomists but more basically on the nature of their species concepts and what kind of populations they admit to the species category.

Students of population evolution and genetics rather than of individual organisms have discovered that a variety of isolating mechanisms are operative in nature and that various degrees of discontinuity exist between populations. They know that biologically different types of organisms can be rightfully admitted to specific status, and that subordinate to the species are other differentiated populations that do not deserve specific rank for one reason or another, primarily because no real line of demarcation separates them from their neighbors. It is the purpose of the remainder of this chapter to examine the nature of "species" in taxonomy and in nature, because the plant geographer must know the real nature of a population that carries a binomial.

Differentiation through gene mutation.—It is perfectly evident that characters which are due to genetic factors that can be segregated at meiosis cannot distinguish species. If, when plants of two species are crossed, the parental species can be obtained again in the F_2 generation, these species cannot maintain their genetic identity in nature. Also, there seems to be no excuse for calling the F_1 plants of such crosses species if their nature is known, but there are cases of both treatments. A character which is usually taxonomically sound may depend sometimes upon simple Mendelian factors or upon linked genes. For example, whether a species has alternate or opposite leaves is usually, but not always, a safe criterion of a species. Henry (340) found that the Huntingdon elm, which at one time had been given specific status, is the F_1 hybrid between *Ulmus glabra* and *U. montana.* The hybrid sets fertile seed and the progeny show segregation in the seedlings for opposite and alternate leaf arrangement in a ratio of 3 : 1. The seedlings also show segregation for small and large leaves in the same ratio. Crane (164) says that *Campanula nitida* is a simple segregate from *C. persicifolia, Urtica dodartii* from *U. pilulifera,* and *Rubus idaeus* var. *obtusifolius* from *R. idaeus,* and that these segregates are recessive forms that result in each case from a single gene difference.

Certain species may differ by more than one character and still show simple segregation if the factors are linked in chromosomes. Winge (705) obtained both parental forms of the cross between *Tragopogon pratensis* and *T. porrifolius* in the F_2 and subsequent generations. Also, from *Verbena tenera* $\times$ *V. Aubletia* it was possible to segregate the parental species in pure form.

Chittenden (141) studied crosses of *Primula acaulis* $\times$ *P. Juliae,* and *P. elatior* $\times$ *P. Juliae,* and found that certain differences are due to single factors. For example, thrum and pin style, pedunculate and non-pedunculate inflorescence, orange and yellow flower-eye color, and degree of hairiness showed simple Mendelian segregation. For other characters, such as "acaulis" leaf shape, it appears that several factors, probably in many chromosomes, are required. These data reveal the important difference between species and named forms that have no right to the specific epithet because their differences segregate out in hybrid progeny. It is probably true that most species show some characters that result from simple genetic difference, but they also have one or more characters that result from the non-segregating action of multiple factors. As Dobzhansky (191) has clearly shown for reproductive isolating mechanisms, they cannot be due to a single factor, and consequently could not have developed at a single step, through one mutation.

Anderson and Sax (22) say that the diploid species related to *Tradescantia virginiana* are apparently differentiated by gene mutation since their genoms are similar in chromosome number, morphology, and structure. These species are interfertile and maintain themselves only because they live in different habitats and geographical areas. One could with considerable reason consider that this group of forms (*Tradescantia virginiana, gigantea, humilis, edwardsiana, paludosa, ernestiana,* and *hirsuticaulis*) composes a single species in which the above-named units are only subspecies.

Two kinds of plants may fulfill the taxonomist's psychic requirements for species and still provide no single sharply discontinuous character difference. This is the case with *Uvularia perfoliata* and *U. grandiflora* (23), which look obviously different; however, the species differentiation is a matter of two combinations of many small tendencies. Such a situation can be very disturbing to the taxonomist, but it should not be surprising to the biologist, who thinks in terms of processes and realizes that germ plasm is made up of a rather enormous number of hereditary units. Whether these uvularias are species depends upon whether they have reproductive isolation, and not upon whether they can be easily diagnosed by the taxonomist.

Within a species, simple genetic differences are important in building up local populations, ecotypes, and subspecies. All these subspecific groups have only partial isolation and differ from one another by characters that can be segregated if combined, but that occur in different frequencies in the different populations. Anderson (15) found that the northern blue flags (*Iris*) occur in more or less isolated colonies that show pronounced statistical differences between colonies which transplant experiments show to be inherent. Through partial isolation, each colony tends to become a breeding unit in which quantitative and qualitative differences can rapidly accumulate. This is just such a situation as hypothesized by Wright (716, 717), in which a continually shifting differentiation among colonies occurs, and which brings about an "indefinitely continuing, irreversible, adaptive, and much more rapid evolution of the species" than is found in larger interbreeding populations. Isolation and the accumulation of Mendelizing differences do not lead to speciation; but when isolated populations that were once panmictic develop linked mutations they may, on overcoming their isolation, be found to have developed reproductive isolation.[4]

Differentiation through structural changes in chromosomes.—In addition to gene or point mutations, a variety of structural changes may occur in chromosomes. Such mutations as changes in chromosome size, position of fiber attachments, inversion of segments, segmental interchange between homologous and non-homologous chromosomes, and fragmentation and loss of segments result in the production of variations which are added to the genetic variability of a species population; they may also result in reduction or loss of fertility between original and mutated forms, thus establishing intrinsic reproductive mechanisms. How much evolution and speciation are

[4] See Chapter 23.

due to structural changes in chromosomes is not well known, but cytological studies are producing considerable evidence. Only one example will be mentioned. Anderson and Sax (22) say that in *Rhoeo* differentiation appears clearly to have been due to segmental interchange. In fact, in the Commelinaceae there is considerable evidence that generic differentiation has resulted from changes in chromosomal ·structure. In *Tradescantia* the chromosomes have approximately median fiber attachments; in *Rhoeo* two chromosomes have developed subterminal attachments; and in *Spironema* and *Callisia* four chromosomes have done so.

Differentiation through changes in chromosome number.—There is no reason for a general discussion of autopolyploidy, allopolyploidy, and aneuploidy at this point, for they are taken up elsewhere, but mention of their role in evolution and speciation may not be out of place.

In autopolyploids there is a duplication of genoms of a fertile species. If *A* represents a genom, the somatic cells of the diploid are *AA,* those of the autopolyploids of *AAAA, AAA,* etc. Autopolyploids when inbred are more or less fertile, but their fertility is less than that of their diploid progenitor because of the difficulty of pairing when three, four, or more homologous chromosomes are present in a cell. When crossed back with diploids, autotetraploids usually produce sterile triploids. Although diploids and their autotetraploids may show little or no morphological differentiation, they have a reproductive isolation which apparently is complete enough to allow them to be recognized as separate species if and when morphological or physiological differentiation has resulted.

Anderson and Sax (22) found both diploid and autotetraploid races in *Tradescantia hirsutiflora, T. occidentalis,* and *T. canaliculata.* They do not consider that these races differ sufficiently for taxonomic recognition, although the tetraploids are larger, have longer blooming periods, are better colonizers, are more northern, and occupy larger areas than the diploids. I believe that the justification of not recognizing the tetraploids as distinct species is entirely practical and is based on the difficulty, if not impossibility, of being certain to which race any single plant belongs. Winge (705) mentions that autopolyploidy frequently results in more or less interrace sterility, but "scarcely anybody would think of designating them [autopolyploids] as different species." As examples he cites *Narcissus Bulbocodium* (with somatic chromosome races of 14, 21, 28, 35, and 42) and *Potentilla argentea* (with 14, 42, and 56 somatic chromosomes).

Unbalanced autopolyploids, such as the triploids *Pyrus minima, Tulipa praecox, T. saxatilis, T. lanata,* and *Nasturtium officinale,* have been preserved and widespread because of their capacity for vegetative reproduction, according to Crane (164). Their assignment to specific status is a practical matter, and if we think back to Du Rietz' definition of a species we see that it includes a population consisting of one strictly asexual and vital biotype.

Differentiation through allopolyploidy.—The two important phenomena distinguishing allopolyploids from autopolyploids are the fact that they result from hybrids, i.e., the genoms are not all homologous, and that they show an increased fertility in comparison with the diploid hybrids. Crane (164) classified allopolyploids according to the part played by cultivation and experimentation. (1) The new species arose in nature and there is good reason for supposing them to have resulted from amphidiploidy, as in *Dahlia variabilis*. (2) The new species arose in nature and subsequent experimentation has shown them to be allopolyploids, as in *Prunus domestica,* which is a hexaploid of tetraploid *P. divaricata* and diploid *P. spinosa*. Crane supposes it probable that *P. domestica* arose independently several times in nature, and it has been produced synthetically from the parents. (3) The new species arose spontaneously under culture from species which are widely separated geographically in nature, as in *Aesculus carnea,* which arose from the European tree *A. hippocastanum,* and the dwarf American *A. pavia,* which was introduced into European arboreta. (4) New species have resulted from spontaneous chromosome doubling after deliberate crossbreeding of sometimes widely related species. In other cases, more or less sterile hybrids have been made into fertile species (or at least potential species) by artificially induced amphidiploidy through the action of colchicine and other agents. For examples, see Chapter 27.

At one step the allopolyploids are reproductively isolated from their progenitors and form a new breeding population which fills all of Emerson's requirements of a species, except that they are not a natural population when experimentally produced. In view of the fact that many species in nature have apparently had this mode of origin, it does not seem to be stretching a good definition too far to include genetically equivalent "artificial" forms under the species concept. Anderson (15), for example, has good morphological, distributional, and genetic evidence that *Iris versicolor* is an allopolyploid of *Iris virginica* and *Iris setosa* var. *interior*. As far back as 1917, Winge hypothesized that constant hybrid species could be produced as a result of chromosome doubling. In 1932 he reviewed 24 clear cases that had been published during the fifteen-year interval. In 8 cases the doubling had occurred in somatic tissues; in 5 the amphidiploids had arisen from a union of unreduced gametes; in the remaining 11 cases how they had arisen could not be determined. These cases are listed in Table 24.

TABLE 24.—Twenty-four cases of known allopolyploids discovered between 1917 and 1932 (Data from 704)

1916. *Primula floribunda* $(g = 9) \times P.$ *verticillata* $(g = 9)$ gave rise to an amphidiploid form of *P. Kewensis* $(g = 18)$. The name *P. Kewensis* was originally applied to the sterile F_1 from which the tetraploid arose spontaneously.

1924. *Rosa pimpinellifolia* $(g = 14) \times R.$ *tomentosa* $(g = 7$ in the pollen) gave rise to *R. Wilsoni* $(g = 21)$.

1924. *Raphanus sativus* $(g = 9) \times Brassica$ *oleracea* $(g = 9)$ resulted in a tetraploid species in the F_2 as a result of the suppression of reduction division in the F_1.

1925. *Nicotiana glutinosa* $(g = 12) \times N.$ *tabacum* $(g = 24)$ gave rise to *N. digluta* $(g = 36)$.

1926. *Fragaria bracteata* $(g = 7) \times F.$ *Helleri* $(g = 7)$ gave rise to one plant which was fertile $(g = 14)$ and was a "new species," according to Winge.

1926. *Aegilops ovata* $(g = 14) \times Triticum$ *dicoccoides* $(g = 14)$ gave rise in some cases to constant hybrids with $g = 28$.

1926. The same is true of the cross of *Aegilops ovata* × *Triticum durum* $(g = 14)$.

1927. *Solanum nigrum* $(g = 36) \times S.$ *luteum* $(g = 24)$ gave a hybrid from which was produced a fertile doubled type $(g = 60)$.

1928. *Digitalis ambigua* $(g = 28) \times D.$ *purpurea* $(g = 28)$ gave rise to *D. Mertonensis* $(g = 56)$.

1929. *Triticum turgidum* $(g = 14) \times T.$ *villosum* $(g = 7)$ gave a hybrid of $g = 21$.

1929. *Nicotiana tabacum* $(g = 24) \times N.$ *silvestris* $(g = 12)$ gave a hybrid of $g = 36$.

1929. *Aesculus hippocastanum* $(g = 20) \times A.$ *pavia* $(g = 20)$ gave *A. carnea* $(g = 40)$.

1929. Crosses of *Aegilops ovata* with five different Triticums (*T. vulgare, T. turgidum, T. compositum-turgidum, T. amyleum* [*dicoccum*], and *T. monococcum*) resulted in fertile constant hybrids with doubled chromosome numbers.

1929. *Triticum vulgare* $(g = 21) \times Secale$ *cereale* $(g = 7)$ gave a constant hybrid of $g = 28$.

1930. *Saxifraga granulata* $(g = 16-22) \times S.$ *rosacea* $(g = 16)$ gave rise to *S. Potterensis* $(g = 32-36)$.

1931. *Primula Bulleyana* $(g = 11) \times P.$ *Beesiana* $(g = 11)$ gave rise to *Primula* "Aileen Aroon" (a horticultural name) of $g = 22$.

1931. *Nicotiana rustica* $(g = 24) \times N.$ *paniculata* $(g = 12)$ gave rise to a hybrid of $g = 36$.

1931. *Triticum dicoccoides* var. *Kotschyanum* $(g = 14) \times Aegilops$ *ovata* $(g = 14)$ gave rise to an amphidiploid of $g = 28$.

1931. *Spartina stricta* $(g = 28) \times S.$ *alterniflora* $(g = 35)$ gave rise to *S. Townsendii* $(g = 63)$.

1932. *Brassica napus* $(g = 19) \times B.$ *campestris* $(g = 19)$ gave rise to *B. napocampestris* $(g = 38)$.

Differentiation through aneuploidy.—The auto- and allo-polyploids are called euploids because whole genoms have been doubled. Aneuploidy is the phenomenon of the duplication of single chromosomes, the duplication of chromosomal segments (471), of the loss of chromosomes; i.e., related aneuploids have irregular chromosome numbers. Many plant series include both euploids and aneuploids (these are called heteroploids), and it appears that aneuploids with one or two chromosomes plus or minus are more likely to develop in polyploid series than in strict diploids. The course of evolution in certain phylogenetic stocks has largely been due to the phenomenon of aneuploidy (Cyperaceae, the California Madiinae). According to Clausen, Keck, and Hiesey (146), the principal difference between aneuploid and euploid series is apparently the fact that certain germ plasms can tolerate single chromosome differences and others cannot. The type of aneuploidy that is due to segmental repeats is important, Muller believes, because the "establishment of these repeats constitutes the only effective means of gene increase in evolution. Thus the whole of the chromosome complement must really represent an accumulation of such repeats, most of which, however, are of such ancient origin that they have become changed beyond recognition."

Aneuploids are frequently diploids plus one chromosome. If the haploid number is 12, as in *Datura,* the diploid may contain 25 chromosomes consisting of 11 pairs and one trisomic which has resulted from non-disjunction of one pair of chromosomes in one gamete. Obviously there might be 12 different kinds of aneuploids in such a case, resulting from the trisomic condition having developed for any one of the chromosomes. Blakeslee and his co-workers have found all 12 aneuploids of this type in *Datura*. The loss of a chromosome always has a more drastic effect than the addition of one. Aneuploids have more or less genetic instability and reduced fertility because the gametes are affected. As a matter of fact, in nature many aneuploid species are largely or wholly apomictic.

Simonet (578) notes that in *Iris* the groups that are morphologically most uniform and taxonomically simplest are those in which the chromosome numbers are stable, but that when the numbers are irregular, taxonomic confusion prevails. Section Oncocyclus is taxonomically well defined and the species have only the chromosome number 10. Section Pogoniris is fairly well defined and the chromosome numbers represent a euploid series, $S = 8$, 12, 16, 20, and 24. Section Apogon, which shows great morphological variation and taxonomic confusion, has a heteroploid series: 8, 10, 11, 12, 14, 16, 17, 19, 20, 21, 22, 36, (41–42), 42, (43–44), and (54–56). In *Iris spuria* of the

Apogon section, Westergaard (681) found somatic numbers of 11, 19, 20, and 22, and in closely allied forms the additional numbers of 8, 16, 17, and 36. Westergaard says, "The uncertainty regarding the taxonomic view of *Iris spuria* is reflected with all desirable distinctness in the variations of the number of chromosomes."

In *Iris* it is rarely possible to cross aneuploid forms with different numbers, but in *Eriophila* (*Draba*) *verna* Winge (706) found it otherwise. In the *E. verna* complex he found the following chromosome numbers: 7, 12, 15, 16, 17, 18, 20, 26, 27, 29, 32; the experimental amphidiploid had 47 chromosomes. Crosses of distinct races with the same chromosome (such as 7×7) behaved exactly as if they were intraspecific crosses, so the chromosomes must be homologous. Crosses between plants with different chromosome numbers resulted in little pairing in the pollen mother cells, meiosis consisting largely in accidental distribution of the chromosomes to the poles, and there was what Winge called a gay segregation in the F_2, with all combinations of parental characters. Plants were selected and inbred for as many as nine generations, with the surprising result that as the generations progressed new stable fertile types segregated out. Variations diminish as the generations increase and finally there results a series of new types, which are quite constant. From various crosses Winge produced constant types with 22, 23, 25, 29, 31, and 34 chromosomes. Stability is apparently reached as the result of the splitting of univalents during meiosis and by a loss of univalents. Winge sees no reason why such types could not compete in nature because they are all vigorously fertile and are in no way vegetatively inferior to natural types.

Winge, along with Du Rietz, Turesson, and others, believes that apomictic species should always be clearly designated as such. From his study of 113 populations of *Eriophila verna* he found many variants, but they were taxonomically disposed of by his recognition that they fall into four morphological and ecogeographical groups: *E. simplex* (S = 7), *E. semiduplex* (S = 12), *E. duplex* (S = 15, 16, 17, 18, 20), and *E. quadruplex* (S = 26, 27, 29, 32). This system of a limited number of "pigeon holes" has a natural basis because plants with low chromosome numbers flower early but not abundantly, are small and compact, and grow in upland dry soils with low, poor vegetation. Plants with high chromosome numbers are large, especially the middle rosette leaves; they have long slender leaves, flower late, and live in low, moist soils with a luxuriant plant cover. Plants with intermediate chromosome numbers are intermediate in characters, but produce more peduncles and fructify vigorously.

Differentiation from hybrid populations.—The word "hybrid" connotes to some biologists mainly that the organism is an F_1 formed by the union of gametes from parents of different species; a "hybrid swarm" consists of F_2 organisms from the selfed F_1 and backcrosses with the parents. At the opposite pole from such a point of view is Darlington's definition (175): "A hybrid is a zygote [or organism developed from it] produced by the union of dissimilar gametes. . . . Whether the gametes come from similar or dissimilar parents does not signify." That, I believe, is equivalent to saying that all organisms are hybrids except those that are completely homozygous and can form only a single type of gamete. There are, then, gene hybrids, sex hybrids, interchange hybrids, and numerical hybrids. To the geneticist, a species population usually consists of hundreds to thousands of genotypes, and all fertilizations, in the sense of Darlington, except those between members of the same genotype result in hybrids. To the ecologist and plant geographer, species consist of local races, ecotypes, and geographical subspecies; they would recognize intraspecific hybrids between members of such groups but not be concerned with the lesser differences within the groups. The taxonomist, however, is usually not concerned except with interspecific hybrids when he encounters occasional individuals (probably F_1) and swarms (subsequent generations and backcrosses).

We have already noted that more or less sterile interspecific hybrids may be raised to specific rank with a striking increase of fertility when amphidiploidy occurs. Without gaining fertility through chromosome doubling, some hybrids have been able to develop extensive colonies and rather wide areas because of parthenogenesis and a highly successful method of vegetative propagation. Several such populations have been given specific rank. It was emphasized in connection with species definitions that species have intrinsic isolating mechanisms which prevent the successful outcrossing of their members, that closely related species have differences which result in their ecological isolation, or that species have geographical isolation. Hybrid swarms cannot develop from crosses between species which have reproductive isolation of the first type. When ecological conditions are changed, as they frequently are by man's activities, such partially isolated species may meet in intermediate ecological situations and produce a hybrid swarm. When geographical isolation is overcome through migration or purposely or inadvertently by man's activities, such species may be found to be fully cross-fertile.

Since hybrids are not all equivalent (gene hybrids, hybrids between subspecies, interspecific hybrids, etc.), and since they may be rare to occasional or present in local swarms, form a boundary population between subspecies, or cover a large area, it is of interest to geographers to know these differences and to be aware of the treatment they have received at the hand of the taxonomist.

Our interest at this point is in more or less freely interbreeding populations because it sometimes is difficult to know whether a polymorphic population has resulted from the hybridization of members of two species, or whether two distinctive populations are in the process of separating out from the polymorphic parental population through reduction of polymorphy. Perhaps the difficulty of interpreting a "hybrid swarm" can be visualized by allowing the palm of the hand to represent the hybrid swarm and the fingers the "species." Was the palm formed by the fusion of the fingers, or were the fingers formed by the separation of the palm? That two species, presumably once sufficiently isolated to have developed their distinctive genetic character, may meet and lose their identity through hybridization is a possibility. For example, Winge (705) found that *Geum rivale* and *G. urbanum* have the same general area but are separated, except for occasional hybridizations, by their growth in different ecological situations. Upon observing the results when he selfed several F_1 hybrids and raised a large F_2 population, he said, "The possibility seems to exist . . . that two species . . . may in thousands of years [in nature] approach one another and perhaps finally be merged into one species." By way of contrast, we may recall Dobzhansky's theory (191), in which it is supposed that hybridization along the boundary between two subspecies may result in the natural selection of factors which tend to reduce the capacity for crossing.

Babcock (35) says that interspecific hybridization, followed by backcrossing or crossing with a third species, may sometimes produce constant new derivative forms which are potential new species, and that this process has been of importance in the evolution of *Crepis*. According to Crane (164), hybrid populations have been given full specific status in *Fragaria*, *Dianthus*, *Streptocarpus*, *Rubus*, *Rhododendron*, *Vitis*, and *Iris*, in cases where there is no question of chromosome doubling or major cytological aberration. He believes that chromosome numbers may sometimes indicate the hybrid origin of species. For example, in *Iris* a group of species with 44 chromosomes (*albicans, germanica, Kochii*) probably arose from species with 40 chromosomes (*chamaeiris, mandshurica, olbiensis, Reichenbachii, subbiflora*) crossed with species with 48 chromosomes (*mesopotamica, trojana*). I do not believe that this has been demonstrated. Allan (9) says that in New Zealand *Hebe elliptica* and *H. salicifolia* are as good species as any "lumper" would wish; yet they hybridize freely and produce fertile offspring. From among these hybrids three forms have found their way into the literature: *Hebe amabilis* Andersen, *H. blanda* Pennell, and *H. divergens* Cockayne.[5] Other examples cited by Allan include *Raoulia Gibbsii*, which includes specimens of the intergeneric cross *Raoulia bryoides* × *Leucogenes grandiceps*. *Celmisia linearis* includes plants from either of two crosses: *C. longifolia* × *argentea*, or *C. longifolia* × *sessiliflora*.

Camp (115), in a study of *Befaria*, has recently pointed out the contrast between constant and highly polymorphic species of a single genus. *B. racemosa*, a species fairly abundant through Florida and with casual outliers on the Gulf coastal plain, has a striking lack of individual variability. "In fact," Camp says, "the specimens are so nearly similar that one might suspect that they all came from the same clone. . . . The stable condition of *B. racemosa* is probably due to its long isolation from any other of the Befarian stocks." He attributes its constancy to a probable high degree of homozygocity. On the other hand, *B. discolor* in Mexico is an extremely variable species which appears to be essentially a hybrid swarm that occurs where *B. glauca* and *B. mexicana* come together and consists of "combinatory and segregate forms apparently derived by hybridization from both species."

Allan (9) says, "That hybrids between species do occur, and that they are sometimes fertile, is now conceded on all hands. Certainly field evidence is circumstantial only, but it is often so overwhelming as to leave no doubt in an unprejudiced mind. . . . But the argument is still repeated that incapability of crossing outside its group is *the* test of a 'good' species, despite the fact that we have definite knowledge of incompatibility within species groups, and of the possibility of species taxonomically widely separated crossing and even producing fertile offspring." Among good species by all other standards than crossibility, he mentions *Nothofagus cliffortioides* and *N. fusca; Cordyline australis* and *C. pumilio; Senecio southlandicus* and *S. Hectori; Coprosma propinqua* and *C. robusta. Hebe salicifolia* is a polymorphic species with several clearly demarcated true-breeding geographical units which are known to hybridize when they meet. *H. elliptica*, however, is distinct and fairly constant, but it hybridizes freely with the preceding species. Allan notes that intergeneric hybrids in New Zealand are found only occasionally and they are usually pauciform and sterile, but the fact that intergeneric hybrids can be produced serves to indicate relationships. He mentions the following genera of Compositae which are linked by hybridization:

[5] Note that the authors of these species are experienced taxonomists.

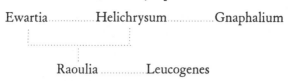

Ewartia............Helichrysum............Gnaphalium

Raoulia............Leucogenes

Burtt and Hill (102), who have made a study of the species of *Gaultheria* and *Pernettya* in New Zealand and nearby lands, have concluded that interspecific crosses are frequent and that intergeneric crosses are not uncommon. No botanist would think of throwing the species that are involved together as one large polymorphic species merely because of their ability to hybridize in nature. These entities constitute good morphological and geographical species except where they meet, and the hybrids do not destroy the species or the species concept. Following are the hybrid connections indicated by Burtt and Hill:

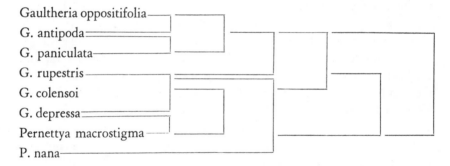

Gaultheria oppositifolia
G. antipoda
G. paniculata
G. rupestris
G. colensoi
G. depressa
Pernettya macrostigma
P. nana

Intergeneric crosses which are indicated include *Pernettya nana* × *Gaultheria rupestris*, and *Pernettya macrostigma* × *Gaultheria rupestris, depressa, antipoda,* and *colensoi.*

An interesting biological as well as taxonomic point results from the naming of hybrids. A large number of citations might be made, but a revision of the genus *Dracophyllum* by Oliver (492) will do. In this genus he recognizes nine hybrids and names them according to the International Code: for example, × *Dracophyllum erectum* (*D. prostratum* × *D. rosmarimifolium*). The geneticist, ecologist, or geographer wonders what kind of hybrids such names represent. Are they occasional F_1 plants? A clone of F_1 plants? Has Oliver observed and drawn a description to fit a few of the possible F_2 and backcross plants? Or does × *D. erectum*, for example, represent all possible variants in a hybrid swarm? We know from experimental crosses between species that the segregating hybrid progeny present a sometimes amazing variety of plants differing both quantitatively and qualitatively in their characters. Some characters may be clearly alternative and others show blending in various degrees. Furthermore, the interaction of genes from different species may result in characters not revealed by either species. As a consequence of these facts, one wonders how a taxonomist can draw up a description to cover a natural hybrid swarm and, if his description is based on one or a few hybrid forms, whether he realizes that he has entered on an almost endless descriptive task.

Introgressive hybridization.—Anderson and Sax (22) have pointed out an interesting phenomenon which has been described as "infection" of one species by genes of another. When one of two species is more abundant in the zone of contact between the two, the majority of crosses will take place between previous crosses, or their descendants, and the commonest parent. Such hybrids, which are three-fourths or more of one parent, will appear not clearly as hybrids but as somewhat unusual or extreme examples of the predominant parent. These two writers cite an example in which *Tradescantia occidentalis,* throughout its zone of contact with *T. canaliculata,* has undergone a mass infection from the latter. Anderson and Hubricht (20) have carried further the analysis of hybridization in these species and have proposed the concept of "introgressive hybridization" for the infiltration of the germ plasm of one species into that of another. In any situation where cross-fertile species meet, conditions are likely to be more favorable to one species and less so to the other, so that the final result of the hybridization depends upon a balance between the deleterious effects of the foreign germ plasm and its advantages in the areas where the hybridization has taken place or to which the hybrids spread.

Two species which have formerly had ecological or geographical isolation may overcome it by migration or, commonly, they may be brought together by disturbance of natural conditions by man's activities. What were formerly two distinct ecological habitats may be broken down, thus providing a wide series of local situations which are in various degrees intermediate between the two major habitat types. A common example is natural grassland and woodland which are modified so that they approach each other. The woodland is opened up by the cutting of trees, and the grassland may be invaded by shrubs and trees as the result of breaking the sod, overgrazing, etc. Two species, formerly kept to their habitats of woodland and grassland, may now meet and hybridize at many points, producing all types of hybrid combinations between the two parent populations. In such a situation, Anderson [6] has said that the hybrids are a better measure of the great

[6] In one of the Jessup Lectures, Columbia University, 1941.

variety of local ecological situations than any possible instrumental measurements the ecologists might make, and that the hybrids which are more
like the woodland species will survive in the situations that are more like
the woodland in just the degree to which they contain germ plasm and the

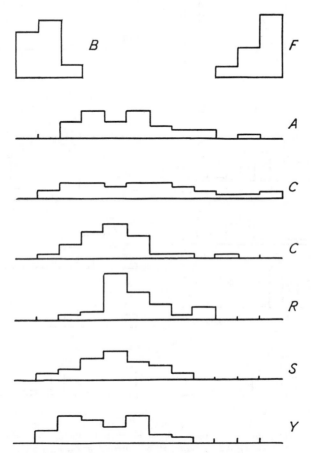

Fig. 47. Frequency distributions for the *Tradescantia virginiana-canaliculata* index. Thirty
plants for each collection. Pure *T. canaliculata* from station B; pure *T. virginiana* from station
F; the remainder are hybrid populations. Data from Anderson (14).

adaptations of the woodland species, and vice versa for the grassland species.

Anderson (14) developed a method which combines both qualitative and
quantitative data concerning species and their hybrids, and allows the scor-
ing of indexes on a frequency basis for different natural populations. This
method, which is illustrated in Fig. 47, helps determine the dynamics of
hybridization and the frequencies as well as the range of variation, and

indicates that in certain cases the resultant variability is so widespread throughout the species that it must constitute the chief raw material for

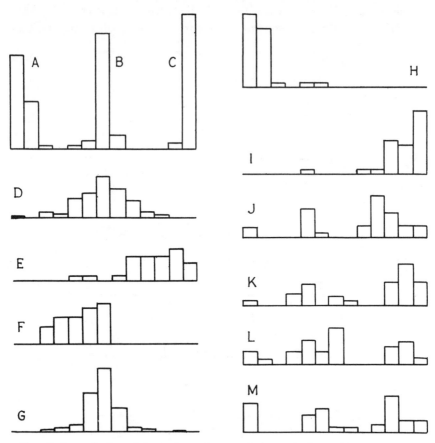

Fig. 48. An application of the Anderson frequency-index method to *Solidago sempervirens,* *S. rugosa,* and hybrids between them, including X *S. asperula.* The histograms of natural populations represent plants from a reclaimed salt marsh at Cambridge, Massachusetts, where the species grow in proximity and hybridize. *A. S. sempervirens. B.* F₁ hybrids. *C. S. rugosa.* *D.* 108 F₂ hybrids. *E.* 23 backcrosses to *S. rugosa. F.* 17 backcrosses to *S. sempervirens.* *G.* Frequency distribution of 98 herbarium specimens labeled X *S. asperula* and recognized as crosses between *S. sempervirens* and *S. rugosa.* Histograms *H–M* are for various populations from one locality. *H* and *I* indicate that the species (compare with *A* and *C*) are not pure but have undergone some introgressive hybridization. *J, K, L,* and *M* indicate that backcrosses are somewhat more frequent with *S. rugosa,* the more numerous species. *Solidago asperula* in nature apparently consists of F₁, F₂, and backcrosses. Data from Goodwin (291).

natural selection. The illustration is drawn from *Tradescantia virginiana,* a shade plant, and *T. canaliculata,* a sun plant. When these species meet in open woods or along roads and forest borders, they hybridize freely. The diagram shows the score for *T. canaliculata* (B) and *T. virginiana* (F) colonies, and the other letters indicate hybrid populations. In this study six well-differentiated morphological characters were employed in producing the indexes. The characters for *T. canaliculata* were valued at zero, and the corresponding characters for *T. virginiana* were given values from 1 to 3. By analyzing their characters it thus becomes possible to give every plant in a colony a total index value that results from the degree to which it has char-

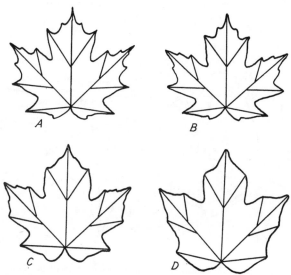

Fig. 49. Average leaves drawn from statistical data. The two top leaves represent *Acer saccharum* from Petersham, Massachusetts (*A*) and Nesbit, Michigan (*B*). The lower-left leaf represents *Acer nigrum* from Michigan (*C*). Leaf *D* is a hypothetical one determined by statistical prediction, and may be called *X*. The hypothesis is that *Acer nigrum* equals *Acer saccharum* plus *X*. The hypothetical leaf *X*, when compared with leaves of other sugar maples, was found to match well with *Acer saccharum* var. *Rugelii*. These leaves are redrawn from (19).

acters of one or the other of the species. The frequency histograms of its index values thus become a measure of the hybridity of the population and its nearness to one or the other of the species.

Goodwin (291) used the Anderson index-frequency method in a study of *Solidago sempervirens* and *S. rugosa* and their hybrids, which are known as *Solidago* × *asperula*. Known stocks of the species, their F_1 and F_2 hybrids, backcrosses, and various wild populations were all analyzed. Goodwin concluded that the polymorphy of *S. sempervirens* and *S. rugosa*, as sometimes encountered in nature, results from the fact that they hybridize where their natural ecological isolation is broken down as the result of disturbance. Also, from a study of natural populations and an analysis of herbarium sheets which taxonomists have designated as *Solidago asperula*, he decided that this "species" includes F_1 plants, backcrosses with either *S. sempervirens* or *S. rugosa*, and possibly F_2 plants. Variable populations develop in nature and result in a sort of reversal of evolution because the distinctiveness of the two species is lost in their fusion.

Other studies of this type are appearing in the literature. Larisey (408) has analyzed *Baptisia leucantha* and *B. viridis* and their hybrids in a region in Texas where they meet. The name *Baptisia fragilis* is assigned to the hybrids. Riley (528) studied *Iris fulva, I. hexagona* var. *giganticaerulea,* and

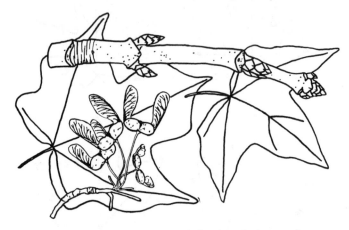

Fig. 50. A reproduction of Trelease's original drawing of *Acer saccharum* var. *barbatum* (which equals *Acer saccharum* var. *Rugelii*). The lower leaf in this drawing compares well with *X*, a putative parent of *Acer nigrum*, the other parent being *Acer saccharum*. Drawing copied from (19).

their hybrids, and later published (529) on the double introgressive hybridization of *Tradescantia canaliculata* (g = 12) into *T. hirsutiflora* (g = 12) and of *T. paludosa* (g = 6) into the tetraploids. Delisle (183) has made a study of species of *Aster* by the index-frequency method.

Introgressive hybridization, such as the mass infection of *Tradescantia occidentalis* by the gene or genes from *T. canaliculata* which result in a tuft of hairs at the sepal tip, tends to change whole populations and break down "specific" characters. The same phenomenon has more drastic effects when the exchange of germ plasm is more complete; it may result in the swamping of the two species. On the other hand, apparently hybridization may not be so complete as to affect most of the parental species populations, and selection may act on the hybrid swarms so as to elevate them to a distinctive and, possibly, a specific status. Woodson (713), who revised the species of

Apocynum of eastern North America, thought that *A. medium* was a hybrid of *A. cannabinum* and *A. androsaemifolium*. This hypothesis was tested by Anderson (13), who grew progeny of the putative hybrid and of its parents. *A. cannabinum* and *A. androsaemifolium* bred true, but *A. medium* produced a variable lot of plants, including some indistinguishable from the parents. Anderson emphasized that for *A. medium* to gain true specific status it would have to be further acted on by natural selection to remove some of its variability and semi-sterility, but that there is no reason why species might not arise in this manner.

An especially interesting case is afforded by the probable hybrid origin of *Acer nigrum*. This result, if true, not only shows that a species may arise through hybridization without amphidiploidy, but also emphasizes the importance of population studies and the appearance of new characters as the result of hybridization. By means of the mass collection technique, Anderson and Hubricht (19) [7] find that estimates of discontinuity between species are provided even when the discontinuity is small, and that a statistical comparison of quantitative characteristics is often useful. Their study of *Acer nigrum* and *A. saccharum* shows that these two sugar maples remain distinct throughout their ranges, although frequently they grow in juxtaposition. From a large series of measurements of many characters of the leaves of these two species, they were able to draw an "average leaf" of each one. Furthermore, their data provided for the construction of an "average leaf" of the hypothetical parent which, with *A. saccharum,* gave *A. nigrum*. Having thus produced a drawing of the unknown putative parent of *A. nigrum,* they found that it compares well with *A. Rugelii*.[8]

I cannot resist the temptation to speculate further concerning these interesting sugar maples. Hybrids occasionally show characters which neither of the parents have, and such characters are sometimes referred to as "reversions." It is accepted that the interaction of genes in a hybrid may result in characters that are not shown by the parents, but whether they are new characters or reversions is usually entirely unknown. The character possessed by *A. nigrum* and not by *A. saccharum* or *A. Rugelii* is foliar stipules; at times these are rather elaborate and almost like distinct pinnae. If this

[7] See (17) and (234) for the mass collection technique and its application.

[8] Anderson and Hubricht employ the following nomenclature for these maples: *Acer saccharum* Marsh, *A. saccharum* var. *nigrum* (Michx. f.) Britt., and *A. saccharum* var. *Rugelii* (Pax) Rehder. In view of the intrinsic barriers to interbreeding between the first two, it is obvious that they should both be accorded specific rank, as already recognized by some authors for *A. nigrum*. Furthermore, they have morphological discontinuity in leaf form, although it is slight, and other characters distinguish them. The evidence is not so detailed, but it appears that *A. Rugelii* is also a good species.

character is a reversion it will be of phylogenetic interest. It could be explained by the following assumptions: (1) that *A. saccharum* and *A. Rugelii* are species that have come from one ancestral stock; (2) that in the evolution of the species and their isolation the complementary factors necessary for the development of foliar stipules became separated so that neither species contained all of them; and (3) that in the production of the hybrid *A. nigrum* the necessary complementary factors sometimes occur together again, with the result that this species, alone of the three, shows foliar stipules. If such an hypothesis should gain cytological support, it would indicate that foliar stipules are vestigial, and possibly that pinnate leaves, as in the *Negundo* section of Acer, are primitive.

Clones.—Biologically, although it may consist of a large number of individuals, a clone is a single plant in the sense that it is genetically constant. By various non-sexual methods of reproduction one organism becomes a colony or a population occupying a more or less large area. At first, one would scarcely think that a clonal population would ever be given specific rank equal to that of variable panmictic populations, but this is the case; when the population is large and distinctive, it may be justified. Several examples will be mentioned.

Whitaker (692) found that *Robinia hispida* and *R. Boyntonii* are sterile triploids. He says, "*Robinia* is one of those genera in which polyploids and hybrids, even though they may be sterile, are factors to be reckoned with in speciation because of the fact that these forms are propagated very readily and rapidly by vegetative means, and are thus able to maintain themselves." *Lilium tigrinum,* according to Chandler, Porterfield, and Stout (124), is a triploid which has functioned successfully on a strictly somatic basis over a known period of at least 130 years. *Oxycoccus gigas* is a newly described allo-hexaploid which, Hagerup says (317), was probably formed in early postglacial time and has survived ever since in certain bogs in Denmark. *Lysimachia Nummularia* is widely distributed through northern and central Europe but rarely produces seeds. By bringing together plants from widely separated places, Dahlgren found that seeds were produced by 28 reciprocal crosses. In nature this plant consists of local colonies, the members of which all belong to the same clone. Being self-sterile because of incompatibilities, the species can produce seed only when members of different genotypes are crossed.

One of the most interesting situations in American taxonomy has resulted from the description of clonal irises from the Mississippi delta area. In 1927 Small described the first of his new irises from Louisiana, *Iris vini-*

color, and in 1929 he published six additional species: *violipupurea, giganti-caerulea, chrysophoenicia, chrysaeola, atrocyanea,* and *miraculosa. Iris mira-culosa* indeed! Two years' continuation of these studies were followed by the publication of 76 additional species by Small and Alexander (595), bringing the total species of *Iris* in southern Louisiana to 85. No wonder Small was led to say (according to Viosca, 661) that southeastern Louisiana is the "Iris center of the universe" and that the "Lower Mississippi Delta natural *Iris* field constitutes the one most spectacular botanical and horti-cultural discovery in North America from the standpoint of a single genus within such a limited area." He offered as a partial explanation of these irises the fact that the ancestors of the species had migrated down from the uplands with the withdrawal of the Mississippi Embayment during the Tertiary, and that the progenitor species had been lost in the highlands and during their migration had become extirpated. Other naïve explanations were not long in appearing; in fact, Cooper [9] suggested that the delta of the Mississippi was apparently the receiving ground for the irises of the whole river system that for countless ages had been floating down from the upper reaches of that river and its tributaries. This may be called the "flot-sam and jetsam" theory; it needs no further comment because of its patent ecological absurdity.

Viosca (662) came to the conclusion that the remarkable colonies of iris were clonal hybrids, and that instead of being ancient species that had come to the area with the Tertiary withdrawal of the Mississippi Embayment, they were probably of recent origin. The species from the area known before Small's epidemic of new species had rather strict habitat requirements. Their ecological isolation was overcome in numerous places by the dis-turbances of local conditions through the formation of drainage ditches, impoundments, etc.; as a result, they intermingled, hybridized, and through vegetative reproduction produced sometimes rather extensive clones. Using the Anderson index-frequency method of studying hybrid population char-acteristics, Riley (528) examined two iris species and the hybrid colonies between them and found that the hybrids represented certain of Small's new species. Thus, in one case at least, Viosca's supposition was proved.

The taxonomic problem in such a case is puzzling. If there were a single clone of F_1 plants, as in *Oxycoccus gigas,* no one would object to the specific designation; but the 83 iris "species" described by Small and Alexander for various clones have by no means disposed of all the forms of iris in this one area. Winge (705) apparently expresses the general opinion when he

[9] Editor of the magazine in which Viosca (661) published.

says that agamospecies should always be distinguished from sexual species, and that it is usually a waste of time to describe and name such unlimited clones.

Panmictic or sexually reproducing populations have a collective genotype and are composed of a large number of individual genotypes. The gene constitutions of the individuals are recombined every generation, and normally no one genotype occurs with a high frequency and certainly not exclusively. The opposite condition characterizes non-sexual organisms or organisms with obligate self-fertilization which, for all genetic effects, are also clones or pure lines and comparable to asexual populations. In many species of plants the asexual condition has been arrived at through loss of sexuality. One common form is parthenogenesis, in which the egg develops without fertilization. Although apomixis is well known in several groups of the Compositae (*Hieracium, Crepis,* the Madiinae, *Taraxacum* [10]), only *Poa* will be discussed. Tinney and Aamodt (628) have found that apomixis is the predominant type of seed production in the bluegrass *Poa pratensis,* and the progeny closely resemble the female parents. When progeny were grown from open pollinated plants, they found that only 1 or 2 per cent· varied from the characters shown by the mother plant. These variants were attributed to one or more of the following processes: (1) occasional sexual reproduction, (2) haploid parthenogenesis, (3) fertilization of a diploid egg, (4) mutation. Tinney and Aamodt believe that the normal constancy results from diploid parthenogenesis, i.e., that the embryo sac develops from a somatic cell of the nucellus, and the diploid egg develops an embryo without fertilization.

Müntzing (478) described in *Poa alpina* the first case of apomixis known for grasses, and Kiellander (390) reported apomixis for *Poa palustris.* Engelbert (213) has shown that four species of *Poa* are apomictic by pseudogamy. In pseudogamy pollen is apparently necessary as an activator, but fertilization does not take place. Experimenting with *P. arctica, P. alpina,* and *P. alpigena* from Greenland, and *P. alpina* and *P. pratensis* from Canada, and trying artificial self- and interspecific cross-pollinations, he found that in all cases the progeny resembled the mother plants. He emphasized the fact that the apomictic races of *Poa* are really clones and that the initial natural selection can be the final one. He also was aware that the taxonomy of *Poa* cannot be worked out in a satisfactory manner until their types of speciation are understood. Brown (97) found certain species of *Poa* to have regular chromo-

[10] Christiansen (143) has recently added to the already burdened genus *Taraxacum* the description of 22 new species from the group *Vulgaria,* and they all reproduce apomictically!

some numbers: $S = 28$ in *P. Wolfii, P. cuspidata,* and *P. sylvestris,* and $S = 42$ in *P. arachnifera.* But *Poa pratensis* had an irregular series of numbers ranging from about 41 to 64. *Poa pratensis* apparently contains some euploids that have sexual reproduction as well as a variable series of apomictic aneuploids, so it is no wonder that bluegrass is highly variable and taxonomically puzzling. Some of the aneuploids have a greater agronomic value than others and it may be expected that by means of colchicine amphidiploids that are fertile may be formed.

Ecads.—An ecad is a habitat form; it is a modification from the usual vegetative structure induced by the action of an unusual environment on the soma. Such a modification comes strictly within the inherent limits of tissue plasticity, has no genetic status, and should be given no taxonomic recognition. That such forms have been called species may be surprising, but when taxonomy is done in the herbarium by persons who have not seen the plants in the field, such mistakes are relatively easy if species lines are finely drawn. No taxonomist would retain an ecad as a species if he knew its real nature; however, such modifications are sometimes difficult to detect and they constitute one of the constant annoyances for taxonomists. In many cases the problem of whether a certain form is an ecad or has a genetic basis can be determined only by transplant and genetic studies (146). One example will suffice. In a study of *Rubus squarrosus,* Allan (9) found that var. *pauperatus* is an inconstant habitat modification, but that var. *subpauperatus* has a genetic basis and is a true-breeding unit that is allied, however, to *R. schmidelioides,* not *R. squarrosus.*

23.

Isolation

It is necessary to recognize a distinction between evolution, which is interpreted to include all development of qualitative and quantitative genetic differences, and speciation, which is also evolution but consists of the attainment of the specific status. Although there are several concepts of the species, only one is capable of strictly scientific delimitation: species are naturally evolved populations with distinct genetic constellations the members of which interbreed but do not usually crossbreed with the members of other specific populations. As is pointed out in the discussion of the species problem, many species in the literature do not meet this criterion of specificity, and, for a variety of reasons of which many are practical, the practice of taxonomy in several fields is still something of an art and cannot everywhere employ the strictly scientific method.

When hereditary traits are of such a nature that segregation takes place in the progeny of a hybrid—as in single gene and chromosome changes—the differences, as pointed out by Dobzhansky (191), cannot be swamped by crossing. When hereditary characters result from the cooperation of several genes—races and species usually differ from one another by several genes and chromosomal alterations—the preservation of distinctive genetic constellations can result only from isolation because interbreeding breaks down these systems. Genes and chromosomes, of course, are fully preserved as such and do not lose their identity in the hybrid, although specific patterns may be disarranged.[1] It follows, then, that isolation plays no role in the lowest stage of evolution where hereditary variability is increased by genic and chromosomal change.

In the development of subspecific status, as in ecotypes and geographical subspecies, partial isolation is important. Mutation, migration, population-size restriction, genic drift, selection, and partial isolation interact in the

[1] Excluded are changes such as segmental loss or interchange which may occur.

attainment of subspecific status. Such groups, in order to be distinct, must have an allopatric distribution with geographical or habitat isolation, for they have not developed reproductive isolation. When reproductive isolation has been attained and the specific level has been reached, related species can have a sympatric distribution because there is little or no exchange of genes between the groups. It will be seen that barriers to interbreeding are partial or complete and temporary or permanent. It is only relatively recently that adequate attention has been given to the important question of isolation types and mechanisms. A brief review of the various classifications follows.

In a cytological monograph on *Tradescantia* (22), Anderson and Sax mention that the species are able to maintain themselves because of such interbreeding barriers as differences in geographical distribution, in habitat preference, in blooming time, in chromosome numbers, and in cytological characteristics other than number. Huxley (371) listed four groups of barriers to interbreeding: geographical and ecological isolation, physiological and genetic isolation. Hogben (348) added psychological isolation to these four and pointed out that they are fundamentally either extrinsic or intrinsic in their mechanism. Extrinsic species distinctness is guaranteed by topographic barriers to their intermingling. It is the usual opinion of those whom Hogben characterizes as "traveler-naturalists," as exemplified by Darwin, that the operation of natural selection upon separated populations produces peculiar hereditary characteristics and, in time, speciation. Intrinsic barriers to interbreeding are the barriers that originate within the genetic make-up of the organism. Following Hogben's concepts, we can make the following outline:

    A. Extrinsic isolating mechanisms:
        1. Geographical isolation.
    B. Intrinsic isolating mechanisms:
        2. Genetic isolation, per se.
        3. Ecological isolation.
        4. Physiological isolation.
        5. Psychological isolation.

Worthington (715) emphasized that isolating mechanisms sometimes have a joint action, and that as a consequence investigators should not have a too specialized approach to isolation. His classification of isolating mechanisms follows:

    A. Geographical isolation.
    B. Ecological isolation.

C. Reproductive isolation.
  1. Reproductive behavior (psychological isolation).
  2. Isolation resulting from changes in the morphology of reproductive organs (lock-and-key theory).
  3. Incompatibility of gametes.
D. Changes of chromosomal mechanism (genetic isolation).
E. Physiological isolation.
  1. Physiological races as in parasites, phytophagous insects, etc.
  2. Types C and D above.

Dobzhansky (191) organizes his consideration of the subject as follows:

A. Geographical isolation.
B. Reproductive isolation.
  1. Incompatibility of parental forms.
    a. Parental forms usually do not meet.
      (1) Ecological isolation.
      (2) Seasonal or temporal isolation.
    b. Parental forms occur together but hybrids do not occur or do not reach maturity.
      (1) Psychological isolation.
      (2) Mechanical isolation.
      (3) Failure of fertilization.
      (4) Inviability of hybrids.
  2. Hybrids develop but do not reach sexual maturity, or gametes give rise to inviable zygotes.

These authors agree that geographical isolation is the only mechanism that is wholly extrinsic. Interbreeding does not exist because populations live in different territories. Such species when brought together may or may not display reproductive isolation. In the following pages geographical isolation will be given first attention because historically it was the first to be recognized and stressed.

Geographic isolation.—Jordan's Law (381) is as follows: "Given any species in any region, the nearest related species is not likely to be found in the same region nor in a remote region, but in a neighboring district separated from the first by a barrier of some sort." Jordan did not claim originality for the idea, and he used as a text for his paper the following quotation from Moritz Wagner written in 1868: "For me, it is the chorology of organisms, that is to say, the study of all the important phenomena embraced in the geography of animals and plants, which is the surest guide to the study of the real phases in the process of the formation of species."

Jordan was aware that a barrier between two parts of a species population provides the necessary isolation to prevent free interbreeding and the swamping of incipient lines of variation. Regardless of how a portion of the species happens to be on one side of the barrier away from the bulk of the species, the isolated population is subjected to a different selection and can develop along lines which are independent of the other portion. The adaptive characters that arise (in either population) are due to natural selection acting on random mutation; the non-adaptive characters that in time come to distinguish the two populations develop because of the isolation. Because of the vicissitudes of environmental change and migration and the necessity for isolation, Jordan believed that the two most closely related species do not live in the same area, but in adjacent areas separated by a barrier. Furthermore, the two *most* closely related species will be found in contiguous territories, because when species become more widely disjunct, the time necessary for the separation to develop is sufficient also, for further evolution to take place.

Jordan's paper opened up a debate which found abundant participants among botanists and zoologists. Lloyd (422) immediately replied that he thought the case would receive scant support from the botanists and that it was easier to find exceptions to the rule than facts in support of it. After citing several cases of related species in the same region, Lloyd said, "The general law as stated by Jordan . . . would be more in harmony with the facts in the case as understood by the botanists if stated in the converse form." Lloyd obviously was not speaking for all the botanists, for Abrams (2) wrote, "Whatever the cause [of the origin of species], we do maintain that the evidence in favor of isolation as an important factor in the *perpetuation* of closely related species is almost overwhelming in plants as well as in animals. Any theory of evolution which will not allow for this fact cannot possibly prevail." Abrams claimed that Lloyd's examples must stand up under two questions: (1) Are we dealing with the *most closely* related species? (2) Are the two species growing associated under the same conditions? When viewed in this way, some of Lloyd's examples offer excellent support of Jordan's Law, rather than refutation.

Leavitt's criticism of Jordan's Law (410) brings out clearly one aspect of the argument that is suggested by the papers of Lloyd and Abrams. According to Leavitt, in North American Orchidaceae there are so many cognate pairs of species with uniform or widely coincident ranges that Jordan's Law loses all force. He lists such pairs as *Habenaria ciliaris* and *H. Blephariglottis,* *H. psycodes* and *H. fimbriata, H. orbiculata* and *H. macrophylla, Spiranthes cernua* and *S. odorata, S. laciniata* and *S. vernalis, S. Beckii* and *S. gracilis, Cypripedium pubescens* and *C. parviflorum, Pogonia verticillata* and *P. affinis.* When we realize that Leavitt considers only whether the two species of a pair live in the same state, it is clear that he missed the real meaning of Jordan's Law and of Abrams' criticism of Lloyd. Geographical isolation does not require a great disjunction, and ecologic isolation can occur within a very local area.

In 1926 Jordan returned to the topic of isolation and speciation. By this time biologists were pretty well agreed that his law was a satisfactory statement of the conditions for extrinsic isolation. At this time Jordan said, "We know nothing of evolution *in vacuo,* of change in life unrelated to environment. . . . We know of no way in which organisms become adapted to special conditions except by the progressive failures of those which do not fit. No organism has escaped or can escape the grasp of selection. In a like manner, the world being full of physical barriers, no organism escapes biological friction which prevents uniformity in breeding. There must be some degree of 'raumliche Sonderung' even in a drop of water. As Wagner truthfully observes: 'Ohne Isolirung keine Arte.' " In an earlier paper (382) Jordan had elaborated his ideas of evolution. He was fully aware that habitat modifications have nothing to do with speciation, and he said that a large part of the work of students of species would have to be concerned with purging the system of species and subspecies founded on characters resulting from reaction to the environment. He claimed that a species can change on its own grounds little by little with the lapse of time and the slow alteration of conditions of selection. Under migration, or with the erection of barriers within the area of a species, "species are torn apart by obstacles as streams are divided by rocks. . . . Nothing is secure against the tooth of time." He knew that topographic segregation can bring about the separation of species or subspecies in precisely the same manner as other methods of geographical isolation. In a similar manner, lines of segregation may be set up within species on the physiological basis of a certain type not breeding freely with those of other physiological types. So we come finally to the conclusion that Jordan intended to include all forms of isolation when he said that the two most closely related species of a genus do not live in the same region, and that by region he apparently meant an interbreeding population.

In addition to considering Jordan's Law, it is useful to examine certain aspects of geographical isolation and to present a few examples. Discontinuity can be defined as a distance between organisms that is greater than the maximum chance for crossbreeding. Obviously, such a distance is relative. Practically, effective partial isolation may occur at much shorter distances than the one defined. If seed dissemination can take place across a wider distance than crossbreeding, the populations are nonetheless discontinuous, for each group of organisms constitutes a separate breeding unit. Among insect-pollinated plants, for example, discontinuity occurs when one group of plants is more widely separated from another than the flying range of the insects. For wind-pollinated plants the average distance is probably much greater, but the chances of two plants being cross-pollinated must decrease approximately with the square of the distance, except under the action of regular wind currents. Among certain animals, such as snails, opposite ends of a hedgerow may provide sufficient isolation for a genetic difference to appear and persist. For certain strong-flying birds or wide-ranging mammals, hundreds of miles may not produce isolation. Distance in such cases, however, depends on the activity range during the mating season and not during the migrating period.

Oceans, lakes, and rivers are more or less effective barriers for land organisms. Land is generally a barrier to aquatic organisms. The continents separate the oceans; even a narrow isthmus may be an effective barrier. Land-locked lakes are for aquatic organisms what islands are for terrestrial organisms. Land barriers to land organisms are usually thought of in terms of major physiographic features, but they obviously grade into barriers that, because of their smaller extent, are thought of as edaphic. Water barriers to water organisms are numerous. Within the length of a stream there is a change of water types from the headwaters to the mouth that may form a series of barriers. The development of a falls may provide an upstream but not a downstream barrier for certain forms even within a certain water type. Fresh, brackish, and saline waters, when a stream enters the sea, provide barriers, and a fresh or brackish water type may become isolated in a series of estuaries along a sinking coast. In deeper bodies of water, the depth of activity of an organism may isolate it from related forms living at different depths.

Distance as a barrier to interbreeding has been emphasized by several authors. Huxley (371), summarizing some of the evidence, concluded that distance promotes geographical differentiation and that full speciation probably requires a more complete or a different type of isolation. A widespread and continuous population may develop two or more gene centers. Partial biological discontinuity results in geographical subspecies; this is well known among mammals and birds. In a continuous although wide population, there can theoretically be a complete gene exchange; but when the area is great the periphery may show differences from the center of the population—in both qualitative and quantitative aspects—or there may be two or more centers for certain gene patterns in a geographical replacement arrangement. This may be the result of distance, differentiation taking place in a partial population more rapidly than the gene flow between this part and the rest of the population. Such cases, however, are usually not just a matter of distance but are associated with differences in selection. If selection results from an environmental gradient, the concept of the cline is applicable; if such a difference is associated with areas of peculiar microclimate or other ecological conditions, the subspecific groups are called ecotypes; if the differences are on a regional basis they are usually associated with regional climates and the groups are called subspecies.

A clear example of differentiation that seems to be related solely to distance between populations is reported by Thompson (624) for the Johnny Darter, a small fish found in Illinois. It occupies small tributary streams, and headwater populations that are only short distances apart overland but may be tens, hundreds, or even thousands of miles apart by water. Distance, of course, is essentially linear for river organisms such as these fish. Thompson noticed that large, strong-swimming fish apparently migrate enough to keep their populations uniform through interbreeding, but that small, weak-swimming fish show differentiation with distance. In the Johnny Darter, differences in mean fin-ray numbers apparently are due not to selection, for the habitats are strictly comparable, but to distance with which the differences can be correlated. Although this phenomenon is probably fairly common, the literature contains almost no evidence of a quantitative nature.

A botanical example is provided in the genus *Ajuga,* which contains about 90 species. Turrill (653) has found that *A. chia* and *A. Chamaepitys* are morphologically more closely related to each other than they are to any other member of the genus. These two species occupy separate yet contiguous areas; the transitional population has intermediate characters. Turrill says that the complex arose in the Levant and Aegean areas and spread westward and northward, becoming less diversified by purification in the process of extension. By purification of the population, he apparently means that the plants which have migrated farthest from the center of origin of the group have become less variable because of the inability to mix back thoroughly with the parent population, *A. chia.*

Although numerous examples of discontinuity between closely related species are presented in the section on areography, two cases will be discussed here. One of the most interesting situations on record in which isolation has existed for a long time and close relationship still remains is found in the *Eucalyptus* forests of Australia. According to Wood and Baas-Becking (712), the native genera probably arose in the southwestern portion of the continent, and distribution occurred eastward from that center. After this dispersal had taken place, the eastern and western portions of the continent were virtually separated in the late Cretaceous by a vast mediterranean sea;

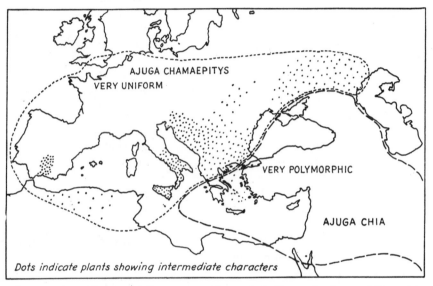

FIG. 51. *Ajuga chia* is essentially an eastern plant and *A. Chamaepitys* is essentially western. Intermediate populations consist of individual characters and character combinations. Apparently the group arose in the Aegean and Levant areas and spread eastward, but especially westward and northward, becoming less diversified by purification in the process of migration, which was completed only after the Ice Age. Map redrawn from (653).

this provided two centers in which endemic species could originate. The gulf regions in South Australia were uplifted in Pleistocene times; hence migration from the eastern and western centers has provided the present flora of South Australia. However, this intermediate country is neither climatically nor edaphically suited to *Eucalyptus* forest species, so they remain isolated today although a broad land connection exists. The high, sclerophyllous *Eucalyptus* forests of New South Wales and southwestern Australia live under conditions of climate and soil that are remarkably similar, and the vegetation is similar both physiognomically and floristically. The characteristic plants of the two regions, playing the same roles in the structure of the two forests, not only belong to the same genera but are closely allied, vicarious species-pairs. Table 25 lists only the abundant plants; the list could

TABLE 25.—Pairs of species that are abundant in the separated areas of the *Eucalyptus* forest of Australia (Data from 712)

Blue Mountains, *New South Wales* *In Southeast*	*Mount Lofty Ranges,* *South Australia* *In Southwest*
Eucalyptus piperita	E. obliqua
Persoonia salicina	P. juniperina
Leptospermum flavescens	L. scoparium
Banksia spinulosa	B. marginata
Hakea dactyloides	H. ulicina
Isopogon anemonifolius	I. ceratophyllus
Acacia discolor	A. myrtifolia
Pultenaea scabra	P. daphnoides
Phyllota phylicoides	P. pleurandroides
Lissanthe sapida	L. strigosa
Tetratheca ericifolia	T. pilosa
Epacris grandiflora	E. impressa

be extended to include other pairs among occasional and rare species that occur separated in the two regions.

This example is useful because it is typical of a common situation in botany. Botanists have no inclination to treat such geographically isolated forms, despite their obvious close relationship, as anything but good species when there is morphological distinctness. So far as I know, there are no data concerning their reproductive isolation, but even if certain pairs proved to be crossable on transplanting, they would still be regarded as species by the botanists.

The other example concerns one of the tarweeds, *Hemizonia paniculata*, in which isolation has played an interesting role. This species has an interrupted range on the California coast from above Santa Barbara to Riverside. Population *A*, south of the barrier, and population *B*, north of the barrier, constitute good morphological subspecies, according to Hall (321), and they retain their characteristics when grown together in the same garden. The southern subspecies consists of a slender, small-flowered type in Baja California, and a more robust form northward in California that is differentiated without discontinuity. In the northern subspecies there has been a differentiation into 8-rayed and 13-rayed populations, with hybrid swarms at the spots marked ✕ on the map. Hall explains the evolution of *Hemizonia paniculata* as consisting first of the separation of the species through the

FIG. 52. The distribution of the forms of *Hemizonia paniculata*. The discontinuity between the southern (*A*) and northern (*B*) subspecies is found in both area and morphology, the morphological differences persisting when the plants are grown together in the same plot. In subspecies *A* there can be detected two hereditary groups (*a* and *b*) which differ principally in size. In subspecies *B* (detailed map in the upper right corner) the hollow circles locate stations for flowers with 8 rays, and the black disks mark stations for plants with 13 rays. In these areas there is no intergradation, but at the stations marked ✕ the ray numbers range between 8 and 13, sometimes approaching one number more than the other.

The history of this group appears to be as follows: (1) the development of a barrier producing the discontinuity between subspecies *A* and subspecies *B;* (2) further evolution in each group; (3) a segregation of 8-rayed and 13-rayed populations in subspecies *B;* and (4) subsequent hybridization between 8-rayed and 13-rayed plants. The illustration is redrawn from (321).

development of a barrier and geographical discontinuity, with subsequent differentiation into subspecies *A* and *B*. Since then the southern subspecies has developed a sort of cline, with types *a* and *b* intergrading because they have only partial isolation. In the northern subspecies there developed distinct 8- and 13-rayed populations under temporary isolation; these now show hybrid swarms where they subsequently came together. From the taxonomic point of view, it would be just as reasonable to call the northern and southern populations species and their internal spatially arranged populations subspecies.

The problem as to the specific or subspecific nature of certain closely related forms is well illustrated by Klauber's study of the snakes of the genus *Rhinocheilus* (397). Ranging from California eastward to Kansas and Texas and southward into Mexico, these snakes consist of four forms which Klauber has designated as subspecies. Two of them have geographical isolation (*antonii*, in Mexico along the eastern coast of the California gulf, and *tessellatus*, centering in Texas) and two of them occupy the same area in southern California and Arizona (*clarus* and *lecontei*). Smith (599) has pointed out that if Klauber's hypothesis is correct—that *lecontei* and *clarus* were once territorially separated—and they are now spreading through each other's territories without merging into an intermediate pattern, the forms must have been sufficiently well differentiated so that interbreeding is now largely inhibited. Since Klauber shows that *clarus* and *lecontei* are clearly separated and hybridization is infrequent, Smith is perfectly logical in concluding that the forms represent species and not subspecies, as stated by Klauber. If the species should be hybridizing to the extent that they are blending into one race, then, as Smith says, "the transition is not from a specific status of each form toward a sub-specific status, but rather toward a complete loss of practical identification and separation." The subspecific concept is useful only for allopatric forms that blend at their junctions but do not swamp each other.

Partial isolation, with respect to the phenomenon of local populations and close inbreeding, deserves special consideration. Wright (717) says that one possibility of a change in gene frequency is exhibited in the tendency toward fixation of one or another chance combination of genes in closely inbred strains. He notes that the effect depends upon the accidental rather than selective factors which determine the individuals that become the parents of the next generation. As Wright expresses this, it requires $2N$ gametes to reconstitute a population of N mature individuals. When N is reasonably large there would appear to be only a slight amount of variability. However, he says, variance of sampling is accumulative, and in the course of geologic time these chance variations may be expected to bring one allele or the other to fixation, even in large populations and without the action of other factors. The rate of fixation is $1/2N$ per generation, and this process, occurring independently in all loci, may bring about any degree of differentiation of isolated populations with respect to indifferent characters. In large species the condition of random mating among all individuals is never realized. Although there is continuity throughout the range of the species, there is restriction of probable mates to a small surrounding territory. Parents, grandparents, and more remote ancestors trace to everwidening territories so that for large populations, when the effective local population in an elementary territory numbers thousands, the results are almost those of random mating. If the effective population of an elementary territory numbers only a few hundred, Wright continues, there may be considerable fluctuating local differentiation such as Anderson (15) has noted for *Iris*. But to approach fixation of different alleles in different regions, the elementary population number must be in the tens rather than in hundreds.

Distinctive local populations can develop within a species because of space factors entirely independently of habitat differences. When a species spreads over several habitat types, however, much of the differentiation is the result of selection of adapted types, and gene combinations increase their frequency, forming adaptive peaks for the different environmental situations.

Ecological isolation.—Within a geographical area there are a number of local habitat types that are due to the action of local topography, microclimate, soils, vegetation, etc. As a consequence of the mosaic-like or alternating pattern of ecological differences, closely related species frequently occur in the same geographical area. It is a comon assumption, when such species hybridize only where they meet at ecological boundaries, that they are kept apart by the spatial pattern which results from their different ecological requirements. Although this phenomenon approaches geographical isolation, it is on a smaller scale and results fundamentally from intrinsic differences that control the habitat requirements. As well known as this phenomenon is, it has received little attention; Dobzhansky (191) notes that "investigations directed towards ascertaining to what extent the ecological and seasonal isolations are actually responsible for the maintenance of separation between species are . . . very rare," and that a genetic analysis of this type of difference between species or races has not been made as far as he knows.

Anderson and Sax (22) and Anderson and Hubricht (20) say that ecological isolation is one of the mechanisms which keep interfertile species of *Tradescantia* from being submerged through hybridization. For example, *T. humilis* is characteristic of shady places and *T. canaliculata* of open places; their hybrids are confined to a narrow zone where such habitats are contiguous, such as near the borders of woods. *T. canaliculata* grows characteristically in hot, dry places, often on limestone outcrops in the northern middle western states; throughout the same region *T. subasper* var. *typica* is for the most part confined to acid soils in shade or semi-shade. Over a wide area these two species are found within short distances of each other, but only rarely are they close enough for hybrids between them to occur.

Salisbury (541) published several cases of pairs of closely related species living in the same general area and rather sharply isolated because of habitat requirements. He mentions that *Quercus robur* occurs on heavy calcareous soils, whereas *Q. sessiliflora* is native on the London clay and on light types of soil deficient in carbonates and exchangeable bases. When a woodland extends across a boundary between two such soils and both species are present, there occurs a series of hybrids between the two species. When the soil transition is abrupt, the band of hybrids is narrow; when the transition is gradual, the zone of hybrids is broad but it does not blur the distributional limits of the parent species. Hogben (348) has a somewhat different interpretation of the oak situation in England. According to him, *Quercus robur* is a polymorphic species consisting of three interfertile races (which have been named *Q. pedunculata, Q. sessiliflora, Q. lanuginosa*). Each has its own focal region and there is much overlapping, so that the case is a borderline one between geographical and ecological isolation. Other pairs of species isolated ecologically, according to Salisbury, include *Galium sylvestre* of calcareous soils and *G. saxatile* of soils poor in bases; *Gentiana excisa* of non-calcareous soils and *G. Clusii* of lime-rich soils; *Veronica spicata* of continental regions and *V. hybrida,* its counterpart in regions with oceanic climate. Hogben also mentions other cases of supposed ecological isolation. In one region *Dianthus deltoides* grows on sandy uplands and *D. armeria* on loamy soil; *Cardamine edule, C. exiguum,* and *C. ovale* are kept separate by the fact that they flourish at different levels of salinity. Dobzhansky (191) adds some examples from observations of Epling. *Quercus Douglasii* and *Q. Garryana* belong to different forest types in California, but meet and hybridize in Lake County; and *Pinus ponderosa* and *P. Jeffreyi* are generally distinct as to range and forest types, but meet and hybridize in the Sierra Nevada.

A few examples of supposed ecological isolation can also be drawn from zoological literature. Two subspecies of the mouse, *Peromyscus maniculatus,* have been reported living in a single region without hybridization (186), although they ordinarily have geographical isolation. *P. m. bairdii* is widely distributed over the Mississippi valley, where it lives in prairies and cultivated fields; *P. m. gracilis* is a northern form that lives in heavy forests. In Charlevoix County, Michigan, *bairdii* is confined to lake beaches, and in Washtenaw County, Michigan, it occurs in open fields. In the latter county it is accompanied by *gracilis;* however, the latter was taken from a woodlot, and no hybrids are known. The reason for considering these forms subspecies is that westerly they form a chain of geographical races (*gracilis, borealis, osgoodi, nebrascensis, bairdii*), with interbreeding at their borders. Dice cites a similar case from *Peromyscus* that was described by Osgood. Along the eastern border of Glacier National Park, Montana, *Peromyscus maniculatus artemisiae* lives in forests and *P. m. osgoodi* lives in sagebrush; the forms do not hybridize. They are of interest because they point to ecological isolation for subspecies which usually have only geographical isolation. When the ends of such species chains meet, as they sometimes do in nature, their lack of crossbreeding may be due to one of several forms of intrinsic isolation. In addition to habitat separation, there may be reproductive isolation on a psychological, structural, or physiological basis. According to Thorpe (626), there are many cases of habitat isolation among crickets. An interesting instance is *Nemobius fasciculatus,* which consists of two races without morphological distinction but which have different songs that are easily recognized. In nature each cricket remains in its own ecological niche. They can be crossed experimentally, and there results a new song never heard in nature; hence there is no genetic isolation.

Physiological isolation of the type found in parasites, phytophagous insects, etc., can be classified as a form of ecological isolation. Thorpe (626) believes that host selection can effectively split a population and that there are many cases of physiological differentiation unaccompanied by observed morphological change. On blueberry there is a maggot that is apparently structurally identical with *Rhagoletis pomonella,* the apple maggot, but neither can be raised on the host of the other. There are two races of the apple-sucker, *Psylla mali;* the race on the hawthorn (*P. peregrina*) cannot be made to oviposit on the apple, nor the apple race on the hawthorn. Neither will they cross in captivity, although their structural characteristics are exceedingly minor. Also, there are certain parasites on the hawthorn race, but the apple race is free of them.

Host or food selection probably originates because of intrinsic factors, just as other ecological isolations do, but there is some evidence that insects can be food-conditioned. Thorpe mentions the conditioning of larvae of the Ichneumonid, *Nemeritis canescens,* to a new food; there followed in adults a tendency to respond to the olfactory stimuli of the new food. Such conditioning may occur in nature and result in splitting a population into two or more parts attached to different hosts or foods, and thus effectively prevent crossbreeding. This is a non-hereditary physiological isolation that favors subsequent differentiation.

The occurrence of physiological races in a termite species reported by Emerson (209) is of interest because it is one of the few cases of host speciation on a physiological basis, and it is also the first reported for termites. *Nasutitermes guayanae* can be divided into two groups because of the species of beetles which the termite harbors. Structural differences between these two are relatively minor and were not noticed or analyzed previously. The soldiers of the two groups show a bimodal variation in head length which corresponds with the differences in termitophile beetles. One race, which retains the name *N. guayanae*, harbors four beetles: *Termitophya amica, Xenopelta cornuta, Thyreozenus major,* and *Eburniola leucogaster.* The other race, named *N. similis* by Emerson, harbors two different beetles: *Termitophya punctata and Xenopelta tricornis.* Because the termites are neither geographically nor ecologically isolated, Emerson believes that the differences in termitophile fauna are a reflection of physiological races.

Darlington (176) considers that the intrinsic differences which result in eclogical isolation may be the first step in an isolation sequence. This may arise through the selection in different habitats of genotypes differing in groups of genes that are favorable for the particular habitats. This ecological diversity may be followed by various genetic changes which result in intrinsic barriers to successful interbreeding.

Seasonal isolation.—Separation of interfertile species that live in the same region may result from differences in the season of reproductive activity. Anderson and Sax (22) state that *Tradescantia hirsuticaulis* and *T. subasper* var. *montana* may live in the same region without crossing because the former blooms early and for a short period and seldom overlaps the blooming period of the latter. Incidentally, it is known in *Tradescantia* that autopolyploidy has overcome seasonal isolation because it results in retardation and extension of the blooming period, thus bringing the autotetraploid into coincidence with other species. Palmer (495) believes that many of the species of hawthorn (*Crataegus*) are prevented from crossing because of slight physiological differences which control the time the pollen ripens and the receptivity of the stigma. A difference of only a day or two is sufficient to make cross-pollination in a wild state unlikely or impossible. By experimental retardation or forcing, it is sometimes possible to bring two blooming times to coincidence and thus demonstrate cross-fertility when hybrids are unknown in nature because of seasonal isolation. Several cases of seasonal isolation are known among animals. The toad, *Bufo Fowleri,* ordinarily breeds later than *B. americanus,* but when their seasons overlap they are further separated by habitat differences and mating calls. The isolation is not complete, however, because hybrids are occasionally found in nature.

Other forms of reproductive isolation.—Non-random mating, or sexual selection, is a phenomenon that produces a form of isolation. This has also been called psychological isolation. Scents, flash signals, dance patterns and antics, songs and calls may act as stimuli to mating within groups and as barriers between slightly divergent groups because mutual attraction does not exist. Apparently the phenomenon may have a physiological, structural, conditioned, or intuitive basis, and it is known for several animal types, including fish, toads, birds, mammals, butterflies, etc. Hogben (348) says that the preference of certain insects for certain flowers of a species may constitute an analogous case for plants that can produce non-random cross-pollination within the plant species. That insects have a basis for non-random selection of flowers when there is no geographical, ecological, or morphological differentiation must lie in the physiology of the flowers and be expressed, for example, through subtle differences of perfume. These differences may have no immediate significance for speciation, but they allow for close inbreeding and the increase of any genes thus isolated or subsequently evolved. It is conceivable that a close relationship might be built up between insects and flowers that would allow for their concomitant speciation. This analogy for plants must not be carried too far, because psychological isolation is frequently a much more complicated phenomenon.

Among plants there is also a case for mechanical isolation that results from the morphological "fit" of flowers and the insect visitors necessary for cross-pollination. Such a role is possible only in a few highly specialized families such as the Orchidaceae, and no satisfactory data concerning the origin of such intimate relations are available, so far as I know. Among several animal groups, however, the peculiar structures of the genitalia have resulted in what may be called the "lock-and-key" hypothesis of copulatory isolation. The structure of the genitalia forms a basis for the taxonomy of certain insects, spiders, mollusks, fish (forms with gonopodia), and mammals (especially bats and rodents), according to Dobzhansky (191), and several observers believe that even minor differences between closely related species may prevent copulation. However, Goldschmidt (282) says that the assumption of the impossibility of mating because of differences in genital armature cannot be argued but must be demonstrated, because there are many cases in which crosses have been accomplished when they would appear to be unlikely or impossible. He sees no reason why these characters cannot vary among geographical races just as others do, and he concludes that they are not of special significance unless copulation is made impossible. In Lepidoptera innumerable species hybrids have been produced, even some that are generic; according to Goldschmidt, "there is no fact available to indicate that the differences in genital armature found in geographic races are such that physiological isolation is affected, or even would be affected by further variation in the same direction." Corresponding situations are known for beetles, and Goldschmidt concludes that "within a single rassenkreis differences in armature may be small or nonexistent, or they may be more considerable. In the latter case all transitions exist within the series. This indicates that the features show the same type of variation as do all other variable characters within a rassenkreis without leading to physiological isolation, because only unimportant details are involved."

In both psychological and mechanical isolation the related forms live in the same general region without successful crossbreeding. In the first case the attempt is not made, and in the second case it fails at an early stage. Failure of the gametes to form a zygote may occur at later stages in the reproductive process. The length of the pollen tube may be inadequate to penetrate the entire style to the ovules, or the tube may burst before it has penetrated the embryo sac to the egg. The gametes may successfully meet and unite to form the zygote, and still the successful interchange of genes between races may fail to be accomplished because the new generation sporophyte of a plant may fail at any time from the zygote to the sexually mature hybrid. Furthermore, a hybrid may reach sexual maturity without being able to produce viable gametes from which the succeeding generation must come. All these failures which may occur at any time constitute forms of reproductive isolation.

Pollen tubes are haploid, and stylar and ovular tissues are diploid. Embryos are diploid and endosperm is triploid, as a result of double fertilization. There is some evidence that normal chromosomal balance is necessary for the functioning of the pollen tube and the development of the embryo. Consequently, abnormal differences in chromosome number may be a cause of reproductive isolation. This has been shown in *Datura* pollen tubes (70) and in *Linum perenne* × *L. austriacum* embryos, which can be raised by removing them from the seed (cited by Dobzhansky, 191).

Anderson and Sax (22) state that there is practically a complete lack of crossing between members of three groups of American species of *Trades-*

cantia: T. micrantha, T. rosea, and *T. virginiana* and its relatives. Cytological evidence for these barriers lies in the fact that 7 of the 12 chromosomes of *T. rosea* have subterminal fibers, all of the chromosomes of *T. micrantha* have them, and all of the chromosomes of *T. virginiana* have median fiber constrictions. There is further separation of species on the basis of chromosome numbers. Diploid species apparently do not cross with tetraploids; such hybrids, if formed, would probably be sterile triploids. In addition to changes in chromosome number (autopolyploidy, allopolyploidy, and aneuploidy), isolation and speciation may result from gene mutation and structural changes in chromosomes (segmental interchange between non-homologous chromosomes, fragmentation and loss, inversion or translocation, changes in chromosome size, and position of the fiber attachments), according to Anderson and Sax (22). They say, "Some of these numerical and structural changes may be directly effective in differentiation, but their chief function is providing the initial isolation permitting independent development of genic changes." Such changes are called genetic because they produce more or less sterility and isolation between the mutant forms and their parents or the non-mutant population.

The origin of isolating mechanisms.—Among the various isolating mechanisms mentioned in the preceding paragraphs there are some which present a difficult problem. Where the isolation results from inviability of the heterozygote or some later ontogenetic stage, or from sterility of the mature hybrid, the homozygotes must be unaffected. As a consequence of this fact, Dobzhansky says, "it follows . . . that a great difficulty is encountered in the establishment of any isolating mechanism in a single mutational step. Since mutants appear in the populations at first as heterozygotes, inviable and sterile heterozygotes are eliminated, regardless of how well adapted might be the corresponding homozygotes. . . . Forms of isolation other than hybrid inviability and sterility fare scarcely better if they arise in a single step. For example, let us suppose that a mutant and the ancestral form reach sexual maturity at different seasons, or that the germ cells of a mutant are incompatible with those of the original type. Since the mutation rates for most genes are known to be low, the number of the mutants produced in any one generation would be so small that they could hardly find mates among masses of unchanged relatives."

When isolating mutations do not result in inviable zygotes or sterility, the initial disadvantage of a low mutation frequency and the concomitant unlikelihood of the mutant finding a mate may be overcome through apomictic modifications of the reproductive method. Thus, interfertile forms with a sexual generation that is independent, as in the mosses *Hypnum* and *Bryum* and the ferns *Aspidium* and *Asplenium*, can produce a local colony of mutant plants (protememata and prothalli). Among higher plants the same result may be accomplished by parthenogenesis, forms of vegetative multiplication, or facultative self-pollination. As Dobzhansky points out, cross-fertilization may then be resumed in some of these colonies. Many polyploids, both euploid and aneuploid, seem to have methods of reproduction other than obligatory cross-pollination. However, such an explanation does not hold in cases of obligatory cross-fertilization, and another mechanism must be found.

Dobzhansky concluded that isolating mechanisms must for the most part result not from a single mutation but from the building up of systems of complementary genes.[2] This conclusion for obligatory cross-fertilizing organisms rests upon the following considerations: (1) that the genes producing isolation are multiple and complementary, their minimum number being 2; (2) that isolating mutations which produce hybrid inviability or sterility are dominant and act in the heterozygotes; and (3) that the development of reproductive isolation is as a rule predicated on the previous territorial separation of the populations later to be reproductively isolated. Dobzhansky illustrates this theory, which is supported by some established facts and opposed by none, by the following hypothetical case: "Let it be assumed that the ancestral population from which two new species are to be evolved has the genetic constitution *aabb*, where *a* and *b* are single genes or groups of genes, and that this population is broken up into two parts temporarily isolated from each other by secular causes—for example by inhabiting territories separated by a geographical barrier. In one part of the population, *a* mutates to *A* and a local race *AAbb* is formed. In the other part, *b* mutates to *B*, giving rise to a race *aaBB*. Individuals of the constitution *aabb*, *Aabb*, and *AAbb* are able to interbreed freely with each other, and hence there is no difficulty in establishing in the populations the gene, or group of genes, *A*. The same is true for the genes *B*, since *aabb*, *aabB*, and *aaBB* are fully capable of interbreeding. But the cross *AAbb* $\times$ *aaBB* is difficult or impossible, because the interaction of *A* and *B* produces one of the physiological isolating mechanisms. If the carriers of the genotype *AAbb* and *aaBB* surmount the barriers separating them, they are now able to coexist in the same territory, since interbreeding is no longer possible."

[2] Dobzhansky says, "As far as the present writer is able to see, this consideration is fatal to Goldschmidt's theory [282] of evolution by 'systematic' mutations, since these mutations are supposed to induce at once a complete isolation of the newly emerged species from its ancestor."

Dobzhansky does not agree with the common idea that isolation results when enough gene differences have developed, or that a certain proportion of genic mutations, by chance, cause isolation; but he believes that the origin of reproductive isolation is probably a separate process from the origin of other differences. It is a well-known fact that populations which are·strongly differentiated morphologically—including vast numbers of good botanical species and even some genera—and which may have been geographically isolated for vast periods of time have not developed reproductive isolation. Also, Dobzhansky says that strains may differ by many genes without any limitation to interbreeding, and that as many as five inversions in *Drosophila melanogaster* do not interfere with fertility. He hypothesizes, then, that races tend to become species not because of indiscriminate differences that tend to build up more at the center of area, but because of separate isolating mutations which tend to have selection value at the margin of area where the race meets and hybridizes with another race. According to him, the adaptive value of a complex is a product not of a few genes but of the whole genotype, and where hybridization jeopardizes the integrity of two or more adaptive complexes, genetic factors which would decrease the frequency or prevent interbreeding would thereby acquire a positive selective value, even though these factors by themselves might be neutral.

From the building-up of morphological and physiological variability through mutation and from the development of adaptive complexes through selection, there results race formation or intraspecific evolution. When these adaptive complexes are exposed to disintegration through interbreeding, isolating mechanisms tend to arise and bring the race to the specific level.

24.

Causes of Species Stability

There is a natural tendency in taxonomy, and in the sciences which contribute to it (cytology, genetics, phylogeny, geography), to emphasize divergence and discontinuity. Yet the causes of relative stability and the factors which tend to prevent speciation are also interesting and important. Emerson[1] has emphasized the aspects of stability and has enumerated the following contributing factors:

1. A low mutation pressure.
2. A limitation of mechanisms.
3. Gene correlation or balance.
4. Character correlation.
5. Panmixy or free interbreeding.
6. Perfection of adaptation or high specialization.
7. Competition.

Huxley (371) states that in the comparatively few cases intensively investigated the mutation rates are all of the same general order of magnitude. There are indications, however, that different types of organisms and different phylogenetic stocks may differ in mutation rate. Certain genes, segments, chromosomes, and genoms are apparently more likely to undergo mutation than are others. Consequently, within a phylogenetic stock, certain characters may remain stable because of the stability of that portion of the gene complex. The rate of mutation, Huxley suggests, may be a limiting factor that condemns some forms to stability, stagnation, or even to eventual extinction. There is some indication that the rate of mutation in certain species may vary cyclically and that it may be influenced by environmental

[1] I wish to thank Professor Alfred Emerson for calling my attention to the importance of species stability and for allowing me to use his notes on the factors of importance in causing species stability.

conditions also. For example, some evidence indicates that temperature may influence the rate of mutations. This influence may be direct or indirect, as Worthington (715) points out; the rate of mutation will vary with temperature because for all except warm-blooded organisms the number of generations tends to increase more in a warmer climate within a unit of time than in a colder climate, other things being equal. It must be admitted, however, that there are few data on this point and that it is extremely difficult in practice to separate mutation rate from other factors that cause variation or stability.

With respect to limitation of mechanisms, Emerson suggests that the number of respiratory pigments are probably extremely limited. Certain other functional mechanisms may also have a limited capacity for successful modification. Although the chemistry of photosynthesis is complicated and not completely understood, it appears that there are and always have been only a rather limited number of physico-chemical systems capable of fixing organic carbon.

Emerson's third and fourth factors, gene correlation and character correlation, are conveniently discussed together. There is ample evidence that some characters, if not most, result from the action of multiple factors, and that at the same time certain factors affect several different characters. According to Muller (471), it would seem that most present-day organisms are the result of a long process of evolution in which at least thousands of mutations must have taken place. Since each gene is effective within a closely integrated and mutually affecting system, each new mutant in turn must have derived its survival value from the effect it produced upon the reaction system that was already in existence. Muller says, ". . . Many of the characters and factors which, when new, were originally merely an asset finally became necessary because other necessary characters and factors had subsequently become changed so as to be dependent on the former. It must result, in consequence, that a dropping out of, or even a slight change in, any one of these parts is very likely to disturb fatally the whole machinery; for this reason we should expect very many, if not most, mutations to result in lethal factors, and of the rest, the majority should be 'semilethal' or at least disadvantageous in the struggle for life, and likely to set wrong any delicately balanced system, such as the reproductive system." Selection cannot operate merely in favor of factors for advantageous characters; it must favor the preservation of all the factors that have a necessary association with the character being selected. Selection thus tends toward species stability through the elimination of all variations (mutations) which adversely affect important characters. The so-called vestigial structures and many stable, apparently non-adaptive characters probably result from their association genetically with derived and adaptive characters and factors.

Salisbury (541) points out the ecological consequences of character correlation. Whereas the taxonomist is usually concerned with adult structures, the ecologist is forced to recognize important juvenile characters. According to Salisbury, "Even in one and the same partial habitat we encounter diversity of leaf-form that it is difficult to believe can bear any direct relation to competitive success. Yet it may well be that these adult distinctions are necessary concomitants of features more subtle that have ecological importance for the species concerned."

Embryologists have to face character correlations between ontogenetic stages more often than the ordinary biologist does. Selection can occur with respect to the properties of organisms in any stage of their ontogeny. DeBeer (180) points out that there may be larval characters that are highly adapted to their milieu and that are correlated with adult characters which show no adaptation to the adult milieu. The correlation of characters together with selection at different stages in ontogeny tends toward species stability. The interrelations between the appearance of novelties and ontogenetic and phylogenetic consequences were first indicated by him (179): ". . . Phylogenetic effects of largest systematic importance seem to be associated with evolutionary novelties which have either made their first appearance and exerted their main effects in early stages of ontogeny, or which have resulted in the retention of juvenile characters in the adult." DeBeer has introduced the term paedomorphosis for mutations that affect early ontogenetic stages and result in the production of a new type of adult organization with high potentialities for further evolution. Gerontomorphosis, on the other hand, is applied to evolutionary novelties which exert their main effects at later stages of the life histories of organisms. Such mutations are less likely to give rise to large changes; rather, they tend toward ever-increasing specialization and progressive loss of potential for further evolution. DeBeer says, "It is concluded that, as evolution proceeds, paedomorphosis is succeeded by gerontomorphosis which actualizes the further evolutionary potentialities opened up by paedomorphosis and exhausts them. The group then lingers or becomes extinct unless a new bout of paedomorphosis supervenes." It is easy to overemphasize his point because there are many instances of characters fundamental to large systematic groups that are manifest only in adult stages and, as far as can be told, function only in adults.

There is some support for the idea that groups with potentially higher systematic value are formed rapidly through paedomorphic processes and that the subsequent phases of evolution result in "adaptive radiation." Paleontologists have independently come to the conclusion that there is evidence for alternate spates of large and small evolution. In the early Tertiary, evolution produced the main lines of mammal phylogeny, but adaptive radiation did not proceed very far. Perhaps a parallel case is afforded by flowering plants. The early records of flowering plants in the Cretaceous and early Tertiary testify to the existence of all of the principal lines of evolution within the group. Yet, since our first knowledge of them, they have undergone an intensive adaptive radiation in most groups. Certain groups have reached their ultimate and become extinct; others are now represented only by relic forms; but many lines are still expanding in a multitude of relatively closely related species. We must not overlook the fact, however, that the evidence of the fossil record may be seriously faulty. Flowering plants may have had a long evolutionary history (in tropical regions or warm temperate plateaus) of an essentially gerontomorphic type prior to their apparently sudden diversified burst in the temperate regions whence come the majority of the fossil records. Be all this as it may, the evidence for the correlation of genes and characters in the tendency toward species stability is fairly conclusive.

Panmixy, in addition to its effect on variation through recombination, has a conservative influence, according to Emerson. Sewell Wright (717) has given mathematical demonstration of the fact that large interbreeding populations separated by partial discontinuities into subgroups provide the greatest reservoir of evolutionary potentiality. This is true with respect both to the number of mutations probable and to the recombinations possible. Isolated and numerically small populations, on the other hand, tend toward uniformity and anomaly. Under selection certain gene frequencies can be reached, and any mutations which are selectively more or less neutral can spread through the population with rapidity. There is an opportunity for accidental non-selective changes to reach almost any extent. Timofeeff-Ressovsky (627) says that "an infinitely large panmictic population consisting of a mixture of genes having equal biological value will, in the absence of mutation and under constant environmental conditions, be stabilized in a certain state of equilibrium." This is a theoretical situation, however, for in large populations mere distance becomes an isolating factor which opposes panmixy. Panmixy and the stability of a population are also opposed by the evolutionary factors of mutation pressure, selection pressure, and the accidental isolation of certain genotypes through population waves or spatial segregation. The fact remains that wherever there is free and complete, or almost complete, interbreeding the result is relative taxonomic constancy or species stability. It appears, then, that although sex is usually thought of as a producer of variability it also serves to integrate a population and produce a certain stability.

Kinsey (394) recognizes two types of species with respect to gall wasp population size and areal magnitude: continental species, which occur over a wide area and consist of large numbers of individuals, and insular species, which are limited in area and in number of individuals.[2] It should first be noted that the occurrence of gall wasps is limited in two ways: most of these wasps are highly specific with respect to their hosts, being confined to a single species of oak, or sometimes to a few closely related species; limitations to the spread of the host oak species is the second limiting factor. Kinsey found that about 76 per cent of the gall wasp species are classifiable as insular (being confined mainly to certain desert mountain ranges of western America), with an average area of 4600 square miles. Continental species have an average area of about 300,000 square miles. A close correlation between area and numbers of individuals being assumed, the largest continental species populations exceed the smaller insular species populations by several thousand times. Kinsey concludes that "these great differences in the sizes of specific populations are the primary bases of the differences in their evolutionary history." He arrives at this conclusion because his observations indicate that 80 per cent of the insular species are more constant in their characters than are any of the continental species. On the other hand, about 80 per cent of the continental species can be classified as distinctly variable. So, says Kinsey, ". . . the uniformity usually found among the insular forms appears to depend upon the smaller sizes of the populations involved." Not only may the larger populations contain more mutations, but panmixy is more difficult of realization than in small populations which are freely interbreeding.

Although the tendency toward population stability exists in all sexual populations, the impossibility of its complete attainment in large populations

---

[2] Goldschmidt (282) has pointed out that according to Kinsey a mutation within a population increases the variation within the species, but the same mutation if isolated or selected results in a new species. The result of this point of view is that Kinsey recognizes a large number of "species" arranged in a geographical pattern of branching chains, and where "genera" are joined the transition is no different in magnitude from that of one species to another in a chain. Kinsey's subgenera would be called species by adherents of the artenkreis concept, and Goldschmidt would call them groups of subspecies of a single species; i.e., Kinsey's gall wasp species are equal to Goldschmidt's subsubspecies.

is indicated by Kinsey's data. He found that 94 per cent of the continental species show intergrading populations between related species, whereas this was true of only 12 per cent among insular species. The latter species, although potentially panmictic at first, cannot long exchange genes with their close relatives, in most cases because of isolation. They develop discontinuities between each other, and a high degree of constancy within each population. Continental species originate in the same way as insular species, but in attaining wide area and numerous individuals they encounter ecological diversity, physical barriers, and distance (probably most important of all); the result is partial isolation and the limitation of the ready spread of new factors. In conclusion, then, we see that panmixy tends always towards species stabilization, and is rather fully attained in small, freely interbreeding populations but only incompletely attained in larger populations.

Perfection of adaptation was Emerson's sixth factor in species stability. He says that highly adapted types are so perfectly adjusted that any mutation is likely to be eliminated by selection, thus keeping the species stable. If the environment does not change, these species tend not to evolve; if the environment changes, they tend to become extinct. We have already noticed the conclusion of embryologists and paleontologists that gerontomorphic variations tend toward specialization and extinction. From the general to the particular may not be a biologically retraceable path. We may look to ontogeny for an analogy. Differentiation and division of labor into tissues and organs, in the ontogeny of an organism, can proceed to the point where cells become so far changed that they cannot regenerate, i.e., return to the embryonic and generalized condition. Also, in evolution it is possible that specialization may be carried to the point where a form is adapted only for a very narrow and peculiar set of environmental conditions. Without a return, fortuitously, to paedomorphic mutation, highly specialized organisms have not the ecological amplitude or the genetic variety to adapt to changing conditions. Salisbury (541) points out that selection for a particular environment imposes a certain stability, especially upon subdivisions of the Linneon. This process of genetic simplification (reduction in number of biotypes) of a population is accomplished not only through selection for particular environmental niches, but also through the chance isolation, as with insular species, of a portion of the genetic stock of the species as a whole.

Competition can also be considered a factor tending to produce species stability. We can think of a region as consisting of a limited number of environmental types or biological niches, for both plants and animals. We can think of these life areas also as being normally saturated in the sense that they are completely occupied by adapted forms. The attainment of the saturated condition, however, is the result of migration and evolution under the limitations of time and chance. In one region there may have been a great burst of adaptive radiation within a certain phylogenetic line so that many of the niches became occupied by related but morphologically widely divergent types. In another region an entirely different phylogenetic stock has been able through adaptive radiation to accomplish the same range of adaptations to the available niches. It follows, then, that a potentially variable phylogenetic stock in a saturated region will be unable to profit from its potentialities because of the competition pressure provided by forms which have already occupied the niches. According to Worthington (715), "In a group capable of rapid differentiation, the environmental conditions are all-important in determining the amount of differentiation which can take place. Of these conditions, two are specially important—the existence of numerous unoccupied ecological niches, and the absence of predators." In this connection Salisbury (541) says, "The geneticist who studies wild species in the protected conditions of the breeding-ground often finds variation far wider than that which the species appears to exhibit in a wild state, probably resuscitating many of Nature's *rejectamenta*. It is, indeed, not without significance that species of open habitats where competition is least severe often display a variety of races that is in marked contrast with the comparative uniformity of those which occupy 'closed' communities where competition is severe."

25.

Rate of Evolution
and Speciation

The rate at which evolution takes place can seldom be determined directly. In fact, the only case is that in which a major mutation, such as polyploidy, at one step produces a reproductive barrier between the mutant and parent organisms, starting a new interbreeding population or an agamic one. In such cases a new species is formed abruptly. Many such species are known in nature, but there are only indirect geological or geographical data to suggest when they occurred. Under experimentation and cultivation it is sometimes known just when such a species arose, as in the case of certain amphidiploids. Most species, however, are assumed to have had their origin through a series of minor mutations, and certain genic conditions have become stabilized through natural selection or under isolation, so that it is impossible to say exactly when the species arose. In this sense the origin of a species is a continuing process that has no particular time of origin. This statement is similar to that of Kinsey: a species has no center of origin, but results from the gradual addition of mutations in a chain of changes along a highway of migration.[1]

In some cases it appears that a few hundred years have been sufficient for elaborate speciation. In others a few million years have been insufficient for the production of interbreeding barriers, except those of a geographical nature. Geological and geographical knowledge often allow the dating of isolation; this, in turn, marks the maximum time during which an observable differentiation has developed between isolated stocks. We shall first consider certain evidences for a relatively rapid evolution (excluding

---

[1] All such discussions are complicated by the existing differences in species definitions. Kinsey's species of *Cynips* are considered to be only subspecies or subsubspecies by Goldschmidt (282).

that known to be due to chromosome number changes), and follow this by data which indicate that speciation is usually a very slow process.

Evidences for relatively rapid evolution.—According to Huxley (371), Jameson recorded an island population of house mice which showed distinct mean adaptive color differences from those of the adjacent mainland within 100 to 200 years. As Wright has shown mathematically (717), small inbreeding populations can develop differentiation within a short period, but in large populations changes in gene frequencies must be slow. Such changes do not necessarily bring a population to the species level.

The hawthorn genus *Crataegus* offers an exceptional opportunity to inquire into the question of the rate of speciation. Marie-Victorin (447), who has discussed this problem, suggests that somewhat parallel to that of *Crataegus* are the problems of *Oenothera* and *Senecio* in eastern Canada, *Rosa, Hieracium,* and *Rubus* in Europe, *Acacia* in Africa, *Sorbus* in the Orient, and *Hebe* in New Zealand. All these genera with wholesale discontinuous evolution suggest "a very important and rather unexpected generalization: that under favorable circumstances, a period of two or three hundred years is sufficient to produce, in some genera at least, by mutation or otherwise, a marvelous outburst of species."

Gray's *Manual of Botany,* covering the northeastern United States, listed 7 species of *Crataegus* in the first edition (1848), 10 species in the sixth edition (1890), and 65 species in the seventh edition (1907). According to Brown (93), between 1896 and 1910 a total of 866 species and 18 varieties were published. Of these newly described forms, 144 came from the pen of C. D. Beadle, 165 from that of W. W. Ashe, and 524 from that of C. S. Sargent. What is the meaning of such an epidemic of descriptions of species within a genus? Was it due to the drawing of finer lines in the species concept and to more intensive field work? Were the species overlooked by earlier botanists? Or, perhaps, did something happen to *Crataegus* rather than to the taxonomists? Are these forms of *Crataegus* hybrids or mutants? Are they really species? Brown says that Ezra Brainerd was the first to suggest that changed conditions, since civilized man entered the scene, had given *Crataegus* its great opportunity. "The genus *Crataegus* . . . has vastly increased in individuals and in 'forms' in the northeastern United States since the forests were cut off; specimens are rarely found in the original forests of this region. But the plants rapidly take possession of neglected pastures, fence-rows, and untilled ledges." The biological interpretation given by Brainerd and Brown was that the clearing away of the forests allowed the spread of the original species until proximity led to extensive cross-pollination. In other words, the new species of *Crataegus* were interpreted as being largely of hybrid nature.

Palmer (495) concurs in the opinión that the clearing away of the forests gave these heliophilous plants a chance to spread and ‚intermingle. He admits that many of the new forms may be hybrids or "divergent forms," but he thinks that many of them are good species. He points out that many of the species are apparently local in occurrence—some are known only from the type locality or from a single tree—and that their endemism can be explained on the basis of two interpretations. Either they are newly formed species which have not had sufficient time to spread widely, or they are relics of ancient and disappearing types. Palmer favors the idea that they are new, for, he reasons, if they were relics they would likely occur in several widely discontinuous spots rather than in only one.

Marie-Victorin (447) says that *C. Victorinii* is a well-defined endemic of the Montreal district and that *C. suborbiculata, C. Jackii, C. Brunetiana, C. laurentiana,* and a host of other species of the St. Lawrence valley seem likewise to be extremely localized. As to the validity of the species, he believes that the characters of the different species of a region, if studied at flowering time, are better and the species far more easily recognized than those of *Epilobium, Aster, Antennaria, Rubus, Carex, Poa,* and other critical polymorphous genera. He agrees with the concept that most of the species of *Crataegus* are very recent in origin. He says that before the advent of the white man, the whole of the St. Lawrence valley was heavily forested and that hawthorn colonies must have existed as sparse individuals scattered along streams in a condition which is still prevalent on the northern boundary of the genus (the Témiscamingue-Abitibi district, Lake St. John, the rivers of Anticosti and Gaspé). Within historical time, the clearings around villages and the old fields of agriculture have produced sunny sites. The "ecological vacuum" became filled with European herbaceous adventives and a few woody heliophytes from the native flora. Among the latter was *Crataegus,* which even today has its great population centers around the older settlements such as Quebec, Montreal, Rochester, and the sites of ancient Indian villages and Hudson Bay forts. According to Marie-Victorin, "It seems safe to assume that the astonishing development of the genus in northeastern America, and particularly in the St. Lawrence Valley, is an immediate biological response to the ecological disturbance or reduced compression brought about by deforestation and the settling of the land. . . . The migration and multiplication of forms must have taken place during the short colonial period." This author also points out that the *Acacia* problem (*Acacia giraffae, A horrida,* etc.) of the N'Gong uplands of Kenya in equatorial Africa has a striking similarity to the *Crataegus* problem in eastern America. They have both exhibited a similar power to produce an enormous number of species.

Crataegus has been shown to be very complicated cytologically. There are hybridization, polyploidy, secondary polyploidy, and apomixis. Many of the new "species" are good even by the strictest definition, but others are not; order remains to be established in this taxonomically confused and difficult group. Whatever the status of many forms, *Crataegus* has undoubtedly produced numerous species within historical time.

Evidences of a slow rate of speciation.—In general, evolution is such a slow process that it can be measured only by geological timetables. For example, according to MacBride (432), Woltereck found that Crustacea in the glacial lakes of Bavaria differed only slightly, but in a constant way, from their nearest allies outside the limits of alpine glaciation. It is estimated that the glacial recession began about 16,000 years ago, and this time has been sufficient only for the differentiation of races or varieties, not for species, despite the high degree of isolation which the lake forms have enjoyed.

Fernald (244) called attention to a situation in which two coastal plain floras have been separated for from 25,000 to 30,000 years, but the speciation has been practically nil. Peninsular southwestern Nova Scotia must have obtained its flora in post-Pleistocene time, because it was scoured by the ice of the Wisconsin glaciation. Its flora was derived partly from New Brunswick and partly from the southern coastal plain by way of a formerly elevated but now submerged continental shelf. The present coastal plain flora of New Jersey has nearly 200 representatives in Nova Scotia, yet during the long time of the latter's isolation no species has been differentiated from its southern counterpart except one weed-like annual, *Agalinis neoscotia.*

Another species in which it is possible to estimate the slowness of speciation is *Bidens hyperborea.* This species occurs exclusively on fresh tidal mud in estuaries from eastern Massachusetts to Quebec. It is found only in the mouths of fresh rivers, and consequently the colonies are widely isolated by the intermediate brackish and saline shores. According to Fernald, all its colonies come from the narrow area once covered by the Champlain Sea. The species probably once had a continuous distribution when great volumes of water from the melting ice sheet freshened the shallow margin of the inner Gulf of Maine and Gulf of St. Lawrence, which were almost landlocked by the elevation of the continental shelf outside. The plants of the present estuarian situations, which have been isolated in their respective regions since the withdrawal of the Champlain Sea, perhaps 15,000 years ago, have undergone some differentiation. The striking phenomenon, however, is that none of these colonies has differentiated enough, despite its rigorous isolation, to have lost its fundamental specific character—a unique achene which immediately distinguishes it from all other species. *Bidens* is a notoriously plastic genus, but all that *Bidens hyperborea* has been able to produce is a series of varieties based mainly on habit and foliage.

The case of *Bidens hyperborea* received detailed attention from Fassett (232, 233), who lists the following occurrences by varieties and river estuaries: var. *typica* (Rupert House, James Bay), var. *cathancensis* (Kennebec River estuarine system, Maine), var. *Svensoni* (Rimouski River, Quebec), var. *arcuans* (Miramichi River, New Brunswick), var. *gaspensis* (St. John and Dartmouth Rivers, Quebec), var. *laurentiana* (Cap-Rouge, St. Lawrence, Boyer, and Bonaventure Rivers, Quebec; Restigouche, Eel, Jacquet, Tetagouche, Tabusintac, Miramichi, and Kouchibouguac Rivers, New Brunswick), and var. *colpophila* (Buctouche and Shediac Rivers, New Brunswick; River Philip, Nova Scotia; Pleasant, Narragaugus, Harrington, Union, Penobscot, Souadabscook, Reed, Sheepscot, Black, Winnegance, Kennebec, Androscoggin, and Mousam Rivers, Maine; Salmon Falls River, New Hampshire; Merrimack River and Mill Creek, Massachusetts).

Fassett believes that this differentiation into varieties, in contrast to the stability of the longer-isolated Nova Scotian plants mentioned above, is due to the extreme conditions under which they grew. I think, however, that *Bidens hyperborea* probably illustrates the Sewell Wright effect in the isolation of small populations, rather than some hypothetical extreme ecological condition that differs for each variety.

Another case of post-Pleistocene differentiation or lack of it is reported by Jordan (382). In early postglacial time Lake Bonneville, a great glacial lake, occupied the basin of Great Salt Lake. At this time it is probable that the same species of fish and insects were found in all its tributaries. Now these streams flow separately into a lifeless lake and their fresh-water inhabitants are widely disjunct. Today one species of sucker (*Catostoma ardens*) and one chub (*Leuciscus lineatus*) are found unaltered throughout this region and in the Upper Snake River above Shoshone Falls into which Lake Bonneville once drained. In other cases certain species have been locally isolated and have undergone no differentiation; only one small minnow of the clay bottoms (*Agosia adobe*) can be shown to have undergone alteration. In contrast to the conservatism of the fish is the plasticity of the tiger beetles

(*Cicindelae*), which have produced a large number of species in the isolation which has resulted from the shrinking of Lake Bonneville.

Worthington (715) has made a comparison of the fish of tropical African lakes, such as Lake Edward, and those of temperate waters of the British Isles and has noted that since the last interglacial period speciation has gone on much more rapidly in the warm than in the cool region. He says, "There is good evidence that where a niche exists, vacant for reasons of isolation, some species will fill it rapidly, even though considerable structural alterations are involved in the process." He also concludes that organisms which have no means of regulating their body temperature will tend to evolve more rapidly in warmer than in cooler regions because more generations will develop in the warmer region in a given length of time. "Looked at from this point of view," he says, "it is less surprising that many endemic species and subspecies of fish have come into being in the African lakes during and since the pluvial epoch, while in temperate regions the same lapse of time has produced relatively little differentiation." If it is assumed that the growth and reproductive rates of fish are accelerated in tropical regions to the extent that a generation is passed through twice as rapidly as in cool temperate regions for organisms of the same general size and evolutionary potentialities, it is obvious that changes in structure should appear in half the time.

Faegri (229) has discussed the relationships between the alpine flora of Scandinavia and corresponding populations of middle Europe. He says that although these floras have been separated 50,000 to 100,000 years the fact remains that vicarious species pairs are surprisingly infrequent. Endemic species in the Scandinavian mountains are few except in such genera as *Taraxacum* and *Hieracium* (which are largely apomictic). Faegri mentions one vicarious pair, *Euphrasia salisburgensis* in the middle European Alps and *E. lapponica* of the Scandinavian mountains. In *Papaver,* a genus which is considered to have shown relatively rapid evolution, 50,000 years or more have been sufficient to produce only a few closely related species or subspecies. The genus *Poa* is a notably critical and polymorphous group; but *Poa flexuosa,* according to Faegri, has shown no perceptible differentiation in Great Britain, Iceland, and Scandinavian regions, where the populations must have survived the last glaciation in separate refugia. The *Poa laxa* group has differentiated into separate species in middle Europe, Scandinavia, and America, but this differentiation must have taken place before the last interglacial period. These members of the genus must be considered taxonomically conservative; this is remarkable, considering that most of the species within the genus *Poa* appear to be in a dynamic condition.

We may now turn our attention to cases in which the development of discontinuity apparently dates to pre-Pleistocene time. For one reason or another it frequently is not possible to say definitely whether the isolation developed in early, middle, or late Tertiary time; but the principal point remains—the disjunction has existed for a few to several million years. We shall note some cases of transatlantic, Euro-Asian, and Asio-American affinities. In a large number of instances the two most closely related species of a genus have just such widely disjunct occurrences. Frequently these pairs of species are so close to each other morphologically that early taxonomists placed them together as a single species; only within recent decades have their differences been discovered. It is of interest to learn from cytogenic studies that such pairs of species, despite their long period of isolation, are sometimes still so little changed that, when brought together, they can produce fertile hybrids.

Such a case is found in the genus *Platanus*. This genus is the sole member of the Platanaceae, which was one of the most widespread of the Cretaceous dicotyledons. The Old World species, *Platanus orientalis,* and the American, *P. occidentalis,* have been isolated populations since no later than the Pliocene. Whether one accepts a land-bridge or continental drift hypothesis, it appears that any effective highway for the migration of such trees as sycamore could not have existed for a few million years. According to Sax (548), *Platanus acerifolia* is a hybrid between the European and American species in which seedlings from the hybrid are variable and segregate for characters of the parents. There is here no question of allopolyploidy, for the species and the hybrid each have the same number of chromosomes (S = 42). The compatibility of the parent species in the first-generation hybrid indicates that they have undergone no fundamental changes since their isolation. The mutation rate in *Platanus* must be very low.

A great quantity of biological evidence points with certainty to a former easy land connection between North America and Europe. These two land areas must have been linked during much if not all of the time from the Cambrian to some time in the Tertiary. However and whenever a disjunction developed in the Tertiary, cold oceanic currents flowed southward and strong climatic and ecological differences developed along with the isolation of the common biota. In the time which has elapsed since the separation of the once remarkably uniform floras of Tertiary time, a number of developments have occurred. Marie-Victorin (447) enumerates five classes of change: (1) Europe has lost a number of present-day American trees—

Pinus Strobus, Liriodendron tulipifera, Taxodium distichum, Juglans nigra, Carya alba, Tsuga canadensis, Acer rubrum. (2) America has lost or nearly lost a number of characteristically European plants—*Trapa natans, Phyllitis Scolopendrium, Atriplex maritima, Sparganium glomeratum, Habenaria albida, Polygonum acadiense, Carex Hostiana, Hieracium groenlandicum.* (3) Common American plants persist as relics in scattered places in the British Isles, the Baltic regions, or Scandinavia—*Botrychium virginianum, Stellaria longipes, Montia lamprosperma, Carex marina, Sisyrinchium angustifolium, Spiranthes Romanzoffiana, Eriocaulon septangulare, Juncus macer, Naias flexilis, Galium trifidum, Spartina alterniflora, Spartina patens.* (4) Most of the tree genera of Europe and America are the same, but not a single species is common to the two areas. (5) Since isolation, however, a remarkable parallel evolutionary development has resulted in the production of numerous pairs of species which stand close together morphologically and ecologically. Following are a series of such pairs listed by Marie-Victorin which illustrate slight but definite divergent evolution:

Ostrya virginiana	O. carpinifolia
Ulmus fulva	U. campestris
Ulmus americana	U. pedunculata
Sorbus americana	S. Aucuparia
Pinus Banksiana	P. sylvestris
Pinus Strobus	P. Peuce
Viburnum lantanoides	V. Lantana
Viburnum trifidum	V. Opulus
Oxalis montana	O. Acetosella
Vallisneria americana	V. spiralis
Maianthemum canadense	M. bifolium
Rumex Hydrolapanthum	R. Britannica
Polypodium virginianum	P. vulgare

Marie-Victorin writes (447): "It seems that this bicentric micro-evolution has been so general that we may well accept the phytogeographical law proposed by Fernald: Species which in America are confined chiefly to the Alleghenian region will be found to differ in very fundamental characters from their nearest allies of continental Europe." With respect to the rate of speciation of the Tertiary, pan-Atlantic, boreal flora on the two sides of the Atlantic, one can only conclude that in general the changes, although definite, are small and that evolution has progressed relatively slowly.

Western Europe and eastern Asia have some species in common which have been isolated since Pre-Pleistocene time. The European larch, *Larix decidua,* and the Japanese larch, *L. Kaempferi,* constitute one pair. According to Sax and Sax (551), *Larix eurolepis* is an F_1 hybrid of these two species; the hybrid shows almost complete chromosome pairing, more than 90 per cent of the pollen is good, and the chiasma frequency at meiosis is as high as that of the parents. Similarly, *Taxus media* is a hybrid between the Japanese yew, *T. cuspidata,* and the English yew, *T. baccata.* Species differentiation in the conifers is caused primarily by gene mutation and position effects—polyploidy, aneuploidy, and other gross changes being exceedingly rare—but despite long periods of geographical isolation many species have developed no barriers to crossbreeding.

Eastern American and eastern Asian affinities have interested botanists since Asa Gray first pointed out the striking relationships in the middle of the last century.[2] Fernald (241) says that the plants of our Alleghenian forest which we share with Japan and China are among the oldest now living on the North American continent. He points out that many of the genera are diatypes, with one American and one Asian species. Even when a genus has two or a few species in one center or another, they are quite distinct. The taxonomist can make a prompt and explicit determination of the species of *Caulophyllum, Dicentra, Panax, Cryptotaenia, Epigaea, Chiogenes,* and *Mitchella.* "Why should he not?" asks Fernald. "These are genera of the eastern Asiatic-eastern American forests; and they are absolutely segregated genera with no complications of freely intergrading species . . . they are old veterans."

The close relationship between certain species of wide disjunction is well illustrated by the taxonomic treatment they have received. In some cases these plants have long been known as the same species, only recently being recognized as sufficiently distinct to warrant separation as varieties or full species. On the other hand, for a long time certain paired populations have been considered different and been given different names, but recent study has shown them to belong to the same species, or to deserve only varietal rank at the most. In either case our point concerning the close relationship of the floras is emphasized, as well as the fact that during vast periods of isolation only relatively slight evolution has occurred.

Two grasses may be mentioned.[3] A species which has long passed as *Beckmannia erucaeformis* in American floras has been shown to be identical with the eastern Asian *B. syzigachne. Festuca scabrella* of American floras

[2] These relationships are discussed in detail in Chapter 17.
[3] I wish to thank Dr. H. A. Gleason for calling my attention to them.

is now known to be identical with *F. altaica* of northern Asia. Hara (330), who has published on the relation between North American and eastern Asian plants, notes that it has been common practice to refer plants of a certain buttercup complex to *Ranunculus Gmelini* when they are Asian and to *R. Purshii* when they are American, but he fails to discover any constant characters by which they could be separated into two distinct species. He says, "I found that some American specimens, for example those from Minnesota and Michigan, agree exactly with the Asiatic plant." The American material is reduced to varietal status. He shows that the enchantress nightshade, known as *Circaea lutetiana* to eastern Asian and American authors, is distinct from the European species. The Asiatic material is referred to *C. quadrisulcata* and the American plant to var. *canadensis* of that species. Hara finds that the circumpolar bedstraw species, *Galium trifidum,* has three geographical centers: the species proper occurs in Europe and North America, var. *pacificum* in North America and eastern Asia, subspecies *tinctorium* in eastern North America. These are too closely related to be maintained as separate species. *Erigeron angulosus* is a species described from the European Alps. He has shown that the common form in North America is identical with the form in eastern Asia known as *E. kamtschaticus,* and he has reduced the latter to varietal status under *E. angulosus.* The pearly everlasting *Anaphalis margaritacea,* is widespread in North America and eastern Asia, but none of the several forms which have been described is worthy of specific status.

In order further to emphasize the relationship between the Asian and American floras and to point out that certain groups have shown very slow evolution, I have assembled from Small's *Manual* (594) the following data concerning taxonomy and distribution. The plants mentioned can be confidently referred to as phylogenetic stocks that were members of the Arcto-tertiary forest vegetation which flourished through much of the Tertiary in America, Europe, and Asia. For the circumboreal distribution of many species of this forest there is corroborative fossil evidence, but even without this their present disjunct occurrences argue for a once continuous area either simultaneously or through migration.[4] Discontinuity must have developed no later than the onset of the glacial age, although it is likely that for many forms it developed even earlier. Furthermore, such far-flung syngameons must have developed differentiation even before their areas were disrupted.

---

[4] The only other alternative would be the assumption of a polytopic origin for these species and species pairs. Such an assumption rapidly loses probability as the number of cases increases, because it involves also the assumption of orthogenetic series in every phylogenetic stock.

In this connection certain families are notable. Iteaceae contains only the genus *Itea*. In southeastern United States there is one species, *I. virginica;* four other species occur in Asia. *Phryma* is the sole genus of Phrymaceae, a segregate from Verbenaceae. One species, *P. leptostachya,* occurs in eastern North America, and the same species or a very closely related one is found in Asia. In the Galacaceae there are four genera, of which two are American and three are Asian. The monotype *Galax aphylla* occurs in eastern United States. *Shortia (Sherwoodia) galacifolia* is a rare species of the southern Blue Ridge; the only other member of the genus occurs in Asia, as do the two other genera of the family. Podophyllaceae consists of four genera monotypic in eastern America. *Caulophyllum thalictroides* is matched by a single Asian species. The same situation exists for *Diphylleia cymosa* and *Jeffersonia diphylla,* but *Podophyllum peltatum* is represented by about three species in Asia.

Following is a list of eastern American genera, each of which has a single species there but from one to a few species in Asia. These species are all unrepresented in Europe and are assumed to belong to the Arcto-tertiary forest.

Trautvetteria carolinensis	Ranunculaceae	1 sp. in Asia
Liriodendron tulipifera	Magnoliaceae	1 sp. in Asia
Hydrastis canadensis	Ranunculaceae	1 sp. in Japan
Mitchella repens	Rubiaceae	1 sp. in Asia
Epigaea repens	Ericaceae	1 sp. in Japan
Hugeria erythrocarpa	Vacciniaceae	1 sp. in Asia
Bignonia radicans	Bignoniaceae	1 sp. in Japan
Sassafras Sassafras	Lauraceae	1 sp. in Asia
Saururus cernuus	Saururaceae	1 sp. in Asia
Croomia pauciflora	Rhoxburghiaceae	1 sp. in Japan
Arethusa bulbosa	Orchidaceae	1 sp. in Japan
Tovara virginiana	Polygonaceae	1 sp. in Japan
Pyrularia pubera	Santalaceae	2 spp. in Asia
Stylophorum diphyllum	Papaveraceae	3 spp. in Asia
Mitella diphylla	Saxafragaceae	3 spp. in Asia
Hamamelis virginiana	Hamamelidaceae	4 spp. in Asia
Schizandra coccinea	Magnoliaceae	5 spp. in Asia
Pieris floribunda	Ericaceae	5 spp. in Asia
Cephalanthus occidentalis	Rubiaceae	5 spp. in Asia

Other genera which have the same type of occurrence but have a few species in eastern America as well as in eastern Asia, include the following:

Illicium (Magnoliaceae) contains two eastern American species, *I. parviflorum* and *I. floridanum,* and five species in Asia. *Gleditsia* (Cassiaceae) consists of *G. triacanthos* and *G. aquatica* of the eastern United States, and four species in Asia. *Stuartia* (Theaceae) consists of *S. Malachodendron* and *S. pentagyna* of the eastern United States, and three Japanese species. *Gelsemium* (Spigeliaceae) consists of *G. sempervirens* and *G. Rankinii* of the eastern United States, and one Asiatic species. *Panax* (Hederaceae) has *P. trifolium* and *P. quinquefolium* in eastern North America, and about five species in Asia. To these cases can be added several genera in which there are somewhat more species: *Aletris, Amsonia, Aralia, Astilbe, Bicuculla, Catalpa, Clintonia, Disporum, Dodecatheon, Glycine, Gordonia, Hydrangea, Leptilon, Magnolia, Nabalus, Nyssa, Osmorrhiza, Opulaster, Parthenocissus, Sapindus, Thermopsis, Tiarella, Toxicodendron, Tracaulon, Trillium,* and *Vagnera.* Some of these genera are represented in western America, but none occurs in Europe. Some occur in the south, but none lacks north temperate disjunction of a wide nature—the fact that is used to support the contention that they are ancient phylogenetic stocks that have survived, in the main, in eastern Asia and eastern America, with relatively slight speciation. In no case can the size and biological condition of these genera be compared with those of *Aster, Crataegus, Draba, Hieracium, Oenothera, Poa, Rubus, Solidago,* and *Veronica.*

The conclusion from these data is relatively simple and direct. The exact length of time which these stocks have been isolated is not known—it may vary from a few to many million years. Although the American and Asiatic species are frequently recognized as distinct, many vicarious pairs exist that are exceedingly close to each other morphologically and genetically. In these generic stocks the rate of speciation has been very slow. In this discussion it has been tacitly assumed that the widely disjunct temperate floras of the northern hemisphere represent segregated floras from the originally circumboreal Tertiary woodland vegetation which was relatively homogeneous. Some of the Asiatic-American groups that are absent from Europe and western America, and largely from western Asia, are undoubtedly of boreal origin. In these cases it can be fairly assumed that the widely disjunct elements had a single, although now bicentric, center of origin. On the other hand, some of the bicentric groups mentioned have tropical affinities today, and it may be that their disjunction is in no way related to boreal conditions, especially glaciation, but represents parallel evolution of temperate members from essentially tropical floras of wide occurrence. In such cases it may be reasoned that primitive pantropical syngameons have given rise to essentially similar genera and species in eastern Asia and eastern America, and that the species of a vicarious nature have never had any connection with each other, and only a remote connection with the same ancestral syngameon. Such a theory of the polytopic origin of a species or of two or more closely related species requires also the assumption of orthogenesis, for otherwise the parent syngameon could never have resulted in essentially the same kinds of organisms except by the rarest chance. Du Rietz (203) has reached approximately this conclusion in discussing bipolar distributions with broad transtropic disjunctions. This topic is discussed elsewhere.

One more example must suffice to round out the indication that in some groups of organisms the rate of speciation is slow. Jordan (382) discusses the relationship between the fishes on the two sides of the Isthmus of Panama. On the basis of studies by Gilbert and Starks, he states that of 374 species of fish from the Bay of Panama, 11 per cent are unchanged on the Pacific side and about 70 per cent consist of pairs of species, one in Pacific and one in Atlantic waters. In most of these cases the paired species are more closely related to each other than to any other species of their genera. Jordan says, "Comparing the fish faunas separated by the isthmus, we find the closest relation possible so far as families and genera are concerned. In this respect the resemblance is far closer than between Panama and Chile, or Panama and Tahiti, or Panama and southern California. On the Atlantic side, similar conditions maintain." All the ichthyological evidence favors a former open communication between the two oceans which became closed some time prior to the Pleistocene—according to Jordan, probably in the early Miocene.

With respect to general similarities between modern and ancient floras as determined by paleontology, Thomas (623) says that the majority of the few hundred species of flowering plants from the Cretaceous in different parts of the world can be assigned to living families, although the approximate age of the Cretaceous has recently been given as around ninety million years. Coming down to the Tertiary, Thomas claims that Eocene seeds and fruits can be assigned to living genera in about one-fourth of the cases, but that the remaining cases are so close to living genera in leaves, seeds, and fruits that their general affinities cannot be questioned. While there has been morphological change during the sixty million years or more since the Miocene began, it seems clear that it has been very slow in many plants which occur in our records. The generic identification of Miocene flowering plants is usually clear, but many of the species are no longer referable to living specific groups. A number of fossils at the close of the Tertiary in the Pliocene can be referred to modern species, and Pleistocene species are nearly always considered to belong to living groups. In some cases the fossil material has been preserved in an excellent state, as in the Oligocene ambers where floral structures can be studied in detail. It is indeed remarkable how little floral structure has evolved in many lines.

PART FIVE

Significance of Polyploidy in Plant Geography

26.

Some Preliminary Considerations
of Polyploidy

A quarter of a century ago Winge (703) summarized the then existing knowledge concerning chromosome numbers[1] and came to the conclusion that a comparative study of this subject would result in important contributions to systematics and phylogeny. About a decade later Tischler (631, 633) correlated what was then known about cytology and taxonomy and concluded that "the value of chromosome morphology and chromosome number is universally recognized as an important factor in systematic botany." He recognized four type-groups of plants with respect to chromosome number: (1) the *Pinus* type, in which all species of the genus have the same chromosome number; (2) the *Chrysanthemum* type, in which the species have euploid (multiple) chromosome numbers; (3) the *Carex* type (aneuploid or dysploid), in which the chromosome numbers differ, but not in a simple polyploid series; (4) the *Antirrhinum* type, in which the number shows only small variations of one or two chromosomes plus or minus the regular number. A type is not always characteristic of a genus or larger taxonomic group; two or more types may occur within the same genus. About this time Hagerup (312) suggested probable phylogenetic relationships between species of Ericales[2] based on sequences of chromosome numbers. Since taxonomic and phylogenetic phenomena frequently cannot be dissociated from geographical considerations, Hagerup, Tischler, and Müntzing in various contributions have made suggestions or proposed rules with respect to the geographical and ecological relations between diploids and polyploids for groups of related species or intraspecific chromosomal races. These ideas will be examined in detail later.

[1] General catalogues of chromosome numbers have been published by Gaiser (259, 261, 262, 263) and Tischler (630, 634, 637, 638, 641), and several authors have made lists for such families as Leguminosae (568) and Ranunculaceae (301).

[2] Hagerup used the group name Bicornes.

Some useful terms and symbols.—In each life cycle higher plants have an alternation of generations which consists of a sporophytic generation terminating in spore production and a gametophytic generation terminating in gamete production. Exceptions result from the loss of the sexual process and the substitution, in the formation of seeds, of reproduction by one of several kinds of non-sexual processes. In other cases seeds may not be formed, plants reproducing only by vegetative means.

The sporophytic generation—which is also the somatic generation in higher plants—begins with the union of two gametes in the process of fertilization subsequent to pollination, and hence with the zygote, or fertilized egg. The zygote has a double complement of nuclear materials, one complement from each gamete. The normal diploid cell—the zygote or its derivatives in the sporophytic tissues—consists of two complete homologous sets of chromosomes which may or may not be homozygous. If A equals one set of chromosomes, each gamete carries a set; the zygote, or any diploid cell, consists of two sets, AA. The sporophytic generation is terminated by the meiotic divisions which result in spores and eventually in the male and female gametophytes. In each nucleus of these cells there is a single set of chromosomes.

It is customary to let n stand for a set of chromosomes or a genom; that is, n stands for the basic haploid number of chromosomes in a set. This number may be 7, 8, or 9, to select common numbers for examples. With the same system of notation, $2n$ stands for the basic diploid condition, that is, for two sets of n chromosomes. For the above examples, these numbers would be 14, 16, 18. It frequently happens in plants that there are races within a species or species within a genus in which the chromosome numbers of some of the kinds are in excess of the basic diploid number just described. When the larger numbers are larger by cardinal multiples of n, there exists what is called a polyploid series of the euploid type. Deviations from the numbers of a regular euploid series may also occur; these numbers are called aneuploid (dysploid) numbers.

The following terminology is applied to euploid series:

Balanced Types	Unbalanced Types
$2n$ = diploid	$3n$ = triploid
$4n$ = tetraploid	$5n$ = pentaploid
$6n$ = hexaploid	$7n$ = heptaploid
$8n$ = octoploid	$9n$ = enneaploid
$10n$ = decaploid, etc.	

These designations all refer to the number of sets of chromosomes in cells of the sporophytic generation. If $n = 8$ chromosomes, the cells of a tetraploid will contain $4 \times 8 = 32$ chromosomes in their somatic nuclei. At reduction division an equal division into two nuclei is possible, each nucleus containing two sets of eight chromosomes. In the basic condition the alternation of generations and the chromosome condition can be symbolized as in Example 1 below. The gametophytic nuclei are spoken of as haploid, and the sporophytic nuclei are called diploid, n and $2n$ respectively. Examples 2 and 3 symbolize the alternation of generations and chromosome conditions in tetraploids and hexaploids, respectively.

	Gametophyte *(Haploid)*	*Sporophyte* *(Diploid)*	*Gametophyte* *(Haploid)*	
1.	n ..... n	.... $2n$	..... n .... n	..... Diploid
2.	$2n$ ..... $2n$	.... $4n$	.... $2n$ .... $2n$	..... Tetraploid
3.	$3n$ ..... $3n$	.... $6n$	.... $3n$ .... $3n$	..... Hexaploid

Confusion arises because the term haploid sometimes is used for both the n number of chromosomes and the gametophytic number, whether it represents one set (n) or more. Likewise, the term diploid is sometimes used for the basic sporophytic chromosome number $(2n)$ and for sporophytic conditions, regardless of whether the number of sets is greater than two.

To make the point clearer, let us suppose that the chromosome number is known for only one species in a genus or that only one number is known for several species of a genus, and that the alternation is between 16 and 32 chromosomes for the gametophyte and the sporophyte, respectively. These are the haploid and diploid numbers, and $n = 16$ chromosomes. We may suspect that the 16 chromosomes include two or four sets, but we do not know; therefore $n = 16$. Let us suppose further that later another species in the genus is found to have a gametophytic number of 8. Its alternation of generations would be between nuclei of 8 and of 16 chromosomes; hence it is clear that the first numbers do not represent the basic chromosome numbers for the genus. They would now be referred to as tetraploids because we know that the alternation is between $2n$ and $4n$ chromosomes.

These examples have dealt with balanced polyploids, such as tetraploids, hexaploids, octoploids, etc., in which the chromosomes can be divided into an equal number of sets at meiosis. The remaining polyploids are called unbalanced (triploid, pentaploid, etc.); they cannot halve their chromosome numbers in reduction division by reassembling them into an equal number of sets.

Polyploids can also be considered to belong to two types depending upon the origin of their chromosome sets. When A equals a set of chromosomes characteristic of a species ($n = A$, in this case), the ordinary alternation of generations of a basic diploid is shown in Example 4.

Gametophyte (*Haploid*)	*Sporophyte* (*Diploid*)	*Gametophyte* (*Haploid*)	
4. $\left.\begin{array}{l} A \dots \\ A \dots \end{array}\right\}$	$\dots AA \dots$	$\left\{\begin{array}{l} \dots A \dots \\ \dots A \dots \end{array}\right.$	$\left\{\begin{array}{l} \\ \end{array}\right.$ $\dots AA \dots$ Diploid
5. $\left.\begin{array}{l} A \dots \\ A \dots \end{array}\right\}$	$\dots AA \dots$	$\left\{\begin{array}{l} \dots AA \dots \\ \dots AA \dots \end{array}\right.$	$\left\{\begin{array}{l} \\ \end{array}\right.$ $\dots AAAA \dots$ Tetraploid
6. $\left.\begin{array}{l} A \dots \\ A \dots \end{array}\right\}$	$\dots AA \dots AAAA \dots$	$\left\{\begin{array}{l} \dots AA \dots \\ \dots AA \dots \end{array}\right.$	$\left\{\begin{array}{l} \\ \end{array}\right.$ $\dots AAAA \dots$ Tetraploid

Two things may happen to such a plant to produce a tetraploid. The reduction division may fail for some reason, with the resultant production of diploid or unreduced gametes (Example 5), or the sets of chromosomes may become doubled in the soma of the sporophyte (Example 6). Such tetraploids are called autotetraploids because there has been a reduplication of similar sets of chromosomes; that is, there has been a doubling of sets of one species.

In contrast to autopolyploids, some polyploids are of hybrid origin and result from the duplication of dissimilar sets of chromosomes. These polyploids are called allopolyploids, in accordance with the term introduced by Kihara and Ono (391). If A still equals a set of chromosomes (genom) of one species, and B equals a set from another related species, we can symbolize certain conditions as shown on page 403.

In Example 7 the non-homologous sets A and B frequently result in a sterile diploid; at least there can be no segregation by sets. In Example 8 there has been a somatic doubling of chromosomes, with the result that the

F_1 is frequently more fertile than the diploid hybrid or completely fertile. Example 9 supposes the union of unreduced gametes which also results in increased fertility of the hybrid.

Gametophyte (*Haploid*)	*Sporophyte* (*Diploid*)	*Gametophyte* (*Haploid*)	
7. $\begin{cases} A\ldots \\ B\ldots \end{cases}$	$\ldots\ldots AB\ldots\ldots$	$\begin{cases} \ldots\ldots ? \\ \ldots\ldots ? \end{cases}$	Usually a sterile F_1 diploid
8. $\begin{cases} A\ldots \\ B\ldots \end{cases}$	$\ldots AB\ldots AABB\ldots$	$\begin{cases} \ldots AB\ldots \\ \ldots\ldots AB\ldots \end{cases}$	$\ldots AABB\ldots$ Allotetraploid (Amphidiploid)
9. $\begin{cases} AA\ldots \\ BB\ldots \end{cases}$	$\ldots\ldots AABB\ldots$	$\begin{cases} \ldots\ldots AB\ldots \\ \ldots\ldots AB\ldots \end{cases}$	$\ldots AABB\ldots$ Allotetraploid (Amphidiploid)

A genus will sometimes contain diploid species and both autoploids and alloploids. We can represent the diploid species as *AA, BB, CC,* etc., and the autotetraploids as *AAAA, BBBB, CCCC,* etc. Other forms, such as *AABB, AACC, BBCC,* and *AABBCC,* are allopolyploids. All these examples are of the balanced type.

Of the polyploid types Stebbins (605) says, "I believe that for practical purposes the best definition of an autopolyploid is a polyploid of which the corresponding diploid is a fertile species. An allopolyploid is a polyploid containing the doubled genoms of a more or less sterile hybrid." These definitions avoid the use of cytological behavior characteristics employed in the definitions formulated by Müntzing (480) and Darlington (174). In horticulture, according to Hurst (363), autopolyploids are called polyploid *varieties,* whereas allopolyploids are called polyploid *species.* Consequently in this branch tetraploid forms such as *AAAA* or *BBBB* are given only varietal rank in nomenclature, but such forms as *AABB* or *AACC* are given full specific status. It appears, however, that this practice has not been strictly adhered to because it is entirely arbitrary.

We have already seen from the preceding discussion that the ordinary nomenclature and symbolism are inadequate and even confusing at times. For example, if we find that the sporophytic nuclei of a plant contain 32 chromosomes, and this is all we know about it, we can write only that $2n = 32$. If we then find a related species of $2n = 16$, the first-mentioned species can be written $4n = 32$. We do not know whether the sets are *AAAA,*

BBBB, AABB, AAAB, etc.; all we know is that the $4n$ plant is some kind of tetraploid.

Various systems of notation have been proposed. When chromosome numbers have been determined from gametophytic cells (usually pollen grains) the author may write, as in different species of *Rumex, n* = 8, 12, 16, 24, 40. When these numbers have been determined from sporophytic cells (usually root tips) he may write, as in different species of *Senecio, N* = 10, 20, 30, 40, 50 . . . 180. Sometimes chromosome numbers are obtained from gametophytic or sporophytic cells but not consistently from one; in such cases it is common practice to reduce sporophytic numbers by $N/2 = n$. We can thus write:

$N/2 = 8$ (1 set of 8 chromosomes in the gametophyte) = diploid.

$N/2 = 16$ (2 sets of 8 chromosomes in the gametophyte) = tetraploid.

$N/2 = 24$ (3 sets of 8 chromosomes in the gametophyte) = hexaploid.

This system, however, is useless for unbalanced polyploids and for aneuploids. For example, in a triploid $N = 3$ sets of chromosomes (3×8, in our example), but the gametophytic number cannot be found by $N/2$ because the segregation is not predictable. Nor do modifications of this system help us. $N/n = 16/8 = 2$ (diploid), or $N/n = 32/8 = 4$ (tetraploid), and $N/2 = 8$, $N/3 = 8$ (triploid), and $N/4 = 8$ (tetraploid), when $N = 16$, 24, 32, respectively, are of no help.

Perhaps a solution can be found by employing a noncommittal set of symbols: $S =$ the sporophytic nuclear condition without reference to more than the number of chromosomes, and $g =$ the gametophytic nuclear condition. For example, in *Solanum* the somatic numbers are $S = 12$, 24, 36, 48, 60, 72, 108, 144, and all are balanced. In *Chrysanthemum* there are both balanced and unbalanced conditions: $S = 18$, 27, 36, 45, 54, 90. In *Hyacinthus* $S = 18$, 19, 21, 22, 23, 24, 25, 27, 28, 30, 36. Gametophytic numbers can also be indicated directly by $g = 6$, 12, 18, 24, 30, 36, 54, 72 for *Solanum*, or whatever is observed for such aneuploids as *Hyacinthus*. The basic number of chromosomes in a gametophytic set is symbolized by n; in *Hyacinthus*, for example, $S = 18 = 2n$, if $n = 9$; or $S = 19 = 2n$ plus 1, etc.

In wheats it is known that $S = 14$, 28, 42; Einkorn and Aegilops are wild types with $S = 14$, in Emmer $S = 28$, and in *Triticum vulgare* cultivated wheats $S = 42$. These designations are noncommittal, but it happens that considerably more is known about wheats than the above would show. Einkorn $S = 14$ consists of two sets of 7 ($2n$) which are symbolized as *AA*. In Aegilops also $S = 14$, but the two sets of 7 are different from those in

Einkorn and are symbolized as CC. The Emmer wheats $S = 28$ have four sets of chromosomes ($4n$), two of which are like Einkorn (AA) and two are different, but not like Aegilops; hence the Emmer wheats are symbolized as $AABB$. In *Triticum vulgare*, $S = 42 = 6n = AABBCC$.

In roses several important characters can be associated with particular sets of chromosomes; a set called A carries factors for compound inflorescence, a B set determines small single flowers, a C set determines crooked stems and prickles. In the Tea Rose "Lady Wellington" $S = AAA$; in Tea Rose "Gloire de Dijon" $S = AAAA$; Perpetual Roses are $AACC$; Hybrid Tea Roses are $AAAC$, etc.

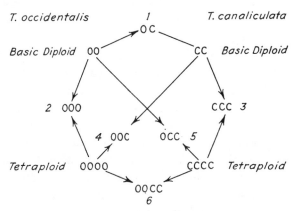

FIG. 53. A diagram illustrating certain relationships in *Tradescantia*. *T. occidentalis* and *T. canaliculata* exist as diploids (OO and CC) and as autotetraploids ($OOOO$ and $CCCC$). Nos. *1* and *6* are semi-fertile interspecific hybrids. Nos. *2, 3, 4,* and *5* are sterile triploids. Nos. *2* and *3* are intraspecific hybrids, and Nos. *4* and *5* are interspecific hybrids. After (24).

In euploids the heredity, the sterilities, and the taxonomy can often be correlated with the character of entire chromosome sets. In these cases the use of the symbol S for the sporophytic condition is an advantage. The following notes indicate the general relations:

Diploid species: $S = 2n = AA, BB, CC$, etc.

Diploid hybrids: $S = 2n = AA', BB', CC'$, etc. Such hybrids will be fertile if the differences between A and A' or B and B' are few or not of the sort to produce sterility.

Diploid hybrids: $S = 2n = AB, AC, BC$, etc. These hybrids will probably be sterile because of the non-homologous sets A and B or B and C.

Autotriploids: $S = 3n = AAA, BBB, CCC$, etc. Autotriploids are usually sterile in spore formation because of the multiple pairing of homologues.

Allotriploids: $S = 3n = AAB, ABB$, etc. Such plants are F_1 hybrids of

$2n \times 4n$ species. As a rule they are sterile in spore formation because one set has nothing to pair with.

Autotetraploids: $S = 4n = AAAA, BBBB, CCCC$, etc. These are usually sterile in spore formation because of the interference of multiple pairing. The homologous chromosomes may develop differences which result in the formation of bivalents, in which case they become fertile because a 4-seriation has given way to a 2-seriation.

Allotetraploids: $S = 4n = AABB, AACC, BBCC$, etc. Such amphidiploids are usually highly fertile because through the process of doubling they have become homozygous for each chromosome type.

The euploid series could be continued. In general it can be said that allopolyploids tend to be more fertile and that autopolyploids tend to be more sterile than their progenitor diploids.

The frequency of polyploidy in plant groups.—Winge (703) concluded that "in the case of higher plants it is a rule that the species in a more or less systematical group have 'related' chromosome numbers, i.e., values which are all single multiples of one and the same cardinal number, and which enter into arithmetical progression." At this early date sufficient species had been investigated for him to assemble such series as those listed below, but he certainly generalized far beyond what later investigations have supported. The following numbers are all for the gametophytic generation: Anthemideae (Compositae): 9 (9 species), 18, 27, 36, and 45 (1 species each); Heliantheae (Compositae): 8 (3 species), 16 (2 species), and 32 (3 species); Solanaceae: 12 (3 species), 18, 24, 36 (1 species each); *Rumex* (Polygonaeae): 8 (4 species), 12, 16, 24, 40 (1 species each); Chenopodiaceae: 6 (1 species), 9 (8 species), and 18 (1 species). Winge went on to say, "In my opinion it will be clearly seen that most families are characterized by quite definite series of values for the number of chromosomes. . . . Nature proceeds according to rules which do not appear to be hopelessly inscrutable, and which may be put to good use in the study of plant systematics. . . . It is by no means possible to construct the natural plant system from a knowledge of the chromosome numbers alone. But in critical cases, when seeking to decide the question of possible relationship between minor systematic units, the chromosome number will at times afford the final weighty argument pro or contra." This prognostication frequently has been borne out.

Something of the widespread occurrence of polyploids is attested to by Hurst (363), who said that autopolyploids are known in apples, strawberries, mulberries, bananas, sugar cane, maize, tomatoes, citrus, ornamental cherries, roses, tulips, hyacinths, dahlias, cannas, daturas, campanulas, evening primroses, petunias, catchflies, buttercups, and Iceland poppies. At the time of his paper, allopolyploids were known from *Rosa, Rubus, Fragaria, Crataegus, Prunus, Potentilla, Alchemilla, Chrysanthemum, Senecio, Hieracium, Tulipa, Triticum, Aegilops, Avena, Hordeum, Populus, Salix, Vaccinium, Linum, Draba, Trifolium, Digitalis, Solanum, Viola, Primula, Nicotiana, Rumex, Polygonum, Rheum,* and many other genera. According to Hurst, the great majority of polyploids are of the hybrid or allopolyploid type. In roses, for example, where both kinds are present, out of 1006 forms examined, 629 (62.5 per cent) were found to be polyploid, and of these 608 (96.6 per cent) were allopolyploid.

It will be interesting to see the extent of polyploidy in some genus which has been thoroughly studied. *Crepis* (Compositae) serves well for this purpose. Hollingshead and Babcock (349) established 11 as the gametophytic number that is basic for the endemic western American species of *Crepis.* They believe, however, that the American diploid species ($S = 22$) are themselves of hybrid origin, having come through chromosome doubling, from species of eastern Asia which had gametophytic numbers of 4 and 7. The details of chromosome numbers for western American species are given by Babcock and Stebbins (39) and are presented in Table 26. In the cases in this table, the chromosome numbers have been determined by cytological examination; in addition, these authors have made a much wider survey from herbarium specimens. They find that they can predict with considerable accuracy the polyploid condition of related plants by a statistical survey of stomatal sizes and frequency.[3]

As has already been noted, the frequency of polyploidy in certain families is lower than the average for all families and much lower than that for certain families. For the Leguminosae, which has been referred to as a family with little polyploidy, Senn (568) assembled all the known cytological data. He found reports on 434 species scattered through 71 genera; 23 per cent are polyploid. Among 296 species of 68 genera in the Cruciferae, Manton (441) found 41.5 per cent to be polyploid. Polyploidy is not uniformly present throughout a family in most cases. Senn found 42 genera in the Leguminosae that are completely diploid and only 10 genera that are completely polyploid. Furthermore, the polyploid genera are much localized taxonomically. The subtribes Aeschynomeninae and Stylosanthinae are entirely polyploid, and Spartiinae and Cytisinae are entirely polyploid except for one species. Gregory (301) found that in 452 known species, in the Ranuncu-

[3] This technique is discussed in Chapter 27.

TABLE 26.—Chromosome numbers in the western American species of *Crepis*
(Data from 39)

Species (Excluding C. nana, C. elegans, and C. runcinata)	Sporophytic Chromosome Numbers						
	22	33	44	55	–	77	88
C. pleurocarpa	1		1				
C. monticola	1		1	1			
C. occidentalis							
subsp. typica	3	6	2				
" costata			1				
" pumila		1				1	
" conjuncta			1				
C. Bakeri							
subsp. typica			1				
" Cusickii	2						
C. nodocensis							
subsp. typica			1				
" subacaulis			2				
C. exilis							
subsp. originalis	1		1	1			
C. acuminata	2	2	1				
C. intermedia		3	2	2			1
C. barbigera			1				4

laceae the relative frequency of diploids is much greater than of polyploids. Within this family there are extremes of stability and instability in chromosome members. In *Clematis* 38 of 40 cytologically known species are simple diploids; a similar situation holds in *Aquilegia*, with 19 of 20 known species diploid. Not only is polyploidy very rare in *Aquilegia*, but interspecific hybrids show little sterility even though the crossed species come from opposite sides of the world. *Ranunculus* and *Thalictrum* are a contrast. *Ranunculus* has run the gamut of chromosome number variation, according to Gregory, with gametophytic numbers among its 250 species of 7, 14, 21, 28, and 8, 16, 24, 32, 40, 48, and 64, and with polyploids predominating. Polyploidy is the rule in *Thalictrum* (about 75 species), with gametophytic numbers of 7, (12), 14, 21, (24), 28, 35, 42, 56, and 77, with *T. polygamum* ($S = 154$) the highest polyploid of the family.

Müntzing (480) suggested that more than half of the species of flowering plants are polyploid. It is difficult to compare this estimate for species with Stebbins' estimate for genera. Stebbins (605) says:

Since allopolyploids usually have different geographical ranges from those of their diploid parents, they are subjected to a different selective activity of the environment, and might therefore be expected to give rise to new morphological types through the process of mutation and natural selection. The frequency with which this has taken place can be estimated from the proportion of genera and larger plant groups for which an allopolyploid origin can be inferred, either because their basic chromosome numbers are multiples or sums of lower numbers existing in related groups or because their basic numbers are so high that polyploidy is the best *a priori* explanation for their origin. The number of these genera is not large; they form about 16 per cent of the genera of Angiosperms which are well enough known cytologically so that any inference can be made concerning them. Most of these genera, moreover, are in complex families like the Rosaceae and the Malvaceae, and have close relatives with lower chromosome numbers. Probably the largest and most diverse single group of plants of which the allopolyploid origin is clearly established is the subfamily Pomoideae of the Rosaceae. . . . The basic haploid number 19, found in the poplars, willows, magnolias and grapes, and the number 23 found in *Fraxinus,* and 40, the basic number in *Tilia,* are very likely of polyploid origin, but such cases are not the rule in the higher plants.

Obviously, polyploidy is an important phenomenon that appears unevenly in different natural groups of plants.

27.

Origin and Characteristics
of Polyploidy

---·»»«‹‹‹·---

The origin of polyploidy.—As early as 1929 Gates said that the evidence indicated that triploids may arise through (1) dispermy, the union of two male gametes with one female gamete; (2) the fusion of a haploid (normal) sperm with a diploid (unreduced) egg; (3) the fusion of a diploid sperm with a haploid egg. It had been thought that tetraploids arise through the union of unreduced gametes, but at the time Gates wrote, the consensus was that they most likely arise through a reduplication of sporophytic sets. Winge (704) reviewed 24 cases of tetraploids reported in the literature and found that in eight there had been chromosome doubling in the sporophytic tissues of the F_1 hybrids, and in five the tetraploids had resulted from the union of unreduced gametes of the F_1 plants. The remaining eleven cases were not sufficiently clear to indicate how the doubling occurred. According to Gates, pentaploids have their origin in crosses between diploid and triploid plants, and hexaploids could arise from crosses between triploid and triploid or diploid and tetraploid, producing sterile plants which in rare cases double their chromosomes and become fertile. Octoploids appear to arise through hybridization of diploids and hexaploids followed by doubling: $6n \times 2n = (3n \& n)\ 2 = 8n$. Belling (50, 51) attributed triploidy to the union of a haploid with a diploid gamete, the diploid gamete apparently often originating from the action of unsuitable temperature in the course of the reduction division. Tischler (635, 636) thought that polyploidy frequently results from plants being exposed to severely cold climates. Böcher (75) also says that temperature may be important in causing cytological irregularities in nature. For example, *Ranunculus reptans* in Denmark was found to be cytologically regular, but in Greenland it is highly irregular. Apparently this is a parallel case to that reported by Medwedewa (460) for Italian hemp which was cultivated in the Caucasus and at Moscow.

In the Caucasus no meiotic irregularities of any importance appeared, but at Moscow there were great cytological irregularities. In the buttercup and the hemp it appears that temperature extremes to which the plants are not originally adapted are likely causes of these irregularities. In fact, Böcher makes the following statement: "It is not unlikely that the degree of meiotic irregularities (and thereby variations) is in certain cases largest along the area limits of the species,. where they meet with unfavourable temperature conditions."

In connection with the natural origin of polyploids, Müntzing (477) believed that certain quantitative relations between embryo, endosperm, and mother tissue are necessary for the successful development of an embryo; hence double fertilization is important in setting up an endosperm-ovular tissue complex which serves as the environment for the young embryo. Consequently, the lack of development of endosperm as in Orchidaceae, or its poor development as in Leguminosae, would account for the low frequency of polyploidy in these families. Blakeslee (70), who has summarized recent work done at Cold Spring Harbor, points out that periclinal chimeras are frequently produced by colchicine treatment. Thus, if three germ layers are recognized in the apical primordium, they may be $2n\text{-}2n\text{-}2n$, $8n\text{-}2n\text{-}2n$, $2n\text{-}8n\text{-}2n$, $2n\text{-}2n\text{-}8n$, etc., numbered from the outer to the inner layer. Stigmatic surfaces of the pistil form from the outer layer and pollen grains form from the middle layer. Thus Buchholz found an explanation for the setting of capsules. "When a tetraploid female ($4n$) is pollinated by a $2n$ male the $1n$ pollen tubes grow well in the $4n$ conducting tissue [sic] of the style. In the reciprocal cross, however, when the $4n$ is the male parent, the $2n$ pollen tubes burst in the conducting tissues of the $2n$ styles. . . . Buchholz had earlier shown that pollen-tube behavior is a block to the crossing between certain species of *Datura*. His recent study of selfs and inter se pollinations among tetraploids of ten species of the genus indicates that pollen-tube growth is generally deleteriously affected by doubling chromosome numbers." Thus, another barrier in the way of successful ploidization is demonstrated. Darlington (174) suggests that there must be a certain balance between chromosome size and number and cell size. He thinks that the tendency toward gigantism in polyploids must be counterbalanced in certain polyploids to enable their survival. The failure to accomplish the necessary dwarfing or to establish the necessary volume-surface relationships in certain plant groups may account for the low frequency of polyploidy in them. His assumption is that all groups probably tend to form polyploids, but that in some groups the tendency is eliminated.

Several investigators, for example Nebel and Ruttle (485), have shown that chromosome doubling in somatic tissues may arise from the action of some factor that prevents the formation of spindle fibers. The result is that the sister chromosomes organize a single nucleus, rather than two daughter nuclei. The work of these authors was based on colchicine treatment, to be discussed later.

Heilborn (338), after assembling the various hypotheses that have been put forward to explain the origin and preservation of polyploids, and their absence in most animal and many plant groups, concludes, "A prolific development of polyploidy, such as is found in many Angiosperm families, obviously requires a certain amount of co-operation of several simultaneous circumstances: (1) the occurrence of somatic doubling or the production of unreduced gametes; (2) only a slight degree of [required] cell-constancy; (3) possibilities for self-fertilization; (4) capacity of enduring a change from separate sexes to hermaphroditism; (5) double fertilization, parthenogenesis or other incompatibility barriers which prevent swamping of the newly established polyploids by crossing with the diploid parents; (6) favourable chromosome conditions that enable a regular meiosis in the polyploids. If one of these conditions fails, polyploidy becomes rare or may be altogether lacking."

Artificially induced polyploids.—Considerable information concerning the differences between diploids and polyploids has been obtained through studies of artificially produced polyploids. For that reason a brief consideration of experimental work follows. Although the production of polyploidy by various types of treatment is not a new field of investigation, considerable impetus to such studies has followed the discovery of the action of colchicine. According to Randolph (515), the Marchals in 1908 were the first to produce new polyploid forms, experimenting with the regeneration of mosses, and Winkler in 1916 was the first to obtain tetraploid strains experimentally in higher plants, using various species of *Solanum*. Following is a list of the principal types of treatment that have resulted in chromosome doubling.

Treatment with chloral hydrate.—Nemec, according to Blakeslee and Avery (71),[1] caused the development of tetraploid segments in roots by the use of chloral hydrate. Later efforts, however, failed to show the value of this chemical in modifying shoots. Other chemicals employed more than a quarter of a century ago with little success include ether and chloroform, according to Randolph (515).

[1] In some instances the citation is to the first published work of the kind, in others it is merely to a convenient reference.

Callus shoots following decapitation.—Winkler (707) found that shoots which regenerated in connection with callus formation following injury, as in grafting, were tetraploid in 1 to 2 per cent of the cases. Later work has combined this method with the application of chemicals to the wound.

Heat treatment.—Heat treatment has taken various forms, but it usually involves shock by an alternation of hot and cold. Randolph (514) can be referred to for this type of investigation; this process was the most practical method of obtaining tetraploids until 1937. Sax (549), on the basis of his review of the pre-colchicine literature on polyploid induction, concluded that temperature appears to affect the synchronization of nuclear and cytoplasmic activities. The development of the chromosomes is accelerated in relation to other nuclear and cytoplasmic activities, according to him, so that the univalents are ready to divide at the first instead of the second meiotic division. If they divide before nuclear division is possible the chromosome number is doubled. Neither constant heat nor cold seems to be effective; it is extreme and sudden changes of temperature that are effective in most genera.

X-ray treatment.—Radiation treatments apparently have been most useful in producing gene mutations, and cases of tetraploidy are reported (288).

Centrifuging.—Kostoff (400) reported the production of tetraploids through the action of centrifugal force.

Insertion of orchid pollinia into wounds.—Povoločko (512) found that the production of polyploid shoots in regeneration from wound tissue was increased by inserting pollinia from tropical orchids into the wounds. This treatment is based on the known high hormone content of the orchid pollinia (250).

Use of bacteria.—According to Kostoff and Kendall (404), the presence of bacteria, probably through the action of some excreted material, caused an increased production of polyploid shoots from wound tissue.

Aging of seeds.—Cartledge and Blakeslee (123) found that certain seed-aging techniques resulted in a higher percentage of polyploid plants.

Colchicine application.—Blakeslee and Avery (71) reported on the production of tetraploids in a number of different plants by the application of colchicine in various ways to various plant organs. Their paper has resulted in a veritable flood of research papers on the colchicine technique and its results.[2] The chromosome doubling with colchicine treatment results from

---

[2] Havas (335), it appears, somewhat preceded Blakeslee and Avery in the use of the alkaloid. Be that as it may, the American paper is to be credited with opening a significant and popular field of investigation.

the inhibiting action of the drug on the formation of spindle fibers, according to Nebel and Ruttle (485), and this has been verified by Derman (185), O'Mara (493), and others. Not only may tetraploids result from the action of colchicine, but cells may continue to increase in chromosome number and size. In *Rhoeo* ($S = 12$, normal diploid) Derman (185) found some stamen hairs with 96 chromosomes, anther cells with about 192 and 384 chromosomes, and ovular tissue that had as many as 48. Levan (416) reported *Allium* cells in which the chromosomes numbered between 500 and 1000.

Acenaphthene.—Kostoff (402) reported the production of polyploids by the use of acenaphthene. Blakeslee (70) considers it a very poor agent for this purpose.

Heteroauxin.—Greenleaf (298) appears to be the first to have used heteroauxin (indol-3-acetic acid) successfully in the production of polyploids. On applying it to decapitated plants, he obtained about 14 per cent polyploid shoots.

Phenyl urethane.—Blakeslee (70) stated that phenyl urethane is effective on *Datura* when used in concentrations near saturation.

Sanguinarine hydrochloride.—Little (420) found that sanguinarine hydrochloride had an effect similar to that of colchicine when tried on *Antirrhinum majus,* the snapdragon.

In the following sections some structural, physiological, and ecological attributes of polyploids are compared with those of diploids. Since natural and artificial polyploids seem to differ from diploids in about the same way, they are discussed together. No attempt has been made to cite the literature completely. Furthermore, the distinction between structural and functional changes is not always clean cut, for of course it cannot be so in fact; hence the grouping of topics is based on convenience.

Morphological comparisons of diploids and polyploids.—From a survey of the literature, Müntzing (480), in pre-colchicine days, found that in every case where intraspecific chromosome races are known there are morphological differences between diploids and polyploids, and that there is a marked correlation between chromosome number and gigas characters. In fact, he concluded that the gigas characters were caused by the higher chromosome numbers and were attributable to increased cell size. Müntzing found that tetraploids differ from the diploids from which they were derived in the following easily observed ways: stems are thicker and longer; leaves are larger, thicker, shorter, broader, and darker green; floral parts are bigger and more showy, and seeds are larger. That the differences are not always like those described above, and that there are other differences, is shown by the following discussion.

Tetraploid flowers are usually larger than diploid flowers; this is mainly attributable to corolla differences. Larger flowers are reported for tetraploids of *Gossypium* (184), *Canna indica* and *Datura stramonium* (72), *Lilium formosanum* and *Antirrhinum* vars. (212), and *Petunia* (70). This is the general rule and it has been observed in many other plants. There are exceptions, however; thus in *Portulaca grandiflora* and *P. parana* (70) the flower size was not changed with polyploidy. New floral types may result from chromosome doubling. In *Portulaca* (70) there are two types of double corollas among diploids (*DD* and *Dd*), and among the tetraploids there are new-type doubles (*DDDd* and *Dddd*). The *DDDd* type is partly fertile; selfing it produces high-grade doubles. In tetraploids the petals may be thicker or changed in form, as in snapdragons where the corolla is more ruffled (212). The inflorescence may be changed, as in snapdragons where the spikes are shorter, and there is a more pronounced tendency to produce double flowers.

Other flower parts are affected by chromosome doubling. Thicker stamens were reported by Blakeslee (70) for the amphidiploid of *Datura metel* × *D. meteloides*. Attention has been directed especially to the pistil, fruit, and seed. In this cross the style was shorter; in daturas generally the capsules become shorter and broader as the *n*-number increases. Cotton boles are larger for tetraploid than for diploid species, according to Denham (184). In gourds the effects of polyploidy on fruit shape are variable; apparently what happens depends upon the genic constitution of the chromosomes that are doubled (70). It appears that generally there is an increase in seed size with tetraploidy; Blakeslee (70) mentions such results in *Rudbeckia hirta, Cannabis sativa, Stellaria media, Cosmos sulphureus, Bidens leucantha, Lychnis dioica,* and *Datura stramonium*. Higher polyploids do not result in still greater seed size, and seed setting is greatly reduced or there is none at all for hexaploids and octaploids.

Foliage leaves are generally broader and thicker in tetraploids than in diploids, and frequently are roughened. The increased size is apparently due to increased cell size rather than to cell number, but the roughened condition is sometimes due to mixoploidy.

In conjunction with the phenomenon of cell-size increase, two features are worthy of special notice. Stomata are affected in such a way that their frequency per unit area is less and their pores and guard cells are larger. For the same reason, epidermal hairs are more widely spaced, but in *Matthiola incana* glandular hairs are also larger and more branched in polyploids (212). Larger cells also result in coarser venation (less total vein length per unit area) that tends to become fasciated (298). The other important feature is pollen grain size; this appears to be double volume in tetraploids. Blakeslee (70) says that there is no exception to the rule that doubling chromosome number increases pollen size. These two features, stomatal frequency and pollen size, are both useful in permitting a quick check of diploid and polyploid races of a species. Wettstein (683, 684, 685) found an increase of moss spore and capsule size with polyploidy and, after several

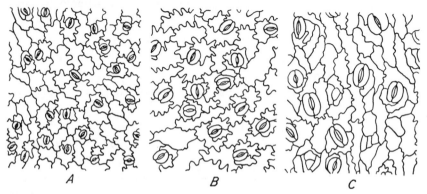

Fɪɢ. 54. Drawings of the epidermis of *Nicotiana* showing the difference between diploids (*A*), colchicine-induced tetraploids (*B*), and octoploids (*C*). Redrawn from (298).

generations, a return to diploid size. No such size regression is known among flowering plants according to Blakeslee (70). Two $4n$ lines of *Datura stramonium* have been inbred for 13 and 14 generations and the large pollen size is still maintained. In *Cuthbertia graminea,* autotetraploids are conspicuously larger than diploids (Table 27), and hexaploids have an even greater gigantism (270).

With respect to the general growth of plants with doubled chromosomes, it is usually observed that they are stockier and coarser because of thickened stems and shorter internodes. What is described as increased vigor frequently occurs in newly formed tetraploids. Clausen (147) says that monobasic *Nicotiana* species respond to tetraploidy in the characteristic manner, with enlarged growth features, but that dibasic (amphidiploid) species show decreased vigor and total growth, and ploidy higher than the first doubling has adverse effects in both auto- and allo-polyploids. Emsweller and Ruttle

TABLE 27.—Comparative measurements of diploids and autotetraploids of
Cuthbertia graminea Small (Data from 270)

	Tetraploids		Diploids	
Character	Mean (mm.)	No. of Plants Measured	Mean (mm.)	No. of Plants Measured
Height of plant	205.0 ± 4.4	53	158.0 ± 4.6	18
Length of longest leaf	144.0 ± 3.2	53	117.0 ± 3.2	18
Width of widest leaf	4.1 ± 0.1	53	2.2 ± 0.1	18
Corolla diameter (average)	20.0	20	16.8	6
Sepal length (average)	5.3	20	4.3	6
Seed length	1.95 ± 0.03	50	1.45 ± 0.02	50
Seed weight (average)	1.48 mg.	162	0.73 mg.	176

(212) found that in snapdragons hybrids between tetraploids were vigorous
and maintained good size, as did diploid hybrids. There are some reports
of natural populations in which the autopolyploid races are structurally
similar to the diploid races. This is true of *Tradescantia* species, according
to Anderson and Sax (22), and of *Galax aphylla* (45). In the latter species
there is a tendency toward larger size in the tetraploids, but it is insufficient
to be detected under field conditions. Fabergé (228) found that tetraploid
tomato embryos were about 30 per cent heavier than diploid, but this initial
advantage was overcome after about two weeks' growth and the chromosomal types ended up about alike in growth experiments. According to
Sears (560), tetraploids from certain grass hybrids were no larger than
diploids. Smith (598) reported increased growth for amphidiploids (tetraploids from sterile hybrids) in some cases but not in others.

Apparently contradictory evidence exists in several published accounts
concerning size and vigor of tetraploid and diploid species in nature. Gaiser
(260) found no correlation between chromosome number and plant size in
Anthurium, and pointed out that it did not exist for such genera as *Triticum, Rosa, Rubus,* and *Crataegus.* Erlanson (226) examined over 100 forms
of *Rosa* and found that the tetraploids were on the average smaller than
the diploids and that their growth form was related to their habitat, the
prairie forms being predominantly tetraploid. According to Blackburn (68),
there was a complete lack of gigas characters for the Spanish races of *Silene
ciliata* in which very similar plants were 2*n*, 4*n*, or even 16*n*. Stockwell (613)
found the northern race of *Opuntia polycantha* (S = 66) to be smaller than
more southern races (S = 44). Clausen, Keck, and Hiesey (146) observed

that *Potentilla diversifolia* was more dwarfish than the alpine *P. gracilis*, despite its chromosome number being larger. Likewise diploid *Artemisia Suksdorfii* is larger than its tetraploid and hexaploid close relatives.

By way of contrast, the general observation that induced tetraploids are more vigorous and larger than their progenitor diploids is supported by field studies in several cases. Müntzing (475) found that in *Potentilla argentea*, with 2*n*, 6*n*, and 8*n* races, there is a direct increase of vigor with chromosome number. Blackburn (68) observed that the Italian *Silene ciliata* (= *S. Graefferi*) consists of diploids and tetraploids, with some gigantism in the latter race. Nakamura (483) reported that the autohexaploid *Solanum nigrum* in northern Japan is markedly larger in both vegetative and reproductive organs than the southern diploid which has recently been named *S. photeinocarpum*.

The contrast between the extreme views of Müntzing (480) ("Intraspecific chromosome races are characterized by quantitative differences and positive correlation between chromosome number and gigas characters") and of others (there is no correlation) may be resolved by the following consideration. From the literature (such as 473, 313, 413, 269, 449, 568, and many others), it appears that there is a well-established increase in cell size with chromosome number, and that this results in some gigantism of tetraploids, especially monobasic or autotetraploids. This size contrast may disappear as the longer-established autotetraploids take on more characteristics of diploids through mutations of one kind or another. If that is true— Darlington (174) believes that there must be a restoration of cell volume/surface ratios through the segregation of factors for dwarfing—then it is not to be expected that long-established polyploid series of species would show a positive correlation between plant size and vigor and chromosome number. Closely connected with Darlington's theory is Wettstein's direct evidence, referred to above, concerning cell-size regression with time, and Blakeslee's negative evidence for *Datura*.

The literature gives some evidence that a change in chromosome number may result in a change in life form. There are considerable data to support the claim that tetraploids develop more slowly than diploids and that this retarded rate of development may change annuals to biennials or perennials. Longley (423) noted that *Euchlaena mexicana* (S = 20) is annual, whereas the closely related *E. perennis* (S = 40) is perennial. He found a similar relation in sorghum (425). A number of species are annual and diploid, as in *Sorghum sudanensis*, whereas the closely related *S. halapensis* (S = 40) is perennial and tetraploid. Both Longley and Huskins and Smith

(366) believe that the tetraploid sorghums were derived from the diploids and that the life-form change was an accompanying phenomenon. Hagerup (313) found in another grass genus that the diploids are annual and the polyploids perennial. Annual *Eragrostis cambessediana* (S = 20) is diploid; the closely related perennial species are *E. albida* (S = 40) and *E. pallescens* (S = 80). Randolph (514) reported that diploid *Zea mays* is annual and that the tetraploid is perennial. Fagerlind (230, 231) found seasonal dimorphism in *Galium palustre*, with the summer form diploid and the autumnal form octoploid.

Such cases as these led Tischler (639) to conclude that in many cases a perennial form has originated from an annual form as the result of a retarded growth rate caused by the slowing down of cell division following polyploidy. This opinion is similar to that held by Müntzing (480), who also contributed cytological material to the old debate of whether woody plants are more primitive than herbaceous plants, and perennials more primitive than annuals. He tabulated the chromosome number of 582 species in 44 European genera in which both annuals and perennials occur. The average chromosome numbers for annuals was 10.65; biennials, 15.50; perennials, 16.95. In 34 of the 44 genera, the average chromosome number was higher for perennials than for annuals by 59 per cent, a statistically significant amount.

Senn (568), after assembling all the data on chromosome numbers of the Leguminosae, concluded that they support Müntzing's contention that a large number of perennial species have arisen from annual types with lower chromosome numbers. The data also suggest that woody legumes may have originated from herbaceous legumes with lower chromosome numbers. Senn found the following average chromosome numbers: annuals 11.05 and perennials 12.95; all herbaceous plants 10.57 and woody plants 15.46. These averages from a single family compare favorably with Müntzing's, which were taken across family lines. Although the direction of evolution in the Ranunculaceae seems to have been the same as in the Leguminosae—from annual, to herbaceous perennial, to woody types—Gregory (301) found that polyploidy apparently had nothing to do with it. *Clematis* represents the maximum of woodiness in the family; although it is the third largest genus and is geographically widely distributed, only two polyploids have been discovered among the nearly 40 species examined. Stebbins (604) concluded from his statistical study of chromosome numbers of 151 herbaceous and 51 woody genera that "the tendency toward polyploidy characteristic of angiosperms is manifest chiefly in perennial herbs, and less marked in woody plants as well as in annuals." The average number in herbaceous genera is 9.0 and 13.3 in woody genera in the groups which he thought were well enough known cytologically. Stebbins did not consider the difference in polyploidy between herbaceous and woody genera to be statistically significant, but he believed that it would become significant when more herbaceous genera become cytologically well known.

It is interesting that these conclusions contribute .a new type of data to the debates concerning evolutionary history, and that they are on the unorthodox side—that herbaceous plants are primitive and woody plants derived, and that perennial plants have had their origin from annuals, in numerous cases. Although there is some good evidence that the most primitive Angiosperms were herbaceous perennials with indeterminate growth from which woody perennials and annuals have evolved, it seems probable that the herbaceous-woody and the woody-herbaceous changes have occurred many times in different families, as have the perennial-annual and the annual-perennial changes. Any relationship to polyploidy is exceedingly difficult to establish because of the fact that all life forms exist in strictly diploid species and also in polyploid species. It would seem that the most significant data are obtained when comparisons are made within single phylogenetic stocks such as the Leguminosae and Ranunculaceae mentioned above.

Another of the interesting suggestions that have been made is that there is frequently an introduction of vegetative methods of reproduction with polyploidy. Turesson (648, 650) reported vivipary associated with higher chromosome numbers in *Festuca ovina*. According to Larter (409), there is a similar relationship for *Ranunculus Ficaria,* in which the polyploid forms bulblets. Levan (415) reported the formation of stolons by polyploid *Tulipa* and of bulblets by polyploid *Lilium*. Maude (458) said that the species of onion which are hexaploids ($S = 48$) are *Allium Babingtonii* and *A. roseum* var. *bulbiliferum,* and that both species reproduce in nature by bulbils. This author's conclusion is that polyploidy, and particularly autopolyploidy, is correlated with an increased importance of vegetative reproduction. In the often cited case of *Biscutella laevigata,* Manton (442) said that the tetraploids, in spite of their cytological instability, are more successful colonizers and cover a wider area than the diploids of this species. She explained this phenomenon in part by the capacity of the polyploids to develop adventitious buds on their roots, a behavior not shown by the diploid races and species of *Biscutella;* it likely compensates for the reduction in seed fertility which probably accompanies the gametic irregularity. Hagerup (316) reported that *Deschampsia alpina* is the only species of the genus with vivipary. The other species are diploids or tetraploids; *D. alpina* has high and irregular chromosome numbers. Hagerup found S = 56 for this species and Flovik (252) reported S = 39, 41, and 49. Hagerup thinks that vivipary in *Deschampsia* is associated with and probably related to the high and irregular chromosome numbers. Emsweller and Brierley (211) induced tetraploidy in *Lilium* by the colchicine method and found that the tetraploid had stem bulblets although the diploid progenitor is characteristically bulbless. This case does not offer good evidence, however, because it is known that bulbs can be caused to form in normally bulbless species of lily by simple decapitation without the introduction of polyploidy.

One more case will be cited because of its special interest. Goodspeed and Crane (290) reported that the coastal redwood, *Sequoia sempervirens,* is apparently tetraploid (S = ±50), and that the Sierra redwood, *Sequoia (Sequoiadendron) gigantea,* is diploid (S = 24). The tetraploid nature of the coastal redwood is almost unique among cone-bearing plants, and it is associated with a remarkable sprouting ability. One of the impressive sights of the redwood forest of the California coast is a family circle of these giant trees that has superseded a preexisting tree and that has been derived from it by sprouts. In contrast, the diploid Sierra redwood is apparently seldom if ever able to reproduce vegetatively.

The argument that polyploids frequently set seed without fertilization, or reproduce vegetatively by bulbs, rhizomes, root sprouts, etc., because of polyploidy—that polyploidy is somehow causal in the development of vegetative reproduction—impresses me as being weak reasoning. Autopolyploids have considerable sterility because of multiple pairing of chromosomes, and unbalanced polyploids are frequently sterile when the unbalance is by either chromosome sets or individual chromosomes, as in triploids and dysploids. It would seem that vegetative reproduction is correlated with various forms of sterility rather than with polyploidy per se, because in such cases only those forms that had vegetative reproduction would survive in the absence of seed setting and they would lose out in competition when only few seeds are formed.

Physiological comparisons of diploids and polyploids.—Some of the differences between diploids and polyploids that have been discovered are more purely physiological than morphological, although, as stated earlier, the distinction is sometimes more one of convenience than of reality. Noguti, Oka, and Otuka, according to Clausen (147), found that autotetraploids of *Nicotiana Tabacum* and *N. rustica* have higher nicotine content than diploids. It is generally reported that autotetraploids have darker green leaves than diploids; this would seem to indicate a more abundant chlorophyll formation, whether the appearance is due to more or larger plastids. Schlösser (554) reported that autopolyploid races in *Lycopersicum cerasiforme, L. racemigerum,* and *Brassica Rapa* had a decreased osmotic concentration, apparently because of a greater water content than in the diploids. As the result of this condition, tetraploids were more cold-sensitive than diploids. Hesse (344) confirmed Schlösser's results, but Fabergé (228) found that diploid and tetraploid tomatoes do not have significantly different water contents, and Kostoff (401) concluded that tetraploid tomatoes are more cold-resistant than diploids. With respect to these experimental data we can take our choice, but there is other information concerning hardiness.

Hagerup (313) and Tischler (635) have upheld the theory that there is a higher percentage of polyploid species in regions that are climatically unfavorable, such as hot and dry or cold regions. From this idea has arisen the concept that polyploids, on the average, are hardier than diploids. This contention has received some support from other sources, such as the work of Müntzing (480) who compared 12 genera in which intraspecific chromosomal races existed, of Anderson (16) on American *Tradescantia,* and of Hagerup (315) on *Orchis maculatus.* Bowden (77) has attempted to measure the hardiness of plants by growing them together at the Blandy Experimental Farm at Boyce, Virgina. He selected plants from tropical families that have some temperate representatives, and from temperate families that have some tropical representatives. He developed an arbitrary scale of hardiness based upon the ability of the plants to withstand certain degrees of cold, and compared the results of survival with chromosome number and the temperature conditions of the native home of the species. Bowden pointed out that among strictly diploid species of certain genera there is a wide range in hardiness, as in *Ruellia, Passiflora, Salix, Helianthemum, Jasimum, Tamarix, Ficus, Periploca, Euphorbia, Lobelia,* and *Itea.* He also reported that in certain families the most northern species is diploid, as in *Asimina triloba* of the Annonaceae, *Pontederia cordata* of the Pontederiaceae, and *Poncirus trifoliata* of the Rutaceae. His general conclusion was that the degree of winter hardiness is not positively correlated with chromosome number. Tetraploids may be hardier, equal to, or less hardy than diploids. He thinks that genic mutations and inter- and intra-specific hybridization are apparently more important processes than polyploidy in the development of hardiness. Many of Bowden's comparisons do not involve really closely related species. When species differ in a number of fundamental ways other than chromosome number, and when they have long maintained the specific status, it is not surprising that no correlation could be found between chromosome number and hardiness. This does not prove, however, that a correlation does not exist for recently formed autopolyploids or amphidiploids.

Hagerup was the first to express the opinion that polyploids could occupy more unfavorable conditions than diploids, but by 1933 he had altered the formulation of this theorem to read: "Polyploid forms may be ecologically changed so as to grow in other climates and formations where the diploid form will not thrive." A number of investigators have agreed with this conclusion. Clausen, Keck, and Hiesey (146) found that polyploids differ in tolerances from diploids, but that they are not necessarily hardier. Babcock and Stebbins (39) say that by their distributions allopolyploid species of *Crepis* show the combination of physiological as well as morphological characteristics and that their different distributions are a result of their changed reactions. Müntzing (480), summarizing the earlier information on the physiology of chromosome races, concluded that polyploids are generally ecologically different from diploids because they have (1) an altered developmental rate, (2) an altered assimilation energy, (3) an altered hardiness, (4) an altered vitamin content, and (5) an altered osmotic concentration.

Ecological characteristics of natural polyploids.—By drawing upon the foregoing data and other sources, it is possible to suggest some ways in which the phenomenon of polyploidy may have significance in ecological and geographical studies. Although this topic will be dealt with in the next chapter, it is useful to make certain points now. Some of the morphological changes may be of importance in changing the survival or competitive capacity of polyploids. Changes in leaf size, plant stature, venation, stomatal size and frequency, hair frequency, size, and form may all have some relationship with water economy because they may affect transpiration. It is not a foregone conclusion, however, that such structural changes affect water balance—as assumed in many ecological studies—and it remains to be shown what significance they may have. Changes in water content, chlorophyll content, anthocyanins, vitamins, or other chemical characteristics may have certain effects favorable to the life of the tetraploid.

Rohweder (530) found that the diploid species of *Dianthus* have poor adaptability and cannot stand soils rich in lime or nitrogen, and that the tetraploids and hexaploids have progressively broader tolerances and adaptabilities. When he attempted to extend his study of soil relations and polyploidy with respect to lime content (531), he encountered some conflicting evidence. In the marshes of the Elba where the ditches are filled with lime-charged water, 95 per cent of 64 species are polyploid, and in the nearby moors, where the roots of the plants can reach lime-charged waters, 79 per cent of 89 species are polyploid; but on the young soils of the seacoast that are flooded by sea water every year this is true of only 39 per cent of the species.

The literature contains many references to the fact that tetraploids have a retarded rate of development, and this may sometimes have considerable ecological significance. Not only is vegetative development slower in tetraploids than in diploids, but flowering is later. Emsweller and Ruttle (212) reported that tetraploid *Lilium formosanum* was two to four weeks later in blooming than diploid forms. Blakeslee (70) found that tetraploid petunias were slower in flowering. Manton (443) noted in *Nasturtium officinale* that polyploid races have much slower growth rates than diploids and that they flower later in the season. Anderson[3] stated that tetraploid tradescantias flower somewhat later and stay in flower much longer than diploids. The data do not uniformly support this relationship. Müntzing (480) found a direct relationship between chromosome number and length of the vegetative period for the following species of *Nymphaea: N. stellata* ($S = 28$), 12 weeks from sowing to flowering; *N. rubra* and *N. lotus* (both $S = 56$), 16 weeks to flowering; *N. gigantea* ($S = 224$), 52 weeks to flowering. The series is broken, however, by *N. alba* ($S = 64$) and *N. candida* ($S = 112$), which require three years from sowing to flowering. Clausen, Keck, and Hiesey (146) found that diploid *Artemisia Suksdorfii* is earlier than its tetraploid and hexaploid relatives when the plants are grown together in the same garden, but that octoploid *Artemisia Rothrockii* is earlier than its close diploid relative *A. Bolanderi,* and that *Potentilla diversifolia* is earlier than its close relative *P. gracilis,* although its chromosome number is higher.

It does not appear that chromosome doubling affects all forms alike with respect to retarding the rate of development. Although the experimental induction of polyploidy points in that direction, such relationships do not persist because of further evolutionary changes, and long-established tetraploids may be either later in development, or the same, or earlier than diploids. Geographically, slowness of flowering may work a hardship on polyploids—in regions with a short growing season because of temperature or moisture —or it may be an advantage or it may have no significance. One point deserves mention. A change in flowering time or length of the flowering sea-

[3] The Jessup Lectures, Columbia University, 1941.

son may permit hybridization between species that in the diploid condition were seasonally isolated.

Life-form changes, as from herbaceous to woody or from biennial to annual, may be of ecological importance, as may, in fact, any change with polyploidy that affects competitive ability. Huskins (364) has reported that allopolyploid *Spartina Townsendii* is successfully replacing the native British *S. stricta,* one of its parents. In any discussion of ecological characteristics of organisms the phenomenon of interspecies competition should be emphasized. Complicated problems arise on two sides. Analyses of chemical, structural, or functional differences obtained under laboratory or other controlled conditions are difficult of ecological interpretation. On the other hand, environmental factors are numerous and interrelated in their action and are difficult to evaluate. Hutchinson (367) has expressed this situation succinctly: "In laboratory experiments clear-cut results can often be obtained; in nature a bewildering number of possibilities present themselves, and the further analysis is carried the more difficult it is to isolate controlling physicochemical variables. This difficulty should be admitted as one of the relevant facts to be considered; it is indeed one of the most important of the data, for it strongly suggests that in many cases modification of the dominance exhibited among numerous competing species is the major role of the environmental factors, rather than the direct transgression of limits of tolerance, so easily studied in laboratory experiments with pure cultures." This situation lies at the heart of all problems in ecology and physiological plant geography, but is frequently unrealized by investigators.

Some other effects of polyploidy.—The recently acquired ability to induce polyploidy by colchicine and other forms of treatment has facilitated genetic research and its application. Not only may autotetraploids be obtained from diploids, but higher forms of ploidy may develop. According to Blakeslee (70), species differ with respect to the number of times their chromosomes can be doubled without loss of vigor. In *Datura* $6n$ and $8n$ conditions have been induced, but no seedlings have been obtained. In *Portulaca parana* $8n$ plants have been obtained, although with some chromosome deficiencies. Clausen (147) found a tendency toward chromosome unbalance in amphidiploid *Nicotiana* forms, with plus or minus chromosomes. Blakeslee also reported chromosome unbalance for *Datura* tetraploids. Emsweller and Ruttle (212) noted that, in addition to causing polyploidy, colchicine treatment also caused an increase of somatic mutations through the loss or gain of chromosomes by otherwise diploid cells.

As a general condition, according to these two authors (212), autotetraploids have a reduced fertility of 50 to 100 per cent when compared with the diploids from which they were obtained, but autotetraploid varieties set seed on being crossed with other related autotetraploid varieties. Clausen (147) finds that amphidiploidy usually results in fertility and in breeding true for hybrid features, especially when the chromosomes of parental species are sufficiently distinct to cause little or no conjugation or ring formation. The amphidiploids are not always completely fertile, however, for in *Nicotiana* there is sometimes a peculiar female sterility in which the embryo sac fails to develop or the embryo aborts. Blakeslee (70) states that pollen sterility is about 85 per cent for *Datura metel* $\times$ *D. meteloides,* but that when the chromosomes of this hybrid are doubled it drops to about 25 per cent. The spontaneous doubling of the chromosomes of sterile species of hybrids to form multiple diploids apparently has been the origin of many wild plants and of the best varieties of oats, wheat, tobacco, cotton, etc.

The phenomenon of mixoploidy is known from nature but is frequently observed in plants treated to induce polyploidy. This may consist only of scattered cells of various degrees of chromosome or genom doubling, or it may involve segmental or periclinal chimera formation. Because differential periclinal ploidy in growing tips affects the subsequently differentiated organs (stamens, pistils, etc.) and tissues (microsporogenous, stigmatic surface, etc.), it may lead to sterility barriers.

Finally, it may be mentioned that ploidy increases breeding possibilities; for example, it allows the transfer of mosaic resistance from *Nicotiana glutinosa* to *N. Tabacum,* according to Holmes (351), and in nature it permits the combination of characters of diploid species through the doubling of chromosomes in sterile hybrids, with a restoration of fertility. Ultimately changes may occur in amphidiploids that transform them in the direction of making them more effectively diploid in genetic behavior.

Non-chromosomal indications of polyploidy.—A cytological survey of sufficient extensiveness to allow the mapping of chromosomal races within a species or among a group of related species is a laborious task. The problem of the preliminary survey in a group would be greatly facilitated if there were some easier, practical way of predicting polyploidy than by chromosome counts—verification of critical material to be made by ordinary cytological procedure. The possibility of working out some technique for rapid survey based on the widespread phenomenon of gigantism in polyploids seems likely. In fact, both Sax and Sax (552) and Babcock and Stebbins (38) have reported such techniques.

Sax and Sax compared diploid and tetraploid races of *Tradescantia canaliculata*. The forms of this species are not easily or certainly distinguishable on the basis of gross morphology alone. Between the races, however, the following anatomical relations exist:

Ratios of Diploid to Tetraploid

Pollen mother cell volume	1	:	1.7
Microspore volume	1	:	1.6
Microspore nuclei volumes	1	:	2.0
Chloroplast volume	1	:	1.8
Length of stem spicules	1	:	1.5
Stomatal length	1	:	1.3
Stomatal frequency per sq. mm.	1	:	0.5

Among these ratios the data likely to be most useful in making a survey of herbarium materials are those relating to stomata. Because stomatal counts are easier to make than stomatal measurements, and since this difference is greater between diploids and tetraploids, this promises to be a fruitful technique. Sax and Sax, after examining species of *Secale, Staphylea,*

TABLE 28.—Stomatal frequencies in relation to chromosome numbers (Data from 547)

Species	Chromosome Complements	Average Number of Stomata per sq. mm.
Betula nigra	S = 2n	82
Betula papyrifera	S = 5n	37
Betula lutea	S = 6n	43
Ulmus laevis	S = 2n	177
Ulmus americana	S = 4n	79
Fraxinus pennsylvanica	S = 2n	162
Fraxinus chinensis	S = 6n	76
Crataegus punctata	S = 2n	105
Crataegus crus-galli	S = 3n	71
Crataegus rotundifolia	S = 4n	65

Deutzia, and the Caprifolium section of *Lonicera,* concluded that an inverse relation exists between stomatal frequency and chromosome numbers for diploid, tetraploid, and hexaploid species within a genus. Sax (547) published more results the next year, some of which are included in Table 28. In certain genera, such as *Malus, Acer,* and *Tilia,* the differences among diploid species and among polyploid species were found to be so great that they obscured any relationship between diploids and tetraploids. In this connection it should be remembered that any tendency toward a gradual loss of cellular gigantism on the part of polyploids will eventually destroy any diploid-tetraploid contrast of this nature. Perhaps, also, only closely related pairs or series of species within a genus may be compared in this way. Differences due to genetic character other than polyploidy would obscure a relationship. Hence, although stomatal frequency cannot be used as an absolute index of polyploidy, it should in many cases be a very helpful criterion in preliminary surveys; in many genera, however, stomatal size and distribution are of no help in indicating the chromosomal condition.

Babcock and Stebbins (38) published the following size relations in races of *Youngia paleaceae,* a very polymorphic species: The long axis of the guard cells, measured in microns, was, for tetraploids, 31–34; hexaploids, 36–37; octoploids, 40.5–41.5. The pollen grain sizes were, for tetraploids, 29–30; hexaploids, 32–34; octoploids, 34.5–35. In their publication on *Crepsis* (39), these two authors presented a series of data on the length of the guard cells of diploid and polyploid species and subspecies; these are shown in Table 29. They found that an application of these size relations allowed

TABLE 29.—Length of the stomata in forms of *Crepis* of which the chromosome number has been counted (Data from 39)

Species	Length of Guard Cells of Basal Leaves	
	Range in Microns	Average
Diploid (S = 22)		
C. pleurocarpa	31–36	33.9 ± 0.3
C. monticola	34–41	37.0 0.4
C. occidentalis ssp. typica	31–40	35.8 0.5
C. Bakeri ssp. Cusickii	35–42	38.1 0.4
C. exilis ssp. originalis	33–42	37.0 0.3
C. acuminata	32–41	36.3 0.5
Triploid (S = 33)		
C. occidentalis	38–47	43.3 0.6
C. acuminata	38–47	43.3 0.5
Tetraploid (S = 44)		
C. monticola	43–54	48.2 0.6
C. occidentalis ssp. costata	45–57	51.1 0.5
C. Bakeri ssp. typica	43–54	48.6 0.6
C. nodocensis ssp. subacaulis	46–55	49.6 0.5
C. exilis ssp. originalis	40–52	45.5 0.7
C. intermedia	41–47	44.9 0.4

Species	Length of Guard Cells of Basal Leaves	
	Range in Microns	*Average*
Pentaploid (S = 55)		
C. monticola	45–58	51.3 0.8
C. exilis ssp. originalis	47–59	51.1 0.6
C. intermedia	42–54	48.7 0.7
Heptaploid (S = 77)		
C. occidentalis ssp. pumila	43–57	51.1 0.9
Octoploid (S = 88)		
C. barbigera	49–60	55.0 0.7

a herbarium survey which facilitated the preparation of distribution maps of chromosome races. They say (39) that several predictions of chromosome numbers made on this basis have been verified by actual counts, and that all errors in prediction have arisen merely in determining the degree of polyploidy. The polyploids could always be told from the diploids.

Burns (101) found that *Saxifraga pennsylvanica* consists of tetraploid (S = 56) and octoploid (S = 112) races, and that a number of morphological characteristics are correlated with the higher chromosome number: gigas characters in var. *crassicarpa,* stoutness of scape in var. *crassicarpa,* shorter and broader leaves in vars. *crassicarpa* and *congesta,* and larger seeds, shorter and broader follicles, and increased cell size in vars. *crassicarpa, congesta,* and *winnebagoensis.* He offers statistical proof that geometric leaf-shape changes are correlated with autopolyploidy, as are cell areas. He found that the tetraploid epidermal cells had a range of mean area from about 1150 to 1900 square microns, and that the octoploids ranged in area from about 2300 to 3400 square microns. He says, "In view of this consistent relationship, it is felt that careful areal measurements of leaf epidermal cells in *Saxifraga pennsylvanica* will serve to indicate to which chromosome race any puzzling forms might belong. A number of such forms were so measured . . . and were found to fit into the previously established ranges for known tetraploids and octoploids."

28.

Principles Concerning
Polyploidy and Related Topics

Since the findings of genetics and cytology have some importance for ecology and geography, the following statements are an attempt to epitomize their value. Each statement is followed by a brief discussion. The bulk of the discussion and the materials which provide the basis for the statements are to be found in the accompanying chapters. The sequence of the principles has no significance.

1. *Polyploids nearly always differ from their diploid progenitors with respect to structure and function. Quantitative and qualitative differences in structure, and new reaction norms cause polyploids to have a tendency toward a different ecology and to occupy habitats and areas different from those of their diploid ancestors.*

New autopolyploids show differences in morphology and physiology from their progenitors, although no apparent change has occurred other than the doubling of the chromosomes. In allopolyploids there is a mingling of the morphological and physiological characteristics of the parents. In both types of polyploidy the new plants, created at one stroke, have structural and physiological characteristics that make them different from the diploids. These differences may not be advantageous, as is sometimes the case for gigas characters, but frequently they apparently give the polyploids a tremendous advantage over their diploid progenitors. Ecological advantages may arise from the competitive ability of the polyploids that allows them to associate favorably with or even to replace their progenitors, or from the capacity of the polyploids to occupy new climatic or edaphic situations, and hence areas in which they are not confronted with competition from their close relatives.

Reports of physiological experimentation on polyploids are not abundant nor are they always conclusive, but there are apparently changes in growth rate, anabolic processes, tolerances, etc. Almost without exception it appears that recently derived tetraploids have a slower rate of development than related diploids; they flower and fruit later in the season, or even change from an annual to a perennial habit. Polyploids have much longer blooming seasons in *Tradescantia,* for example, thereby increasing the chances for inter-specific crossing when sympatric species have only seasonal isolation. Limits of tolerance may be extended in polyploids with respect to heat, moisture, and chemical factors of the soil. It is not clear that tetraploids are more cold-resistant, drought-resistant, acid-tolerant, etc., than diploids as an invariable rule, although such correlations have been suggested; but they are changed and, through fortuitous circumstances, the change may have survival value. Structurally, an increase in woody tissues, vegetative reproduction, massive-ness of organs (flowers, fruits, seeds), etc., may have ecological significance. Accompanying the gigas characters of the cells of the polyploids are dif-ferences in stomatal size, stomatal frequency, hairiness, osmotic concentra-tion, etc., which must have important but as yet unanalyzed effects on the water economy of such plants.

2. *Within a small phylogenetic group, polyploids tend to occupy more extensive areas than related diploid species. Allopolyploids, having their origin from hybrids and combining the tolerances and variability of both parents, frequently have a foundation for a wider range of tolerance which allows them to occupy wider ranges, both ecologically and geographically, than either their diploid progenitors or related autopolyploids.*

Within a group, certain diploids have wider areas than other diploids and certain polyploids have wider areas than other polyploids. Sometimes a diploid has a wider area than a related polyploid. Also, when two groups are compared, certain diploids may range wider than the polyploids of the other group. But in those instances where systematic complexes have been studied in their entirety, the polyploids tend more frequently to occupy a wider area than the diploids.

3. *In complexes containing diploid anad polyploid species, the latter tend to predominate in geographical regions which have recently been sub-jected to great climatic or other environmental changes. In these cases the diploids tend to occupy older areas and more stable habitats.* Cf. No. 8.

This generalization appears to be based upon the fact both that polyploid species have new reaction norms which allow them to occupy new areas, and that they escape the competition of closely related forms by their ca-pacity to migrate into new territory. New territory results from geological processes such as glaciation, from other forms of physiographic instability, and from climatic changes, whatever their origin. It appears likely that many of the geographical contrasts, such as those of north-south areas, alpine-lowland areas, coastal-interior areas, etc., are coincidental in part, and that the real cause of the preponderance of polyploids in northern regions, etc., is the fact that these regions are geologically new and hence more recently available for occupancy by the more recently evolved, more variable, and more broadly tolerant polyploids of a phylogenetic stock.

The present evidence indicates that the percentage of polyploid species in high northern regions is definitely greater than in more southern regions, and apparently polyploidy has provided many new forms for the colonization of such areas. That increased chromosome number per se signifies increased adaptability to cold and unfavorable conditions is by no means proved. The abundance of polyploids in cool temperate and in arctic or alpine regions may result from the geological history of such regions (glaciation), their physiographic instability, their vegetational instability and youth, and the fact that wide areas have only recently become available for colonization. Since allopolyploids tend to have wide tolerances, it is reasonable to assume that relatively more polyploids than diploids have found their way northward from the old lands and phylogenetic stocks of more temperate regions, and from Pleistocene refugia.

With respect to the effects of glaciation, we may suppose a history somewhat as follows: In the general vicinity of the glacial boundary old diploid species, formerly well isolated, are brought together because of the vicissitudes of forced migrations. In such a region, especially after glacial recession has commenced, there are numerous new, variable, and closely associated habitats in which populations of a variety of species can live in rather close proximity. The result of this intermingling of species may be the production of hybrids, followed sometimes by amphidiploidy. With continued glacial recession, the polyploids and backcrosses are in a position to expand their area tremendously. Some of the diploids also may extend far onto the glacial plain, but most of them will probably have only a limited expansion. The chances of such polyploids spreading into unglaciated territory to any considerable extent seems unlikely because penetration of closed communities is more difficult.

4. *The center of a polyploid complex lies in the region of the old diploid species where, through autoploidy, hybridization, allopolyploidy, and dysploidy, new elements may be added to the complex which later extend the boundaries of the group as a whole.*

The position of the center of origin of a complex may be northern or southern, alpine or lowland, coastal or interior, in moist or dry regions. It will tend, however, to be in a relatively old region. The polyploid derivatives will tend to spread toward the periphery of the area of the complex as a whole and extend it, as well as occupying individually wider areas than the diploids. This does not signify that the relative number of forms, or even of diploids, will be low at the center of origin of the group, for there may be more species and races there than in any other area of equivalent size; however, since polyploids tend to migrate into new territory (at least different territory), a greater number of them will tend to be peripheral. But any single peripheral area may contain only one or two polyploid forms. These situations, of course, do not refute the assumption that disjunct peripheral populations (whether of diploid or polyploid species or varieties) may be old. Allopolyploid complexes can arise only when two or more diploid species are brought together in one region through the vicissitudes of migration, usually dependent upon climatic reversal. The center of origin of a polyploid complex is a secondary center and may bear no close relationship spatially to the original center in which the old diploid species began. Finally, although the term center has been used in the singular, there can, of course, be two or more centers of polyploid formation in a genus or section.

5. *Within a section, genus, or other closely related phylogenetic group, there is a tendency for polyploid races and species to be more frequent at the periphery of range for the group.*

Inasmuch as recently formed polyploids tend to have increased vigor and hardiness with respect to closely related diploids, and collectively tend to have a broader ecological amplitude as measured by types of habitats occupied and area covered, it follows that they will tend to be more frequent at the periphery of the range for a group.

6. *Allopolyploidy permits the combination of genes which may have been isolated because of sterility barriers in the two or more diploid species of a phylogenetic stock.*

The F_1 hybrids of related species are frequently highly sterile, but the allopolyploids which sometimes arise from these hybrids are usually more or less fertile. Furthermore, when autopolyploids and allopolyploids exist together in the same complex of related species, hybridization and secondarily derived polyploids are almost certain to occur. Thus allopolyploids may come to contain genes from as many as three or four diploid species, and a complicated series of interrelated species including aneuploids may spread through many habitat types of a large area. The morphological and ecological characteristics of the species of such a complex can be explained on the basis of the recombination of the various morphological and ecological characteristics of the diploid species, and through the effects of autopolyploidy. The phenomenon of apomictic reproduction which frequently accompanies polyploidy also accounts for the preservation of numerous variants that otherwise would be lost because of their cytological irregularity and their inability to reproduce by seeds. The whole tendency, however, is toward the production of numerous forms—with ecological, geographical, or reproductive isolation—which represent the recombinations of factors present in the diploid progenitors. These processes are of great importance in evolution within certain phylogenetic stocks, but they seem to be of little or no importance in the origination of new phylogenetic stocks.

7. *Polyploids seldom if ever originate new phylogenetic stocks because of the statistical improbability of the same mutations occurring in all the allelomorphic genosomes.*

A polyploid complex may result in almost endless new forms, but they all tend to be variations of the original genetic themes of the diploid progenitors. A polyploid complex tends to be a closed system, as compared with diploid species, because a genic mutation, in order to gain expression if it is recessive, would have to be reduplicated in all the homologous chromosomes of the somatic cell. It is statistically highly improbable that the same genic mutation would occur as many times as would be necessary in a tetraploid or higher polyploid. At the same time, it would seem that polyploids would tend to become great reservoirs of recessive mutations with little immediate chance for their expression as characters. Mutations are probably as frequent in polyploids as in diploids, but many of them remain hidden.

According to Anderson (15), *Iris versicolor* is probably an amphidiploid hybrid (S = 108) between *I. virginica* (S = 72) and *I. setosa* var. *interior* (S = 36). Also, *I. virginica* is probably itself an ancient amphidiploid hybrid between two species, each with 36 chromosomes. These blue flags are important in connection with the concept that polyploidy cannot initiate new lines of evolution but can only result in variations (combinations) on the preexisting themes (genic materials). For example, albinism is not uncommon in many species no more variable than these blue flags; it is fairly frequent in *I. missouriensis*. The factor for albinism is recessive. Albinos cannot appear in amphidiploids unless the recessive factor develops in the genoms of all the constituent species. Anderson makes this point graphic by assuming a frequency of albinos of one in 5000 plants for *I. setosa* and for each of the two unknown putative parents of the amphidiploid *I. virginica*. If this is the frequency of the recessive gene, then one might expect an albino in *I. virginica* once in every 25,000,000 plants, and only once in every 125,-000,000,000 plants in *I. versicolor*. Anderson says, "The frequency of albinism in *I. virginica* and its even greater rarity in *I. versicolor* is therefore in strict accord with theoretical expectation."

8. *Systematic complexes, it has been hypothesized, tend to undergo a cyclic development. Beginning in a diploid, endemic, and monotypic condition, genera tend to become wide, polytypic, continuous, and in many cases polyploid. In old age, the range of the genus becomes disrupted and the genus may end with only relic polyploid species, endemic in widely disjunct areas.*

The above principle embodies a combination of Good's Theory of Generic Cycles and Stebbins' concept of the rise and decline of polyploid complexes. These theories do not account for the origin of monotypic generic stocks, but they do provide a basis for the interpretation of some types of range—juvenile endemic, relic endemic, wide continuous, wide discontinuous—and of the chromosome conditions found in various plant groups. When two or more diploid species are brought together and when hybridization and polyploidy (and possibly aneuploidy) occur, a genus tends to become complex, widely distributed, and continuous; this constitutes its mature stage. As time goes on, the vicissitudes of climatic change and migration cause the elimination first of the less plastic diploid species, with the resultant development of a strictly polyploid genus. With a greater lapse of time, the polyploids themselves may be decimated and the genus may end as it started, endemic and monotypic. One great difference remains; the juvenile species seem to be filled with evolutionary potency, whereas the old polyploid species are apparently incapable of initiating a new line of evolution. An exception to the latter statement may be found in cases of intergeneric crosses followed by amphidiploidy. In this connection it should not be overlooked that many genera, such as *Quercus, Heuchera, Rhododendron, Pinus,* etc., have little or no polyploidy, and that in many cases diploid species are old and occupy relic situations. This theory therefore is on rather uncertain ground.

9. *Differences in chromosome number may furnish new diagnostic characters for the recognition and taxonomic treatment of species and their sub-*

divisions. Cytogenetic information can be used to advantage in conjunction with morphological and geographical data in the systematization of genera or higher taxonomic units.

When two closely related but morphologically distinct chromosomal races bear the relationship of species and variety, it is obviously unsound and misleading to dispose of them by referring to the diploid of the pair as the variety and the polyploid as the species, for polyploids are derived from diploids phylogenetically. When cytological investigation reveals that a form is a polyploid derivative, and when the rules of nomenclature do not permit a revision of the rank of the forms within a complex, it is nevertheless necessary that the true phylogenetic relationship be made clear and that the implication of the nomenclatural disposition be corrected. Fundamental cytological differences within a family or other higher category, as in the Ranunculaceae, take precedence over morphological differences for purposes of natural classification when it can be shown that the morphological characters were probably obtained by parallel evolution within the cytologically different groups.

10. *Certain diploid species, without benefit of chromosomal races, and genera composed only of diploid species, have attained all the ecological adaptations and types of geographical range that other species with polyploid races and polyploid genera have been able to attain.*

When closely related diploid and polyploid species or chromosomal races within a single species are compared, there frequently arise important contrasts in ecological adaptation or in geographical range. These contrasts, however, are also known from among closely related strictly diploid species, and from among ecotypes of diploid species. That is to say, genic or larger mutations, without changes of chromosome number, can build up within a population ecological tolerances of great variation in degree and kind. Chromosomal differences, as in polyploidy and aneuploidy and the frequently accompanying condition of apomixis, are not necessary in the building up of ecological and geographical contrasts between related plants. For example, circumpolar species frequently appear to have acquired their wide boreal range because of the presence of chromosomal races, but at the same time other strictly diploid species are also circumpolar. Plants of strongly characterized climates or edaphic situations frequently are polyploid, and their closely related diploids seem unable to occupy these severe situations. Nevertheless, it also occurs that some diploids are originally adapted to severe conditions and the derived polyploids have moved into the more favorable situations. In view of these facts, we may be sure that the world would be fully clothed with vegetation and the great fundamental patterns of plant geography and plant morphology would exist even if there were no such phenomenon as polyploidy.

Cytogenetic investigations have recently produced evidence that widespread diploid species are not homogeneous, in spite of their constant chromosome number, but consist of biotype groups with more or less ecological and geographical isolation. The morphological and physiological differences between the ecotypes, ecospecies, and geographical races of a widespread diploid taxonomic species seem to be based firmly on genetic differences such as inversions and translocations, as well as gene mutations. More and more it appears that what have been called genic mutations are many times due to changes in chromosome morphology. Be that as it may, selection pressure in the various environments has resulted in the "breaking up" of a wide population into many more or less local races. Ultimately, then, the contrast between the widespread diploid species and the widespread species with chromosomal races is not a contrast between genetically homogeneous and genetically heterogeneous populations. Both types are complex and the complexity is of two kinds.

I do not consider that the principle which has just been discussed negates the preceding statements. I have been careful not to make these statements as if they were laws to which no exceptions are known, but to formulate them as working rules because they happen to be true more often than not. Such contrasts as have been described are attributable to polyploidy, but changes in chromosome number alone are only one way in which evolutionary materials are provided.

29.

Geographical Aspects
of Polyploidy

The abundance of polyploids in certain floras.—Only a few studies have been made of the percentage of a flora which is composed of polyploids. Tischler (635) gathered the cytological information concerning the flora of Schleswig-Holstein and compared that region with certain others. Sokolovskaja and Strelkova (602) investigated the high alpine regions of Pamir and Altai from this point of view, and Flovik (252) published on the chromosome numbers of the plants at Isfjorden, Spitzbergen. These results are summarized in Table 30.

TABLE 30.—The percentage of polyploids among the species of certain floras

Author and Reference	Region	Number of Forms for Which the Chromosome Number Is Known	Percentage Polyploids
Tischler (635)	Schleswig-Holstein, total flora	714	44
	Species of the northern element		60
	Species of the southern element		27
	Iceland	359	54.5
	Faroes	267	49.4
	Sicily	689	31.3
Sokolovskaja and Strelkova (602)	Pamir	150	85
	Altai	200	65
Flovik (252)	Isfjorden, Spitzbergen	68	80

438

The most conspicuous fact revealed by these comparisons is the high percentage of polyploidy in the floras of northern and high-altitude regions. The small flora from Isfjorden was 80 per cent polyploid. This compares favorably with the 85 per cent of polyploids in the Pamir region and 65 per cent in the Altai. The Russian authors point out that the climatic and growth conditions of Pamir are very unfavorable, and that in comparison the conditions in Altai are milder. If we employ the rather large flora of Schleswig-Holstein, for which chromosome numbers are known, as a temperate flora to be used for comparison, we see that the more northern floras of Iceland and the Faroes have somewhat more polyploidy. On the other hand, the more southern flora of Sicily has strikingly fewer polyploids. Perhaps the strongest contrast is revealed when Tischler divides the flora of Schleswig-Holstein into those species which range northward and those which range southward from this province. The northern element in the flora, including some species that are circumpolar, is 60 per cent polyploid, whereas the southern element is only 27 per cent polyploid.

It is of further interest to note that 49 species of the flora of Schleswig-Holstein are known to have intraspecific polyploidy (a total of 7 per cent), and that 22 of these species have such races within Schleswig-Holstein itself (about 3 per cent). Tischler considers this region of special interest because it represents a meeting of species of northern and southern distributions.

Flovik (252) interprets these comparisons as evidence of a tendency toward the fact that increased chromosome number increases adaptability to extreme habitats, such as arctic and alpine conditions, and that polyploidy appears to have played an important role in the origin of forms which have been able to colonize cold regions. The point should not be overlooked, however, that polyploidy is not necessary for successful life in high arctic and alpine regions. Many diploids are equally successful in such regions. This topic is continued in the discussion of north-south contrasts in the occurrence of related diploids and polyploids.

In this connection it should be remembered that all such comparisons will be relatively inaccurate and inconclusive until many different counts have been made for many more species throughout their ranges. One count of a species is inconclusive, for in many cases (*Tradescantia, Galax,* etc.) further counts have revealed intraspecific chromosome races without accompanying morphological differences of sufficient extent to have attracted the notice of taxonomists.

North-south distributional contrasts.—An early paper by Hagerup (312) is of interest in a study of distributional contrasts between diploids and polyploids. He examined cytologically about 30 species of Ericales and discovered the following four pairs of related species in which the one with the more northern distribution was tetraploid:

Species Pairs	Gametophytic Chromosome Numbers
Arctostaphylos diversifolia	13
A. uva-ursi	26
Empetrum nigrum	13
E. hermaphroditum	26
Clethra arborea	8
C. alnifolia	16
Kalmia latifolia	12
K. glauca	24

Hagerup said, "It is worth noting that among these four pairs of species, those with the higher chromosome number are always the ones growing farther north, and thus more exposed to extremes of temperature." Although he apparently did not propose this statement as a general rule, several subsequent investigators have treated it as one.

It is not within the province of this book to attempt an exhaustive survey of the literature, but a cross section of it will be useful in examining the possible significance of diploid-polyploid distributional contrasts. The following discussion includes several cases in which it has been reported that the polyploid has a more northern distribution than the related diploid.

Erlanson (226) found that in the section Cinnamomeae of *Rosa* the hexaploids extend farthest to the north in America. Manton (441, 443) reported that in *Nasturtium officinale* the chromosome race $S = 64$ is more northern than the race $S = 32$. According to Mangelsdorf and Reeves (439), *Tripsacum dactyloides* consists of chromosome races in which the higher number ($S = 72$) is found in more northern regions than the $S = 36$ race. Håkansson (318) found that subspecies of *Pimpinella saxifraga* differ as to chromosome number and geographical distribution. The subspecies *eusaxifraga* ($S = 36$) occurs in northern Sweden and the subspecies *nigra* ($S = 18$) in continental Sweden. Clausen (145) reported for *Pentstemon azureus* chromosomal differences between subspecies; *parvulus* ($S = 32$) is more southern than the subspecies *typicus* and *gutissimus,* both of which are $S = 48$. Stockwell (613) found that in the common prickly-pear cactus *Opuntia polycantha* of the North American grasslands the northern race from Peace River, Alberta, has 66 sporophytic chromosomes, whereas the plants from farther south in Alberta and on down to Colorado are $S = 44$. Müntzing (479) reported that *Phleum alpinum* consists of a diploid and a tetraploid race, of which the latter ($S = 28$) is more northern. Nakamura (483) found that hexaploid *Solanum nigrum* ($S = 72$) is generally distributed through the northern part of Japan and that only the diploid ($S = 24$) *S. photeinocarpum* occurs in the south of Japan in subtropical regions. According to Fagerlind (231), the diploid *Galium verum* is southern, centering in the Balkans, and the tetraploid race of the species is more northern, occurring in middle Europe, Scandinavia, and England. *Galium Mollugo* presents a parallel case. Turesson (652) reports a diploid race of *Sedum telephium* ($S = 24$) in the vicinity of Vienna, Budapest, and Moscow, and a tetraploid race ($S = 48$) in Sweden and Siberia. Flovik (252)

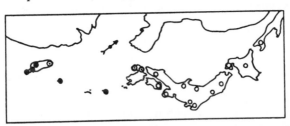

Fig. 55. The natural range of *Solanum nigrum* ($S = 36$) in the Japanese Islands is indicated by hollow circles. The range of the newly described diploid, *S. photeinocarpum* ($S = 12$), indicated by black disks, is distinctly more southern. The species occur together in the areas indicated by the circle with a dot. Map redrawn from (483).

has contributed a number of similar cases to the record in connection with his studies of the plants of Spitzbergen. *Saxifraga oppositifolia* in Spitzbergen was found to be $S = 52$, whereas Skovsted (591) had reported $S = 26$ for it from Norway. *Dupontia Fisheri* ($S = 88$) and *D. Fisheri* var. *psilosantha* ($S = 44$) are cases of intraspecific chromosome races in which both forms are typically arctic but in which *D. Fisheri,* with the higher chromosome number, extends farther to the north and otherwise occupies more unfavorable regions. *Cardamine pratensis* was reported by Manton (441) to consist of two chromosome races in England ($S = 32$ and 64). Flovik found only the race with the higher number in Spitzbergen. *Saxifraga stellaris* is $S = 28$, but its var. *comosa* is $S = 56$ (591, 76, 252). According to Flovik, "The two forms differ in respect to ecology and geographical distribution. Variety *comosa,* having the higher number of chromosomes, is circumpolar, high-arctic and high-alpine, whereas the species is subarctic, temperate and alpine." *Saxifraga nivalis* ($S = 60$) has a wider distribution than the diploid var. *tenuis* ($S = 20$). Flovik says that the polyploid form goes farther down into the lowland as well as higher up on the mountains than the diploid. Furthermore, both forms occur in Spitzbergen, but the polyploid is found somewhat farther north than the diploid. Hagerup had early reported that *Empetrum hermaphroditum* is tetraploid and *E. nigrum* is diploid. The tetraploid is the sole form in East Greenland and is now known in Spitzbergen, whereas the diploid is generally more southern.

Although the first known cases and many subsequent ones indicated that the more northern member of a pair of related species was often polyploid and the more southern member had a lower chromosome number, it was soon discovered that there are exceptions to this relationship. Shortly after his original discovery of the more northerly distribution of tetraploids in four pairs of species in the Ericales, Hagerup (314) reported a contrary case concerning *Vaccinium uliginosum*. He writes, "In general it may be said that the northern limit of the tetraploid [f. *genuina*] lies slightly more to the north than the southern limit of the diploid [f. *microphylla*, g = 12]. Only in those regions [around the polar circle] can these two forms be found together, the diploid form being of a more markedly arctic distribution than the tetraploid. . . . Among the polyploids previously investigated it was the form with the greatest number of chromosomes which was most resistant in unfavorable climates. In this connection *Vaccinium* is especially of value as showing that this is not a universal rule."

Peterson (503) found *Stellaria neglecta* to consist of two races (S = 22 and 44), of which the diploid is more northern. Matsuura (456) has shown that the diploid race of *Fritillaria camschatcensis* is more northern than the tetraploid race. Anderson (15) found that *Iris setosa* (S = 36) is more northern than the related *I. virginica* (S = 72). The allopolyploid *I. versicolor* (S = 108), incidentally, occupies an area between its supposed parents mentioned above. Böcher (74) reports in *Campanula rotundifolia* that the arctic form (known as var. *uniflora*) is diploid (S = 34), whereas a subarctic and a Danish race are each tetraploid (S = 68). Turesson (652) found in *Galeobdolon luteum* a tetraploid race (S = 36) in Bavaria and a diploid race farther north in Sweden and Latvia. Flovik (252) reported that *Cochlearia officinalis* var. *groenlandica* from Spitzbergen was diploid (S = 14). Crane and Gairdner (166) had reported the species from Wales as tetraploid (S = 28) and Böcher (76) had found the divergent but approximately tetraploid number S = 26 in the Faroes. According to Hagerup (317), *Oxycoccus microcarpus* (S = 24) occurs farther north than the tetraploid *O. quadripetalus* (S = 48).

Closely related to the contrasts in north-south distribution of diploids and polyploids is their alpine occurrence. Griesinger (302) reported that *Arenaria serpyllifolia* is a complex in which the diploid subspecies is alpine and the tetraploid subspecies are subalpine and lowland. Matsuura (456) likewise found that the diploid race of *Fritillaria camschatcensis* is alpine and the tetraploid race is lowland. An opposite case is *Biscutella laevigata* reported by Manton (442, 444), in which the tetraploids are alpine and the diploids lowland. Clausen, Keck, and Hiesey (146) found instances in the California flora of both types; sometimes the polyploids and sometimes the diploids were characteristic of higher altitudes, within groups of related forms. Sakai (539) reported on the chromosome numbers of 28 Japanese alpine plants; in 1935 he added 38 more species and varieties. His data reveal that two-thirds of these alpine species are diploid and one-third polyploid. These percentages are about the opposite of what would be expected on the basis of the high percentages of polyploids in far-northern regions.

Relations of polyploids to glaciated territory.—The question of whether polyploids bear any relation to glaciated areas is obviously somewhat similar to the previous discussion of their relations to northern and alpine occurrence. If any relationship exists it may be tied up with temperature and other climatic conditions and consequently be the same as for northern-alpine occurrences, or it may concern only the opportunities for colonization which result from the appearance of bare land following glacial recession. This point carries the inference that polyploids are peculiarly fitted to occupy new territory, or at least are more aggressive than diploids. Another aspect of the problem is the influence of glaciation, through the forced intermingling of floras, on the production of hybrids from which occasional allopolyploids might develop which would be well fitted to occupy the new territory resulting from glacial recession. Diploid species that had become geographically isolated prior to a glacial advance might be brought into proximity by enforced migrations caused by climatic changes. These diploids might hybridize freely and, even though their crosses were sterile, they would enter a new phase of development, with chromosome doubling (amphidiploidy) and the development of seed production. Furthermore, through the development of double diploids (autoploidy) the way would also be open for hybridization of previously isolated species. Thus hybrid complexes may develop, to which may be added dysploidy and the stabilization of such forms through apomixis. In this way glaciation, and its attendant plant migrations and commingling, may cause the development of new forms with a higher chromosome number that are in a position to take advantage of the new variable territories made available by glacial recession. The studies on *Crepis* (39) and *Vaccinium* (117) substantiate this type of development. Three cases will be discussed in which the phenomenon of glaciation is of importance in connection with the polyploids of a group.

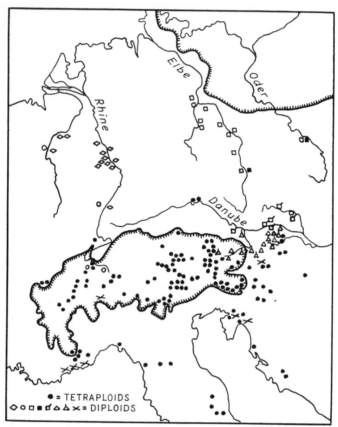

Fig. 56. The occurrence of *Biscutella laevigata* in central Europe. All forms occur south of the continental terminal moraine (hatched line); the tetraploids are found mainly within the alpine glaciated area. The other symbols indicate different diploid relics which occur in sheltered valleys of the Rhine, Elbe, Oder, and Danube Rivers still confined largely to Pleistocene refugia. Map redrawn from (442).

Biscutella laevigata is a variable species which has caused taxonomists considerable trouble; in a recent revision numerous subspecies and varieties were recognized. Manton (444), who made a careful cytological study of the species, discovered the presence of several races, some of them diploid and some tetraploid, thus confirming cytologically the complexity of the group. The diploids and tetraploids do not occupy the same territory and only in a few places do their areas make contact. The tetraploids have essentially a continuous range and occupy nearly all the central European mountain ranges, colonization being extended to the southern slopes of the Carpathians, the Transylvanian Alps, and the mountains of Italy and the northern Balkans. Tetraploid *Biscutella laevigata* is obviously a successful type. Diploid races, in contrast, occupy areas which are limited in extent and highly discontinuous. The three main centers of the diploid population are situated in the river valleys of the Rhine, the Austrian Danube, and the Elbe. Each of the centers is characterized by its own assemblage of morphological types to which subspecific or varietal status has been given. According to Manton, the evidence is strong that the diploid races are preglacial or interglacial—at least older than the last ice advance—and that as relics they occupied approximately the same locations where they are now found. Tetraploid races are absent from these relic areas; they occupy extensive regions which were once covered by alpine glaciers. It is obvious that the alpine and montane tetraploids must be postglacial immigrants to most of their areas, and they are probably still spreading. They seem to have a capacity for taking advantage of the soil exposed by the retreat of the mountain ice, whereas the diploids apparently have no capacity to spread beyond their anciently occupied stations where they survived the last ice age.

Another interesting case is described in the report by Hagerup (317) on the recently discovered hexaploid *Oxycoccus gigas* (S = 72). This species appears to be a hybrid between a diploid and a tetraploid species. Tetraploid *O. quadripetalus* lives in the vicinity of *O. gigas,* but no diploid is now found in the region. Diploid *O. microcarpus* (S = 24) lives in regions north of Denmark and could have been the other parent of *O. gigas* during glacial times when the diploid, along with other northern species, was forced southward into the range of the tetraploid. Hagerup suggests that these vegetative populations of *O. gigas*[1] became established, as in the bog near Copenhagen, in early postglacial time when a diploid species lived in the same bog. The latter would have retreated northward with glacial recession, and the vegetative colony of the allopolyploid has presumably persisted ever since.

Babcock and Stebbins (39) consider that an important role in the development of western American species of *Crepis* has been played by the migrations which were attendant on glacial advance and recession. They think that these species of *Crepis* are basically a group of distinct diploids (S = 22) that immigrated into the northwest from Asia in pre-Pleistocene time, and

---

[1] *O. gigas* is a sterile clone. It is questionable whether it should be given specific rank.

that they in turn are allopolyploids which originated in Asia from species with gametophytic numbers of 4 and 7. The vicissitudes of climatic change and migration allowed the species to stabilize through isolation and the development of partial reproductive barriers. Later changes have resulted in bringing certain of the diploid species together, with some resultant hybridization and the origin of a host of polyploid and apomictic forms which effectively combine the morphological and ecological characteristics of the original diploids. In general, the diploids have remained relatively local and the polyploids have shown considerable ability to enter new territory and attain an extensive distribution.

American species of *Iris* also show the same tendency. *Iris versicolor,* the highest polyploid, is found almost wholly in glaciated territory.

Relations of polyploids to coastal and interior distribution.—Shimotomai's report (574) that in Japan the polyploid species of *Chrysanthemum* are coastal and the diploid species are interior has caused some other investigators to comment on the possible relationship. Tischler (639, 640) reached the conclusion that the flora of the Halligen Islands have an increased percentage of polyploidy over that of the continental flora of Schleswig-Holstein. Although these islands in the North Sea are overflowed by the sea every year and must have plants with some tolerance for salt, they are not completely occupied by halophytic plants. A cytological investigation of all the species revealed divergent conditions in the various ecological groups, as seen in the data in Table 31. About half of the true halophytes and the

TABLE 31.—The percentage of polyploids in various maritime ecological types of the Halligen Islands

Vegetation Type	Number of Species	Percentage of Polyploids
The grassland flora	19	47.4
The true halophytes	36	55.6
The non-halophytic sand flora	6	100.0
The flora of ponds and their shores	4	100.0
The weed flora:		
Completely naturalized species outside of the grasslands	15	86.7
The spontaneous flora	53	33.9

grassland flora are composed of polyploid species. The non-halophytic sand flora and the pond flora are completely polyploid. Among the weeds there are two groups. Those that are completely naturalized are highly polyploid, but those that occur repeatedly but sporadically are only about one-third polyploid. Tischler concludes that there appears to be increased polyploidy among coastal plants over interior floras, but he warns his readers that the situation in the Halligen Islands may be obscured by the "northern factor" and that comparable investigations should be made in other regions, especially in tropical islands and maritime floras.

Warburg (675) found *Erodium cicutarium* to consist of diploid and tetraploid races, in which the diploids are maritime and the tetraploids inland in distribution. Simonet (579) reported two chromosome races in *Agropyrum junceum*, both of them coastal. On the French west coast there is a race of $S = 28$ and on the Mediterranean coast there is a race of $S = 42$. According to Clausen, Keck, and Hiesey (146), among the California plants they investigated there was no consistent tendency for the polyploids to be coastal in groups of closely related forms. They found frequently that the polyploid race of a species or species of a group are interior and the diploids coastal. For example, in the coenospecies *Zauschneria* the coastal *Z. cana* is diploid and the tetraploid *Z. californica* extends from the coast to subalpine altitudes. *Z. cana* is monotypic, but the tetraploid is differentiated into ecotypes, of which subspecies *angustifolia* is maritime, subspecies *typica* occurs in the Coast Range, and subspecies *latifolia* is mid-altitude in the Sierra Nevada. In *Artemisia* the diploid *A. Suksdorfii* is strictly maritime and tetraploid *A. ludoviciana* is interior. The *Achillea millefolium* complex contains hexaploid *A. borealis*, which has a distinctly maritime ecotype, and tetraploid *A. lanulosa*, which is interior. Tetraploid *Horkelia* (*Potentilla*) *californica* and diploid *H. cuneata* are both maritime.

The emphasis given by Shimotomai, Tischler, and Rohweder to the abundance of polyploids in maritime regions will not allow the formulation of a general rule. It seems that if the basic diploid happens to be interior, the derived polyploid may also be interior in a different area or it may become coastal. If the basic diploid happens to be coastal, the derived polyploid may find another area which is also coastal or it may find an inland area. That is to say, the polyploids usually occupy different areas from the diploids, but there is no special tendency for polyploids to be coastal and diploids to be interior. It merely happened that in certain early cases that were investigated the polyploids were coastal and the diploids interior, wherefore, under the general impression that polyploids occupy less favorable situations, it was assumed that the relationship might be general.

The relation of polyploids to xerophytic and non-xerophytic regions.— Hagerup (313) appears to be the first investigator to suggest that hot and dry regions have a relatively high number of polyploids. He had already found a tendency in certain paired species for the polyploid member of the pair to occupy the colder region. Furthermore, certain laboratory investigations at this time had shown that polyploidy could sometimes be induced by heat and cold shock. From his studies of some plants of Timbuktu, Hagerup decided that the tendency for polyploids to occupy unfavorable conditions was a phenomenon extending to xerophytic situations. For instance, *Portulaca oleracea* var. *gigas* (S = 54) grows in hotter and drier situations than the species proper (S = 18). He found his best example, however, among species of the grass *Eragrostis*. *E. cambessediana* (S = 10) grows in mud flats of lake shores, *E. albida* (S = 40) grows in medium moist soils at the foot of dunes, and *E. pallescens* (S = 80) grows on the high dunes where the soil is very dry and the soil temperature may reach 80° C.

The investigations of Clausen, Keck, and Hiesey (146)—to select an example from the literature—seem not to support Hagerup's findings as a general rule. *Aster adscendens* is a complex in which the diploid species occurs in the climatically extreme Great Basin region of the western United States and the tetraploid is found in the Sierra Nevada. *Viola purpurea* is a parallel case. Likewise, *Artemisia Bolanderi* is a diploid of desert flats; its close relative, octoploid *A. Rothrockii,* grows in the high mountains. Although *Potentilla pectinisecta* is a species of the arid Great Basin, its relatives in the coenospecies *P. gracilis,* with high and irregular chromosome numbers, are found in the mesophytic climates of the mountains.

Once again we find that there is no rule covering a specific geographical relationship with polyploidy.

Relations of polyploids and diploids to extent of area occupied.—To the frequently expressed idea that polyploids can occupy more unfavorable regions than diploids, such as colder, hotter, drier, or maritime regions, we have found numerous exceptions which prevent the enunciation of any strict rules. There remains the more general question of whether polyploids merely tend to occupy different ecological and geographical regions and, perhaps, wider areas than their related diploids.

In an early paper on *Anthurium* Gaiser (260) reported that among the polyploid species there were some which had the widest ranges in the genus, but there were also others which had the smallest ranges. Manton (442), we remember, found that in *Biscutella laevigata* the diploids were of very restricted range, whereas the polyploids enjoyed a wide and continuous range. In *Crepis* Babcock and Stebbins (39) found that the polyploids have much greater areas than the diploids, with the exception of the homoploid species,

C. nana and *C. runcinata,* which have wide areas. In general, however, they believe that the data warrant the conclusion that both polyploidy and hybridization (allopolyploidy) are important in allowing such species of *Crepis* to attain strikingly wider ranges than most diploids.

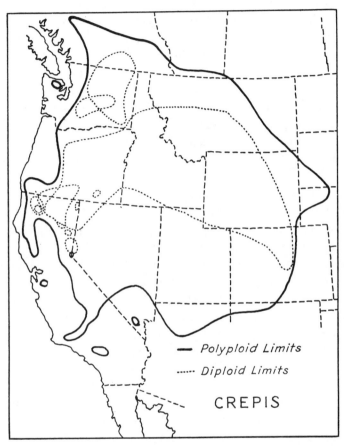

—— *Polyploid Limits*

······ *Diploid Limits*

CREPIS

FIG. 57. In the western American species of *Crepis* the range of the genus is greatly extended by the polyploids. The diploids are relatively local in occurrence, except for one species. Map redrawn from (39).

What appears to be the situation in *Crepis,* and one that is probably true in other groups, is that different diploids have different capacities for occupying territory which are due to their genic composition. One diploid species may be wide and another may be narrow in ecological amplitude, and consequently in geographical range. With the introduction of polyploidy, and especially allopolyploidy, it appears likely that the new form would have the capacity of attaining a wider area than the progenitor diploids. In a later paper (606) emphasizing the role of polyploidy and apomixis in *Crepis,* Stebbins and Babcock write that "within the group seven diploid, purely sexual forms have been found. These are all strikingly different from each other both in their outward appearance and their geographic distribution, and unquestionably represent seven distinct species. With one exception, *C. acuminata,* these diploids are restricted in their geographic distribution to a small portion of the range of the group as a whole, and some are very rare. The polyploids occur abundantly, for the most part, along with the diploid forms, and in addition extend far beyond the limits of the ranges of these diploids." In a still later paper on *Crepis* (605), Stebbins notes that the strictly autopolyploid forms are almost as restricted in distribution as the diploids, and that the farther edges of the ranges of the group are occupied by allopolyploid forms of complex origin. There is considerable evidence to indicate that these allopolyploids have been able to occupy a wide range because they have acquired through hybridization a favorable new combination of the different physiological characteristics of their diploid ancestors.

In a monographic study of the cytology of the large family Leguminosae, Senn (568) makes some pertinent observations on the question of the area occupied by polyploids. The relative distribution of diploids and polyploids varies greatly from genus to genus. In ten genera (*Trigonella, Trifolium, Lotus, Astragalus, Indigofera, Robinia, Oxytropis, Lespedeza, Vicia,* and *Lathyrus*) the diploids have a wider distribution than the polyploids. In three genera (*Acacia, Amorpha,* and *Caragana*) the distribution of polyploids is wider than that of diploids. After mentioning some other aspects of distribution, such as relations to cold or hot regions, Senn concludes that in general the data from the Leguminosae do not support the concept that polyploids have a wider distribution or a more northern distribution than diploids. He says, "Instead, the whole situation seems to be more complex, involving the origin of the species in question and its geographic relationships. That is, polyploids in the juvenile endemic stage occupy a small area, which may not as yet reach the boundaries of their specific tolerance. The changes in specific tolerance resulting from polyploidy may enable the plant to live under very different conditions from those in which its diploid ancestor lived. The new conditions are not necessarily linked with cold and drought but may run the whole gamut of climatic and edaphic conditions."

Baldwin (43) notes that diploid *Shortia galacifolia* (S = 12) is extremely localized and rare in the Southern Appalachian Mountains, whereas tetraploid *Galax aphylla* (S = 24) is relatively widespread and aggressive, extending from the mountains onto the Piedmont and even to the coastal plain of the southeastern states. He suggests that perhaps the greater ecological adaptability of *Galax* results from its polyploid nature. Later he found that *Galax aphylla* consists of two races, S = 12 and 24, and that the tetraploid occupies less area than the diploid (45). Although these two species are the only members of the old, small, and discontinuous family Diapensiaceae in the Southern Appalachians, it would seem that they are not closely enough related for the comparison to be of much significance. For example, *Diapensia lapponica* of the same family is a diploid species which is circumpolar, extending from northeastern America through Europe to northern Asia and Japan. Moreover, cytological investigations of this species, made from material from America, Sweden, Greenland, and Japan and including its two varieties (*obovata* and *genuina*), show that throughout this wide area it remains constant as to chromosome number. This case again indicates that certain diploids can attain a wide range and morphological and ecological variability. In *Arisaema* Bowden (77) found that *A. triphyllum* (S = 56) is hardier and has a much wider area in North America than *A. quinatum* (S = 28). Eigsti (208) reported that tetraploid *Polygonatum canaliculatum* ranges more widely, is more abundant, and occupies different habitats than diploid *P. biflorum*.

In some widely distributed species it appears that the presence of chromosome races have probably assisted the attainment of the wide distribution. For example, *Potentilla fruticosa* is a circumpolar species which, according to Turesson (652), is known to be diploid (S = 14) in Alberta, Canada, and tetraploid (S = 28) near the type locality in the islands of the Baltic. Clausen, Keck, and Hiesey (146) have shown that the greatest variability and ecological adaptability can be developed in either the diploid or the polyploid ecospecies of a complex. Thus, in the cenospecies *Zauschneria* the diploid *Z. cana* is monotypic and the tetraploid *Z. californica* has developed three ecotypic subspecies that are differentiated ecologically and geographically, as well as morphologically. On the other hand, *Viola purpurea* has a monotypic tetraploid ecospecies; it is the diploid ecospecies that is differentiated into three ecotypic subspecies. Generally speaking, of course, the ability to develop ecotypes results in an expansion of range, whether the form is diploid or polyploid. There are some data to indicate that "weediness" is associated with polyploidy, as in American *Tradescantias* related to *T. virginiana*, but there are not enough to allow an opinion whether weeds tend to average higher in chromosome number than non-weeds. A couple of instances can be quoted. Bhaduri (66) reported diploid, tetraploid, and hexaploid races in India for *Solanum nigrum,* which is an almost cosmopolitan weedy vine. Tischler (640) found only the hexaploid form of this species in Europe and America, but Nakamura (483) found both diploid and hexaploid races in Japan. Maude (458) reports that *Chenopodium album,* which is a widespread and variable weed, has sporophytic numbers of 18 from England, 18 and 54 from Sweden, and 36 from North America.

Among the species of *Tradescantia* allied to *T. virginiana,* Anderson and Sax (22) found eight species known only as diploids (*edwardsiana, ernestiana, gigantea, paludosa, bracteata, hirsuticaulis, humilis,* and *subacaulis*), which have an average area of 83,475 square miles. Nine species of the same

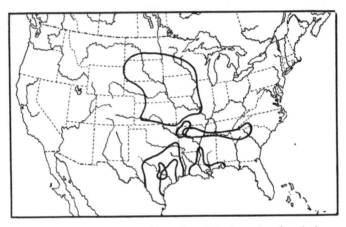

Fig. 58. Areas of the diploid species of *Tradescantia,* after (22).

group of *Tradescantia* (*subasper, ozarkana, virginiana, hirsutiflora, tharpi, canaliculata, longipes, roseolens,* and *occidentalis*) are known only or predominantly as tetraploids, and their average area is 376,300 square miles, or about four and one-half times as great as that of the diploids. If these areas are changed to average diameters of the areas, which they say more nearly represent the spreading power of a species, the ratio between tetraploids and diploids is about 2 to 1. These data seem to be especially illuminating because the species that are compared are closely related to one another.

The work of Giles (270) on *Cuthbertia graminea,* which is rather closely related to the spiderworts of the *Tradescantia virginiana* complex, presents an unusually clear case in which the autotetraploid race occupies a much larger area than the diploid race. The diploids occur in the Cretaceous sand hills along the western edge of the coastal plain in North Carolina, and the tetraploids are spread through much of the south Atlantic coastal plain, having colonized the newer suitable habitats as they became available. Giles thinks that the spread of the tetraploids has not been so much a matter of correlation with climate (they are more southern than the diploids) as of their ability to colonize new territory because of their greater variability, ecological amplitude, and vigor. The autotetraploids appear to have been better able to take advantage of the opportunities afforded by the expanding coastal plain.

One of the most fruitful studies published, which has a bearing on these problems, is a cytological survey of *Veronica* by Lehmann (411). He pays special attention to the question of the respective areas of diploids and polyploids, and to the Tischler-Hagerup theory. This work is further reliable

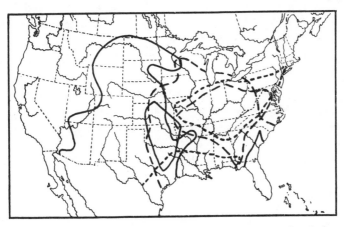

Fig. 59. Areas of the autotetraploid species of *Tradescantia*, after (22).

because the comparisons are made on the basis of the closest taxonomic relationship. If polyploidy plays a role in range extension it should be observable when closely related species are compared, not merely two species of the same genus. It apparently does not signify much if comparisons are made between widely related species unless the parallel cases are sufficiently numerous to assume statistical significance.

In Lehmann's studies we shall first note cases in which there is a strong contrast in area between two or more closely related species with the same chromosome number. *Veronica fruticulosa* and *V. fruticans*, in Group Fruticulosae of Section Veronicastrum, are both S = 16. The former has a narrow area and the latter a wide area. *Veronica Gouani* (S = 32, 34, 36) and *V. Schmidtiana* (S = 34) both belong to Group Gouani of Section Veronicastrum; the former is narrow and the latter wide in area. *V. ticinen-*

sis, V. longifolia var. *japonica,* and *V. maritima* belong to Group Longifoliae of Section Pseudolysimachia, and all are S = 34. The last species is
of wide area and the others are narrow. *V. spuria, V. foliosa,* and *V.
Komaronii* all have the chromosome number S = 34 and belong to Group
Spuriae of Section Pseudolysimachia. The first species is wide and the
other two are narrow in area. *V. syriaca* and *V. acinifolia* are both S = 14
and belong to Group Acinifoliae of Section Alsinebe. The former is narrow
and the latter is somewhat wider in area. *V. scutellata* and *V. montana* are
both S = 18 and belong to Group Scutellatae of Section Chamaedrys. The
former has a much wider range than the latter. *V. officinalis* and *V. Onoei*
are both S = 36 and belong to Group Officinalis of Section Chamaedrys.
The former is wide and the latter narrow in area. *V. incana* and *V. crassi-
folia* are both S = 68 and belong to Section Pseudolysimachia. The former
is wide and the latter narrow in area.

Thus, in several groups and sections, and with sporophytic chromosome
numbers of 14, 16, 18, 32, 34, 36, and 68, we find that closely related species
with the same chromosome number can diverge widely with respect to their
areas. That is to say, narrow and wide areas constitute frequent contrasts
between closely related species, whether the pairs or triplets are diploid or
polyploid.

Let us next notice cases in which there are reversals of the expected relationship, that is, cases in which the diploid species range more widely
than the polyploid species. In Group Souani of Section Veronicastrum
Veronica Stelleri (S = 18) has the widest area, *V. Schmidtiana* (S = 34) has
the next widest, and *V. Gouani* (S = 32, 34, 36), *V. gentianoides* (S = 48),
and var. *variegata* (S = 64–68) have narrower ranges. *V. laxa* and *V.
Chamaedrys* of Group Euchamaedrys and Section Chamaedrys have sporophytic numbers of 46 and 32, respectively. The tetraploid has a wider range
than the hexaploid. In these cases it can be reasoned that the forms with the
higher chromosome numbers have narrower ranges because of the time
factor—they are too recent to have attained their potential areas.

Finally, there is a larger series of cases in which, according to Lehmann,
the Tischler-Hagerup principle seems to apply. *Veronica longifolia* is typically S = 68, but its varieties diverge from this number: var. *typica* and
var. *subsessilis* are both S = 48, and var. *japonica* is S = 34. The species
proper (nomenclaturally, that is) has a range which encloses the smaller
areas of the varieties and also the other species of Group Longifoliae of
Section Pseudolysimachia (*V. ticinensis* and *V. maritima,* which are S = 34).
The latter species, however, has a range nearly as large as that of *V. longi-*

folia. Group Spicatae of Section Pseudolysimachia is a good example of the "expected" results. *V. spicata* consists of two races (S = 34 + 68). Three other species are S = 34 (*V. euxina, V. orchidea*, and *V. Barrelieri*). All the forms with 34 sporophytic chromosomes have a narrow area and are included by the 68-chromosome race of *V. spicata*. In Group Bilobae of Section Alsinebe the polyploid species have much larger areas than the diploid species *V. cardiocarpa* (S = 14, 16), except for *V. arguteserrata* (S = 42), which is also very local. In Group Megaspermae of Section Alsinebe the polyploids have larger areas than the species with low chromosome numbers. The cytologically known species in this group are *V. panormitana* (S = 18), *V. sibthorpioides* (S = 30), *V. Cymbalaria* (S = 36), and *V. heredifolia* (S = 56). The species in Group Austriaca of Section Chamaedrys also seem to fit the rule well. *V. prostrata* consists of two chromosome races (S = 16 + 32), in which the tetraploid is wide and the diploid narrow in range. Likewise, in *V. austriaca* the hexaploid subspecies have wider areas than the tetraploid subspecies (subsp. *orbiculata* S = 32, subspp. *dentata* and *Jacquini* S = 48). *V. Teucrium* (S = 64), which belongs to the same group, has the widest range of all, is the most variable morphologically, and has the largest size, especially in flowers and capsules.

When closely related species are compared, we find three types of relationship between chromosome number and area: (1) forms with the same chromosome number can have strikingly different total areas, whether diploid or polyploid; (2) diploids and low polyploids may have wider areas than tetraploids and higher polyploids; (3) forms with higher chromosome numbers have larger areas than those with lower chromosome numbers. In the majority of cases the latter type, in accordance with the Tischler-Hagerup principle, seems to prevail. An intensive study of the history of the species of a group or section, however, might account for some of the departures from the expected results. For example, a newly developed polyploid may have a narrow range merely because it has not had sufficient time to spread to its potential area. On the other hand, ancient diploids or polyploids may be relics and may now occupy only a fragment of their original area. Furthermore, polyploidy seems to have different effects on different phylogenetic stocks, even within a single genus, so that chromosome duplication may not always result in a broader ecological amplitude and a potentially greater range. The behavior of a plant goes back ultimately to its genic condition; chromosome number is only one of the factors in genetic composition.

Gregory's cytological monograph on Ranunculaceae (301) gives some data relating to areas of genera. This family includes genera that represent extremes of chromosomal number, stability, and instability, and a sharp division on the basis of two types of chromosomes. Among genera with large chromosomes, *Clematis* (170 species) is highly stable with respect to chromosome number and is about 95 per cent diploid. It occurs on every large land mass of the earth except Antarctica, and also on Madagascar, New Zealand, the Fiji Islands, etc. *Ranunculus* (250 species), in contrast, has gametophytic numbers of 7, 8, 14, 16, 21, 24, 28, 32, 40, 48, and 64. Its geographical distribution is wide but is practically limited to the northern cool temperate. Among the Ranunculaceae with small chromosomes, *Aquilegia* (50 species) and *Thalictrum* (76 species) present a similar contrast, the former being 95 per cent diploid and the latter highly polyploid, with gametophytic numbers of 7, 14, 21, 28, 35, 42, 56, and 77. *Aquilegia* covers a tremendous territory in the northern hemisphere, and *Thalictrum* is also widely distributed. We have nothing to choose from concerning the effects of polyploidy on areas when genera such as these are compared.

The relation of polyploidy to the peripheral area of a group.—We have already seen that in *Crepis* the polyploids occupy the peripheral portions of the area of the group as a whole. Sax (550), who has made a similar observation for a variety of types, says that in those genera of the Old World which are also represented in the New World one might expect a greater portion of polyploids in the latter. Such a relation is found in *Spiraea,* for example, and in certain other genera of the Rosaceae which seem to be of Old World origin. The reason for this frequent phenomenon he attributes to the greater vigor, hardiness, and adaptability of polyploids which result in their extension of range, at least in certain genera. Another study by Sax (549) shows that a relation similar to *Spiraea* exists for *Malus, Rosa, Acer, Staphylea,* and *Ulmus,* but that the reverse is true for *Lonicera* and *Fraxinus*. The clarification of such contrasts results from consideration of the origin of these groups which seem to fit the hypothesis stated by Sax. Many genera, however, such as *Rhododendron,* have Old and New World distributions and are entirely or almost completely diploid in both hemispheres. Sax notes that the common interpretation is that extreme temperature changes (for a particular form) may cause polyploidy as a diploid race extends its area by migration into unfavorable territory. To this he adds the interesting statement, " . . . But it is also possible that the polyploids have originated before the extension of the range, and because they are hardier, they extend their range into more extreme environments." We are thus confronted with questions of area, range extension, speciation, and speciation at the periphery of range—questions that deserve further examination.

The problem of the origin and migration of polyploid races is not different in some respects from that of the origin and migration of ecotypes or incipient species of whatever nature.[2] We may start with the well-grounded assumption that most species are divided into local populations and ecotypes or subspecies, each of which has a distinct genetic basis. Furthermore, an ecotype that is adapted to a certain habitat—an ecotype exists because the factors in a certain environment have selected the particular biotypes which characterize the ecotype—cannot migrate into another habitat to which it is not adapted. Exceptions to this statement exist only in the occasional propagules of a form which may sporadically germinate out of the native habitat of the kind and, although depauperate, may produce some seed and survive for a while in the face of mild competition, but be unable to establish permanently. At the same time, a complex of environmental factors which constitute a habitat type cannot select out a group of biotypes constituting a new ecotype in harmony with the habitat until the species is represented by plants in the area of the habitat type. We seem to have encountered a biological impasse. An ecotype, being adapted to a certain habitat, cannot migrate out of its habitat type into a new and different one because of the inherent physiological limitations of the type; a new ecotype can arise only under the action of the environmental factors of a new habitat.

This dilemma has the usual two horns. We can assume that species tend to have broad ecological amplitudes in their early beginning, and that evolution within a species is partly a matter of selection out of populations of narrower ecological amplitudes. In other words, microevolution is in part a matter of genetic and ecological simplification. Basic species are broad in their adaptations and derived forms are narrower in this respect as the result of a progressive selection of biotypes. It is true that partial populations of a species are genetically simpler than the whole population of the species. Partial populations, with isolation, become ecotypes as panmixy is reduced and selection eliminates the less fit; they may, perhaps, become species with the establishment of reproductive barriers. Biologists in several fields—paleontologists, geneticists, taxonomists—have reasoned in this way. Thus the problem of migration and the above dilemma are obviated. There is at least one fundamental difficulty in such a concept. Primi-

[2] On the debatable assumption that ecotypes, geographical races, subspecies, etc., are incipient species, see 282.

tive organisms would have to be genetically richer than derived organisms. To carry the contrast to an absurd degree, we could say from amoeba to man, or from bluegreen algae to daisies, by simplification. In the second place, we would have to assume that original adaptations are broad. Third, in order to avoid the assumption of some kind of biological entropy, that

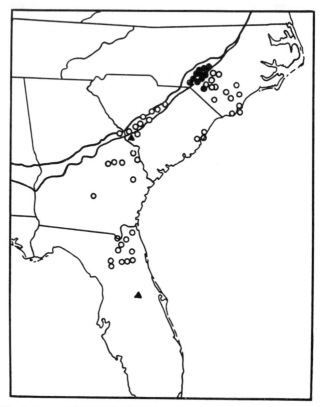

Fig. 60. Distribution of diploid (black disks), tetraploid (circles), and hexaploid (triangles) forms of *Cuthbertia graminea*. The diploids seem to be confined to Cretaceous sand areas and the tetraploids are widely spread through more recent sand areas of the south Atlantic coastal plain. Map redrawn from (270).

evolution is running down, we would have to conceive of a local population or an ecotype as also being pregnant with evolutionary possibility so that it, in its turn, could become a broad and basic species. In this connection, Goldschmidt's contrast (282) between microevolution (within the species) and macroevolution (forming species and higher categories) is of considerable interest. The paleontological record makes it appear that in many genera the fossil species have had wide distributions and that derived species have had more and more narrow areas and, presumably, increasingly narrower ecological amplitudes. The appearance of the phenomenon may be due in part to the inadequacy of the fossil record. On the other hand, it may be due to the fact that at certain geological times broad areas of the

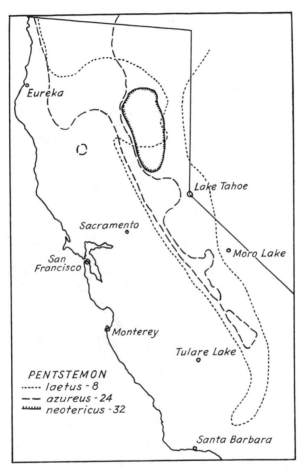

FIG. 61. The allopolyploid *Pentstemon neotericus* (g = 32) occupies an intermediate area between *P. laetus* (g = 8) and *P. azureus* (g = 24) that is largely unoccupied by the progenitor species. Map simplified from (145).

earth were characterized by equable and not very heterogeneous climates, and that, for example, with the progress of the Tertiary and the onset of the Pleistocene the breakup into many divergent climatic areas was accompanied by a phylogenetic breakup into many more narrow species. At any rate, today it appears that ecotypes, of which in many cases there are from a few to many within the boundaries of a taxonomic species, are not formed *in situ* from a preexisting and different population that already occupied the area of the ecotypes.

The other approach to the dilemma seems to lie in the variability of a population at its periphery. Preadapted biotypes, whether polyploid or not, can migrate into a new habitat. Dissemination of propagules of non-adapted

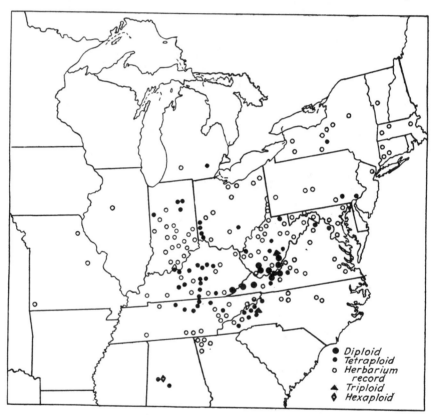

FIG. 62. Polyploidy and area in *Sedum ternatum*. Map redrawn from (46).

biotypes does not result in migration, because the non-adapted types cannot establish themselves. The weight of argument, then, is apparently in favor of Sax's second alternative, that polyploids arise, and because of their wider or different adaptability are enabled to extend the range of their group, especially if they arise near the periphery of an area.

Speciation can be accomplished only when variations can become "consolidated" through some form of isolation. It is probable that all species populations are constantly undergoing mutation with greater or lesser frequency. Mutations which fail to carry with them some form of automatic sterility barrier have no chance of characterizing a new species unless they have some means of escape from panmixy with the parent population. It follows that speciation can take place most easily at the periphery of range for a species population. Here plants carrying new genic mutations, or other types, have a chance to go into territory more or less isolated from the original range of the species population. Isolation would seem to be the first necessity for speciation, although evolution, in the sense of increased variability, can occur without speciation.

At this point it may be well to describe more specifically what is meant by certain terms. Periphery of range consists of two types. In the first place, it refers to the boundary of the total area of the species in a geographical sense. In the second place, it refers to ecological boundaries within the area of the species as a whole. A shade-requiring species meets ecological boundaries wherever open places occur within the whole area. This peripheral region or boundary provides spatial isolation, however broad in actual distance, when suitable habitats are separated from each other by distances which are greater than the disseminative capacity of the propagules of the plant under question, or greater than the range of activity of the agents of cross-pollination for certain sexually reproducing species.

It would seem to follow logically that speciation usually takes place at the periphery of range, along a geographical or ecological boundary, where a certain amount of isolation is provided.

To return to the question of the development of ecotypes, it follows that an ecotype cannot be formed *in situ* from a preexisting population unless, of course, it is assumed that species are originally of broad ecological adaptations and wide area. An ecotype can develop only as certain biotypes of a population that exists at the periphery of the range migrate into a new ecological situation and are selected by that environment because they happen to be preadapted, i.e., genetically better fitted for survival in the new situation. In other words, if these assumptions are accepted, a new ecotype cannot develop within the area of an old ecotype; neither can a new ecotype be selected in a new area from plants of the old ecotype which have migrated into an area of a different nature. With respect to the last point, it is conceivable that a diploid plant, let us say of a shade habitat, may now and then appear sporadically in a sunny situation. In such a place, subjected to greater and more rapid temperature changes, it may give rise to an autotetraploid adapted to the open situation. In this case, migration precedes speciation.

Ordinarily, ecotypes must arise at the transition region between two areas, geographical or ecological, simultaneously with the dissemination into the new region of biotypes from the population of the old region. It would appear that it cannot arise far within either area. Migration, selection, and

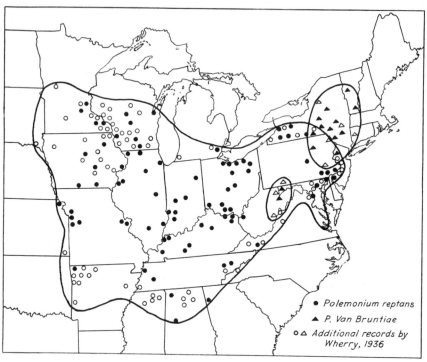

• *Polemonium reptans*
▲ *P. Van Bruntiae*
○△ *Additional records by Wherry, 1936*

Fig. 63. *Polemonium reptans* has the somatic chromosome number of 18. *P. Van Bruntiae* was reported by Flory (251) as having the same number, but Dr. C. A. Berger, S.J., of Fordham University, has established that it is a tetraploid (S = 36); his data are unpublished. The tetraploid may have been formed during the Pleistocene or merely survived in its southern area during that time. It is a reasonable hypothesis that *P. Van Bruntiae* extended its area northeastward during early postglacial time, only to have it dissected by the xerothermic period into the two disjunct modern areas.

the origin of an ecotype are apparently simultaneous and interdependent processes.

With respect to polyploid races or species, the same type of reasoning can be employed. A tetraploid can arise only from a diploid. A tetraploid, whatever its present area, must have arisen somewhere within the range of the progenital diploid population. Having arisen, the tetraploid, with its new characteristics, can migrate into new territory if its ecological attributes make it suited for life there. Tetraploids have new reaction norms and apparently they generally have wider ecological amplitudes. Consequently, they usually do migrate into new and wider territories and occupy the boundaries of the range of the group to which they belong.

To carry the thesis further, we may note that habitats frequently change gradually—one habitat type merges into another—and we can recognize a central habitat area that is typical. With respect to this typical habitat, which is "normal" or near the mean conditions of the habitat type, we could expect selection of biotypes adapted to the normal or mean conditions. Toward the periphery of a habitat type conditions would be somewhat different and the biotypes would be expected to reflect these differences in variation from the mean or normal structural and physiological types. This variation would progress across the ecotone into a second habitat, and a second center of biotypes different from the first would appear. Ecotypes would therefore be expected to show transition just as habitats do. Such, in part, are Haldane's clines. But evolution of one ecotype from another could result only from the simultaneous migration to and selection from the periphery of the preexisting ecotype.

Within a population, the greatest total variation would be expected at the margin of range because at the various parts of the margin the populations live in environments that are variable with respect to the average conditions of the main area of the type. Hence again we can return to the statement that speciation is usually a population-periphery phenomenon which gains expression through the migration that allows isolation and selection.

30.

Polyploidy and
Phylogenetic Relationships

In the early days of this phase of cytology Winge (703) pointed out that chromosome numbers have some relationship to natural plant groups. For example, in the complicated Compositae, which had been thought to be very irregular with respect to chromosome numbers, he showed that the Anthemideae have 9 as the basic number and that the Heliantheae have 8 as their low number, indicating a fundamental phylogenetic divergence between these two tribes. A decade later Hagerup (312) said that the chromosome numbers of the Ericales present a handsome sequence that can be taken as an indication of cytological alterations and is very suggestive as to the relationship among the forms. He found the following gametophytic numbers which he arranged in a pattern of their probable origin:

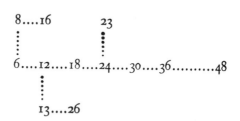

The species of Ericales for which Hagerup knew chromosome numbers at that time are listed below according to their gametophytic numbers.[1]

g = 6 *Phyllodoce coerulea, Diapensia lapponica.*

g = 12 *Kalmia latifolia, Loiseleuria procumbens, Leiophyllum buxifolium, Erica tetralix, Erica cinerea, Erica arborea, Erica carnea,*

---

[1] No attempt is made to add the considerable quantity of later data.

Vaccinium vitis-idaea, Polycodium stamineum, Gaylussacia baccata.

g = 18 *Bruckenthalia spiculiflora.*

g = 24 *Kalmia glauca, Andromeda polifolia, Cassiope hypnoides, Vaccinium* spp.

g = 30 *Vaccinium* (1 hybrid).

g = 36 *Vaccinium* spp., *Oxycoccus palustris.*

g = 48 *Gaultheria shallon.*

g = 8 *Calluna vulgaris* var. *pubescens, Clethra arborea.*

g = 16 *Clethra alnifolia.*

g = 13 *Rhododendron lapponicum, Ledum groenlandicum, Empetrum nigrum, Arctostaphylos diversifolia, Arbutus andrachne, Arbutus canariensis, Epacris impressa.*

g = 26 *Empetrum hermaphroditum, Arctostaphylos uva-ursi.*

g = 23 *Pyrola grandiflora, Pyrola rotundifolia, Pyrola minor.*

Hagerup did not go into a detailed analysis of the relationships of the Ericales as determined through ordinary comparative morphology and geographical distribution in comparison with the suggestions of the cytological condition, but he drew a few conclusions. He pointed out four pairs of related species in which it was assumed that the tetraploid was derived from the diploid by chromosome doubling. One pair exists in the 6-series: *Kalmia latifolia* and *K. glauca.* Two pairs are in the 13-series: *Arctostaphylos diversifolia* and *A. uva-ursi,* and *Empetrum nigrum* and *E. hermaphroditum.* The fourth pair, in the 8-series, is *Clethra arborea* and *C. alnifolia.* He noted that *Ledum groenlandicum* is an arctic species that corresponds well in both morphology and biology with *Rhododendron lapponicum,* and that in both g = 13. Furthermore, the small family Empetraceae (*Empetrum,* g = 13) is known to be closely related to the Rhododendraceae (Ericaceae). *Leiophyllum* is said to be closely related to *Loiseleuria,* and both are g = 12. Also, *Bruckenthalia* is very closely related to *Erica,* and they too are g = 12. The 13-series he suggested had its origin from some tetraploid member of the 6-series through the addition of a chromosome, probably by duplication. The Pyrolaceae (g = 23) were thought to have been derived from the 6-series through the loss of a chromosome by some tetraploid. The Pyrolaceae have been considered as forming a primitive family in the Ericales, but their chromosome number does not support this view.

In another early paper (362) Hurst reviewed the mechanism of heredity and evolution and pointed out the significance of autopolyploidy, allo-polyploidy (then called differential polyploidy), and aneuploidy for the interrelations within taxonomic groups. At this time chromosome numbers were known for 2845 species of plants and animals, representing 1326 genera and 417 families. The numbers found in these species range among animals from 1 pair of chromosomes in the nematode *Ascaris* to more than 100 pairs in the crayfish *Cambarus,* and for plants from 2 pairs in the fungus *Eumycetes* to more than 100 pairs in the horsetail *Equisetum* and the ferns *Ophioglossum* and *Ceratopteris.* With respect to aneuploidy, Hurst cites the work of Heilborn, who in 44 species of *Carex* found 22 different chromosome numbers varying from 9 to 56 pairs. In this genus the different numbers and sizes of chromosomes in each species clearly have resulted in characters of taxonomic significance, and the species belonging to the same section have numbers which are alike or nearly so. It had already been demonstrated in *Datura, Oenothera* and *Viola* that one method of origin of aneuploid numbers was the non-disjunction of a pair of chromosomes, increasing the basic number by 1. In *Datura,* with 12 pairs of chromosomes, such a non-disjunction (24 plus 1) had been found for each of the chromosomes; this resulted in 12 distinct varieties. Incidentally, this demonstrates that each of the chromosomes carries different genes. Autopolyploid varieties were known in *Rosa* (S $= 14$, 21, 28), *Oenothera* (S $= 14$, 21, 28), *Datura* (S $= 12$, 24, 36, 48), and *Solanum* (S $= 24$, 72, 144); and among animals in *Ascaris* (S $= 2$, 4, 8), *Artemia* (S $= 42$, 84), and *Drosophila* (S $= 8$, 12, 16). Allo-polyploids, first recognized in wild species of *Rosa* in 1923, were also known in *Triticum;* this was confirmed in 1926.

A most illuminating recent paper on the relations of cytology to phylogeny and systematics is that by Gregory (301) on the natural family Ranunculaceae. There has long been confusion regarding this family because of the inability of taxonomists to select and agree upon fundamental characters on which to base a natural classification. As Gregory says, "The taxonomic characters of follicular versus achene fruits group together genera which not only look very different vegetatively but behave differently under cultural conditions. . . . However, no sooner had Langlet made his observations on the chromosomes [the existence of the large *Ranunculus*-type and the small *Thalictrum*-type chromosomes] than the situation resolved itself into a matter of parallel fruit development in two very different groups of the Ranunculaceae." Intermediate conditions between follicular and achene fruits are known, so there is little difficulty with respect to this evolutionary change, and, according to Gregory, there is every reason to believe that chromosome type is of a more fundamental nature than fruit type within this family. This does not mean that fruit types are ignored; rather, they are relegated to a subordinate position in the arrangement of the genera, thus allowing a more reasonable grouping according to gross morphology, chromosome type, and basic chromosome number within the family.

Other cases in which cytological data have been useful in supplementing the taxonomist's ideas concerning relationship, or lack of it, may be mentioned. For example, certain species were assigned to *Anchusa* by Gürke which were later placed in separate genera by Johnston. The latter shifted *Anchusa sempervirens* to *Caryolopha* and *Anchusa myosotidiflora* to *Brunnera*. On a cytological basis, Smith decided that Johnston's treatment was warranted. The species of *Anchusa* proper have chromosome numbers based on an 8-series: 16, 24, 32. In *Caryolopha sempervirens* (*Anchusa sempervirens*) $S = 22$, and *Brunnera macrophylla* (*Anchusa mysotidiflora*) has the same number. In addition to these number differences, there are other cytological features of the karyotype, such as chromosome size and position of fiber attachments, that make these plants different from *Anchusa*. In another case (see Ranunculaceae) cytological data have shown that parallel evolution has caused taxonomic confusion. According to McKelvey and Sax (435), "*Yucca* and *Agave* are similar in many taxonomic characters [2] although one genus is placed in the Liliaceae and the other in the Amaryllidaceae. *Yucca* and the closely allied genera *Hesperoyucca, Hesperaloe,* and *Samuela* have 5 pairs of large chromosomes and 25 pairs of small chromosomes at the meiotic divisions. Exactly the same chromosome constitution is found in *Agave* and in at least one species of the closely related genus *Furcraea*. The similarity in taxonomic characters and chromosome constitution indicates that these genera have had a common origin and are closely related." The occurrence of 5 pairs of large and 25 pairs of small chromosomes is so unusual that it can scarcely be attributed to chance; therefore *Yucca* is properly removed from the Liliaceae despite its superior ovary.

Foster (255) has discovered cytological details in a large number of species and varieties of *Acer* and *Staphylea* that afford evidence that the two families to which they belong had a common origin. Thirteen is the basic number in each genus and each genus shows a polyploid series of 13, 26, 39. Whitaker (691), who investigated the Magnoliales cytologically, found that there is a

---

[2] Both *Yucca* and *Agave* have a six-parted perianth and six stamens, and *Yucca* is highly suggestive of *Agave* in general appearance. *Yucca* has a superior ovary, which has caused its inclusion in the Liliaceae, and *Agave* has an inferior ovary.

basic division of the order into two groups. Not only are the groups cytologically distinct, but there are morphological correlations which call for a systematic revision that will result in a more natural treatment of these important primitive plants. Group I, with 19 as the basic chromosome number, includes *Magnolia, Liriodendron, Cercidiphyllum, Drimys, Trochodendron,* and *Tetracentron.* Group II, with 14 as the basic number, includes *Illicium, Schizandra, Kadsura,* and *Euptelea.*

Anderson and Sax (21) found that the Hamamelidaceae are a polyphyletic family. The tribe Hamamelidoideae (*Hamamelis, Corylopsis, Parrotiopsis, Fothergilla,* and *Sinowilsonia*) has 12 as the basic chromosome number, and the tribe Liquidambaroideae (*Liquidambar*) has 15 as the basic number. These authors believe that the tribe Liquidambaroideae was probably derived by reticulate evolution (hybridization and ampidiploidy) from Rosales stock with 7 and 8 chromosomes. They say, "While the anastomoses of the main trunks of the Rosales stock represent supposed true-breeding allopolyploid hybridizations, they do not necessarily indicate a cross between families as such. On any evolutionary hypothesis, related families derive, ultimately, from forms no more differentiated than present-day genera or species. All that need be hypothesized for these hybrids is that they are between forms as diverse morphologically as certain hybrids which have been experimentally obtained, those between Zea and Tripsicum, for example. . . . The general conception, however, of a more or less webbed net-tree for the Rosales is strongly supported by cytological evidence. In some groups of flowering plants (the Tubiflorae, for instance) the webbing would be so much more complex that one would scarcely use the word tree in describing it. In the Cyperaceae, on the other hand, there would be few if any anastomosing branches. The cytogenetic evidence shows with increasing force that the actual pattern of evolutionary progress has been different in different plant groups."

The polyphyletic origin of a group is usually supposed to mean simply that convergent evolution has occurred and that systematists now place together plants which are similar, although they may have had separate phylogenetic origins. This type of polyphyletic condition is certainly widespread among taxonomic groups. The condition of amphidiploid hybrids, however, is an entirely different case. In allopolyploids an actual anastomose of phylogenetic stocks does occur, and the descendants of such a hybridization have a dual ancestry.

Chromosome number reaches its greatest stability, according to Anderson (16), in the gymnosperms, where the great majority of the genera belong to the 12-chromosome type (Taxaceae and Abieteae) or to the 11-chromosome type (Taxodieae and Cupresseae). Sax and Sax (551) gathered information of a cytological nature concerning the conifers and concluded that the similarities in chromosome numbers and meiotic configuration are remarkable, considering the condition in Angiosperms. Differentiation of genera often seems to be associated with changes in chromosome morphology, presumably caused by segmental interchange. Flory (251) later added to the cytological information concerning gymnosperms and discussed their probable phylogenetic relationships on the basis of morphology and cytology. Only two polyploids (*Sequoia sempervirens* and *Juniperus chinensis Pfitzeriana*) are known for sure and there is one probable case (*Pseudolarix*); these are autotetraploids. Hybridization, however, is not infrequent, and most authors believe that geographical separation is the principal factor responsible for the maintenance of species within a genus. But hybridization seems never to have been followed by amphidiploidy.

Some other modern detailed work with phylogenetic implications, combining the practices of morphology in taxonomy with the geographical and cytogenetic principles, is being published by such investigators as Clausen, Keck, and Hiesey (146) and Babcock and Stebbins (38, 39) on western American plants, by Turesson (645 and later publications) on Scandinavian and European plants, and by Turrill and others in England. Stebbins (605) has recently expanded his ideas on the role of polyploidy in evolution and has developed a concept of the rise and decline of systematic complexes. He says:

Since the polyploid members of a complex are more numerous and widespread than the diploids, one would naturally expect that as a polyploid complex becomes older and as conditions cease to be favorable for the type of plant represented by that particular complex, its diploid members would be the first to go. An old or senescent polyploid complex, therefore, is one that consists only of polyploids. With increasing age, the polyploids also begin to die out, so that in the last stages of its existence a polyploid complex is simple once more, and is a monotypic or ditypic genus without any close relatives. Examples of such vestigial polyploid complexes, that is, of isolated monotypic or small genera with high chromosome numbers, are scattered throughout the plant kingdom. Perhaps the most striking ones are the two living genera, *Psilotum* and *Tmesipteris,* which are the only survivors of the most ancient order of vascular plants, the Psilotales. Both of these genera are frequently considered to be monotypic; their species have more than a hundred chromosomes in their sporophytic cells. They may represent the remnants of polyploid complexes which flourished hundreds of millions of years ago in the Paleozoic era. We know from fossil evidence that this order formed a dominant part of the earth's vegetation at that time. Other vestigial polyploid complexes are probably the redwood, *Sequoia sempervirens,* and the genera *Lyonothamnus* and *Fremontia,* familiar relic species of our California flora which have high chromosome numbers. The evidence from the plant kingdom as a whole, therefore, suggests that polyploidy has been most important in developing large, complex and widespread genera; but that in respect to the major lines of evolution, it has been more important in preserving relics of old genera and families than in producing new ones.

Stebbins' concept of the rise and decline of polyploid complexes calls to mind immediately Good's Theory of Generic Cycles (286). Senn (568), as far as I know, is the only cytologist who has published on the relationship of Good's theory to the cytological history of genera. Briefly, Good's theory calls for the following stages in the history of a genus: (1) the genus is juvenile, monotypic, and endemic; (2) it becomes mature, polytypic, and continuous over a more or less wide area; (3) through extinction and radical evolution the genus becomes senile and discontinuous, and may end by being relic-monotypic. Of this theory Senn (568) says, "These concepts may be associated with the concept of chromosome evolution to provide for estimating the relative phylogenetic positions of closely related genera. Genera may arise as a result of aneuploidy, polyploidy, or accumulation of genic changes. In groups in which the basic number is well established, endemic genera with the basic number could be distinguished at once as relic endemics, whereas genera with derived polyploidy or aneuploid numbers would be juvenile endemics. Thus a genetic basis for distinguishing 'old' and 'new' endemics may be established."

Good (287) also proposed a theory of tolerance which postulates that each species has a specific tolerance that represents the range of environmental factors (climatic and edaphic) within which the species is capable of living and reproducing. Any species population, whether diploid or polyploid, can presumably extend its area eventually to the natural limits set jointly by its specific tolerance and the conditions of the environment. The present area of any species population is the result of the interaction of the above phenomena, together with any modifications in the way of change in size of area, position of area, or disruption of area caused by migrations forced by climatic changes and changes in the configuration of land and sea. Polyploidy is of importance in the theory of specific tolerances because it results in changed reaction norms and thus allows the population to occupy a different area in which the ecological or climatic conditions deviate from those to which the diploids were adapted. In his survey of the cytological condition of the Leguminosae, Senn (568) points out that the Genisteae afford an interesting illustration of the application of the theory of tolerance and of generic cycles. The diploid species of *Crotalaria* form a genus supposedly in the mature stage of the cycle, as they are widespread and continuous. The polyploid species of *Genista* are more widespread than the diploid species; this suggests a change in tolerance accompanying chromosome doubling. The discontinuous distribution of the species of the polyploid genus *Lupinus* may indicate that these are very ancient polyploids which are becoming extinct more rapidly than the progenital stock represented by the diploid *Crotalaria*. The other genera, *Spartium, Laburnum, Cytisus,* and *Ulex* are polyploid and are limited in distribution; this indicates that they are juvenile endemic genera of recent origin. It is interesting that *Ulex,* which has three octoploid species, has the most limited distribution of the whole group. This may be because of their recentness of origin. Their specific tolerances may allow them a potentially greater area but they may not yet have had time to occupy it. To such suppositions should be added the idea that a species may be restricted in area, whatever its cytological condition or however great its ecological amplitude, because of the chance that it arose in an area which is climatically or ecologically circumscribed by barriers to its potential spread.

When Darwin wrote *The Origin of Species,* chromosomes were unknown, and it was only in the last quarter of the last century that their relations to the life cycle were first investigated. The great pioneer student of heredity, Gregor Mendel, worked on the polymorphic *Hieracium* in an attempt to explain its complexity; he was dismayed when his breeding experiments yielded only a few hybrids and a large number of offspring exactly resembling the maternal parent. He had no way of knowing that the anomalous behavior of *Hieracium* was due to apomixis. At the turn of the century came the realization that the basis of Mendel's law of heredity lay in the behavior of chromosomes at meiosis. This discovery was shortly followed by evidence that certain mutations in *Oenothera* could be related to changes in chromosome number.

The present century has seen rapid progress in cytology and genetics and in the interpretation of nuclear conditions with respect to evolution. Rosenberg (532) discovered that *Hieracium aurantiacum,* one of Mendel's puzzling species, is a tetraploid (S = 36), and that the related sexual species *H. auricula* is diploid. Several apomictic forms, some of them aneuploid, soon became known. According to Rosenberg (533) and Stebbins and

Babcock (606), about 30 apomictic genera are now known, and apomictic species are almost exclusively polyploid (euploid or aneuploid), whereas their nearest relatives which are panmictic are diploid. Ernst (227) first pointed out that most apomictic plants appear to be of hybrid origin. Some autopolyploids, however, have also been reported to be apomictic, according to Bergman (54). Gates (265) emphasized the fact that unbalanced polyploids (such as triploids and pentaploids) can be maintained only by apogamous or vegetative reproduction and that these unbalanced species must have been apogamous since their origin.

Let it be admitted that much of the evidence on the importance of polyploidy in evolution is circumstantial; there are some instances of the experimental proof of the course of evolution in the synthesis of good Linnean species. The most famous case is the synthesis of *Galeopsis Tetrahit* by Müntzing (474). He crossed *G. pubescens* and *G. speciosa,* Linnean species of S = 16 chromosomes. From this cross he obtained a highly sterile triploid which he called *"pseudo-Tetrahit,"* in which S = 24 chromosomes. A backcross in which pollen of one parent, *G. pubescens,* was used resulted in a single seed and this gave rise to a tetraploid plant which resembled in every respect the natural *G. Tetrahit,* S = 32. This plant, which must have arisen from an unreduced triploid ovule and a haploid male gamete, is *"artificial-Tetrahit."* *Galeopsis Tetrahit, artificial-Tetrahit,* and *G. bifida* are all tetraploids and are freely interfertile, but they are sterile with the diploid parents of *artificial-Tetrahit, G. pubescens,* and *G. speciosa.* This was the first case of a Linnean species being synthesized by means of species hybridization and chromosome summation. The relations just described are shown in Table 32.

TABLE 32.—A diagrammatic representation of the synthesis of *Galeopsis Tetrahit*

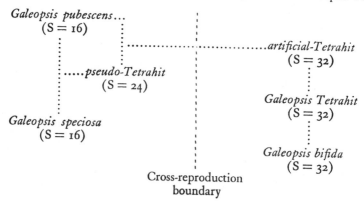

Other cases in which species known in nature have been synthesized include artificial *Salix cinerea* (g = 38), obtained by Nilsson (488) from *S. viminalis* and *S. caprea,* both g = 19; and *Brassica napus* (g = 19) from *B. campestris* (g = 10) and *B. oleracea* (g = 9), obtained by U (655). G = 19 is a new basic number in the genus. The case is interesting in showing that tetraploidy may result from a sterile cross in which the parents have different gametophytic numbers. These synthetically produced species, which are also known in nature, represent a clear verification of Winge's theory (703) on the origin of polyploidy in nature. Also, it can be said with complete confidence that new species can arise by hybridization, followed by doubling, and be perfectly stable, as was claimed by Lotsy (427).

Polyploidy and taxonomic nomenclature.—In a discussion of the interrelations between chromosome numbers, phylogeny, systematics, and plant ecology and geography, there is one technical situation in taxonomic nomenclature that should be remarked on. Ecologists and plant geographers, in considering the relations of populations, are influenced in their interpretations of plant history by the systematic position of a form. For example, one naturally assumes that a variety is subordinate to a species and that it is derived from the species, and consequently that its ecological and geographical features are best understood in terms of difference from the species—such differences having resulted from the vicissitudes of evolution, environmental or selection pressure, migration, etc. That the exact reverse is sometimes the case is well known to systematists. It frequently happens that the first known form of a certain group, and therefore, according to the rules of nomenclature, the one that receives the specific name, is atypical and that the "real species," discovered later, has to be assigned to a varietal status. In plant exploration the first members of a complex to be encountered are often those near the periphery of range for the complex, as when a coastal region is explored earlier than the interior. The peripheral population frequently represents a form subordinate to and derived from the interior population. Nevertheless, the first described form in a group is the species and the later forms are varieties, nomenclaturally. The concept of a species, however, includes any subspecific units taxonomically recognized.

Cytological investigations have revealed numerous cases of the unfortunate conservative effect of the rules of nomenclature, cases which call for revision of nomenclatural status of the forms concerned for the sake of phylogenetic, ecological, and geographical clarity, but which cannot be made under the rules. The following examples are cited from Flovik (252): *Cochlearia officinalis,* from the southwest coast of Wales, was found by

Crane and Gairdner (166) to be tetraploid (S = 28). Flovik found that the form known as *C. officinalis* var. *groenlandica* from Spitzbergen was diploid. *Dumontia Fisheri* is S = 88, whereas the so-called variety *psilosantha* is S = 44. Flovik suggests that *D. Fisheri* originated from *D. Fisheri* var. *psilosantha* by a simple doubling of chromosomes. *Saxifraga nivalis* is S = 60 and *S. nivalis* var. *tenuis* is S = 20. In each of these cases the taxonomic variety has a smaller chromosome number than the species. Phylogenetically the relationship between these paired forms is exactly the reverse of that suggested by the taxonomic disposition of them. It is not surprising that such situations should occur, but it is unfortunate that the condition cannot be corrected in the nomenclature. The lesson for ecologists and geographers is always to inquire into the probable phylogenetic relations between forms and not take the taxonomic disposal at its face value—the so-called variety may be the parent population and the so-called species the derived material.

Glossary

―――――――――――――――――――‑>>><<<‑――――――――――――――――――――

Since plant geography utilizes several more specialized sciences, all the terminology of which may not be familiar to specialists in any one science, an effort has been made to include in the Glossary such terms as may be generally unfamiliar or which may have special meanings in the separate sciences.

Acquired character. A non-heritable environmental modification. *See* Modification.

Action. The effect of environmental factors on organisms. *See* Coaction; Reaction.

Adaptive radiation. The evolution of several closely related but morphologically and ecologically divergent forms. *See* Non-adaptive radiation.

Agamospecies. Species which do not have sexual reproduction, as in parthenogenetic aneuploids.

Age-and-area. The common designation for Willis' hypothesis that the older a species the larger its area.

Aggregation. The process resulting in the grouping of organisms either through active movement or as a result of dissemination.

Allele (allelomorph). One of a pair of alternative genes that occurs in the same relative position in homologous chromosomes. *See* Multiple allelomorphism.

Allogenous flora. Relic plants of an earlier prevailing flora and environment; epibiotic plants.

Allopatric. When species or subspecies areas do not occur together; the forms that occupy such areas.

Allopolyploid. A polyploid from a more or less sterile hybrid. A polyploid containing a complement of chromosomes derived from two or more organisms of dissimilar origin, usually different species. Equivalent to amphidiploid.

Allosyndesis. The pairing of non-homologous chromosomes from the sets contributed by the different parents of an allopolyploid.

Allotetraploid. Synonym of amphidiploid.

Amphidiploid. An allotetraploid; a polyploid of a more or less sterile hybrid between two organisms of dissimilar origin.

Amphimictic population. A population with free-crossing and vital and fertile descendants. Equivalent to panmictic population.

Amphimixis. Cross-fertilization.

Anastomosis. A joining, as in ordinary interspecific hybridization, amphidiploidy, and polyphylesis.

Anemochore. An organism whose disseminules are wind-borne, as in many seeds, spores, etc.

Anemophilous. Pertaining to air-borne bodies, especially pollen and spores.

Aneuploid. An organism having a chromosome number that differs from the diploid or some euploid number by more, or less, than a genom or single set of chromosomes; e.g., a trisomic $(2n +1)$ or a monosomic $(2n -1)$.

Apomixy. The phenomenon of limited or no cross reproduction. The opposite of panmixy.

Area. 1. A geographical unit smaller than a region. 2. The total range of a taxonomic unit, such as the area of a species or genus.

Areography. The description of areas and their causes.

Artenkreis. A circle of related species that show a geographical replacement pattern.

Association. 1. A climax community that is the largest subdivision of a climax, biome, or formation. 2. Loosely, any stable community.

Association-individual. A stand; a concrete example of an association.

Association-segregate. An association that has differentiated out of a mixed association under the historical influence of climatic differentiation.

Associes. Any seral community below the climax association.

Autecology. The study of the ecological relations of individuals or kinds, in contradistinction to the study of communities. *See* Synecology.

Autochthonous. Self-produced.

Autopolyploid. A polyploid of a fertile organism; a polyploid arising by somatic doubling of a non-hybrid organism or by the union of diploid gametes of one species and having multiple genoms of the same type.

Autosyndesis. The pairing of two or more homologous chromosomes from the sets contributed by one parent of an allopolyploid.

Balanced polyploids. Euploids whose genoms are an even number, such as tetraploids, hexaploids, octoploids, etc.

Barrier. 1. A topographic, climatic, edaphic, or biological condition which separates a form from an area with a suitable environment for the form, the breadth of the barrier being greater than the normal dispersal capacity of the form. 2. Any condition that effectively reduces or prevents cross-breeding.

Bicentric. When species, genera, etc., have two centers of evolutionary development.

Binomial. The generic and specific name of an individual.

Biome. A major climax community composed of plants and animals. Equivalent to climax or formation.

Biotope. The smallest natural area or space characterized by a particular environment.

Biotype. The smallest morphological unit within a species whose members are usually genetically identical. Most species consist of thousands of biotypes. *See* Genotype.

Bipolar distribution. The distribution of a form whose area is discontinuous between the northern and southern hemispheres, not necessarily within the polar regions.

Bivalents. Homologous chromosomes that pair.

Bulbil. A small bulb or modified bud that accomplishes vegetative reproduction.

Calcicole. An organism usually found on calcium-rich substrata. Synonym of calciphile; antonym of calcifuge.

Calcifuge. An organism usually found on acid substrata.

Calciphile. *See* Calcicole.

Calciphobe. *See* Calcifuge.

Center of origin. The more or less local area where a phylogenetic stock arose.

Chaparral. A type of vegetation dominated by shrubs with small, broad, hard, evergreen leaves, and occupying a region with a "Mediterranean" type of climate.

Character. Any characteristic of an organism, either structural or functional. Genes, not characters, are inherited. Environment may modify a character only within inherited limits.

Chiasma. A visible change of pairing between two of the four chromatids of a bivalent at meiosis which is evidence that crossover has occurred.

Chimera. An organ with segments or concentric zones differing as to chromosome number or other genetic characters.

Chorology. The science of areas and their development.

Chromatid. One-half of a split chromosome, or one of the four strands of a bivalent at heterotypic division. Chromatids separate at anaphase and enter opposite nuclei.

Chromosome. One of the rod-shaped bodies of the nucleus along which the genosomes are linearly arranged.

Chromosome conjugation. Synapsis; the coming of homologous chromosomes into intimate association.

Chromosome duplication. The doubling of one or more chromosomes through failure of reduction division or of the spindle in mitosis.

Circumboreal. Species of the northern hemisphere whose areas include stations in America, Europe, and Asia.

Climax. The terminal community of a sere which is in dynamic equilibrium with the prevailing climate. The major world climaxes are equivalent to formations and biomes. The term is also used in connection with any subdivision, such as a climax association.

Cline. A series of form changes; a gradient of biotypes along an environmental transition.

Clisere. A series of climaxes following one another in any area as a result of climatic change. Contiguous climaxes move together in a common direction because of a widespread climatic change that induces regional parallelism.

Clone. Members of the same genotype that have descended from one plant by vegetative reproduction.

Coaction. The interaction of organisms; the reciprocal effects of plants and animals.

Codominants. The few species that together dominate or control a community.

Coenospecies. A group of species which may not all be able to exchange genes directly, but which have the possibility of doing so at least to a limited extent through various hybridizations.

Commiscuum. All individuals that can successfully exchange genes; species in the usual sense.

Community. An organized group of plants or animals, or both. The term is employed when it is not necessary or desirable to use a more specific designation such as association, associes, etc.

Comparium. All individuals held together because of the possibility of crossing; equals a coenospecies and taxonomically may be a whole section or genus.

Compatibility. The ability of two organisms to crossbreed successfully.

Compensation. The condition when one or a group of factors usually limiting to the life of an organism is counteracted by an excess or difference of other factors.

Competition. The struggle for existence that results when two or more organisms have requirements in excess of the supply. Competition is largely responsible for the characteristic structure of phytocoenoses.

Complementary genes. Genes that produce a character as a result of their joint action, the character not appearing unless the complementary genes occur together in an organism.

Congeners. Members of the same generic stock.

Conjugation. The lateral association of homologous chromosomes.

Continental bridge hypothesis. The concept that present land masses now separate were once bridged by continents where oceans now lie. An alternative hypothesis to the continental drift concept.

Continental drift hypothesis. Several hypotheses hold that the continents have not always had their present relative positions but have split off from an original land mass and have drifted on the magma. *See* Continental bridge hypothesis.

Convergent evolution. The process whereby phylogenetic stocks that are not closely related produce similar-appearing forms. Such forms are not as closely related genetically as they seem to be.

Convivium. A population differentiated within the commiscuum and isolated by geographical influences; subspecies in the usual sense; also probably equal to ecotypes.

Crossing over. An interchange of segments between chromatids of homologous chromosomes, resulting in a change of linkage relation.

Cryptogam. A lower form of plant life that does not bear seeds.

Cytogenetics. The science combining the methods and findings of cytology and genetics.

Cytology. The study of the structure and function of cells, particularly the protoplasmic bodies of the cytoplasm and nucleus.

Davalloid. Like *Davallia,* a large genus of ferns chiefly of the Old World tropics.

Dendrochronology. The science of dating and investigating historical climates through the study of differences between successive annual rings of trees that result especially from the correlation between ring growth and climatic rhythms.

Diaspore. Any spore, seed, fruit, bud, or other portion of a plant that consists of its active dispersal phase and is capable of reproducing a new plant.

Diastrophism. Deformation of the earth's crust resulting in the major physiographic features.

Dibasic species. Amphidiploid or allopolyploid species.

Diploid. An organism containing two genoms or chromosome sets derived from the egg and sperm; the sporophytic generation. *See* Haploid, Polyploid, etc.

Disclimax. A long-enduring subclimax stage that is prevented from attaining the climax condition by human or animal interference.

Disjunction. 1. In areography, the possession of discontinuous areas. 2. In genetics, the separation of paired chromosomes at anaphase.

Dispersal. The transport of diaspores. It does not constitute migration, but is a necessary antecedent to it.

Disseminule.   *See* Diaspore.

Dominance.   1. In ecology, the term used to refer to the extent of area covered, space occupied, or degree of control of a community. 2. In genetics, the term applied to genes that cause the appearance of a character in the phenotype. *See* Recessive.

Dominant.   1. A form that has a high degree of dominance, or control over a community. 2. A character possessed by one of the parents of a hybrid which appears in the F_1 to the exclusion of its allelomorphic recessive character.

Double fertilization.   The union in higher plants of one male nucleus with an egg and another with a primary endosperm nucleus, producing a zygote in the first case and an endosperm nucleus in the second.

Dysploid.   Equivalent to aneuploid.

Ecad.   A habitat form, a modification, a change or difference that is purely somatic and not heritable.

Ecesis.   The germination and establishment of plants.

Ecological amplitude.   Equivalent to breadth of tolerance.

Ecology.   The study of the complex interrelations between organisms and habitats; more broadly, the term used to include much of the fields of geography and sociology.

Ecospecies.   An amphimictic population with vital and fertile descendants, but with reduced fertility or not readily crossed with other members of the coenospecies.

Ecotone.   The transition region between two communities which contains characteristic species of each. Ecotones are narrow where transitions between environmental types are abrupt, and broad where living conditions change gradually.

Ecotype.   The smallest taxonomic group within a coenospecies whose members cross freely and successfully with other ecotypes of an ecospecies and whose individuality is maintained only because of habitat isolation.

Edaphic.   Pertaining to the soil, especially with respect to its influence on organisms; soil factors.

Element.   1. Phytogeographical element: plants typical of a certain natural area, whether growing within that area or extraneously. 2. Taxonomic element: members of a certain taxonomic group which may be variously represented within any particular area.

Endemic.   1. Forms that are confined to a single natural area. 2. Forms that are confined to a single area, whether natural or not, large or small, isolated or continuous.

Endosperm.   A nutritive tissue of many seeds of flowering plants that contains a complement of chromosomes, resulting from double fertilization, that is greater than 2x.

Endozoochore. A seed, spore, etc., that is disseminated by being carried within the body of an animal.

Entomophilous. Insect-borne, especially with reference to pollen.

Environment. The sum total of effective factors to which an organism responds.

Epibiotic species. Endemic species that are relics of a "lost" flora and compose a minor portion of the biota of most regions.

Epizoochore. A seed, spore, etc., that is disseminated by being carried upon the body of an animal.

Euploid. A diploid or a polyploid that has additional whole genoms, in distinction to aneuploids; an exact multiple of the haploid number of chromosomes.

Evolution. The development by a population of qualitative or quantitative differences of an hereditary nature. *See* Speciation.

Extraneous. Nearer the periphery than the center.

Extrinsic isolating mechanisms. Those mechanisms operating externally and presumably without an hereditary basis that may be overcome if the organisms are brought together.

F_1 generation. The first-generation offspring following a particular mating.

F_2 generation. Progeny derived by selfing the F_1 or by crossing *inter se*.

Faciation. A portion of an association characterized by a particular combination of dominants of the association and having an areal basis related to climatic difference within the general climatic type controlling the association.

Factor. Synonymous with gene.

Flora. The plants of an area considered as kinds rather than as composing communities.

Form (forma). The smallest botanical taxonomic category consisting of occasional individuals that differ by a single character.

Formation. Equivalent to biome and climax.

Frutescent. Having the form of a shrub or bush.

Gametophyte. That portion of the life cycle of an organism that results in the production of gametes and has half the number of chromosomes as the sporophyte; the sexual generation in organisms with alternation of generations.

Gene. A single hereditary factor operating alone or with other genes to produce a character of an organism in conjunction with the environment; a unit of inheritance.

Gene flow. The spread of a gene through a sexual population.

Genecology. The ecology of organisms or taxonomic units; the combination of genetic and ecological methods and concepts, particularly in the study of species and infraspecific groups.

Genetic constellation. The variety and frequency of all genes within a species or other population.

Genetics. The science of heredity.

Genom. The set of chromosomes contributed by one gamete in a basic diploid species, the sporophytic cells containing two, and polyploids higher numbers of genoms. The haploid chromosome set.

Genorheithrum. The stream of genes passing down a phylogenetic stock.

Genosome. The protoplasmic material of a chromosome where a gene is located.

Genotype. 1. Organisms that have identical genetic structure. 2. The entire genetic constitution of an organism, in contrast to phenotype.

Geosere. A series of climax formations developed through geological time.

Germ plasm. The part of the protoplasm that contains the hereditary factors.

Gerontomorphosis. Evolutionary radiation as the result of small or superficial mutations. *See* Paedomorphosis.

Growth form. Equivalent to life form.

Gypsophily. The phenomenon whereby certain species are confined to or are more abundant on soils rich in gypsum.

Habitat. The environment at a particular station occupied by a species, community, etc.

Halophytes. Plants of salt marshes and other salty substrata.

Haploid. The single chromosome set characteristic of the gametophytic generation of a diploid organism. The reduced number of chromosomes in contradistinction to the double number resulting from fertilization.

Heliophilous organisms. Forms usually found living in full light; sun plants and animals; "loving" light.

Hermaphroditism. The presence of two sexes in the same individual.

Heteroploid. A complex containing diploid, euploid, and aneuploid members.

Heterozygote. A hybrid.

Heterozygous. Containing both genes of an allelomorphic pair, or two different genes of an allelomorphic series.

Holocoenotic. The nature of the action of the environment; i.e., the factors of the environment do not act separately and independently, but have mutual interactions and a concerted action upon organisms, and are thus holocoenotic.

Homologous chromosomes. The pairs of chromosomes of a diploid that contain genes in the same sequence for the same characters, and that have come one from each parent. In an autotetraploid homologous chromosomes occur in sets of four.

Homozygote. A non-hybrid.

Homozygous. Capable of producing identical gametes, i.e., whose gametes carry the same genes.

Hybrid. A heterozygote. An organism whose parents are unlike.

Hybrid swarm. A considerable number of first and subsequent generation crosses and various backcrosses with the parents; usually applied to species, but also to subspecies.

Hydrochore. An organism whose seeds, etc., are water-disseminated.

Hydrophyte. An organism living under conditions of high environmental moisture; usually an aquatic.

Hygrochase. A seed case that opens in humid air and closes in dry air.

Inbreeding. The mating of closely related organisms with the genetic result of increasing homozygosis.

Interchange. An exchange of segments of non-homologous chromosomes.

Intraneous. Well within the area.

Intrinsic isolating mechanisms. Those mechanisms operating internally and presumably because of hereditary factors.

Introgressive hybridization. The spread of one or more genes of one species into the germ plasm of another species as the result of hybridization.

Inversion. A segmental rearrangement resulting in a change of order of genes along a chromosome; such a rearranged segment.

Isoflors. Lines delimiting regions with equal numbers of species within the circle of affinity, such as a genus or family.

Isogenous. Having the same origin.

Isohyets. Lines connecting areas of the same annual rainfall.

Isopolls. Lines connecting points of equal pollen percentages during equivalent periods of time; applied only to fossil pollen.

Jordanon. A microspecies; a species of slight variability.

Karyotype. The gross morphology of the chromosomes of a type, usually diagrammed so as to show chromosome number, length, breadth, satellites, etc.

Life form. The vegetative form of an organism such as tree, shrub, annual, liana, bunchgrass, broad-leafed sclerophyll, etc. Synonymous with growth form.

Life zone. Latitudinal and altitudinal belts characterized by a certain fauna and flora and supposedly correlated with effective heat during the growing season.

Linked genes. Genes that do not segregate at meiosis because they are located in the same chromosome; consequently the characters they determine appear together in one organism.

Linneon. A large species, usually polymorphic but with well-characterized limits; species in the usual sense of Linnaeus.

Lociation. A local variant of an association which may not differ from the rest of the association or faciation with respect to dominants, but which is characterized by its own subdominants in the inferior layers of the community; usually correlated with microclimate.

Macroevolution. Speciation at one step; large mutations; paedomorphosis.

Macrospecies. Large species; Linneons; polymorphic species sharply discontinuous from their congenors.

Massenerhebung effect. The effect of the size of a mountain mass upon the altitudinal limits of climatic formations; in larger masses altitudinal limits are extended.

Meiosis. Reduction division when the double sporophytic chromosome number is reduced to the single gametophytic number. Meiosis consists of two divisions: the heterotypic or reductional division, followed by the homotypic or non-reductional division.

Melanism. An unusual development of dark pigment, usually due to a mutation.

Mesophyte. An organism living under moderate moisture conditions; intermediate between xerophyte and hydrophyte.

Microclimate. The climatic environment of a very local area such as the north- or south-facing slopes of a hill, or an even smaller space.

Microevolution. Intraspecific evolution; gerontomorphosis; evolution of relatively minor differences based upon small mutations.

Microspecies. Small species; Jordanons, most likely without reproductive barriers with their related neighbors and consequently probably subspecies.

Microsporogenous tissue. The stamen tissue from which microspores and pollen grains develop.

Migration. The culmination of dissemination in ecesis; when the dissemination of a diaspore is followed by establishment of the organism in a new area, migration has occurred.

Migratory tract. 1. A highway of migration, i.e., a series of similar, approximate sites. 2. An outlying tongue of the area of a genus, extending from the center of distribution.

Migrule. Equivalent to diaspore.

Mitosis. Ordinary non-reductional cell division involving the formation of chromosomes whose halves go to the respective poles, thus maintaining a constant chromosome number.

Mixoploidy. The phenomenon of contiguous cells, cell masses, or tissues with different chromosome numbers. *See* Chimera.

Modification. An acquired characteristic that results from the formative effects of the environment and is not heritable; an extreme morphological expression within the inherited limits of tissue and organ plasticity.

Monobasic species. Diploid or autopolyploid species composed of organisms whose genoms are similar.

Monoclimax hypothesis. The theory that within a regional climatic type there will develop, given sufficient time, a single climax formation.

Monophyletic. A taxonomic group, usually above a species, the members of which have had the same immediate ancestors; i.e., belonging to the same phylogenetic stock.

Monoploid. Equivalent to haploid in the sense of either the basic gametophytic number when polyploids are known, or the gametophytic number in any case.

Monosomic. An organism whose chromosomes are $2n - 1$.

Monotopic. Having a single area; used in reference to species that do not have discontinuous areas.

Monotype. Usually a genus that contains only one species; the term is applicable to other categories.

Multiple allelomorphism. The existence of more than a pair of alleles.

Mutation. An abruptly appearing change in the hereditary materials which may involve genes, chromosome structure, or chromosome number.

Non-adaptive radiation. The evolution of several closely related and morphologically divergent forms without apparent ecological diversification.

Non-disjunction. The failure of a conjugated pair of chromosomes to separate; it results, in gametogenesis, in one gamete having one chromosome plus and another having one chromosome less than the regular number.

Nunatak. A driftless area within a region of general glaciation, usually a tableland or mountain.

Ontogeny. The development of an individual from the zygote or propagule to maturity.

Orthogenesis. Evolutionary development that is not at random, but according to a plan. An hypothesis of orthogenesis may be metaphysical, but it may be based on a theory of chemical structure of the genoplasm that permits only certain types of mutations as the result of limited possibilities of molecular rearrangement or growth.

Paedomorphosis. Evolution through large or basic mutations. *See* Gerontomorphosis.

Paleoecology. The study of past biota on the basis of ecological concepts and methods.

Pandemic. Occurring widely in several natural areas. Antonym of endemic.

Panmixy. The condition of free and more or less unlimited cross reproduction.

Parallel evolution. The phenomenon whereby related but distinct phylogenetic stocks develop comparable forms.

Páramo. The common name applied to alpine vegetation in the mountains of northern South America and the Andes.

Parthenogenesis. The development of an egg without fertilization. Haploid parthenogenesis is the development of the usual reduced egg without fertilization. Diploid parthenogenesis is the development of a diploid egg which has arisen in an embryo sac that came from a somatic cell of the nucellus.

Periclinal chimera. A chimera made up of genetically different tissues arranged parallel with the surface, as when the dermatogen, periblem, and plerome differ as to chromosome number.

Phenotype. The appearance of an organism, in contradistinction to its hereditary constitution. The phenotype results from expressed characters. *See* Genotype.

Photoperiod. Length of the period of light; used in connection with plants that flower under long, medium, or short day length.

Phylad. A small phylogenetic stock, usually a few species linearly related.

Phyllode. A flat expanded petiole replacing the blade of a leaf.

Phylogeny. The evolutionary history and relationships of a phylad or larger group.

Phytocoenosis. The total assemblage of plants living at a particular station. A phytocoenosis is usually composed of several lesser communities which are more homogeneous with respect to flora, life form, and environment, i.e., the synusiae.

Phytosociology. In particular, the study of the composition and structure of communities, and their classification, but extended to include much of the field usually considered as ecology.

Pollen analysis. The identification and determination of the frequency of the pollen species accumulated in a sedimentary deposit.

Pollen profile. An expression, in tabular or diagrammatic form, of the changes in the percentage composition of pollen grains of different species from a consecutive series of samples taken vertically through a deposit.

Pollen spectrum. An expression of the percentage composition of pollen grains of different species from a single sample.

Pollinia. Pollen masses, as in Orchidaceae.

Polster. A cushion plant; a low, compact, shrubby perennial.

Polychronism. The independent origin of a species at more than one time.

Poly-climax hypothesis. The theory that within a regional climatic type there will develop more than one climax because the climatic factors cannot completely subordinate the edaphic factors. *See* Monoclimax.

Polymorphism. The phenomenon whereby a population, such as a species, is composed of several forms, varieties, etc.

Polyphyletic. The condition of a taxonomic group, usually a genus or higher category, the members of which have not come from the same phylogenetic stock. Polyphyletic groups represent mistaken classification, according to most concepts, because of convergent or parallel evolution.

Polyploid. Having more chromosomes than the basic diploid number, i.e., having 3 or more monoploid sets of chromosomes; includes also aneuploids.

Polytopic. Having more than one area; used in reference to species that are disjunct.

Postclimax. A climax community requiring more favorable moisture conditions than prevail generally in an area, usually a relic that has survived because of the compensation of favorable edaphic or microclimatic conditions.

Preclimax. A climax community requiring less favorable moisture conditions than prevail generally in an area, usually a relic that has survived because of the compensation of suitable edaphic or microclimatic conditions which, at the same time, are too dry for the prevailing climax.

Primordium. An embryonic region; a region of localized growth.

Progressive equiformal areas. A series from the smallest to the largest areas formed by plants and animals that have radiated from the same center and spread different distances.

Propagule. Synonymous with diaspore.

Prothallus. The thalloid stage in the gametophyte of ferns and their allies.

Protonema. The alga-like stage in the gametophyte of several mosses, ferns, etc.

Pseudogamy. The parthenogenetic development of an egg, but only after activation by a gamete (or after pollination). Fertilization does not occur.

Rassenkreis. A chain or circle of closely related races within a species; it has a geographical replacement pattern.

Ray. The ligulate or strap-shaped corolla characteristic of many Compositae.

Reaction. The effects of organisms living in an area upon the environment of that area.

Recessive. In genetics, the term applied to genes that result in the appearance of the character which they determine only when the dominant allele is absent. *See* Dominance.

Reciprocal crosses. The phenomenon whereby in one cross the pollen parent is of one type and the ovular parent of another; in the other cross the parents are reversed.

Reduction division. The heterotypic division of meiosis.

Refugium. An area that has not been as drastically altered as the region as a whole; generally used with reference to climatic changes. Usually a center for relic species, and for postglacial dispersal.

Regional parallelism. The development of equivalent but not necessarily identical changes in communities in a country as the result of a general climatic change.

Relic. A surviving organism, population, or community characteristic of an earlier time and usually of a different condition.

Ring formation. The result of the attempted conjugation of three or more homologous chromosomes in polyploids; the meeting of chromosome ends to form a circle.

Savanna. Open forest with grass. The trees may be widely spaced or grouped in clumps with grassland between. A transitional type of vegetation between woodland and grassland.

Sclerophyll. A hard-leafed plant, usually an evergreen. It may be broad- or narrow-leafed.

Sectorial chimera. A chimera in which genetically different tissues are arranged in sectors.

Segregation. The separation of alleles in the formation of gametes.

Sere. A succession; a sequence of communities from pioneer to climax condition.

Soil reaction. Soil acidity or pH.

Soma. The vegetative body, in distinction to the reproductive cells and structures.

Speciation. The final stage in the evolution of species from lower categories when two populations develop interbreeding barriers and become specific. *See* Evolution.

Sporophyte. The portion of the life cycle of an organism that results in the production of spores and has the double number of chromosomes as a result of fertilization; the asexual generation.

Station. The location in which an organism or community type occurs; used in a strictly geographical sense without inference as to environment.

Stoma. A pore in the epidermal tissue of leaves of most plants. It is usually surrounded by chlorophyll-bearing guard cells that can regulate the stomatal size.

Subclimax. An associes that endures overly long because of the interference of edaphic or biotic factors with the normal succession favored by the prevailing climate.

Subendemic. When species are nearly confined to a single natural area.

Subspecies. A population of several biotypes that forms a more or less distinct regional facies of a species and interbreeds with other subspecies where they meet.

Succession. The development of a sere; the replacement of one community by another.

Symbiosis. The living together of two organisms of different species with mutual advantages, or at least with some cooperation.

Sympatric. When two or more closely related species occupy the same area.

Synapsis. The union of homologous maternal and paternal chromosomes to form bivalents.

Synecology. The study of communities, especially their environmental relations and structure.

Syngameon. The members of a population that can exchange genes either directly by crossing or indirectly through a series of crosses.

Synusiae. Minor communities within a phytocoenosis that are characterized by relative uniformity of life form, floristic composition, and environment.

Taiga. Boreal forests, usually coniferous, lying south of the tundra.

Termitophile organisms. Organisms cohabitating with termites.

Tetradynamous. Having six stamens, four long and two short ones, as in mustards.

Tetraploid. An organism that has four sets of chromosomes in its sporophytic cells. An autotetraploid has chromosomes in homologous sets of four because it has resulted from chromosome doubling in a normally fertile organism. An allotetraploid has chromosomes in homologous pairs like a diploid because it has arisen from chromosome doubling in a sterile hybrid.

Tolerance.   The ranges of intensity of factors of the environment within which an organism can function.

Transad.   Organisms of the same or closely related species that occur on two sides of a barrier and must, at one time, have extended across it.

Translocation.   The attachment of a fragment of one chromosome to a non-homologous chromosome.

Triploid.   An organism that has three sets of chromosomes in its sporophytic cells.

Trisomic.   An organism whose chromosomes are $2n + 1$.

Tundra.   The vegetation of arctic and alpine regions above the timber line.

Unbalanced polyploids.   Aneuploids, and euploids with an odd number of genoms, such as triploids, pentaploids, etc.

Variety.   A subdivision of a species consisting of individuals of one or more biotypes and forming a more or less distinct local facies of a species.

Vegetation.   The plants of an area considered in general or as communities, but never taxonomically, as in floristics.

Vicariads.   Vicarious species; closely related allopatric species derived from a common ancestral population.

Vicarious areas.   Mutually exclusive areas belonging to closely related species or subspecies.

Xerochase.   A seed case that opens in dry air and closes in humid air.

Xeromorph.   An organism having a morphology typical of xerophytes, but not necessarily xerophytic.

Xerophyte.   An organism that characteristically lives under dry conditions.

Xerothermic period.   An historical warm-dry period; there seems to have been at least one in the northern hemisphere since the last ice age.

Zygote.   The fertilized egg; the cell produced by the union of two gametes.

Literature Cited

—»»>‹‹‹—

The abbreviations employed in the titles conform to the style given in "Abbreviations used in the Department of Agriculture for titles of publications," by Carolyn Whitlock (U. S. Dept. Agr., *Misc. Pub. 337*, 1939). Sources not included in the above list have been handled in the same style on the basis of titles in the "Union List of Serials." A small portion of the literature cited has not been seen by me and has been referred to on the authority of others. Special credit and appreciation are gladly given Mrs. Lazella Schwarten, Assistant Librarian, New York Botanical Garden, who has kindly checked all citations directly from the literature in the libraries of the New York Botanical Garden, the New York Public Library, the American Museum of Natural History, the Geographical Society, and the Arnold Arboretum. Any inaccuracies which may remain are not her responsibility.

1. Aario, L. Pflanzentopographische und paläographische Mooruntersuchung in N.-Satakunta. *Comm. Forest. Fenniae,* 1932, 17:1–179.
2. Abrams, LeRoy. The theory of isolation as applied to plants. *Science,* 1905, 22:836–838.
3. Adams, C. C. Baseleveling and its faunal significance, with illustrations from southeastern United States. *Amer. Nat.,* 1901, 35:839–852.
4. Adams, C. C. Southeastern United States as a center of geographical distribution of fauna and flora. *Biol. Bul.,* 1902, 3:115–131.
5. Adams, C. C. Post-glacial origin and migrations of the life of northeastern United States. *Jour. Geog.,* 1902, 1:303–310, 352–357.
6. Adams, C. C. The post-glacial dispersal of the North American biota. *Biol. Bul.,* 1905, 9:53–71.
7. Adams, C. C. The Coleoptera of Isle Royale, Lake Superior, and their relation to the North American centers of dispersal. In *Ecological Survey of Isle Royale. Mich. Geol. Rept. 1908,* 1909, 157–215.
8. Adams, C. C. The variations and ecological distribution of the snails of the genus *Io. Natl. Acad. Sci. Mem.,* 1915, 12 (2):1–92.
9. Allan, H. H. Natural hybridization in relation to taxonomy. In J. Huxley, *The New Systematics.* Oxford, Clarendon Press, 1940, pp. 515–528.
10. Allee, W. C., and T. Park. Concerning ecological principles. *Science,* 1939, 89:166–169.
11. Allen, J. A. Variations in vertebrated animals. *Amer. Nat.,* 1892, 26:87–89.

12. Allen, J. A.   The evolution of species through climatic conditions. *Science,* 1905, 22:661–668.

13. Anderson, E.   An experimental study of hybridization in the genus *Apocynum. Mo. Bot. Gard. Ann.,* 1936, 23:159–168.

14. Anderson, E.   Hybridization in American Tradescantias. *Mo. Bot. Gard. Ann.,* 1936, 23:511–525.

15. Anderson, E.   The species problem in *Iris. Mo. Bot. Gard. Ann.,* 1936, 23:457–509.

16. Anderson, E.   Cytology in its relation to taxonomy. *Bot. Rev.,* 1937, 3:335–350.

17. Anderson, E.   The technique and use of mass collections in plant taxonomy. *Mo. Bot. Gard. Ann.,* 1941, 38:287–292.

18. Anderson, E., and D. G. Diehl.   Contributions to the *Tradescantia* problem. *Arnold Arboretum Jour.,* 1932, 13:213–231.

19. Anderson, E., and L. Hubricht.   The American sugar maples. I. Phylogenetic relationships as deduced from a study of leaf variation. *Bot. Gaz.,* 1938, 100:312–323.

20. Anderson, E., and L. Hubricht.   Hybridization in *Tradescantia.* III. The evidence for introgressive hybridization. *Amer. Jour. Bot.,* 1938, 25:396–402.

21. Anderson, E., and K. Sax.   Chromosome numbers in the Hamamelidaceae and their phylogenetic significance. *Arnold Arboretum Jour.,* 1935, 16:210–215.

22. Anderson, E., and K. Sax.   A cytological monograph of the American species of *Tradescantia. Bot. Gaz.,* 1936, 97:433–476.

23. Anderson, E., and T. W. Whitaker.   Speciation in *Uvularia. Arnold Arboretum Jour.,* 1934, 15:28–42.

24. Anderson, E., and R. E. Woodson.   The species of *Tradescantia* indigenous to the United States. *Arnold Arboretum Contrib.,* 1935, 9:1–132.

25. Arldt, T.   *Handbuch der Palaeogeographie.* Leipzig, Borntraeger, 1919–22, 2 vols., 1647 pp.

26. Arldt, T.   *Die Entwicklung der Kontinente und ihr Lebenwelt.* Berlin, Borntraeger, 1936, ed. 2, Erster teil, 448 pp.

27. Atkins, W. R. G.   Some factors affecting the hydrogen ion concentration of the soil and its relation to plant distribution. *Roy. Dublin Soc. Sci. Proc.,* 1922, n.s. 16:369–413.

28. Auer, V.   Stratigraphical and morphological investigations of peat bogs of southeastern Canada. *Com. ex Instit. Quaest. Forestal. Finlandiae Editae,* 1927, 12:1–62.

29. Auer, V.   Peat bogs in southeastern Canada. *Canad. Dept. Mines, Geol. Surv. Mem.,* 1930, 162:1–32.

30. Auer, V.   Peat bogs of southeastern Canada. *Handb. d. Moorkunde,* 1933, 7:141–223.

31. Axelrod, D. I. A Pliocene flora from the Mount Eden beds, Southern California. *Carnegie Inst. Wash. Pub.,* 1937, 476: 125–183.

32. Axelrod, D. I. A Miocene flora from the western border of the Mohave Desert. *Carnegie Inst. Wash. Pub.,* 1939, 516: 1–129.

33. Axelrod, D. I. The concept of ecospecies in Tertiary paleobotany. *Natl. Acad. Sci. Proc.,* 1941, 27: 545–551.

34. Babcock, E. B. Cyto-genetics and the species-concept. *Amer. Nat.,* 1931, 65: 5–18.

35. Babcock, E. B. Genetic evolutionary processes. *Natl. Acad. Sci. Proc.,* 1934, 20: 510–515.

36. Babcock, E. B. Systematics, cytogenetics and evolution in *Crepis. Bot. Rev.,* 1942, 8: 139–190.

37. Babcock, E. B., and D. R. Cameron. Chromosomes and phylogeny in *Crepis.* II. *Calif. Univ. Pubs. Agr. Sci.,* 1934, 6: 287–324.

38. Babcock, E. B., and G. L. Stebbins, Jr. The genus *Youngia. Carnegie Inst. Wash. Pub.,* 1937, 484: 1–106.

39. Babcock, E. B., and G. L. Stebbins, Jr. The American species of *Crepis.* Their interrelationships and distribution as affected by polyploidy and apomixis. *Carnegie Inst. Wash. Pub.,* 1938, 504: 1–199.

40. Backer, C. A. *The problem of Krakatao as seen by a botanist.* Java, 1929.

41. Bailey, I. W., and E. W. Sinnott. The climatic distribution of certain types of angiosperm leaves. *Amer. Jour. Bot.,* 1916, 3: 23–39.

42. Baker, H. B. *The Atlantic Rift and its meaning.* Privately printed. Ann Arbor, Edwards Bros., 1932, 305 pp.

43. Baldwin, J. T., Jr. Chromosomes of the Diapensiaceae. *Jour. Hered.,* 1939, 30: 169–171.

44. Baldwin, J. T., Jr. Distribution of *Galax aphylla* in Virginia. *Va. Jour. Sci.,* 1941, 2: 68–69.

45. Baldwin, J. T., Jr. *Galax:* The genus and its chromosomes. *Jour. Hered.,* 1941, 32: 249–254.

46. Baldwin, J. T., Jr. Polyploidy in *Sedum ternatum* Michx. II. Cytogeography. *Amer. Jour. Bot.,* 1942, 29: 283–286.

47. Baur, E. Artumgrenzung und Artbildung in der Gattung *Antirrhinum,* Sektion *Antirrhinastrum. Ztschr. f. Induktive Abstam. u. Vererbungslehre,* 1933, 63: 256–302.

48. Beasley, J. O. The production of polyploids in *Gossypium. Jour. Hered.,* 1940, 31: 39–48.

49. Becker, G. Experimentelle Analyse der Genon- und Plasmonwirkung bei Moosen. III. Osmotischer Wert heteroploider Pflanzen. *Ztschr. f. Induktive Abstam. u. Vererbungslehre,* 1931, 60: 17–38.

50. Belling, J. The origin of chromosomal mutations in *Uvularia. Jour. Genet.,* 1925, 15: 245–266.

51. Belling, J. Production of triploid and tetraploid plants. *Jour. Hered.*, 1925, 16:463–464.

52. Benson, L. The mesquites and screw-beans of the United States. *Amer. Jour. Bot.*, 1941, 28:748–754.

53. Bentall, R. Spores in Tennessee coals and their use in correlation of seams. Thesis, Univ. Tenn. Library, 1940, 140 pp.

54. Bergman, B. Zytologische Studien über sexuelles und asexuelles *Hieracium umbellatum*. *Hereditas*, 1935, 20:47–64.

55. Berry, E. W. The upper Cretaceous and Eocene floras of South Carolina and Georgia. *U. S. Geol. Survey, Prof. Paper*, 1914, 84:1–200.

56. Berry, E. W. The lower Eocene floras of southeastern North America. U. S. Geol. Survey, Prof. Paper, 1916, 91:1–481.

57. Berry, E. W. Upper Cretaceous floras of the Eastern Gulf Region in Tennessee, Mississippi, Alabama, and Georgia. *U. S. Geol. Survey, Prof. Paper*, 1919, 112:1–177.

58. Berry, E. W. *Tree ancestors, a glimpse into the past.* Baltimore, Williams and Wilkins, 1923, 270 pp.

59. Berry, E. W. The Middle and Upper Eocene floras of southeastern North America. *U. S. Geol. Survey, Prof. Paper*, 1924, 92:1–206.

60. Berry, E. W. The flora of the frontier formation. *U. S. Geol. Survey, Prof. Paper*, 1929, 158:129–135.

61. Berry, E. W. A revision of the flora of the Latah formation. *U. S. Geol. Survey, Prof. Paper*, 1929, 154:225–264.

62. Berry, E. W. Revision of the lower Eocene Wilcox flora of the Southeastern States. *U. S. Geol. Survey, Prof. Paper*, 1930, 156:1–196.

63. Berry, E. W. The past climate of the north polar region. *Smithsn. Inst., Misc. Collect.*, 1930, 82 (6):1–29.

64. Bertsch, K. Paläobotanische Monographie des Federseerieds. *Bib. Bot.*, 1931, 26, Hefte 103:1–127.

65. Bessey, C. E. The phylogenetic taxonomy of flowering plants. *Mo. Bot. Gard. Ann.*, 1915, 2:109–164.

66. Bhaduri, P. N. Chromosome numbers of some Solanaceous plants of Bengal. *Indian Bot. Soc. Jour.*, 1933, 12:56–64.

67. Blackburn, K. B. Chromosomes and classification in the genus *Rosa*. *Amer. Nat.*, 1925, 59:200–205.

68. Blackburn, K. B. On the relation between geographic races and polyploidy in *Silene ciliata* Pourr. *Genetica*, 1933, 15:49–66.

69. Blackburn, K. B., and J. W. H. Harrison. Genetical and cytological studies in hybrid roses. I. The origin of a fertile hexaploid form in the *Pimpinellifoliae-villosae* crosses. *Brit. Jour. Expt. Biol.*, 1924, 1:557–570.

70. Blakeslee, A. F. Effect of induced polyploidy in plants. *Biol. Symposia*, 1941, 4:183–201.

71. Blakeslee, A. F., and A. G. Avery. Methods of inducing doubling of chromosomes in plants. *Jour. Hered.,* 1937, 28:393–411. [Other accounts of the same report: *Science,* 1937, 86:408. (*Paris*) *Acad. des Sci. Compt. Rend.,* 1937, 205:476–479.]

72. Blakeslee, A. F., and J. Belling. Chromosomal mutations in the Jimson weed, *Datura Stramonium. Jour. Hered.,* 1924, 15:195–206.

73. Blum, H. F. A consideration of evolution from a thermodynamic viewpoint. *Amer. Nat.,* 1935, 69:354–369.

74. Böcher, T. W. Cytological studies on *Campanula rotundifolia. Hereditas,* 1936, 22:269–277.

75. Böcher, T. W. Cytological studies in the genus *Ranunculus. Dansk Bot. Arkiv.,* 1938, 9 (4):1–33.

76. Böcher, T. W. Zur Zytologie einiger arktischen und borealen Blütenpflanzen. *Svensk Bot. Tidskr.,* 1938, 32:346–361.

77. Bowden, W. M. Diploidy, polyploidy, and winter hardiness relationships in the flowering plants. *Amer. Jour. Bot.,* 1940, 27:357–371.

78. Bowman, P. W. Study of a peat bog near the Matamek River, Quebec, Canada, by the method of pollen analysis. *Ecology,* 1931, 12:694–708.

79. Braun, E. L. Glacial and post-glacial plant migrations indicated by relic colonies of southern Ohio. *Ecology,* 1928, 9:284–302.

80. Braun, E. L. A history of Ohio's vegetation. *Ohio Jour. Sci.,* 1934, 34:247–257.

81. Braun, E. L. The undifferentiated deciduous forest climax and the association-segregate. *Ecology,* 1935, 16:514–519.

82. Braun, E. L. The vegetation of Pine Mountain, Kentucky. *Amer. Midland Nat.,* 1935, 16:517–565.

83. Braun, E. L. Some relationships of the flora of the Cumberland Plateau and the Cumberland Mountains in Kentucky. *Rhodora,* 1937, 39:193–208.

84. Braun, E. L. A remarkable colony of coastal plain plants on the Cumberland Plateau in Laurel County, Kentucky. *Amer. Midland Nat.,* 1937, 18:363–366.

85. Braun, E. L. Deciduous forest climaxes. *Ecology,* 1938, 19:515–522.

86. Braun, E. L. Mixed deciduous forests of the Appalachians. *Va. Jour. Sci.,* 1940, 1:1–4.

87. Braun, E. L. The differentiation of the deciduous forest of Eastern United States. *Ohio Jour. Sci.,* 1941, 41:235–241.

88. Braun-Blanquet, J. *L'origine et le développement des flores dans le massif central de France.* Paris, Zürich, 1923, 286 pp.

89. Briquet, J. Une Graminée nouvelle pour la flore des Alpes (*Poa Balfourii* Parn.). *Conserv. Jard. Bot. Genève Ann.,* 1901, 5:174–176.

90. Brown, C. A. The flora of Pleistocene deposits in the western Florida

parishes, West Feliciana Parish, and East Baton Rouge Parish, Louisiana. *La. Dept. Conserv. Geol. Surv.,* 1938, Bul. 12:59–96.

91. Brown, F. B. H. Origin of the Hawaiian flora. *Pan-Pacific Sci. Conf. First Proc.,* 1921, 1:131–142.

92. Brown, F. B. H. The secondary xylem of Hawaiian trees. *Bernice P. Bishop Mus. Occas. Paper,* 1922, 8 (6):217–371.

93. Brown, H. B. The genus *Crataegus,* with some theories concerning the origin of its species. *Torrey Bot. Club Bul.,* 1910, 37:251–260.

94. Brown, R. W. The recognizable species of the Green River flora. *U. S. Geol. Survey, Prof. Paper,* 1934, 185 C:45–68.

95. Brown, R. W. Miocene leaves, fruits, and seeds from Idaho, Oregon, and Washington. *Jour. Paleont.,* 1935, 9:572–587.

96. Brown, R. W. Further additions to some fossil floras of the western United States. *Wash. Acad. Sci. Jour.,* 1937, 27:506–517.

97. Brown, W. L. Chromosome complements of five species of *Poa* with an analysis of variation in *Poa pratensis. Amer. Jour. Bot.,* 1939, 26:717–723.

98. Bruun, H. G. The cytology of the genus *Primula. Svensk Bot. Tidskr.,* 1930, 24:468–475.

99. Bruun, H. G. Cytological studies in *Primula* with special reference to the relation between the karyology and taxonomy of the genus. *Symbolae Bot. Upsalienses,* 1932, 1 (1):1–239.

100. Bryan, K. Geologic antiquity of man in America. *Science,* 1941, 93:505–514.

101. Burns, G. W. The taxonomy and cytology of *Saxifraga pennsylvanica* L. and related forms. *Amer. Midl. Nat.,* 1942, 28:127–160.

102. Burtt, B. L., and A. W. Hill. The genera *Gaultheria* and *Pernettya* in New Zealand, Tasmania, and Australia. *Linn. Soc. London, Jour. Bot.,* 1935, 49:611–644.

103. Buxton, B. H., and W. C. F. Newton. Hybrids of *Digitalis ambigua* and *Digitalis purpurea,* their fertility and cytology. *Jour. Genet.,* 1928, 19:269–279.

104. Cain, S. A. Certain floristic affinities of the trees and shrubs of the Great Smoky Mountains and vicinity. *Butler Univ. Bot. Stud.,* 1930, 1:129–150.

105. Cain, S. A. An ecological study of the heath balds of the Great Smoky Mountains. *Butler Univ. Bot. Stud.,* 1930, 1:177–208.

106. Cain, S. A. The quadrat method applied to sampling spruce and fir forest types. *Amer. Midland Nat.,* 1935, 16:566–584.

107. Cain, S. A. Pollen analysis as a paleo-ecological research method. *Bot. Rev.,* 1939, 5:627–654.

108. Cain, S. A. The climax and its complexities. *Amer. Midland Nat.,* 1939, 21:146–181.

109. Cain, S. A. Some observations on the concept of species senescence. *Ecology,* 1940, 21:213–215.

110. Cain, S. A. The identification of species in fossil pollen of *Pinus* by size-frequency determinations. *Amer. Jour. Bot.*, 1940, 27:301–308.

111. Cain, S. A. The Tertiary nature of the cove hardwood forests of the Great Smoky Mountains. *Torrey Bot. Club Bul.*, 1943, 70:213–235.

112. Cain, S. A., and J. D. O. Miller. Leaf structure of *Rhododendron catawbiense* Michx. grown in *Picea-Abies* forest and in heath communities. *Amer. Midland Nat.*, 1933, 14:69–82.

113. Cain, S. A., and J. E. Potzger. A comparison of leaf tissues of *Gaylussacia baccata* (Wang.) C. Koch. and *Vaccinium vacillans* Kalm. grown under different conditions. *Amer. Midland Nat.*, 1933, 14:97–112.

114. Cain, S. A., and J. E. Potzger. A comparison of leaf tissues of *Gaylussacia baccata* grown under different conditions. *Amer. Midland Nat.*, 1940, 24:444–462.

115. Camp, W. H. Studies in the Ericales. A discussion of the genus *Befaria* in North America. *Torrey Bot. Club Bul.*, 1941, 68:100–111.

116. Camp, W. H. Studies in the Ericales: A review of the North American Gaylussacieae; with remarks on the origin and migration of the group. *Torrey Bot. Club Bul.*, 1941, 68:531–551.

117. Camp, W. H. On the structure of populations in the genus *Vaccinium*. *Brittonia*, 1942, 4:189–204.

118. Campbell, D. H. The origin of the Hawaiian flora. *Torrey Bot. Club Mem.*, 1918, 17:90–96.

119. Campbell, D. H. The derivation of the flora of Hawaii. *Stanford Univ. Pub. Univ. Ser. Biol.*, 1919, pp. 1–34.

120. Campbell, D. H. The Australasian element in the Hawaiian flora. *Amer. Jour. Bot.*, 1928, 15:215–221.

121. Candolle, A. de. *Géographie Botanique Raisonnée*. Paris, 2 vols., 1855.

122. Carroll, G. The use of bryophytic polsters and mats in the study of recent pollen deposition. *Amer. Jour. Bot.*, 1943, 30:361–366.

123. Cartledge, J. L., and A. F. Blakeslee. Mutation rate increased by aging seeds as shown by pollen abortion. *Natl. Acad. Sci. Proc.*, 1934, 20:103–110.

124. Chandler, C., W. M. Porterfield, and A. B. Stout. Microsporogenesis in diploid and triploid types of *Lilium tigrinum* with special reference to abortions. *Cytologia Fujii Jubilaei*, 1937, 2:756–784.

125. Chaney, R. W. Quantitative studies of the Bridge Creek flora. *Amer. Jour. Sci.*, 1924, S 5, 8:127–144.

126. Chaney, R. W. A comparative study of the Bridge Creek flora and the modern redwood forest. *Carnegie Inst. Wash. Pub.*, 1925, 349:1–22.

127. Chaney, R. W. The Mascall flora—Its distribution and climatic relation. *Carnegie Inst. Wash. Pub.*, 1925, 349:23–48.

128. Chaney, R. W. Geology and paleontology of the Crooked River Basin with special reference to the Bridge Creek flora. *Carnegie Inst. Wash. Pub.*, 1927, 346:45–138.

129. Chaney, R. W. The succession and distribution of Cenozoic floras around the Northern Pacific basin. In *Essays in Geobotany in Honor of William Albert Setchell.* Berkeley, Univ. Calif. Press, 1936, pp. 55–85.

130. Chaney, R. W. Plant distribution as a guide to age determination. *Wash. Acad. Sci. Jour.,* 1936, 26:313–324.

131. Chaney, R. W. Ancient forests of Oregon: A study of earth history in Western America. *Carnegie Inst. Wash. Pub.,* 1938, 501:631–648.

132. Chaney, R. W. Paleoecological interpretations of Cenozoic plants in Western North America. *Bot. Rev.,* 1938, 4:371–396.

133. Chaney, R. W. The Deschutes flora of Eastern Oregon. *Carnegie Inst. Wash. Pub.,* 1938, 476:185–216.

134. Chaney, R. W. Tertiary forests and continental history. *Geol. Soc. Amer. Bul.,* 1940, 51:469–486.

135. Chaney, R. W. Bearing of forests on the theory of continental drift. *Sci. Monthly,* Dec., 1940, pp. 489–499.

136. Chaney, R. W., and M. K. Elias. Late Tertiary floras from the High Plains. *Carnegie Inst. Wash. Pub.,* 1936, 476:1–72.

137. Chaney, R. W., and H. H. Hu. A Miocene flora from Shantung Province, China. II. Physical conditions and correlation. *Carnegie Inst. Wash. Pub.,* 1940, 507:85–140.

138. Chaney, R. W., and H. L. Mason. A Pleistocene flora from Santa Cruz Island, California. *Carnegie Inst. Wash. Pub.,* 1930, 415:1–24.

139. Chaney, R. W., and H. L. Mason. A Pleistocene flora from the asphalt deposits at Carpinteria, California. *Carnegie Inst. Wash. Pub.,* 1933, 415:45–79.

140. Chaney, R. W., and E. I. Sanborn. The Goshen flora of west central Oregon. *Carnegie Inst. Wash. Pub.,* 1933, 439:1–103.

141. Chittenden, R. J. Notes on species crosses in *Primula, Godetia, Nemophila,* and *Phacelia. Jour. Genet.,* 1928, 19:285–314.

142. Christ, H. *Die Geographie der Farne.* Jena, G. Fischer, 1910, 357 pp.

143. Christiansen, M. P. Nye *Taraxacum*-Arten af Gruppen *Vulgaria. Dansk Bot. Arkiv,* 1936, 9 (2):1–31.

144. Clausen, J. Cyto-genetic and taxonomic investigations on *Melanium* violets. *Hereditas,* 1931, 15:219–308.

145. Clausen, J. Cytological evidence for the hybrid origin of *Penstemon neotericus* Keck. *Hereditas,* 1933, 18:65–76.

146. Clausen, J., D. D. Keck, and W. M. Hiesey. Experimental studies on the nature of species. I. Effect of varied environments on western American plants. *Carnegie Inst. Wash. Pub.,* 1940, 520:1–452.

147. Clausen, R. E. Polyploidy in *Nicotiana. Biol. Symp.,* 1941, 4:95–110.

148. Clements, F. E. The polyphyletic disposition of lichens. *Amer. Nat.,* 1897, 31:277–284.

149. Clements, F. E. *The development and structure of vegetation.* Lincoln, Neb., Woodruff-Collins, 1904.

150. Clements, F. E. Plant succession. *Carnegie Inst. Wash. Pub.,* 1916, 242:1–512.

151. Clements, F. E. Scope and significance of paleo-ecology. *Geol. Soc. Amer. Bul.,* 1918, 29:369–374.

152. Clements, F. E. The relict method in dynamic ecology. *Jour. Ecol.,* 1934, 22:39–68.

153. Clements, F. E. The origin of the desert climax and climate. In *Essays in Geobotany in Honor of William Albert Setchell.* Berkeley, Univ. Calif. Press, 1936, pp. 87–140.

154. Clements, F. E. Nature and structure of the climax. *Jour. Ecol.,* 1936, 24:252–284.

155. Clements, F. E., and R. W. Chaney. Methods and principles of paleo-ecology. *Carnegie Inst. Wash. Yearbook,* 1923, 22:319.

156. Clements, F. E., and R. W. Chaney. Environment and life in the Great Plains. *Carnegie Inst. Wash. Suppl. Pub.,* 1936, 24:1–54.

157. Clements, F. E., and V. E. Shelford. *Bio-ecology.* New York. John Wiley & Sons, 1939, 425 pp.

158. Cockerell, T. D. A. The evolution of species through climatic conditions. *Science,* 1906, 23:145–146.

159. Cockerell, T. D. A. Discontinuous distribution in bees. *Nature,* 1932, 130:58–59.

160. Cockerell, T. D. A. Discontinuous distribution in plants. *Nature,* 1932, 130:812.

161. Cooper, W. S. The broad-sclerophyll vegetation of California. *Carnegie Inst. Wash. Pub.,* 1922, 319:1–124.

162. Cowles, H. C. The physiographic ecology of Chicago and vicinity; a study of the origin, development, and classification of plant societies. *Bot. Gaz.,* 1901, 31:73–108, 145–182.

163. Cowles, H. C. The succession point of view in floristics. *Intern. Congr. Plant Sci., Ithaca, Proc.,* 1929, 1:687–691.

164. Crane, M. B. The origin and behaviour of cultivated plants. In J. Huxley, *The New Systematics.* Oxford, Clarendon Press, 1940, pp. 529–547.

165. Crane, M. B., and C. D. Darlington. The origin of new forms in *Rubus. Genetica,* 1927, 9:241–278.

166. Crane, M. B., and A. E. Gairdner. Species-crosses in *Cochlearia,* with a preliminary account of their cytology. *Jour. Genet.,* 1923, 13:187–200.

167. Cranwell, L. M. Southern-beech pollens. *Auckland Inst. Mus. Rec.,* 1939, 2:175–196.

168. Cranwell, L. M. Pollen grains of the New Zealand conifers. *N. Zealand Jour. Sci. Tech. B,* 1940, 22:1–17.

169. Cranwell, L. M., and L. von Post. Post-Pleistocene pollen diagrams from the southern hemisphere. *Geogr. Annaler, Årg,* 1936, 18:308–347.

170. Cretzoiu, P. *Rhododendron ponticum* L. *Die Pflanzenareale,* 1938, 4 Reihe, Heft 6, Karte 53.

171. Dahl, O. Floraen i Finnmark fylke. *NYT Mag.,* 1934, 69:1–430.

172. Danser, B. H. Über die Begriffe Komparium, Kommiskuum und Konvivium und über die Entstehungsweise der Konvivien. *Genetica,* 1929, 11:399–450.

173. Dark, S. O. S. Chromosomes of *Taxus, Sequoia, Cryptomeria* and *Thuya. Ann. Bot.* (London), 1932, 46:965–977.

174. Darlington, C. D. *Recent advances in cytology.* Philadelphia, Blakiston, 2nd ed., 1937, 559 pp.

175. Darlington, C. D. What is a hybrid? *Jour. Hered.,* 1937, 28:308.

176. Darlington, C. D. Taxonomic species and genetic systems. In J. Huxley, *The New Systematics.* Oxford, Clarendon Press, 1940, pp. 137–160.

177. Darwin, C. R. *On the origin of species by means of natural selection, or the preservation of favoured races in the struggle for life.* London. J. Murray, 1859, 502 pp.

178. Davies, D. Correlation and palaeontology of the coal measures in East Glamorganshire. *Roy. Soc. London, Phil. Trans.,* 1929, B 217:91–153.

179. DeBeer, C. R. *Embryology and evolution.* Oxford, Clarendon Press, 1936, 116 pp.

180. DeBeer, C. R. Embryology and taxonomy. In J. Huxley, *The New Systematics.* Oxford, Clarendon Press, 1940, pp. 365–393.

181. Deevey, E. S., Jr. Studies on Connecticut lake sediments. *Amer. Jour. Sci.,* 1939, 237:691–724.

182. Degelius, G. Das ozeanische Element der Strauch- und Laubflechten-flora von Skandinavien. *Acta Phytogeog. Suecica,* 1935, 7:1–411.

183. Delisle, A. L. Cytogenetical studies on the polymorphy of two species of *Aster. Gen. Program, Amer. Assoc. Adv. Sci.* 101st Meeting, 1937, 121 pp.

184. Denham, H. J. The cytology of the cotton plant. Chromosome numbers of Old and New World cottons. *Ann. Bot.* (London), 1924, 38:433–438.

185. Derman, H. A cytological analysis of polyploidy induced by colchicine and by extremes of temperature. *Jour. Hered.,* 1938, 29:211–229.

186. Dice, L. R. The occurrence of two subspecies of the same species in the same area. *Jour. Mammal.,* 1931, 12:210–213.

187. Diels, L. Vikariierende Formen. In Schneider, *Illust. Handwörterb.* Bot. Leipzig, 2nd ed., 1917, 753 pp.

188. Digby, L. The cytology of *Primula kewensis* and of other related *Primula* hybrids. *Ann. Bot.* (London), 1912, 26:357–388.

189. Diver, C. The problem of closely related species living in the same area.

In J. Huxley, *The New Systematics*. Oxford, Clarendon Press, 1940, pp. 303–328.

190. Dobzhansky, T. *Genetics and the origin of species*. New York, Columbia Univ. Press, 2nd ed., 1941, 364 pp.

191. Dobzhansky, T. Speciation as a stage in evolutionary divergence. *Amer. Nat.*, 1941, 74:312–321.

192. Domin, K. Grundzüge der Pflanzengeographischen Verbreitung und Gliederung der Lebermoose. *Soc. Roy. Sci. Bohème Mem. Nr. 8*, Prague, 1923, 1–74.

193. Dorf, E. Pliocene floras of California. *Carnegie Inst. Wash. Pub.*, 1933, 412:1–112.

194. Dorf, E. A late Tertiary flora from southwestern Idaho. *Carnegie Inst. Wash. Pub.*, 1936, 476:73–124.

195. Dorsey, E. Chromosome doubling in the cereals. *Jour. Hered.*, 1939, 30:393–395.

196. Drude, O. *Handbuch der Pflanzengeographie*. Stuttgart, J. Engelhorn, 1890, 582 pp.

197. Drummond, A. T. The distribution of plants in Canada in some of its relations to physical and past geological conditions. *Canad. Nat.*, 1867, Ser. 2, 3:161–177.

198. Duncan, W. H. Ecological comparison of leaf structures of *Rhododendron punctatum* Andr. and the ontogeny of the epidermal scales. *Amer. Midland Nat.*, 1933, 14:83–96.

199. Du Rietz, G. E. Studien über die Vegetation der Alpen, mit derjenigen Skandinaviens vergleichen. *Geobot. Inst. Rübel, Zürich, Veröffent.*, 1924, 1:31–138.

200. Du Rietz, G. E. The discovery of an Arctic element in the lichen-flora of New Zealand and its plant geographical consequences. *Australasian Assoc. Adv. Sci. Rpt. Hobart Meeting*, 1929, 1928:628–635.

201. Du Rietz, G. E. The fundamental units of biological taxonomy. *Svensk Bot. Tidskr.*, 1930, 24:333–428.

202. Du Rietz, G. E. Life-forms of terrestrial flowering plants. I. *Acta Phytogeogr. Suecica*, 1931, 3 (1): 1–95.

203. Du Rietz, G. E. Problems of bipolar plant distribution. *Acta Phytogeog. Suecica*, 1940, 13:215–282.

204. Dyakowska, J. Researches on the rapidity of the falling down of pollen of some trees. *Polon. Acad. Sci. Let. Cl. Sci. Math. Nat. Bul. Internatl. Sér. B*, 1936, (I): 155–168.

205. Eghis, S. A. Hybridization between the species *Nicotiana rustica* L. and *Nicotiana Tabacum* L. *Bul. Appl. Bot. Genet. and Plant Breeding*, 1927, 17, (3):184–189. (Eng. Summary)

502   *Foundations of Plant Geography*

206. Eig, A.   Les éléments et les groupes phytogéographiques auxiliaires dans la flore palestinienne. *Feddle, Repert. spec. nov. regni veget. Beihefte Bd.,* 1931, 63:1–201.
207. Eigsti, O. J.   A cytological study of colchicine effects in the induction of polyploidy in plants. *Natl. Acad. Sci. Proc.,* 1938, 24:56–63.
208. Eigsti, O. J.   A cytological investigation of *Polygonatum* using the colchicine-pollen tube technique. *Amer. Jour. Bot.,* 1942, 29:626–636.
209. Emerson, A. E.   Termitophile distribution and quantitative characters as indicators of physiological speciation in British Guiana termites (Iosptera). *Ent. Soc. Amer. Ann.,* 1935, 28:369–395.
210. Emerson, A. E.   The origin of species: A review of Dobzhansky, *Genetics and the origin of species. Ecology,* 1938, 19:152–154.
211. Emsweller, S. L., and P. Brierley.   Colchicine-induced tetraploidy in *Lilium. Jour. Hered.,* 1940, 31:223–230.
212. Emsweller, S. L., and M. L. Ruttle.   Induced polyploidy in floriculture. *Biol. Symp.,* 1941, 4:114–130.
213. Engelbert, V.   Reproduction in some *Poa* species. *Canad. Jour. Res. Bot. Sci.,* 1940, 18:518–521.
214. Engler, A.   *Versuch einer Entwicklungsgeschichte der Pflanzenwelt I. Die extratropischen Gebiete der nördlichen Hemisphäre.* Leipzig, Engelmann, 1879, 202 pp.
215. Engler, A.   *Ibid. II. Die extratropischen Gebiete der südlichen Hemisphäre.* Leipzig, Engelmann, 1882, 347 pp.
216. Erdtman, G.   Pollen statistics from the Curragh and Ballough Isle of Man. *Liverpool Geol. Sci. Proc.,* 1925, 14:158–163.
217. Erdtman, G.   Literature on pollen statistics published before 1927. *Geol. Fören Stockholm Förhandl.,* 1927, 49:196–211.
218. Erdtman, G.   Peat deposits of the Cleveland Hills. *Naturalist,* 1927, 39–46.
219. Erdtman, G.   Studies in the post-arctic history of the forests of northwestern Europe. I. Investigations in the British Isles. *Geol. Fören. Stockholm Förhandl.,* 1928, 50:123–192.
220. Erdtman, G.   Some aspects of the post-glacial history of British forests. *Jour. Ecol.,* 1929, 17:112–126.
221. Erdtman, G.   Pollen-statistics: A new research method in paleo-ecology. *Science,* 1931, 73:399–401.
222. Erdtman, G.   The boreal hazel forests and the theory of pollen statistics. *Jour. Ecol.,* 1931, 19:158–163.
223. Erdtman, G.   Literature on pollen-statistics and related topics published 1930 and 1931. *Geol. Fören Stockholm Förhandl.,* 1932, 54:395–418.
224. Erdtman, G.   Literature on pollen statistics and related topics published in 1937–1939. *Geol. Fören Stockholm Förhandl.,* 1940, 62:61–97.

225. Erdtman, G. *An introduction to pollen analysis.* Waltham, Mass., Chronica Botanica, 1943, 239 pp.

226. Erlanson, E. W. Cytological conditions and evidences for hybridity in North American wild roses. *Bot. Gaz.,* 1929, 87:443–506.

227. Ernst, A. *Bastardierung als Ursache der Apogamic im Pflanzenreich.* Jena, Fischer, 1918, 665 pp.

228. Fabergé, A. C. The physiological consequences of polyploidy. *Jour. Genet.,* 1936, 33:365–400.

229. Faegri, K. Some recent publications on phytogeography in Scandinavia. *Bot. Rev.,* 1937, 3:425–456.

230. Fagerlind, F. Beiträge zur Kenntnis der Zytologie der Rubiaceen. *Hereditas,* 1934, 19:223–232.

231. Fagerlind, F. Embryologische, zytologische und bestäubungs-experimentelle Studien in der Familie Rubiaceae nebst Bemerkungen über einiger Polyploiditätsprobleme. *Acta Horti Bergiani,* 1937, 11 (9):195–470.

232. Fassett, N. C. *Bidens hyperborea* and its varieties. *Rhodora,* 1925, 27:166–171.

233. Fassett, N. C. The vegetation of the estuaries of Northeastern North America. *Boston Soc. Nat. Hist. Proc.,* 1928, 39:73–130.

234. Fassett, N. C. Mass collections: *Rubus odoratus* and *R. parviflorus. Mo. Bot. Gard. Ann.,* 1941, 28:299–368.

235. Fenton, C. L. Viewpoints and objects of paleoecology. *Jour. Paleont.,* 1935, 9:63–78.

236. Fernald, M. L. The soil preferences of certain alpine and subalpine plants. *Rhodora,* 1907, 9:149–193.

237. Fernald, M. L. A botanical expedition to Newfoundland and Southern Labrador. *Rhodora,* 1911, 13:109–162.

238. Fernald, M. L. The alpine bearberries and the generic status of *Arctous. Rhodora,* 1914, 16:21–33.

239. Fernald, M. L. A calciphile variety of *Andromeda glaucophylla. Rhodora,* 1916, 18:100–102.

240. Fernald, M. L. The geographic affinities of the vascular floras of New England, the Maritime Provinces and Newfoundland. *Amer. Jour. Bot.,* 1918, 5:219–236.

241. Fernald, M. L. Isolation and endemism in northeastern America and their relation to the age-and-area hypothesis. *Amer. Jour. Bot.,* 1924, 11:558–572.

242. Fernald, M. L. Persistence of plants in unglaciated areas of boreal America. *Amer. Acad. Arts and Sci. Mem.,* 1925, 15:241–342.

243. Fernald, M. L. The antiquity and dispersal of vascular plants. *Quart. Rev. Biol.,* 1926, 1:212–245.

244. Fernald, M. L. Some relationships of the floras of the Northern Hemisphere. *Intern. Congr. Plant Sci. Ithaca, Proc.,* 1929, 2:1487–1507.

245. Fernald, M. L. Specific segregations and identities in some floras of eastern North America and the Old World. *Rhodora*, 1931, 33:25–63.
246. Fernald, M. L. Local plants of the inner Coastal Plain of southeastern Virginia. III. *Rhodora*, 1937, 39:465–491.
247. Fernald, M. L. Misinterpretation of Atlantic Coastal Plain Species. *Rhodora*, 1942, 44:238–246.
248. Firbas, F. Ueber die Bestimmung der Walddichte und der Vegetation waldloser Gebiete mit Hilfe der Pollen-analyse. *Planta*, 1934, 22:109–145.
249. Fischer-Piette, E. The concept of species and geographical isolation in the case of North Atlantic Patellas. *Linn. Soc. London Proc. Session 150*, 1938, (4):268–275.
250. Fitting, H. Die Beeinflussung der Orchideenblüten durch die Bestäubung und durch andere Umstände. *Ztschr. f. Bot.*, 1909, 1:1–86.
251. Flory, W. S. Chromosome numbers and phylogeny in the Gymnosperms. *Arnold Arboretum Jour.*, 1936, 17:83–89.
252. Flovik, K. Chromosome numbers and polyploidy within the flora of Spitzbergen. *Hereditas*, 1940, 26:430–440.
253. Forbes, E. On the distribution of endemic plants, more especially those of the British Isles, considered with regard to geological changes. *Brit. Assoc. Adv. Sci. Rpt.*, 1845, 67–68.
254. Forbes, E. On the connexion between the distribution of the existing fauna and flora of the British Isles, and the geological changes which have affected their area, especially during the Epoch of the Northern Drift. *Geol. Survey England and Wales, Mem. 1*, 1846, 336–432.
255. Foster, R. C. Chromosome number in *Acer* and *Staphylea*. *Arnold Arboretum Jour.*, 1933, 14:386–393.
256. Fritel, P. H. Remarques sur quelques espèces fossiles du genre *Magnolia*. *Soc. Géol. France, Bul. 13*, 1913, 277–292.
257. Fuller, G. D. Pollen analysis and postglacial vegetation. *Bot. Gaz.*, 1927, 83:323–325.
258. Gairdner, A. E. *Campanula persicifolia* and its tetraploid form, "Telham Beauty." *Jour. Genet.*, 1926, 16:341–351.
259. Gaiser, L. O. A list of chromosome numbers in Angiosperms. *Genetica*, 1926, 8:401–482.
260. Gaiser, L. O. Chromosome numbers and species characters in *Anthurium*. *Roy. Soc. Canada, Proc. and Trans. III*, 1927, 21:1–137.
261. Gaiser, L. O. Chromosome numbers in Angiosperms. II. *Bibliographia Genetica*, 1930, 6:171–466.
262. Gaiser, L. O. Chromosome numbers in Angiosperms. III. *Genetica*, 1930, 12:161–260.
263. Gaiser, L. O. Chromosome numbers in Angiosperms. IV. *Bibliographia Genetica*, 1933, 10:105–250.

264. Gams, H. Einige homologe Pflanzengesellschaften in der subalpinen und alpinen Stufe der Alpen und Skandinaviens. *Schweiz. Naturf. Gesell. Verhandl.,* 1921, 102:142–143.

265. Gates, R. R. Species and chromosomes. *Amer. Nat.,* 1925, 59:193–200.

266. Gates, R. R. The origin of polyploidy. *John Innes Hort. Inst., Conference on polyploidy,* 1929, pp. 22–26.

267. Gattinger, A. *The flora of Tennessee and a philosophy of botany.* Nashville. Gospel Advocate, 1901, 296 pp.

268. Gelting, P. Studies on the vascular plants of East Greenland between Franz Joseph fjord and Dove Bay. *Meddel. Grønl.,* 1934, 101 (2):1–337.

269. Gershoy, A. Studies in North American violets. III. Chromosome numbers and species characters. *Vt. Agr. Expt. Sta. Bul.,* 1934, 367:1–92.

270. Giles, N. H., Jr. Autopolyploidy and geographical distribution in *Cuthbertia graminea* Small. *Amer. Jour. Bot.,* 1942, 29:637–645.

271. Gilmour, J. S. L. Taxonomy and philosophy. In J. Huxley, *The New Systematics.* Oxford, Clarendon Press, 1940, pp. 461–474.

272. Gleason, H. A. A revision of the North American Vernonieae. *N. Y. Bot. Gard. Bul.,* 1906, 4:144–243.

273. Gleason, H. A. An isolated prairie grove and its phytogeographical significance. *Bot. Gaz.,* 1912, 53:38–49.

274. Gleason, H. A. Vernonieae. *North Amer. Flora,* 1922, 33 (1):47–110.

275. Gleason, H. A. Evolution and geographical distribution of the genus *Vernonia* in North America. *Amer. Jour. Bot.,* 1923, 10:187–202.

276. Gleason, H. A. The vegetational history of the middle west. *Assoc. Amer. Geog. Ann.,* 1923, 12:39–85.

277. Gleason, H. A. Species and area. *Ecology,* 1925, 6:66–74.

278. Goddijn, W. A. On the species conception in relation to taxonomy and genetics. *Blumea,* 1934, 1:75–89.

279. Godwin, H. Pollen analysis. An outline of the problems and potentialities of the method. *New Phytol.,* 1934, 33:278–305, 325–358.

280. Godwin, H. Pollen analysis and forest history of England and Wales. *New Phytol.,* 1940, 39:370–400.

281. Godwin, H. Studies of the postglacial history of British vegetation. III and IV. *Roy. Soc. London Philos. Trans. B.,* 1940, 230:239–303.

282. Goldschmidt, R. *The material basis of evolution.* New Haven, Yale Univ. Press, 1940, 436 pp.

283. Good, R. D'O. The past and present distribution of the Magnolieae. *Ann. Bot. (London),* 1925, 39:409–430.

284. Good, R. D'O. On the phytogeographical distribution of the Stylidaceae. *New Phytol.,* 1925, 24:225–240.

285. Good, R. D'O. The genus *Empetrum* L. *Linn. Soc. London Jour. Bot.,* 1927, 47:489–523.

286. Good, R. D'O. The geography of the genus *Coriaria*. *New Phytol.*, 1930, 29:170–198.

287. Good, R. D'O. A theory of plant geography. *New Phytol.*, 1931, 30:149–171.

288. Goodspeed, T. H. Occurrence of triploid and tetraploid individuals in X-ray progenies of *Nicotiana tabacum*. *Calif. Univ. Pub. Bot.*, 1930, 11:299–308.

289. Goodspeed, T. H., and M. V. Bradley. Amphidiploidy. *Bot. Rev.*, 1942, 8:271–316.

290. Goodspeed, T. H., and M. P. Crane. Chromosome number in *Sequoia*. *Bot. Gaz.*, 1920, 69:348–349.

291. Goodwin, R. H. The cyto-genetics of two species of *Solidago* and its bearing on their polymorphy in nature. *Amer. Jour. Bot.*, 1937, 24:425–432.

292. Gray, A. Analogy between the flora of Japan and that of the United States. *Amer. Jour. Sci. and Arts*, 1846, 52:135–136.

293. Gray, A. Diagnostic characters of new species of phaenogamous plants, collected in Japan by Charles Wright, with observations upon the relations of the Japanese flora to that of North America, and other parts of the northern temperate zone. *Amer. Acad. Arts and Sci. Mem.*, 1859, 6:377–452.

294. Gray, A. Illustrations of the botany of Japan in its relation to that of Central and Northern Asia, Europe, and North America. *Amer. Acad. Arts and Sci. Proc.*, 1860, 4:131–135.

295. Gray, A. Address of Professor Asa Gray, ex-president of the association. (Consists of a comparative study of the floras of Eastern North America and of Eastern Asia.) *Amer. Assoc. Adv. Sci. Proc.*, 1873, 21:1–31.

296. Gray, A. Forest geography and archeology. *Amer. Jour. Sci.*, 1878, 3 S. 16:85–94, 183–196.

297. Gray, A. *Gray's new manual of botany*. Revised 7th ed. by B. L. Robinson and M. L. Fernald. New York, American Book Co., 1907, 926 pp.

298. Greenleaf, W. H. Induction of polyploidy in *Nicotiana*. *Jour. Hered.*, 1938, 29:451–464.

299. Greenman, J. M. The age-and-area hypothesis with special reference to the flora of tropical America. *Amer. Jour. Bot.*, 1925, 12:189–193.

300. Gregor, J. W., and F. W. Sansome. Experiments on the genetics of wild populations. II. *Phleum pratense* L. and the hybrid *P. pratense* L. X *P. alpinum* L. *Jour. Genet.*, 1930, 22:373–386.

301. Gregory, W. C. Phylogenetic and cytological studies in the Ranunculaceae. *Trans. Amer. Phil. Soc.*, 1941, N.S. 31:443–521.

302. Griesinger, R. Über hypo- und hyperdiploide Formen von *Petunia, Hyoscyamus, Lamium* und einiger andere Chromosomenzählungen. *Deut. Bot. Gesell. Ber.*, 1937, 55:556–571.

303. Griggs, R. F. Observations on the behavior of some species on the edges of their ranges. *Torrey Bot. Club Bul.*, 1914, 41:25–49.

304. Griggs, R. F. The edge of the forest in Alaska and the reasons for its position. *Ecology*, 1934, 15:80–96.

305. Griggs, R. F. Timberlines in the Northern Rocky Mountains. *Ecology*, 1938, 19:548–564.

306. Griggs, R. F. Indications as to climatic changes from the timberline of Mount Washington. *Science*, 1942, 95:515–519.

307. Grinnell, J., and H. S. Swarth. An account of the birds and mammals of the San Jacinto area of southern California. *Calif. Univ. Pubs. Zool.*, 1913, 10 (10):197–406.

308. Guillaumin, A. *Les régions florales du Pacifique. IV. Contribution à l'étude du peuplement zoologique et botanique des Iles du Pacifique.* Paris, 1934.

309. Guppy, H. B. *Observations of a naturalist in the Pacific between 1896 and 1899. II. Plant dispersal.* London, Macmillan, 1906, 627 pp.

310. Hagedoorn, A. L. and A. C. *The relative value of the processes causing evolution.* The Hague, Nijhoff, 1921, 294 pp.

311. Hagerup, O. *Empetrum hermaphroditum* (Lge) Haerup. A new tetraploid, bisexual species. *Dansk Bot. Arkiv*, 1927, 5 (2):1–17.

312. Hagerup, O. Morphological and cytological studies of *Bicornes*. *Dansk Bot. Arkiv*, 1928, 6 (1):1–27.

313. Hagerup, O. Uber Polyploidie in Beziehung zu Klima, Ökologie, und Phylogenie. *Hereditas*, 1932, 16:19–40.

314. Hagerup, O. Studies on polyploid ecotypes in *Vaccinium uliginosum* L. *Hereditas*, 1933, 18:122–128.

315. Hagerup, O. Studies on the significance of polyploidy. II. *Orchis. Hereditas*, 1938, 24:258–264.

316. Hagerup, O. Studies on the significance of polyploidy. III. *Deschampsia* and *Aira. Hereditas*, 1939, 25:185–192.

317. Hagerup, O. Studies on the significance of polyploidy. IV. *Oxycoccus. Hereditas*, 1940, 26:399–410.

318. Håkansson, A. Beiträge zur Polyploidie der Umbelliferen. *Hereditas*, 1933, 17:246–248.

319. Haldane, J. B. S. Genetics of polyploid plants. *John Innes Hort. Inst., Conference on Polyploidy*, 1929, pp. 9–12.

320. Hall, H. M. A botanical survey of San Jacinto Mountain. *Calif. Univ. Pub. Bot.*, 1902, 1:1–140.

321. Hall, H. M. Heredity and environment as illustrated by transplant studies. *Sci. Monthly*, 1932, 35:289–302.

322. Hansen, H. P. Pollen analysis of two Wisconsin bogs of different age. *Ecology*, 1937, 18:136–148.

323. Hansen, H. P.   Pollen analysis of some interglacial peat from Washington. *Wyoming Univ. Pub.,* 1938, 5:11–18.

324. Hansen, H. P.   Postglacial forest succession and climate in the Puget Sound region. *Ecology,* 1938, 19:528–542.

325. Hansen, H. P.   Paleoecology of a montane peat deposit at Bonaparte Lake, Washington. *Northwest Sci.,* 1940, 14 (3):60–68.

326. Hansen, H. P.   A pollen study of Post-Pleistocene lake sediments in the Upper Sonoran life zone of Washington. *Amer. Jour. Sci.,* 1941, 239:503–522.

327. Hansen, H. P.   Paleoecology of a bog in the spruce-hemlock climax of the Olympic Peninsula. *Amer. Midland Nat.,* 1941, 25:290–297.

328. Hansen, H. P.   Paleoecology of two peat deposits on the Oregon Coast. *Oregon State Monogr.,* 1941, 3:1–31.

329. Hansen, H. P.   Further studies of post Pleistocene bogs in the Puget Lowland of Washington. *Torrey Bot. Club Bul.,* 1941, 68:133–148.

330. Hara, H.   Some notes on the botanical relation between North America and Eastern Asia. *Rhodora,* 1939, 41:385–392.

331. Harper, R. M.   Coastal Plain plants in New England. *Rhodora,* 1905, 7:69–80.

332. Harshberger, J. W.   The comparative age of the different floristic elements of eastern North America. *Acad. Nat. Sci. Phila. Proc.,* 1904, 56:601–615.

333. Hartshorne, R.   *The nature of geography. A critical survey of current thought in the light of the past.* Lancaster, Assoc. of Amer. Geographers, 1939, 658 pp.

334. Hatcher, J. B.   Discovery of a musk ox skull (Ovibos cavifrons Leidy) in West Virginia, near Steubenville, Ohio. *Science,* 1902, 16:707–709.

335. Havas, L. J.   A colchicine chronology. *Jour. Hered.,* 1940, 31:115–117.

336. Hay, O. P.   Bibliography and catalogue of the fossil vertebrata of North America. *U. S. Geol. Survey Bul.,* 1902, 179:1–868.

337. Heer, O.   *Flora Fossilis Arctica.* Zürich, 1868–1883, 7 vols.

338. Heilborn, O.   On the origin and preservation of polyploidy. *Hereditas,* 1934, 19:233–242.

339. Hemsley, W. B.   Report on the present state of knowledge of various insular floras. *Challenger Expedition. Botany,* 1885, 1:1–75.

340. Henry, A.   On elm-seedlings showing Mendelian results. *Linn. Soc. London Jour. Bot.,* 1910, 39:290–300.

341. Herzog, T.   *Geographie der Moose.* Jena, G. Fischer, 1926, 439 pp.

342. Herzog, T.   Geographie. In F. Verdoorn, *Manual of Bryology.* The Hague, Nijhoff, 1932, pp. 273–296.

343. Hesmer, H.   Die natürliche Bestockung und die Waldentwicklung auf verschiedenartigen Märkischen Standorten. *Zeit. Forst- und Jagdwesen,* 1933, 65:505–540, 569–606, 631–651.

344. Hesse, R.   Vergleichende Untersuchungen an diploiden und tetraploiden Petunien. *Ztschr. f. Induktive Abstam. u. Vererbungslehre,* 1938, 75:1–23.

345. Hiesey, W. M., J. Clausen, and D. D. Keck. Relations between climate and intraspecific variation in plants. *Amer. Nat.,* 1942, 76:5–22.

346. Hinton, M. A. C., and others. A discussion on "subspecies" and "varieties." *Linn. Soc. London Proc. Session 151,* 1939, (2):89–114.

347. Hitchcock, C. H. The distribution of maritime plants in North America: a proof of oceanic submergence in the Champlain Period. *Amer. Assoc. Adv. Sci. Proc.,* 1871, 19:175–181.

348. Hogben, Lancelot. Problems of the origin of species. In J. Huxley, *The New Systematics.* Oxford, Clarendon Press, 1940, pp. 269–286.

349. Hollingshead, L., and E. B. Babcock. Chromosomes and phylogeny in *Crepis. Calif. Univ. Pubs. Agr. Sci.,* 1930, 6:1–53.

350. Holmboe, J. The Trondheim district as a center of late glacial and post-glacial plant migrations. *Norske Vidensk.-Akad. Avh. Oslo I. Mat.-Nat. 1936,* 1937, (9):1–59.

351. Holmes, F. O. Inheritance of resistance to tobacco-mosaic disease in tobacco. *Phytopathology,* 1938, 28:553–560.

352. Hooker, J. D. *The botany of the Antarctic voyage of H. M. Discovery Ships* Erebus *and* Terror *in the years 1839–1843.* Vol. I. *Flora Antarctica,* part I. *Botany of Lord Auckland's Group and Campbell's Island,* part II. *Botany of Fuegia, The Falklands, Kerguelen's Land, etc.* London, 1844–1847.

353. Hooker, J. D. *The botany of the Antarctic Voyage of H. M. Discovery Ships* Erebus *and* Terror *in the years 1839–1843. Flora Novae-Zelandiae,* Vol. II, part 1, *Flowering Plants.* London, 1853.

354. Hooker, J. D. Insular floras. *Gard. Chron.,* 1867, [32]:6–7, 27, 50–51, 75–76, 152, 181–182.

355. Hooker, J. D. The distribution of the North American flora. *Gard. Chron.,* 1878, 44:140–142, 216–217.

356. Houdek, P. K. Pollen statistics for two Indiana bogs. *Ind. Acad. Sci. Proc.,* 1933, 42:73–77.

357. House, H. D. The genus *Shortia. Torreya,* 1907, 7:233–235.

358. Hu, H. H. A comparison of the ligneous flora of China and Eastern North America. *Chinese Bot. Soc. Bul.,* 1935, 1:79–97.

359. Hultén, E. On the American component in the flora of eastern Siberia. *Svensk Bot. Tidskr.,* 1928, 22:220–229.

360. Hultén, E. *Outline of the history of Arctic and Boreal biota during the Quaternary period. Their evolution during and after the glacial period as indicated by the equiformal progressive areas of present plant species.* Stockholm, 1937, 168 pp.

361. Humboldt, Alexander von. *De distributione geographica plantarum, secundum coeli temperiem et altitudinem montium prolegomena.* Paris, 1817.

362. Hurst, C. C. The mechanism of heredity and evolution. *Eugenics Rev.,* 1927, 19:19–31.

363. Hurst, C. C. Polyploidy as a source of species and horticultural varieties. *John Innes Hort. Inst., Conference on Polyploidy,* 1929, pp. 13–21.

364. Huskins, C. L. The origin of *Spartina Townsendii. Genetica,* 1931, 12:531–538.

365. Huskins, C. L. Polyploidy and mutations. *Biol. Symp.,* 1941, 4:133–148.

366. Huskins, C. L., and S. G. Smith. A cytological study of the genus *Sorghum* Pers. I. The somatic chromosomes. *Jour. Genet.,* 1932, 25:240–249.

367. Hutchinson, G. E. Ecological aspects of succession in natural populations. *Biol. Symp.,* 1941, 4:8–20.

368. Huxley, J. Species formation and geographical isolation. *Linn. Soc. London Proc. Session 150,* 1938, (4):253–264.

369. Huxley, J. Clines: an auxiliary taxonomic principle. *Nature* (London), 1938, 142:219–220.

370. Huxley, J. In Hinton, *et al.,* A discussion on subspecies and varieties. *Linn. Soc. London Proc. Session 151,* 1939, (2):105–114.

371. Huxley, J. Introductory: Toward the new systematics. *The new systematics.* Oxford, Clarendon Press, 1940, pp. 1–46.

372. Hyde, J. Variations in the rate of agricultural production and one of their causes. *Science,* 1898, 8:575–576.

373. Ihering, H. von. Das neotropische Florengebiet und seine Geschichte. *Bot. Jahrb.,* 1893, 17, Beiblatt Nr. 42:1–54.

374. Ihering, H. von. Die phyto-geographischen Grundgesetze. *Bot. Jahrb.,* 1928, 62:113–154.

375. Irmscher, E. Pflanzenverbreitung und Entwicklung der Kontinente. Studien zur genetischen Pflanzengeographie. *Hamburg Inst. f. Allg. Bot. Mitt.,* 1922, 5:17–235.

376. Irmscher, E. Pflanzenverbreitung und Entwicklung der Kontinente. II. Teil. Weitere Beiträge zur genetischen Pflanzengeographie unter besonderer Berücksichtigung der Laubmoose. *Hamburg Inst. f. Allg. Bot. Mitt.,* 1929, 8:169–374.

377. Jeffrey, E. C. Polyploidy and the origin of species. *Amer. Nat.,* 1925, 59:209–217.

378. Jentys-Szafer, J. La structure des membranes du pollen de *Corylus,* de *Myrica* et des espèces européennes de *Betula,* et leur détermination à l'état-fossile. *Polon. Acad. Sci. Let. Cl. Sci. Math. Nat. Bul. Internatl.,* 1928, Sér. B:75–125.

379. Johnston, I. M. The floristic significance of shrubs common to North and South American deserts. *Arnold Arboretum Jour.,* 1940, 21:356–363.

380. Johnston, I. M. Gypsophily among Mexican desert plants. *Arnold Arboretum Jour.,* 1941, 22:145–170.

381. Jordan, D. S. The origin of species through isolation. *Science,* 1905, 22:545–562.

382. Jordan, D. S. Concerning variation in animals and plants. *Pop. Sci. Monthly,* 1906, 69:481–502.
383. Jordan, D. S. Isolation with segregation as a factor in organic evolution. *Smithsn. Inst. Ann. Rpt. 1925,* 1926, pp. 321–326.
384. Karpechenko, G. D. The production of polyploid gametes in hybrids. *Hereditas,* 1927, 9:349–368.
385. Karpechenko, G. D. Konstantwerden von Art- und Gattungsbastarden durch Verdoppelung der Chromosomenkomplexe. *Züchter,* 1929, 1:133–140.
386. Kearney, T. H. The pine-barren flora in the East Tennessee mountains. *Plant World,* 1897, 1:33–35.
387. Kearney, T. H. Plant geography of North America. III. The lower austral element in the flora of the southern Appalachian region. *Science,* 1900, 12:830–842.
388. Kearney, T. H. Plants new to Arizona. (An annotated list of species added to the recorded flora of the state or otherwise interesting.) *Wash. Acad. Sci. Jour.,* 1931, 21:63–80.
389. Kerner Von Marilaun, A. *Die Abhängigkeit der Pflanzengestalt von Klima und Boden.* Innsbruck, 1869, 48 pp.
390. Kiellander, C. L. On the embryological basis of apomixis in *Poa palustris* L. *Svensk. Bot. Tidsk.,* 1937, 31:425–429.
391. Kihara, H., and T. Ono. Chromosomenzahlen und systematische Gruppierung der *Rumex*-Arten. *Ztschr. f. Zellforsch. u. Mikros. Anat.,* 1926, 4:475–481.
392. Kinsey, A. C. The gall wasp genus *Cynips.* A study in the origin of species. *Ind. Univ. Studies,* 1930, 16 (84–86):1–577.
393. Kinsey, A. C. The origin of higher categories in *Cynips. Ind. Univ. Publ. Sci. Ser.,* 1936, 4:1–334.
394. Kinsey, A. C. An evolutionary analysis of insular and continental species. *Natl. Acad. Sci. Proc.,* 1937, 23:5–11.
395. Kirikov, S. V. Sur la distribution du hamster noir et ses relations avec la forme normale de *Cricetus cricetus. Zool. Zurn.* (Moscow), 1934, 13:361–368. Russian with French summary.
396. Klages, K. H. W. *Ecological crop geography.* New York, Macmillan, 1942, 615 pp.
397. Klauber, L. M. The long-nosed snakes of the genus *Rhinocheilus. San Diego Soc. Nat. Hist. Trans.,* 1941, 9 (29): 289–332.
398. Knowlton, F. H. Fossil flora of the John Day Basin, Oregon. *U. S. Geol. Survey Bul.,* 1902, 204:1–153.
399. Knowlton, F. H. A catalogue of the Mesozoic and Cenozoic plants of North America. *U. S. Geol. Survey Bul.,* 1919, 696:1–815.
400. Kostoff, D. Changes in karyotypes induced by centrifuging. *Acad. des Sci. U. S. S. R. Compt. Rend.,* 1935, 2:71–76. (English summary.)

401. Kostoff, D.   Studies on polyploid plants. *Current Sci.*, 1938, 6:549–552.
402. Kostoff, D.   Polyploid plants produced by colchicine and acenaphthene. *Current Sci.*, 1938, 7:108–110.
403. Kostoff, D.   Cytogenetic behaviour of the allopolyploid hybrids *Nicotiana glauca* Grah. X *Nicotiana Langsdorffii* Weinm. and their evolutionary significance. *Jour. Genet.*, 1938, 37:129–209.
404. Kostoff, D., and J. Kendall.   Studies on plant tumors and polyploidy produced by bacteria and other agents. *Arch. f. Mikrobiol.*, 1933, 4:487–508.
405. Lagerheim, G.   In H. Witte, *Stratiotes aloides* L. funnen i Sveriges postglaciala aflagringar. *Geol. Fören. Förh.*, 1905, 27:443–445.
406. Lagerheim, G.   In N. O. Holst, Postglaciala tidsbestämningar. *Sveriges Geol. Unders., Ser. C.*, 1909, no. 216:30, 74.
407. Lam, H. J.   Studies in phylogeny. I. On the relation of taxonomy, phylogeny and biogeography. II. On the phylogeny of the Malaysian Burseraceae-Canarieae in general and of *Haplolobus* in particular. *Blumea*, 1938, 3:114–158.
408. Larisey, M. M.   Analysis of a hybrid complex between *Baptisia leucantha* and *Baptisia viridis* in Texas. *Amer. Jour. Bot.*, 1940, 27:624–628.
409. Larter, L. N. H.   Chromosome variation and behaviour in *Ranunculus* L. *Jour. Genet.*, 1932, 26:255–283.
410. Leavitt, R. G.   The geographic distribution of closely related species. *Amer. Nat.*, 1907, 41:207–240.
411. Lehmann, E.   Polyploidie und geographische Verbreitung der Arten der Gattung *Veronica. Jahrb. f. Wiss. Bot.*, 1940, 89:461–542.
412. Lesquereux, L.   Contributions to the fossil flora of the Western Territories. III. The Cretaceous and Tertiary floras. *U. S. Geol. Survey of the Territories*, 1883, 8:1–283.
413. Levan, A.   Cytological studies in *Allium*. II. Chromosome morphological contributions. *Hereditas*, 1932, 16:257–294.
414. Levan, A.   Zytologische studien an *Allium schoenoprasum. Hereditas*, 1936, 22:1–126.
415. Levan, A.   Cytological studies in the *Allium paniculatum* group. *Hereditas*, 1937, 23:317–370.
416. Levan, A.   The effect of colchicine on root mitoses in *Allium*. *Hereditas*, 1938, 24:471–486.
417. Lewis, I. F., and E. C. Cocke.   Pollen analysis of Dismal Swamp peat. *Elisha Mitchell Sci. Soc. Jour.*, 1929, 45:37–58.
418. Liebig, J.   *Chemistry in its application to agriculture and physiology*. Philadelphia. Peterson, 3rd ed., 1843.
419. Lippmaa, T.   Areal und Altersbestimmung einer Union (*Galeobdolon-Asperula-Asarum*-U.) sowie das Problem der Charakterarten und der Konstanten. *Acta Inst. Horti Botan. Univ. Tartuensis*, 1938, 6 (2):1–152

420. Little, T. M. Tetraploidy in *Antirrhinum majus* induced by sanguinarine hydrochloride. *Science,* 1942, 96: 188–189.

421. Livingston, B. E., and F. Shreve. The distribution of vegetation in the United States, as related to climatic conditions. *Carnegie Inst. Wash., Publ.,* 1921, 284: 1–590.

422. Lloyd, F. E. Isolation and the origin of species. *Science,* 1905, 22: 710–712.

423. Longley, A. E. Chromosomes in maize and maize relatives. *Jour. Agr. Res.,* 1924, 28: 673–682.

424. Longley, A. E. Chromosomes in *Vaccinium. Science,* 1927, 66: 566–568.

425. Longley, A. E. Chromosomes in grass sorghums. *Jour. Agr. Res.,* 1932, 44: 317–321.

426. Lotsy, J. P. *Evolution by means of hybridization.* The Hague, Nijhoff, 1916, 166 pp.

427. Lotsy, J. P. *Evolution considered in the light of hybridisation.* Christchurch, New Zealand. Printed for Canterbury College by Andrews, Baty and Company, 1925, 59 pp.

428. Lüdi, W. Die Pollensedimentation in Davoserhochtale. *Ber. Geobot. Forschungsinstitut Rübel, Zürich, 1936,* 1937, pp. 107–127.

429. Lüdi, W. Die Signaturen für Sedimente und Torfe. *Ber. Geobot. Forschungsinstitut Rübel, Zürich 1938,* 1939, pp. 87–91.

430. Lundegårdh, H. G. (Trans. by Eric Ashby.) *Environment and plant development.* London, Edward Arnold, 1931, 330 pp.

431. Lynge, B. On *Dufourea* and *Dactylina.* Three Arctic lichens. *Skrifter om Svalbard og Ishavet no. 59,* 1933, 1–62.

432. MacBride, E. W. Mutations and variations and their bearing on the origin of species. *Linn. Soc. London Proc. Session 150,* 1938, (4): 227–231.

433. MacGinitie, H. D. Redwoods and frost. *Science,* 1933, 78: 190.

434. MacGinitie, H. D. The Trout Creek flora of southeastern Oregon. *Carnegie Inst. Wash. Pub.,* 1933, 416: 21–68.

435. McKelvey, S. D., and Karl Sax. Taxonomic and cytological relationships of *Yucca* and *Agave. Arnold Arboretum Jour.,* 1933, 14: 76–81.

436. McLaughlin, W. T. Atlantic coastal plain plants in the sand barrens of northwestern Wisconsin. *Ecol. Monog.,* 1932, 2: 335–383.

437. Malmström, C. Eine botanishe, hydrologische und entwicklungsgeschichte. Untersuchungen eines nordschwedischen Moorkomplexes. *Mitt. Forst. Versuchs. Schwedens,* 1923, 20: 177–205.

438. Malthus, T. R. *An essay on the principle of population.* London, J. Murray, 1826, 6th ed., 2 vols.

439. Mangelsdorf, P. C., and R. G. Reeves. Hybridization of maize, *Tripsacum* and *Euchlaena. Jour. Hered.,* 1931, 22: 329–343.

440. Mangelsdorf, P. C., and R. G. Reeves. The origin of Indian corn and its relatives. *Tex. Agr. Expt. Sta. Bul. 574,* 1939 (Monogr.): 1–315.

441. Manton, I.   Introduction to the general cytology of the Crucifera. *Ann. Bot.* (London), 1932, 46:509–556.

442. Manton, I.   The problem of *Biscutella laevigata* L. *Ztschr. f. Induktive Abstam. u. Vererbungslehre*, 1934, 67:41–57.

443. Manton, I.   The cytological history of watercress, *Nasturtium officinale* R. Br. *Ztschr. f. Induktive Abstam. u. Vererbungslehre*, 1935, 69:132–157.

444. Manton, I.   The problem of *Biscutella laevigata* L. II. The evidence from meiosis. *Ann. Bot.* (London), 1937, N.S. 1:439–462.

445. Marie-Victorin, Fr.   Le dynamisme dans la flore du Quebec. *Cont. Bot. Lab. Univ. Montreal*, 1929, no. 13:1–89.

446. Marie-Victorin, Fr.   Some evidences of evolution in the flora of Northeastern America. *Jour. Bot.*, 1930, 68: 161–172.

447. Marie-Victorin, Fr.   Phytogeographical problems of Eastern Canada. *Amer. Midland Nat.*, 1938, 19:489–558.

448. Marty, P.   *Magnolia* fossile des Arkoses de Ravel (Puy-de-Dôme). *Soc. Géol. France Bul.*, 1915, 4ᵐᵉ sér., 15:242–259.

449. Marvin, J. W.   Cell size and organ size in two violet species and their hybrid. *Torrey Bot. Club Bul.*, 1936, 63:17–32.

450. Mason, H. L.   Fossil records of some west American conifers. *Carnegie Inst. Wash. Pub.*, 1927, 346:139–158.

451. Mason, H. L.   The Santa Cruz Island pine. *Madroño*, 1930, 2:8–10.

452. Mason, H. L.   A phylogenetic series of the California closed-cone pines suggested by the fossil record. *Madroño*, 1932, 2:49–56.

453. Mason, H. L.   Pleistocene flora of the Tomales Formation. *Carnegie Inst. Wash. Pub.*, 1934, 415:81–179.

454. Mason, H. L.   The principles of geographic distribution as applied to floral analysis. *Madroño*, 1936, 3:181–190.

455. Mason, H. L.   Distributional history and fossil record of *Ceanothus*. In Van Rensselaer and McMinn, *Ceanothus*. Santa Barbara, Santa Barbara Botanic Garden, 1942, pp. 281–303.

456. Matsuura, H.   On Karyo-ecotypes of *Fritillaria camschatcensis* (L.) Kergawler. *Hokkaido Imp. Univ. Faculty Sci. Jour. Ser. V Botany*, 1935, 3, (5):219–232.

457. Matthew, W. D.   *Climate and evolution.* N. Y. Acad. Sci., Special Pub., 1939, 2nd ed., 1:1–223.

458. Maude, P. F.   Chromosome numbers in some British plants. *New Phytol.*, 1940, 39:17–32.

459. Maximov, N. A.   (Trans. by R. H. Yapp.)   *The plant in relation to water.* London, George Allen and Unwin, 1929, 451 pp.

460. Medwedewa, G. B.   The climatic influences upon the pollen development of the Italian hemp. *Ztschr. f. Induktive Abstam. u. Vererbungslehre*, 1935, 70:170–176.

461. Meinke, H. Atlas und Bestimmungsschlüssel zur Pollenanalytik. *Bot. Arch.*, 1927, 19:380–449.

462. Merriam, J. C. *The living past.* New York, Scribner's, 1930, 144 pp.

463. Merrill, E. D. Malaysian phytogeography in relation to the Polynesian flora. In *Essays in geobotany,* Berkeley, Univ. Calif. Press, 1936, pp. 247–261.

464. Merrill, E. D., and M. L. Merritt. The flora of Mount Pulog. *Philippine Jour. Sci.,* 1910, 5:287–403.

465. Miller, A. H. Speciation in the avian genus *Junco. Calif. Univ. Pubs. Zool.,* 1941, 44:173–434.

466. Mirov, N. T. Phylogenetic relations of *Pinus Jeffreyi* and *Pinus ponderosa. Madroño,* 1938, 4:169–171.

467. Mirov, N. T., and P. Stockwell. Colchicine treatment of pine seeds. *Jour. Hered.,* 1939, 30:389–390.

468. Mitscherlich, E. A. Das Gesetz des Minimums und das Gesetz des abnehmenden Bodenertrags. *Landw. Jahrb.,* 1909, 38:537–552.

469. Molinier, R., and P. Muller. La dissemination des espèces vegetales. *Rev. Gen. de Bot.,* 1938, 50:53–72, 152–169, 202–221, 277–293, 341–358, 397–414, 472–488, 533–546, 598–614, 649–670.

470. Montfort, C. Die Xeromorphie der Hochmoorpflanzen als Voraussetzung der "physiologischen Trockenheit" der Hochmoore. *Zeitschr. Bot.,* 1918, 10:257–357.

471. Muller, H. J. Bearings of the "Drosophila" work on systematics. In J. Huxley, *The New Systematics,* Oxford, Clarendon Press, 1940, pp. 185–268.

472. Munns, E. N. The distribution of important forest trees of the United States. *U. S. Dept. Agr. Miscl. Pub.,* 1938, 287:1–176.

473. Müntzing, A. Chromosome number, nuclear volume and pollen grain size in *Galeopsis. Hereditas,* 1928, 10:241–260.

474. Müntzing, A. Über Chromosomenvermehrung in *Galeopsiskreuzungen* und ihre phylogenetische Bedeutung. *Hereditas,* 1930, 14:153–172.

475. Müntzing, A. Note on the cytology of some apomictic *Potentilla*-species. *Hereditas,* 1931, 15:166–178.

476. Müntzing, A. Quadrivalent formation and aneuploidy in *Dactylis glomerata. Bot. Notiser,* 1933, pp. 198–205.

477. Müntzing, A. Hybrid incompatibility and the origin of polyploidy. *Hereditas,* 1933, 18:33–55.

478. Müntzing, A. Apomictic and sexual seed formation in *Poa. Hereditas,* 1933, 17:131–154.

479. Müntzing, A. Cyto-genetic studies on hybrids between two *Phleum* species. *Hereditas,* 1935, 20:103–136.

480. Müntzing, A. The evolutionary significance of autopolyploidy. *Hereditas,* 1936, 21:263–378.

481. Müntzing, A. Genetics in relation to general biology. *Hereditas*, 1938, 24:492–504.

482. Myers, W. M. Colchicine induced tetraploidy in perennial ryegrass, *Lolium Perenne* L. *Jour. Hered.*, 1939, 30:499–504.

483. Nakamura, M. Cyto-genetical studies in the genus *Solanum*. I. Auto-polyploidy of *Solanum nigrum* Linn. *Cytologia. Fujii Jubilaei*, 1937, 1:57–68.

484. Nannfeldt, J. A. Taxonomical and plant-geographical studies in the *Poa laxa* group. *Symbolae Bot. Upsalienses*, 1935, 1 (5):1–113.

485. Nebel, B. R., and M. L. Ruttle. The cytological and genetical significance of colchicine. *Jour. Hered.*, 1938, 29:3–9.

486. Newberry, J. S. The later extinct floras of North America. A posthumous work edited by Arthur Hollick. *U. S. Geol. Survey Monog.*, 1898, 35:1–295.

487. Newton, W. C. F., and C. Pellew. *Primula kewensis* and its derivatives. *Jour. Genet.*, 1929, 20:405–466.

488. Nilsson, H. Synthetische Bastardierungsversuche in der Gattung *Salix*. *Lunds Univ. Arsskr. N. F. Avd.*, 1930, Bd. 27, nr. 4:1–97.

489. Nilsson, H. The problem of the origin of species since Darwin. *Hereditas*, 1935, 20:227–237.

490. Nilsson, H. Die Analyse der synthetisch hergestellten *Salix laurina*. *Hereditas*, 1935, 20:339–353.

491. Nordhagen, R. Versuch einer neuen Einteilung der subalpinen-alpinen Vegetation Norwegens. *Bergens Mus. Aarbog Naturen, 1936*, 1937, (7):1–88.

492. Oliver, W. R. B. A revision of the genus *Dracophyllum*. *New Zeal. Inst. Trans. and Proc.*, 1929, 59:678–714.

493. O'Mara, J. G. Observations on the immediate effects of colchicine. *Jour. Hered.*, 1939, 30:35–37.

494. Ostenfeld, C. H. Meeresgräser I. Marine Hydrocharitaceae. *Die Pflanzenareale*, 1927, 1 Reihe, Heft 3, Karte 21–24.

495. Palmer, E. J. The *Crataegus* problem. *Arnold Arboretum Jour.*, 1932, 13:342–362.

496. Palmgren, A. Chance as an element in plant geography. *Intern. Congr. Plant Sci. (Ithaca) Proc.*, 1929, 1:591–602.

497. Pax, F. Gesamtareal der Gattung *Acer* und einiger Sektionen. Verbreitung einiger Sektionen der Gattung *Acer* zur Tertiärzeit. *Die Pflanzenareale*, 1926, 1927, 1. Reihe, Heft 1, Karte 4, Heft 4, Karte 4, 31–33.

498. Pax, F. Hippocastanaceae DC. *Die Pflanzenareale*, 1928, 2. Reihe, Heft 1, Karte 8.

499. Payson, E. B. A monograph of the genus *Lesquerella*. *Mo. Bot. Gard. Ann.*, 1922, 8:103–236.

500. Peattie, D. C. *Shortia*, the flower that was lost for a century. *Farm and Gard.*, 1922, 9 (12):1–5.

501. Peattie, D. C. The Atlantic coastal plain element in the flora of the Great Lakes. *Rhodora*, 1922, 24:57–70, 80–88.

502. Pennell, F. W. The Scrophulariaceae of eastern temperate North America. *Acad. Nat. Sci. Phila. Monog.*, 1935, 1:1–650.

503. Peterson, D. Some chromosome numbers in the genus *Stellaria*. *Bot. Notiser*, 1935, pp. 409–410.

504. Phillips, J. Succession, development, the climax, and the complex organism: An analysis of concepts. I–III. *Jour. Ecol.*, 1934–35, 22:554–571; 23:210–246, 488–508.

505. Post, L. von. Skogsträdspollen i sydsvenska torvmosselagerföljder. Forhandl. 16 (1916) *Skand. Naturforskermøte*, Kristiania, 1918.

506. Post, L. von. Die postarktische Geschichte der europäischen wälder nach den vorliegenden Pollendiagrammen. *Meddel. Stockholms Högsk. Geol. Inst.*, 1929, 16:1–27.

507. Post, L. von. Problems and working lines in the post-arctic forest history of Europe. *Fifth Intern. Bot. Congr. Rept. Proc.* Cambridge Univ. Press, 1930, pp. 48–54.

508. Potbury, S. S. A Pleistocene flora from San Bruno, San Mateo County, California. *Carnegie Inst. Wash. Pub.*, 1932, 415:25–44.

509. Potzger, J. E. Succession of forests as indicated by fossil pollen from a northern Michigan bog. *Science*, 1932, 75:366.

510. Potzger, J. E., and R. R. Richards. Forest succession in the Trout Lake, Vilas County, Wisconsin area: A pollen study. *Butler Univ. Bot. Studies*, 1942, 5:179–189.

511. Poulton, E. B. The conception of species as interbreeding communities. *Linn. Soc. London Proc. Session 150*, 1938, (4):225–226.

512. Povoločko, P. A. An autotetraploid of *Nicotiana sylvestris* obtained by regeneration effected by growth hormones. *Acad. des Sci. U. S. S. R. Compt. Rend.* (Dok.), 1935, 4:77–80. (English text.)

513. Ramsbottom, J. Linnaeus and the species concept. *Linn. Soc. London Proc. Session 150*, 1938, (4):192–219.

514. Randolph, L. F. Some effects of high temperature on polyploidy and other variations in maize. *Natl. Acad. Sci. Proc.*, 1932, 18:222–229.

515. Randolph, L. F. An evaluation of induced polyploidy as a method of breeding crop plants. *Biological Symposia*, 1941, 4:151–167.

516. Raunkiaer, C. *The life forms of plants and statistical plant geography, being the collected papers of C. Raunkiaer.* Oxford, Clarendon Press, 1934, 632 pp.

517. Raup, H. M. Recent changes of climate and vegetation in southern New England and adjacent New York. *Arnold Arboretum Jour.*, 1937, 18:79–117.

518. Raup, H. M. Botanical problems in Boreal America. *Bot. Rev.*, 1941, 7:147–248.

519. Raup, H. M.   Trends in the development of geographic botany. (In press, *Assoc. Amer. Geogr. Ann.*), 1942.

520. Reeves, R. G., and P. C. Mangelsdorf. Chromosome numbers in relatives of *Zea mays* L. *Amer. Nat.*, 1935, 69:633–635.

521. Reid, C. and E. M.   The lignite of Bovey Tracey. *Roy. Soc. London, Phil. Trans.*, 1910, Ser. B, 201:161–178.

522. Reid, E. M.   A comparative review of Pliocene floras, based on the study of fossil seeds. *Geol. Soc. London, Quart. Jour.*, 1920, 76 (part 2):145–161.

523. Reid, E. M., and M. E. J. Chandler. *The London Clay Flora.* British Mus. (Nat. Hist.), London, 1933, 561 pp.

524. Reinig, W. F.   *Elimination und Selektion. Eine Untersuchung über Werkmalsprogressionen bei Tieren und Pflanzen auf genetisch- und historisch-chorologischer Grundlage.* Jena, Fischer, 1938, 146 pp.

525. Rensch, B.   Some problems of geographical variation and species-formation. *Linn. Soc. London Proc. Session 150,* 1938, (4):275–285.

526. Ridley, H. N.   Endemic plants. *Jour. Bot.*, 1925, 63:182–183.

527. Ridley, H. N.   *The dispersal of plants throughout the world.* Kent, Ashford, 1930, 744 pp.

528. Riley, H. P.   A character analysis of colonies of *Iris fulva, Iris hexagona* var. *giganticaerulea* and natural hybrids. *Amer. Jour. Bot.*, 1938, 25:727–738.

529. Riley, H. P.   Introgressive hybridization in a natural population of *Tradescantia. Genetics,* 1939, 24:753–769.

530. Rohweder, H.   Beiträge zur Systematik und Phylogenie des Genus *Dianthus* unter Berücksichtigung der karyologischen Verhältnisse. *Englers Bot. Jahrb.*, 1934, 66:249–366.

531. Rohweder, H.   Die Bedeutung der Polyploidie für die Anpassung der Angiospermen an die Kalkgebiete Schleswig-Holsteins. *Bot. Centbl. Beihefte.*, 1936, 54A:507–519.

532. Rosenberg, O.   Cytological studies on the apogamy in *Hieracium. Bot. Tidsskr.*, 1907, 28:143–170.

533. Rosenberg, O.   Apogamie und Parthenogenesis bei Pflanzen. *Handb. der Vererbungswiss.*, 1930, Bd. 2, Lief. 12:1–66.

534. Ross, M. N.   Seed reproduction of *Shortia galacifolia. N. Y. Bot. Gard. Jour.*, 1936, 37:208–211.

535. Rudolph, K., and F. Firbas. Paläofloristische und stratigraphische Untersuchungen böhmischer Moore. Die Hochmoore des Erzgebirges. *Beih. Bot. Centralbl.*, 1925, 41²:1–162.

536. Rudolph, K., and F. Firbas. Pollenanalytische Untersuchung subalpinen Moore des Riesengebirges. *Ber. Deut. Bot. Ges.*, 1926, 44:227–238.

537. Rybin, V. A.   Polyploid hybrids of *Nicotiana Tabacum* X *N. rustica* L. *Bul. Appl. Bot. Genet. and Plant Breeding,* 1927, 17 (3):235–240. (Eng. summary.)

538. Saint Hilaire, G. Lectures. Résumé in *Rev. et Mag. de Zool.* 1, January, 1837.

539. Sakai, K. Studies on the chromosome number in alpine plants. I. *Jap. Jour. Genet.,* 1934, 9:226–230.

540. Sakai, K. Studies on the chromosome number in alpine-plants. II. *Jap. Jour. Genet.,* 1935, 11:68–73.

541. Salisbury, E. J. Ecological aspects of plant taxonomy. In J. Huxley, *The New Systematics.* Oxford, Clarendon Press, 1940, pp. 329–340.

542. Samuelsson, G. Über die Verbreitung einiger endemischer Pflanzen. *Arkiv för Bot.,* 1910, 9 (12):1–16.

543. Sando, W. J. A colchicine-induced tetraploid in buckwheat. *Jour. Hered.,* 1939, 30:271–272.

544. Sansone, F. W., and S. S. Zilva. Polyploidy and vitamin C. *Biochem. Jour.,* 1933, 27:1935–1941.

545. Sargent, C. S. [A comparison of eastern Asiatic and eastern North American woody plants.] Introduction in E. H. Wilson, *A naturalist in Western China,* 1913, vol. 1, pp. xvii–xxxvii.

546. Sax, H. J. Chromosome pairing in *Larix* species. *Arnold Arboretum Jour.,* 1932, 13:368–373.

547. Sax, H. J. The relation between stomata counts and chromosome number. *Arnold Arboretum Jour.,* 1938, 19:437–441.

548. Sax, K. Species hybrids in *Platanus* and *Campsis. Arnold Arboretum Jour.,* 1933, 14:274–278.

549. Sax, K. The experimental production of polyploidy. *Arnold Arboretum Jour.,* 1936, 17:153–159.

550. Sax, K. Polyploidy and geographic distribution in *Spiraea. Arnold Arboretum Jour.,* 1936, 17:352–356.

551. Sax, K. and H. J. Chromosome number and morphology in the conifers. *Arnold Arboretum Jour.,* 1933, 14:356–375.

552. Sax, K. and H. J. Stomata size and distribution in diploid and polyploid plants. *Arnold Arboretum Jour.,* 1937, 18:164–172.

553. Schimper, A. F. *Pflanzen-geographie auf physiologischer Grundlage.* Jena. G. Fischer, 1898 (1903), 876 pp. [Eng. transl. by Groom and Balfour, 1903. Oxford.]

554. Schlösser, L. A. Frosthärte und Polyploidie. *Züchter,* 1936, 8:75–80.

555. Schonland, S. On the theory of "age and area." *Ann. Bot.,* 1924, 38:453–472.

556. Schouw, J. F. *Grundzüge einer allegemeinen Pflanzengeographie.* Berlin, G. Reimer, 1823, 524 pp.

557. Schröter, C. *Das Pflanzenleben der Alpen, Eine Schilderung der Hochgebirgsflora.* Zürich, Raustein, 1926, ed. II, 1288 pp.

558. Schuchert, C. The making of paleogeographic maps. Leopoldina, 1929, 4:116–125.

559. Schuchert, C. *Historical geology of the Antillean-Caribbean region.* New York. Wiley, 1935, 811 pp.

560. Sears, E. R. Amphidiploids in the Triticinae induced by colchicine. *Jour. Hered.*, 1939, 30:38–43.

561. Sears, P. B. Common fossil pollen of the Erie Basin. *Bot. Gaz.*, 1930, 89:95–106.

562. Sears, P. B. A record of post-glacial climate in northern Ohio. *Ohio Jour. Sci.*, 1930, 30:205–217.

563 Sears, P. B. Glacial and postglacial vegetation. *Bot. Rev.*, 1935, 1:37–51.

564. Sears, P. B. Types of North American pollen profiles. *Ecology*, 1935, 16:488–499.

565. Sears, P. B. Climatic interpretation of postglacial pollen deposits in North America. *Amer. Meteorol. Soc. Bul.*, 1938, 19:177–185.

566. Sears, P. B. Postglacial migration of five forest genera. *Amer. Jour. Bot.*, 1942, 29:684–691.

567. Sears, P. B., and G. C. Couch. Microfossils in an Arkansas peat and their significance. *Ohio Jour. Sci.*, 1932, 32:63–68.

568. Senn, H. A. Chromosome number relationships in the Leguminosae. *Bibliographia Genetica*, 1938, 12:175–336.

569. Sernander, R. Zur Morphologie und Biologie der Diasporen. *Nova Acta. Reg. Soc. Upsaliensis,* 1927.

570. Setchell, W. A. Pacific insular floras and Pacific paleogeography. *Amer. Nat.*, 1935, 69:289–310.

571. Setchell, W. A. Geographic elements of the marine flora of the North Pacific Ocean. *Amer. Nat.*, 1935, 69:560–577.

572. Sharp, A. J. Taxonomic and ecological studies of Eastern Tennessee bryophytes. *Amer. Midl. Nat.*, 1939, 21:267–354.

573. Shimotomai, N. Chromosomenzahlen und Phylogenie bei der gattung *Potentilla. Hiroshima Univ. Jour. Sci.*, 1930, Ser. B., Div. 2, Vol. 1, Art. 1:1–11.

574. Shimotomai, N. Zur Karyogenetik der Gattung *Chrysanthemum. Hiroshima Univ. Jour. Sci.*, 1933, Ser. B., Div. 2, Vol. 2, Art. 1:1–100.

575. Shreve, F. Vegetation of the northwestern coast of Mexico. *Torrey Bot. Club Bul.*, 1934, 61:373–380.

576. Shreve, F. The plant life of the Sonoran desert. *Sci. Monthly,* 1936, 42:195–213.

577. Shreve, F. Lowland vegetation of Sinaloa. *Torrey Bot. Club Bul.*, 1937, 64:605–613.

578. Simonet, M. Nouvelles recherches cytologiques et génétiques chez les *Iris. Ann. Sci. Nat. Bot.*, 1934, 10 Ser., 16:229–383.

579. Simonet, M. Contributions á l'étude cytologique et génétique de quelques *Agropyrum. Compt. Rendu. Acad. Sci.,* 1935, Paris, 201: 1201–1212.

580. Sinnott, E. W., and I. W. Bailey. Investigations on the phylogeny of the Angiosperms. 5. Foliar evidence as to the ancestry and early climatic environment of the Angiosperms. *Amer. Jour. Bot.,* 1915, 2: 1–22.

581. Sinnott, E. W., and A. F. Blakeslee. Changes in shape accompanying tetraploidy in cucurbit fruits. *Science,* 1938, 88: 476.

582. Širjaev, G. *Ononis L. Die Pflanzenareale,* 1934, 4. Reihe, Heft 2: 11–15.

583. Sissingh, Ir. G. Het exotenvraagstuk en de plantensociologie, speciaal met het oog op Nederlandsche boschgezelschappen en hun vicarieerende associaties in Amerika. *Nederlandsch Boschbouw-Tijdschrift,* 1939 (4): 145–165.

584. Skottsberg, C. Juan Fernandez and Hawaii. A phytogeographical discussion. *Bernice P. Bishop Mus. Bul.,* 1925, 16: 1–47.

585. Skottsberg, C. Pollinationsbiologie und Samenverbreitung auf den Juan Fernandez-Inseln. *The Natural History of Juan Fernandez and Easter Island,* 1928, 2 (18): 503–547.

586. Skottsberg, C. Remarks on the relative independency of Pacific floras. *Third Pan-Pacific Sci. Congr. Proc. Tokyo, 1926,* 1928, 1: 914–920.

587. Skottsberg, C. Pollination and seed dispersal in the Juan Fernandez Islands. *Fourth Pacific Sci. Congr. Proc. Java, 1929,* 1929, 3: 395–399.

588. Skottsberg, C. Remarks on the flora of the high Hawaiian volcanoes. *Götesborgs Bot. Trädgard Meddel.,* 1931, 6 (1930): 47–65.

589. Skottsberg, C. Geographical isolation as a factor in species formation, and its relation to certain insular floras. *Linn. Soc. London Proc. Session 150,* 1938, (4): 286–293.

590. Skottsberg, C. Remarks on the Hawaiian flora. *Linn. Soc. London Proc. Session 151,* 1939, (3): 181–186.

591. Skovsted, A. Cytological studies in the tribe Saxifrageae. *Dansk Bot. Arkiv.,* 1934, 8 (5): 1–52.

592. Small, J., and I. K. Johnston. Quantitative evolution in Compositae. *Roy. Soc. Edinb. Proc.,* 1937–38, 57 (i): 26–54; (iii): 215–227; 58 (i): 14–54.

593. Small, J. K. A new bog-asphodel from the mountains. *Torreya,* 1924, 24: 86–87.

594. Small, J. K. *Manual of the Southeastern Flora.* New York, 1933, 1554 pp.

595. Small, J. K., and E. J. Alexander. Botanical interpretations of Iridaceous plants of the Gulf States. *N. Y. Bot. Gard. Contrib.,* 1931, 327: 325–357.

596. Smith, A. C., and M. F. Koch. The genus *Espeletia:* A study on phylogenetic taxonomy. *Brittonia,* 1935, 1: 479–530.

597. Smith, E. C. Chromosome behavior in *Catalpa hybrida* Spaeth. *Arnold Arboretum Jour.,* 1941, 22: 219–221.

598. Smith, H. H. The induction of polyploidy in *Nicotiana* species and species hybrids by treatment with colchicine. *Jour. Hered.*, 1939, 30:291–306.

599. Smith, H. M. Another case of species versus subspecies. *Amer. Midl. Nat.*, 1942, 28:201–203.

600. Smith, P. Correlations of pollen profiles from glaciated eastern North America. *Amer. Jour. Sci.*, 1940, 238:597–601.

601. Smith, S. G. Cytology of *Anchusa* and its relation to the taxonomy of the genus. *Bot. Gaz.*, 1932, 94:394–403.

602. Sokolovskaja, A. P., and O. S. Strelkova. Polyploidy in the high mountain regions of Pamir and Altai. *Compt. Rendu. (Doklady) Acad. Sci. USSR*, 1938, 21:68–71.

603. Stapf, O. On the flora of Mount Kinabalu in North Borneo. *Linn. Soc. London Proc.*, 1894, 4 (2):69–263.

604. Stebbins, G. L., Jr. Cytological characteristics associated with the different growth habits in the Dicotyledons. *Amer. Jour. Bot.*, 1938, 25:189–198.

605. Stebbins, G. L., Jr. The significance of polyploidy in plant evolution. *Amer. Nat.*, 1940, 74:54–66.

606. Stebbins, G. L., Jr., and E. B. Babcock. The effect of polyploidy and apomixis on the evolution of species in *Crepis*. *Jour. Hered.*, 1939, 30:519–530.

607. Steenis, C. G. G. J. van. Malayan Bignoniaceac, their taxonomy, origin, and geographical distribution. *Rec. des Trav. Bot. Neerland.*, 1927, 24:787–1049.

608. Steenis, C. G. G. J. van. On the origin of the Malaysian mountain flora. Part 1, Facts and statement of the problem. Part 2, Altitudinal zones, general considerations and renewed statement of the problem. Part 3, Analysis of floristic relationships. *Buitenzorg Jard. Bot. Bul. Ser. 3*, 1934, 1935, 1936, 13:135–262, 289–417; 14:56–72.

609. Steffen, H. Beiträge zur Begriffsbildung und Umgrenzung einiger Floren-elemente Europas. *Beih. Bot. Centralbl.*, 1935, Bd. 53, Abt. B:330–404.

610. Steiner, M. Ökologische Pflanzengeographie. Halophyten. *Fortschritte der Botanik.*, 1938, 7:259–262.

611. Steyermark, J. A. A revision of the genus *Menodora*. *Mo. Bot. Gard. Ann.*, 1932, 19:87–176.

612. Stocker, O. Die Transpiration und Wasserökologie nordwestdeutscher Heide und Moorpflanzen am Standort. *Zeitschr. Bot.*, 1923, 15:1–41.

613. Stockwell, P. Chromosome numbers of the Cactaceae. *Bot. Gaz.*, 1935, 96:565–570.

614. Stockwell, P. A revision of the genus *Chaenactis*. *Dudley Herb. Contrib.*, 1940, 4:89–168.

615. Svenson, H. K. Effects of post-Pleistocene marine submergence in eastern North America. *Rhodora*, 1927, 29:41–48, 57–72, 87–93, 105–114.

616. Szymkiewicz, D. Contributions à la géographie des plantes. I. Phyto-

géographie floristique et écologique. *Kosmos,* 1933, 58:405-424. (Polish and French.)

617. Szymkiewicz, D. Une contribution statistique à la géographie floristique. *Soc. Bot. Poloniae Acta,* 1934, 11:249-265.

618. Szymkiewicz, D. Seconde contribution statistique à la géographie floristique. *Soc. Bot. Poloniae Acta,* 1936, 13:271-292.

619. Szymkiewicz, D. Troisième contribution statistique à la géographie floristique. *Soc. Bot. Poloniae Acta,* 1937, 14:215-238.

620. Szymkiewicz, D. Contributions à la géographie des plants. IV. Une nouvelle méthode pour la recherche des centres de distribution géographique des genres. *Kosmos,* 1937, 62:1-15. (Polish and French.)

621. Szymkiewicz, D. Quatrième contribution statistique à la géographie floristique. *Soc. Bot. Poloniae Acta,* 1938, 15:15-22.

622. Taylor, W. P. Significance of extreme or intermittent conditions in distribution of species and management of natural resources, with a restatement of Liebig's law of minimum. *Ecology,* 1934, 15:374-379.

623. Thomas, H. H. Palaeobotany and the origin of the Angiosperms. *Bot. Rev.,* 1936, 2:397-418.

624. Thompson, D. H. Variation in fishes as a function of distance. *Ill. State Acad. Sci. Trans.,* 1931, 23:276-281.

625. Thornthwaite, C. W. The climates of North America according to a new classification. *Geogr. Rev.,* 1931, 21:633-655.

626. Thorpe, W. H. Ecology and the future of systematics. In J. Huxley, *The New Systematics.* Oxford, Clarendon Press, 1940, pp. 341-364.

627. Timofeeff-Ressovsky, N. W. Mutations and geographical variation. In J. Huxley, *The New Systematics.* Oxford, Clarendon Press, 1940, pp. 73-136.

628. Tinney, F. W., and O. S. Aamodt. The progeny test as a measure of the types of seed-development in *Poa pratensis* L. *Jour. Hered.,* 1940, 31:457-464.

629. Tischler, G. Chromosomenzahl, -Form und -Individualität im Pflanzenreiche. *Progr. Rei Bot.,* 1916, 5:164-284.

630. Tischler, G. Pflanzliche Chromosomen-Zahlen. *Tab. Biol.,* 1927, 4:1-83.

631. Tischler, G. Über die Verwendung der Chromosomenzahl für phylogenetische Probleme bei den Angiospermen. *Biol. Zentbl.,* 1928, 48:321-345.

632. Tischler, G. Revisionen früherer Chromosomenzählungen und anschliessende Untersuchungen. *Planta,* 1929, 8:685-697.

633. Tischler, G. Verknüpfungsversuche von Zytologie und Systematik bei den Blütenpflanzen. *Deut. bot. Gesell. Ber. 47 generalversam.,* 1929, 1:30-49.

634. Tischler, G. Pflanzliche Chromosomen-Zahlen. *Tab. Biol.,* 1931, Per. 1 (*Tab. Biol.,* 7):109-226.

635. Tischler, G. Die Bedeutung der Polyploidie für die Verbreitung der

Angiospermen, erläutert an den Arten Schleswig-Holsteins, mit Ausblicken auf andere Florengebiete. *Bot. Jahrb.*, 1935, 67: 1–36.

636. Tischler, G. Die Bedeutung der Polyploidie für Pflanzengeographische Probleme. *6th Intern. Bot. Congr. Proc.*, 1935, 2: 54–56.

637. Tischler, G. Pflanzliche Chromosomen-Zahlen. *Nachtrag Nr. 2*, Teil I, 1935. *Tab. Biol.* Per. 5 (*Tab. Biol.*, 11): 281–304.

638. Tischler, G. Pflanzliche Chromosomen-Zahlen. *Nachtrag Nr. 2*, Teil II, 1936. *Tab. Biol.* Per. 6 (*Tab. Biol.*, 12): 57–115.

639. Tischler, G. On some problems of cytotaxonomy and cytoecology. *Indian Bot. Soc. Jour.*, 1937, 16: 165–169.

640. Tischler, G. Die Halligenflora der Nordsee im Lichte cytologischer Forschung. *Cytologia, Fujii Jubilaei*, 1937, 1: 162–170.

641. Tischler, G. Pflanzliche Chromosomen-Zahlen IV (*Nachtrag Nr. 3*), 1938. *Tab. Biol.*, 16: 162–218.

642. Tokugawa, Y., and Y. Kawada. Cytological studies on some garden varieties of *Canna*. *Jap. Jour. Bot.*, 1924, 2: 157–173.

643. Torrey, J. *A flora of the state of New York, comprising full descriptions of all the indigenous and naturalized plants hitherto discovered in the state, with remarks on their economical and medicinal properties.* (Natural History of New York.) Albany, Carroll and Cook, 1843, 2 vols.

644. Trela, J. Zur Morphologie der Pollenkörner der einheimischen *Tilia*-arten. *Polon. Acad. Sci. Let. Cl. Sci. Math. Nat. Bul. Intern.*, 1928, Sér. B: 45–54.

645. Turesson, G. The plant species in relation to habitat and climate. *Hereditas*, 1925, 6: 147–236.

646. Turesson, G. Contributions to the genecology of glacial relics. *Hereditas*, 1927, 9: 81–101.

647. Turesson, G. Zur Natur und Begrenzung der Arteinheiten. *Hereditas*, 1929, 12: 323–334.

648. Turesson, G. Studien über *Festuca ovina* L. II. Chromosomenzahl und Viviparie. *Hereditas*, 1929, 13: 177–184.

649. Turesson, G. Über verschiedene Chromosomenzahlen in *Allium schoenoprasum* L. *Bot. Notiser*, 1931, pp. 15–20.

650. Turesson, G. Studien über *Festuca ovina* L. III. Weitere Beiträge zur Kenntnis der Chromosomenzahlen virviparen Formen. *Hereditas*, 1931, 15: 13–16.

651. Turesson, G. Die Genezentrumtheorie und das Entwicklungszentrum der Pflanzenart. *Kungl. Fysiogr. Sällsk. i Lund Förhdl.*, 1932, 2 (6): 76–86.

652. Turesson, G. Chromosome stability in Linnean species. *Agr. Coll. Sweden Ann.*, 1938, 5: 405–416.

653. Turrill, W. B. The correlation of morphological variation with distribution in some species of *Ajuga*. *New Phytol.*, 1934, 33: 218–230.

654. Turrill, W. B.   Principles of plant geography. *Kew Bul.*, 1939, pp. 208–237.

655. U, N.   Genome-analysis in *Brassica* with special reference to the experimental formation of B. *napus* and peculiar mode of fertilization. *Jap. Jour. Bot.*, 1935, 7: 389–452.

656. Underwood, L. M.   A preliminary comparison of the hepatic flora of boreal and sub-boreal regions. *Bot. Gaz.*, 1892, 17: 305–312.

657. Vavilov, N. I.   Geographical regularities in the distribution of the genes of cultivated plants. *Bul. Applied Bot. Genet. and Plant Breeding*, 1927, 17 (3): 420–428. (Russian with English summary.)

658. Vavilov, N. I.   *Geographische Genzentren unserer Kultur pflanzen.* Verhandl. V. Intern. Kongr. Vererbungswissenschaft. Berlin, 1928.

659. Vavilov, N. I.   The new systematics of cultivated plants. In J. Huxley, *The New Systematics.* Oxford, Clarendon Press, 1940, pp. 549–566.

660. Vierhapper, F.   Über echten und falschen Vikarismus. *Oest. Bot. Zeitschr.*, 1919, 68: 1–22.

661. Viosca, P., Jr.   Irises of Louisiana. *Flower Grower*, 1932, 19: 386–387.

662. Viosca, P., Jr.   The Irises of southeastern Louisiana. A taxonomic and ecological interpretation. *Amer. Iris Soc. Bul.*, 1935, 57: 3–56.

663. Vogler, P.   Ueber die Verbreitungsmittel der Schweizerischen *Alpen-pflanzen. Flora*, 1901, 89. (Ergänzungsband.)

664. Voss, J.   Preliminary report on the paleo-ecology of a Wisconsin and an Illinois bog. *Ill. State Acad. Sci. Trans.*, 1931, 24: 130–137.

665. Vries, H. de.   *Gruppenweise Artbildung unter spezieller Berücksichtigung der Gattung* Oenothera. Berlin, Bornträger, 1913, 365 pp.

666. Wacław, G.   La relation entre les aires géographiques des plantes et les canyons en Podolie. *Soc. Bot. Poloniae Acta*, 1934, 11 (Supplement): 445–460.

667. Walker, R. I.   The effect of colchicine on microspore mother cells and microspores of *Tradescantia paludosa. Amer. Jour. Bot.*, 1938, 25: 280–285.

668. Walker, R. I.   The effect of colchicine on somatic cells of *Tradescantia paludosa. Arnold Arboretum Jour.*, 1938, 19: 158–162.

669. Walker, R. I.   The effect of colchicine on the developing embryo sac of *Tradescantia paludos. Arnold Arboretum Jour.*, 1938, 19: 442–445.

670. Wallace, A. R.   *The Malay archipelago: The land of the orang-utan, and the bird of paradise. A narrative of travel, with studies of man and nature.* London, Macmillan, 1869, 638 pp.

671. Wallace, A. R.   *Island Life: or, the phenomena and causes of insular faunas and floras, including a revision and attempted solution of the problem of geological climates.* London, Macmillan, 1880, 526 pp.; 2nd ed., 1892.

672. Walter, H.   Die Anpassungen der Pflanzen an Wassermangel. Das Xerophytenproblem in kausal-physiologischen Betrachtung. *Naturwissenschaft und Landwirtschaft*, 1926, Heft 9: 1–115.

673. Walter, H. *Die Hydratur der Pflanze.* Jena, G. Fischer, 1931, 174 pp.

674. Wangerin, W. Florenelemente und Arealtypen. *Beih. Bot. Centralbl.*, 1932, Bd. 49 Erganzbd.: 515–566.

675. Warburg, E. F. Taxonomy and relationship in the Geranales in the light of their cytology. *New Phytol.*, 1938, 37: 189–210.

676. Warming, E. Oecology of Plants. (Trans. by Groom and Balfour.) London, 1909.

677. Warmke, H. E., and A. F. Blakeslee. Induction of simple and multiple polyploidy in *Nicotiana* by colchicine treatment. *Jour. Hered.*, 1939, 30: 419–432.

678. Weaver, J. E., and F. E. Clements. *Plant Ecology.* New York, McGraw-Hill, 2nd ed., 1938, 601 pp.

679. Webber, I. E. Woods from the Ricardo Pliocene of Last Chance Gulch, California. *Carnegie Inst. Wash. Pub.*, 1933, 412: 113–134.

680. Werth, E. Die Vegetation der subantarktischen Inseln Kerguelen, Possesion- und Heard-Eiland. II. Teil. Deutsche Südpolar-Expedition 1901–1903, Bd. 8, Botanik. 1911, Heft. III:223–371.

681. Westergaard, M. Karyotypes of the collective species *Iris spuria* L. *Dansk Bot. Arkiv*, 1938, 9 (5): 1–11.

682. Wetmore, R. H., and A. L. Delisle. Studies in the genetics and cytology of two species in the genus *Aster* and their polymorphy in nature. *Amer. Jour. Bot.*, 1939, 26: 1–12.

683. Wettstein, F. von. Morphologie und Physiologie des Formwechsels der Moose auf genetischer Grundlage. *Zeitschr. f. induk. Abstammungs- und Vererbungslehre*, 1924, 33: 1–236.

684. Wettstein, F. von. Morphologie und Physiologie des Formwechsels der Moose auf genetischer Grundlage. II. *Bibliographia Genetica*, 1928, 10: 1–216.

685. Wettstein, F. von. Die genetische und entwicklungsphysiologie Bedeutung des Cytoplasms. *Zeitschr. f. induk. Abstammungs- und Vererbungslehre*, 1937, 23: 349–366.

686. Wettstein, R. von. Die europäischen Arten der Gattung *Gentiana* aus der Section *Endotricha* Froel. und ihr entwicklungsgeschichtlicher Zusammenhang. *Denkschrift. d. Math.-Naturw. Classe d. Akad. d. Wiss.*, 1896, 64: 1–74.

687. Wettstein, R. von. *Grundzüge der geographisch-morphologischen Methode der Pflanzensystematik.* Jena, G. Fischer, 1898, 64 pp.

688. Wettstein, R. von. Untersuchungen über den Saison-dimorphismus im Pflanzenreiche. *Akad. der Wiss. Wien. Denkschr.*, 1908, 70: 305–346.

689. Wherry, E. T. Divergent soil reaction preferences of related plants. *Ecology*, 1927, 8: 197–206.

690. Wherry, E. T. *Guide to eastern ferns.* Lancaster, Science Press, 1937.

691. Whitaker, T. W. Chromosome number and relationship in the Magnoliales. *Arnold Arboretum Jour.*, 1933, 14: 376–385.

692. Whitaker, T. W. A Karyo-systematic study of *Robinia. Arnold Arboretum Jour.,* 1934, 15:353–357.

693. Willis, J. C. Some evidence against the theory of the origin of species by natural selection of infinitesimal variations, and in favour of origin by mutation. *Peradeniya Roy. Bot. Gard. Ann.,* 1907, 4:1–15.

694. Willis, J. C. Further evidence against the origin of species by infinitesimal variations. *Peradeniya, Roy. Bot. Gard. Ann.,* 1907, 4:17–19.

695. Willis, J. C. The floras of hill tops in Ceylon. *Peradeniya, Roy. Bot. Gard. Ann.,* 1907, 4:131–138.

696. Willis, J. C. *Age and area. A study in geographical distribution and origin of species.* Cambridge, University Press, 1922, 259 pp.

697. Willis, J. C. Some conceptions about geographical distribution and origin of species. *Linn. Soc. London Proc. Session 150,* 1938, (3):162–167.

698. Willis, J. C. *The course of evolution by differentiation or divergent mutation rather than by selection.* Cambridge, University Press, 1940, 207 pp.

699. Willis, J. C., and G. U. Yule. Some statistics of evolution and geographical distribution in plants and animals, and their significance. Nature, 1922, 109:177–179.

700. Wilson, L. R. The Two Creeks forest bed, Manitowoc County, Wisconsin. *Wis. Acad. Sci. Trans.,* 1932, 27:31–46.

701. Wilson, L. R. Further fossil studies of the Two Creeks forest bed, Manitowoc County, Wisconsin. *Torrey Bot. Club Bul.,* 1936, 63:317–325.

702. Wilson, L. R., and E. F. Galloway. Microfossil succession in a bog in northern Wisconsin. *Ecology,* 1937, 18:113–118.

703. Winge, O. The chromosomes. Their numbers and general importance. *Carlsberg Lab. Compt. Rend. des Trav.,* 1917, 13:131–275.

704. Winge, O. On the origin of constant species-hybrids. *Svensk Bot. Tidskr.,* 1932, 26:107–122.

705. Winge, O. The genetic aspect of the species problem. *Linn. Soc. London Proc. Session 150,* 1938, (4):231–238.

706. Winge, O. Taxonomic and evolutionary studies in *Erophila* based on cytogenetic investigations. *Carlsberg Lab. Compt. Rend. Ser. Physiol.,* 1940, 25:41–74.

707. Winkler, H. Über die experimentelle Erzeugung von Pflanzen mit abweichenden Chromosomenzahlen. *Zeits. Bot.,* 1916, 8:417–531.

708. Winkler, H. Geographie. In Verdoorn, *Manual of Pteridology.* The Hague, Nijhoff, 1938, pp. 451–473.

709. Wodehouse, R. P. Tertiary Pollen—I. Pollen of the living representatives of the Green River flora. *Torrey Bot. Club Bul.,* 1932, 59:313–340.

710. Wodehouse, R. P. Tertiary Pollen—II. The oil shales of the Eocene Green River formation. *Torrey Bot. Club Bul.,* 1933, 60:479–524.

711. Wodehouse, R. P. *Pollen grains, their structure, identification and significance in science and medicine.* New York, McGraw-Hill, 1935, 123 pp.

712. Wood, J. G., and L. G. M. Baas-Becking. Notes on convergence and identity in relation to environment. Blumea, 1937, 2:329–338.

713. Woodson, R. E., Jr. Studies in the Apocynaceae. I. A critical study of the Apocynoideae (with special reference to the genus *Apocynum*). *Mo. Bot. Gard. Ann.,* 1930, 17:1–212.

714. Woodward, A. S. Palaeontology and the Linnaean classification. *Linn. Soc. London Proc., Session 150,* 1938 (4):238–241.

715. Worthington, E. B. Geographical differentiation in fresh waters with special reference to fish. In J. Huxley, *The New Systematics.* Oxford, Clarendon Press, 1940, pp. 287–302.

716. Wright, S. Evolution in Mendelian populations. *Genetics,* 1931, 16:97–159.

717. Wright, S. The statistical consequences of Mendelian heredity in relation to speciation. In J. Huxley, *The New Systematics.* Oxford, Clarendon Press, 1940, pp. 161–183.

718. Wulff, E. V. Introduction to the historical geography of plants. *Bul. Applied Bot. Suppl.,* 1932, 52:325–335. (English summary.)

719. Wulff, E. V. *An introduction to historical plant geography.* [Eng. trans. based on the 2nd ed. (1936) revised in 1939.] Waltham, Mass., Chronica Botanica, 1943, 223 pp.

720. Yule, G. U. A mathematical theory of evolution, based on the conclusions of Dr. J. C. Willis. *Phil. Trans. Roy. Soc. London,* 1924, 213:21–87.

Index